POWER SYSTEM HARMONICS AND PASSIVE FILTER DESIGNS

POWER SYSTEM HARMONICS AND PASSIVE FILTER DESIGNS

J.C. DAS

WILEY

Library of Congress Cataloging-in-Publication Data:

Das, J. C., 1934-
 Power system harmonics and passive filter design / J. C. Das.
   pages cm
 ISBN 978-1-118-86162-2 (hardback)
1. Electric power system stability. 2. Harmonics (Electric waves) 3. Electric filters, Passive. I. Title.
 TK1010.D37 2015
 621.31'7–dc23

                                                                    2014034588

10 9 8 7 6 5 4 3 2 1

CONTENTS

FOREWORD

Dr. Jean Mahseredjian[1]

This book on power system harmonics and passive filter designs is a comprehensive resource on this subject, covering harmonic generation, mitigation, measurement and estimation, limitations according to IEEE and IEC standards, harmonic resonance, formation of shunt capacitor banks, modeling of power system components and systems. Harmonic penetration in the power systems, passive filters, and typical study cases, covering renewable energy sources – solar and wind power generation – are included. There are many aspects of harmonics discussed in this book, which are not covered in the current publications.

The following is a chapter-wise summary of the book content.

Chapter 1 forms a background on the subject of power system harmonics with discussions of harmonic indices and power theories. The coverage of nonsinusoidal single-phase and three-phase systems and popular instantaneous power theory of H. Akagi and A. Nabe, much used for active filter designs discussed later on in the book, leads a reader to understand the nonlinearity.

The second chapter on Fourier analysis, though much mathematical, paves the way for the applications to harmonic analysis and measurements with limitations of window functions. The examples given in the chapter help the readers to understand the transformations.

Harmonic generation from conventional power equipment, ferroresonance, and electronically switched devices, converters, home appliances, cycloconverters, PWM, voltage source converters, switch mode power supplies, wind farm generation, pulse burst modulation, chopper circuits, traction and slip recovery schemes, are well described in Chapters 3 and 4. A reader will find an interesting analysis of transformer modeling, third harmonic voltages in generators, and many EMTP simulations. Harmonics due to saturation of current transformers is an added feature. Chapter 4 is fairly exhaustive and includes harmonic generation from many sources of practical importance. The analysis and topologies of ASDs (adjustable speed drives) are well documented. Though the author provides some background, yet a reader must be conversant with elements of power electronics.

Interharmonics is a new field of research, and Chapter 5 is well written so as to provide a reader a clear concept of interharmonic generation and their effects. This is followed by a well-written work on flicker from arcing loads, arcing and induction

---

[1] Dr. Jean Mahseredjian is an IEEE-Fellow and Professor of Electrical Engineering at École Polytechnique de Montréal, Montréal, Québec, Canada. He is world renowned authority on the simulation and analysis of electromagnetic transients. He was also a member of IEEE working groups on Power System Harmonics.

furnaces, and tracing methods of flicker. The control of flicker through the application of a STATCOM followed by torsional analysis due to harmonics in large drives with graphics is one problem that is not so well addressed in current texts. The subsynchronous resonance in series compensated HV transmission lines and drive system cascades, with EMTP simulation results, will be of interest to special readers interested in this field.

Having discussed the generation of harmonics in previous chapters, Chapter 6 is logically placed to discuss the various strategies that can be adopted to reduce the harmonics at source itself, so that harmonic penetration in the power systems is avoided. This covers active filters, combination of active and passive filters, their controls, active current shaping matrix converters, multilevel inverters, THMI inverters and theory of harmonic reduction at source, new breed of matrix and multilevel converters, followed with the theory of the resultant of polynomials. Then, the demonstration of this theory and control of switching angles is demonstrated to reduce harmonic distortion to a very low level. Some sections of this chapter will need a prior understanding of many aspects of converters and their switching, and on first reading the mathematical treatment cannot be easily followed by an average reader. The author provides excellent references at each step for further reading.

The calculations, estimation, time stamp of harmonics are the first step before a model can be generated for study. The relevance of modeling angles of the harmonics, measuring equipment, transducers, analysis of various waveforms will be of interest to all readers, while probabilistic concepts, regression methods, Kalman filtering, and so on will be of special interest. The author provides fundamental aspects leading to these advanced concepts.

The effects of harmonics can be very deleterious on electrical power equipment, Chapter 8. Practically all power system equipment of interest, motors, insulation stresses, and traveling wave phenomena on drive system cables, common mode voltages, bearing currents, protective relaying, circuit breakers, and the like are covered. Of special interest to a reader will be derating of dry and liquid-filled transformers serving nonlinear loads, which at times may be ignored, resulting in overloads.

After this background is grasped, harmonic resonance in various forms is discussed in Chapter 9. The reactance curves, Foster networks, composite resonance, secondary resonance are illustrated, which are commonly missing topics in other texts.

The limits of harmonic distortions in Chapter 10 cover both, IEEE and IEC guidelines, with limits on interharmonics and calculations of effects of notching on harmonic distortions.

In the design of passive filters, formation of shunt capacitor banks and their grounding and protection is an important aspect, Chapter 11. Often failures on harmonic filters occur due to improper selection of the ratings of unit capacitors forming the bank, as well as ignoring their protection and switching transients. The importance of this chapter cannot be overstated for a reader involved in harmonic filter designs.

The next step in harmonic analysis is accurate modeling of power system components and power systems, depending on their nature and extent of study, which is detailed in Chapters 12 and 13. These two chapters form the backbone of harmonic analysis. The modeling described for transmission lines, transformers, loads, cables, motors, generators, and converters in Chapter 12 is followed by system modeling in industrial, distribution, and transmission systems and HVDC, which are the aspects that should be clearly grasped by a reader interested in harmonics.

Study of harmonic penetration discussed in Chapter 14 can be undertaken after the material in the previous chapters is grasped. Apart from time and frequency domain methods, the chapter covers the latest aspects of probabilistic modeling.

It may seem that in the entire book only one chapter, Chapter 15, is devoted to passive filters. However, harmonic filter designs may be called the last link of the long chain of harmonic studies. The chapter describes practically all types of passive filters commonly applied in the industry, with some new technologies such as genetic algorithms and particle swarm theories.

Lastly, Chapter 16 has many real-world studies of harmonic analysis and filters designs, including arc furnaces, transmission systems, solar and wind generation plants. A reader with adequate modeling tools and software can duplicate these studies and it will be a tremendous exercise in learning.

I conclude that the book is well written and should appeal to beginners and advanced readers, in fact, this can become a standard reference book on harmonics. Many solved examples and real-world simulations of practical systems enhance the understanding. The book is well illustrated with relevant figures in each chapter.

PREFACE

The power system harmonics is a subject of continuous research; this book attempts to present the state-of-art technology and advancements. It is a subject of interest of many power system professionals engaged in harmonic analysis and mitigation and the applications in the modern climate when the nonlinear loads in the utility systems are on the increase.

The book provides a comprehensive coverage of generation, effects, and control of harmonics. New harmonic mitigation technologies, detailed step-by-step design of passive filters, interharmonics, and flicker are covered. The intention is that the book can serve as a reference and practical guide on harmonics.

A beginner should be able to form a clear base for understanding the subject of harmonics, and an advanced reader's interest should be simulated to explore further. A first reading of the book followed by a detailed critical reading is suggested. The many real-world study cases, examples, and graphics strive for this objective and provide clear understanding. The subject of harmonics may not form a curriculum even for graduate studies in many universities. In writing this book, an undergraduate level of knowledge is assumed; yet, the important aspects with respect to connectivity of each chapter are not lost sight of. It has the potentiality of serving as advance undergraduate and graduate textbook. Surely, it can serve as continuing education textbook and supplementary reading material.

The effects of harmonics can be experienced at a distance, and the effect on power system components is a dynamic and evolving field. These interactions have been analyzed in terms of current thinking.

The protective relaying has been called "an art and science." The author will not hesitate to call the passive harmonic filter designs and mitigation technologies the same. This is so because much subjectivity is involved. Leaving aside high-technology research tools such as Monte Carlo simulations, the available computer techniques invariably require iterative studies to meet a number of conflicting objectives.

A first reading of the book will indicate that the reader must understand the nature of harmonics, modeling of power system components, and characteristics of filters, before attempting a practical filter design for real-world applications. Chapter 16 is devoted to practical harmonic passive filter designs and case studies including solar and wind generation. A reader can modal and reproduce the results and get a "feel" of the complex iterative and analytical procedures.

The author acknowledges with thanks permission for republication of some work from his book: *Power System Analysis: Short-Circuit Load Flow and Harmonics*, CRC Press.

J.C. DAS

ABOUT THE AUTHOR

J.C. Das is principal and consultant with Power System Studies, Inc. Snellville, Georgia. He headed the Power System Analysis department at AMEC, Inc. for many years. He has varied experience in the utility industry, industrial establishments, hydroelectric generation, and atomic energy. He is a specialist in performing power system studies, including short circuit, load flow, harmonics, stability, arc flash hazard, grounding, switching transients, and protective relaying. He conducts courses for continuing education in power systems and has authored or coauthored about 65 technical publications nationally and internationally. He is the author of the following books:

- *Arc Flash Hazard Analysis and Mitigation*, IEEE Press, 2012.
- *Transients in Electrical Systems: Analysis Recognition and Mitigation*, McGraw-Hill, 2010
- *Power System Analysis: Short-Circuit Load Flow and Harmonics, Second Edition*, CRC Press, 2011.

These books provide extensive converge, running into more than 2400 pages and are well received in the technical circles. His interests include power system transients, EMTP simulations, harmonics, power quality, protection, and relaying. He has published 200 study reports on electrical power system for his clients.

Related to harmonic analysis, Mr. Das has designed some large harmonic passive filters in the industry, which are in successful operation for more than 18 years.

Mr. Das is a Life Fellow of Institute of Electrical and Electronics Engineers, IEEE (United States), Member of the IEEE Industry Applications and IEEE Power Engineering societies, a Fellow of Institution of Engineering Technology (United Kingdom), a Life Fellow of the Institution of Engineers (India), a Member of the Federation of European Engineers (France), and a member of CIGRE (France). He is a registered Professional Engineer in the States of Georgia and Oklahoma, a Chartered Engineer (C. Eng.) in the United Kingdom and a European Engineer (Eur. Ing.) in the Europe. He received meritorious award in engineering, IEEE Pulp and Paper Industry in 2005.

He received MSEE degree from the Tulsa University, Tulsa, Oklahoma, and BA (advanced mathematics) and BEE degrees from the Punjab University, India.

POWER SYSTEM HARMONICS

The electrical power systems should be designed not only for the sinusoidal currents and voltages but also for nonlinear and electronically switched loads. There has been an increase in such loads in the recent times, and these can introduce harmonic pollution, distort current and voltage waveforms, create resonances, increase the system losses, and reduce the useful life of the electrical equipment. Harmonics are one of the major problems of ensuring a certain power quality. This requires a careful analysis of harmonic generation and their measurements and the study of the deleterious effects, harmonic controls, and limitation to acceptable levels. Interest in harmonic analysis dates back to the early 1990s in connection with high voltage DC (HVDC) systems and static var compensators (SVC; Reference [1]). The analytical and harmonic limitation technology has progressed much during this period (see Reference [2] for a historical overview of the harmonics in power systems).

DC power is required for a number of applications from small amount of power for computers, video equipment, battery chargers, UPS (uninterrptible power supplies) systems to large chunks of power for electrolysis, DC drives, and the like. A greater percentage of office and commercial building loads are electronic in nature, which have DC as the internal operating voltage. Fuel and solar cells and batteries can be directly connected to a DC system, and the double conversion of power from DC to AC and then from AC to DC can be avoided. A case study conducted by Department of Electrical Power Engineering, Chalmers University of Technology, Gothenburg, Sweden is presented in [3]. This compares reliability, voltage drops, cable sizing, grounding and safety: AC verses DC distribution system. In Reference [4], DC shipboard distribution system envisaged by US Navy is discussed. Two steam turbine synchronous generators are connected to 7000 V DC bus through rectifiers, and DC loads are served through DC–DC converters. However, this is not a general trend, bulk and consumer power distribution systems are AC; and we will not be discussing industrial or commercial DC distribution systems in this book, except that HVDC converter interactions with respect to harmonics and DC filters are of interest and discussed in the appropriate chapters.

Harmonics in power systems originate due to varied operations, for example, ferroresonance, magnetic saturation, subsynchronous resonance, and nonlinear and electronically switched loads. Harmonic emission from nonlinear loads predominates.

Power System Harmonics and Passive Filter Designs, First Edition. J.C. Das.

1.1   NONLINEAR LOADS

To distinguish between linear and nonlinear loads, we may say that linear time-invariant loads are characterized so that an application of a sinusoidal voltage results in a sinusoidal flow of current. These loads display constant steady-state impedance during the applied sinusoidal voltage. Incandescent lighting is an example of such a load. The electrical motors not supplied through electronic converters also approximately meet this definition. The current or voltage waveforms will be almost sinusoidal, and their phase angles displaced depending on power factor of the electrical circuit. Transformers and rotating machines, under normal loading conditions, approximately meet this definition. Yet, it should be recognized that flux wave in the air gap of a rotating machine is not sinusoidal. Tooth ripples and slotting in rotating machines produce forward and reverse rotating harmonics. Magnetic circuits can saturate and generate harmonics. Saturation in a transformer on abnormally high voltage produces harmonics, as the relationship between magnetic flux density B and the magnetic field intensity H in a magnetic material (the transformer core) is not linear. Yet, the harmonics emissions from these sources are relatively small (Chapter 3).

In a nonlinear device, the application of a sinusoidal voltage does not result in a sinusoidal flow of current. These loads do not exhibit constant impedance during the entire cycle of applied sinusoidal voltage. *Nonlinearity is not the same as the frequency dependence of impedance*, that is, the reactance of a reactor changes in proportion to the applied frequency, but it is linear at each applied frequency if we neglect saturation and fringing. However, nonlinear loads draw a current that may even be discontinuous or flow in pulses for a part of the sinusoidal voltage cycle.

Mathematically, linearity implies two conditions:

- Homogeneity
- Superposition

Consider the state of a system defined in the state equation form:

$$\dot{x} = f[x(t), r(t), t] \tag{1.1}$$

If $x(t)$ is the solution to this differential equation with initial conditions $x(t_0)$ at $t = t_0$ and input $r(t)$, $t > t_0$:

$$x(t) = \varphi[x(t_0), r(t)] \tag{1.2}$$

then homogeneity implies that

$$\varphi[x(t_0), \alpha r(t)] = \alpha\varphi[x(t_0), r(t)] \tag{1.3}$$

where α is a scalar constant. This means that $x(t)$ with input $\alpha r(t)$ is equal to α times $x(t)$ with input $r(t)$ for any scalar α.

Superposition implies that

$$\varphi[x(t_0), r_1(t) + r_2(t)] = \varphi[x(t_0), r_1(t)] + \varphi[x(t_0), r_2(t)] \tag{1.4}$$

That is, $x(t)$ with inputs $r_1(t) + r_2(t)$ is equal to the sum of $x(t)$ with input $r_1(t)$ and $x(t)$ with input $r_2(t)$. Thus, linearity is superimposition plus homogeneity.

1.2   INCREASES IN NONLINEAR LOADS

Nonlinear loads are continuously on the increase. It is estimated that, during the next 10 years, more than 60% of the loads on utility systems will be nonlinear. Also much of the electronic load growth involves residential sector and household appliances. Concerns for harmonics originate from meeting a certain power quality, which leads to the related issues of (1) effects on the operation of electrical equipment, (2) harmonic analysis, and (3) harmonic control. A growing number of consumer loads are sensitive to poor power quality, and it is estimated that power quality problems cost US industry tens of billion of dollars per year. Although the expanded use of consumer automation equipment and power electronics is leading to higher productivity, these heavy loads are a source of electrical noise and harmonics and are less tolerant to poor power quality. For example, adjustable speed drives (ASDs) are less tolerant to voltage sags and swells as compared to an induction motor; and a voltage dip of 10% of certain time duration may precipitate ASD shutdown. These generate line harmonics and a source containing harmonics impacts their operation, leading to further generation of harmonics. *This implies that the nonlinear loads which are a source of generation of harmonics are themselves relatively less tolerant to the poor power quality that originates from harmonic emission from these loads.*

Some examples of nonlinear loads are as follows:

- ASD systems
- Cycloconverters
- Arc furnaces
- Rolling mills
- Switching mode power supplies
- Computers, copy machines, television sets, and home appliances
- Pulse burst modulation
- Static var compensators (SVCs)
- Thyristor-controlled reactors (TCRs)
- HVDC transmission, harmonics originate in converters
- Electric traction, chopper circuits
- Wind and solar power generation
- Battery charging and fuel cells
- Slip frequency recovery schemes of induction motors
- Fluorescent lighting and electronic ballasts
- Electrical vehicle charging systems
- Silicon-controlled rectifier (SCR) heating, induction heating, and arc welding.

The harmonics are also generated in conventional power equipment, such as transformer and motors. Saturation and switching of transformers generate harmonics. The harmonic generation is discussed in Chapters 3–5. The application of capacitor banks for power factor corrections and reactive power support can cause resonance and further distortions of waveforms (Chapter 9). Earlier rotating synchronous condensers have been replaced with modern shunt capacitors or SVCs (Chapter 4).

1.3   EFFECTS OF HARMONICS

Harmonics cause distortions of the voltage and current waveforms, which have adverse effects on electrical equipment. The estimation of harmonics from nonlinear loads is the first step in a harmonic analysis, and this may not be straightforward. There is an interaction between the harmonic producing equipment, which can have varied topologies, and the electrical system. Over the course of years, much attention has been focused on the analysis and control of harmonics, and standards have been established for permissible harmonic current and voltage distortions (Chapter 10). The effects of harmonics are discussed in Chapter 8.

1.4   DISTORTED WAVEFORMS

Harmonic emissions can have varied amplitudes and frequencies. The most common harmonics in power systems are sinusoidal components of a periodic waveform, which have frequencies that can be resolved into some multiples of the fundamental frequency. Fourier analysis is the mathematical tool employed for such analysis, and Chapter 2 provides an overview.

The components in a Fourier series that are not an integral multiple of the power frequency are called *noninteger harmonics* (Chapter 5).

The distortion produced by nonlinear loads can be resolved into a number of categories:

- A distorted waveform having a Fourier series with fundamental frequency equal to power system frequency and a periodic steady state exists. This is the most common case in harmonic studies. The waveform shown in Fig. 1.1 is synthesized from the harmonics shown in Table 1.1. The waveform in Fig. 1.1 is symmetrical about the x-axis and can be described by the equation:

$$I = \sin(\omega t - 30°) + 0.17 \; \sin(5\omega t + 174°) + 0.12 \; \sin(7\omega t + 101°) + \ldots$$

  Chapter 4 shows that this waveform is typically of a six-pulse current source converter, harmonics limited to 23rd, though higher harmonics will be present. The harmonic emission varies over wide range of distorted waveforms. Figure 1.2 shows a typical waveform for HVDC link, DC drives, and a six-pulse voltage source inverter (VSI) ASD, Ref. [1]. Chapter 4 studies typical waveforms and distortions from various types of power electronic

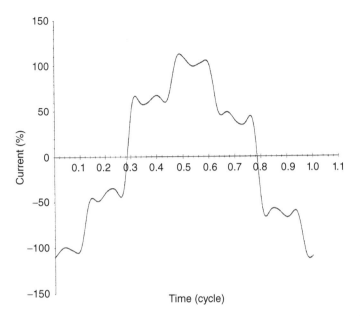

Figure 1.1   Simulated waveform of the harmonic spectrum shown in Table 1.1.

TABLE 1.1   Harmonic Content of the Waveform in Fig. 1.1

h	5	7	11	13	17	19	23
%	17	12	11	5	2.8	1.5	0.5

h = harmonic orders shown in percentage of fundamental current.

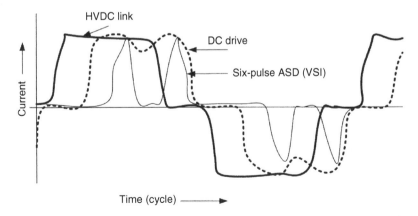

Figure 1.2   Typical line current waveforms of HVDC, DC drive, and six-pulse ASD.

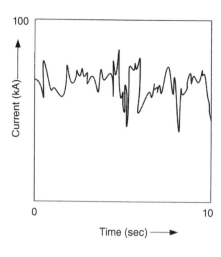

Figure 1.3   Erratic current signature of an electric arc furnace during scrap melting.

switching equipment. This is the most common situation in practice, and the distorted waveforms can be decomposed into a number of harmonics. The system can usually be modeled as a linear system.

- A distorted waveform having a submultiple of power system frequency and a periodic steady state exists. Certain types of pulsed loads and integral cycle controllers produce these types of waveforms (Chapters 4 and 5).

- The waveform is aperiodic, but perhaps almost periodic. A trigonometric series expansion may still exist. Examples are arcing devices: arc furnaces, fluorescent, mercury, and sodium vapor lighting. The process is not periodic in nature, and a periodic waveform is obtained if the conditions of operation are kept constant for a length of time. Consider the current signature of an arc furnace during scrap melting (Fig. 1.3). The waveform is highly distorted and aperiodic. Yet, typical harmonic emissions from arc furnace during melting and refining have been defined in IEEE standard 519 [5].

The arc furnace loads are highly polluting and cause phase unbalance, flicker, impact loading, harmonics, interharmonics, and resonance, and may give rise to torsional vibrations in rotating equipment.

1.4.1   Harmonics and Power Quality

Harmonics are one of the major power quality concerns. The power quality concerns embrace much wider concerns such as voltage sags and swells, transients, under and overvoltages, frequency variations, outright interruptions, power quality for sensitive electronic equipment such as computers. Table 3.1 summarizes some power quality problems. A reference of importance is IEEE Recommended Practice for Emergency and Standby Power Systems for Industrial and Commercial Applications, [6]. This book is not about power quality; however, some important publications are separately listed in References for the interested readers.

1.5 HARMONICS AND SEQUENCE COMPONENTS

The theory of sequence components is not discussed in this book and references [7–10] may be seen. In a three-phase balanced system under nonsinusoidal conditions, the hth-order harmonic voltage (or current) can be expressed as

$$V_{ah} = \sum_{h \neq 1} V_h(h\omega_0 t - \theta_h) \tag{1.5}$$

$$V_{bh} = \sum_{h \neq 1} V_h(h\omega_0 t - (h\pi/3)\theta_h) \tag{1.6}$$

$$V_{ch} = \sum_{h \neq 1} V_h(h\omega_0 t - (2h\pi/3)\theta_h) \tag{1.7}$$

Based on Eqs. (1.5–1.7) and counterclockwise rotation of the fundamental phasors, we can write

$$V_a = V_1 \sin \omega t + V_2 \sin 2\omega t + V_3 \sin 3\omega t + V_4 \sin 4\omega t + V_5 \sin 5\omega t + \dots$$

$$V_b = V_1 \sin(\omega t - 120°) + V_2 \sin(2\omega t - 240°) + V_3 \sin(3\omega t - 360°) + V_4 \sin(4\omega t - 480°)$$
$$\quad + V_5 \sin(5\omega t - 600°) + \dots$$
$$\quad = V_1 \sin(\omega t - 120°) + V_2 \sin(2\omega t + 120°) + V_3 \sin 3\omega t + V_4 \sin(4\omega t - 120°)$$
$$\quad + V_5 \sin(5\omega t + 120°) + \dots$$

$$V_c = V_1 \sin(\omega t + 120°) + V_2 \sin(2\omega t + 240°) + V_3 \sin(3\omega t + 360°) + V_4 \sin(4\omega t + 480°)$$
$$\quad + V_5 \sin(5\omega t + 600°) + \dots$$
$$\quad = V_1 \sin(\omega t + 120°) + V_2 \sin(2\omega t - 120°) + V_3 \sin 3\omega t + V_4 \sin(4\omega t + 120°)$$
$$\quad + V_5 \sin(5\omega t - 120°) + \dots$$

Under balanced conditions, the hth harmonic (frequency of harmonic $= h$ times the fundamental frequency) of phase b lags h times 120° behind that of the same harmonic in phase a. The hth harmonic of phase c lags h times 240° behind that of the same harmonic in phase a. In the case of triplen harmonics, shifting the phase angles by three times 120° or three times 240° results in cophasial vectors.

Table 1.2 shows the sequence of harmonics, and the pattern is clearly positive–negative–zero. We can write

$$\text{Harmonics of the order } 3h + 1 \text{ have positive sequence} \tag{1.8}$$

$$\text{Harmonics of the order } 3h + 2 \text{ have negative sequence} \tag{1.9}$$

$$\text{Harmonics of the order } 3h \text{ are of zero sequence} \tag{1.10}$$

All triplen harmonics generated by nonlinear loads are zero sequence phasors. These add up in the neutral. In a three-phase four-wire system, with perfectly balanced

TABLE 1.2 Harmonic Order and Rotation

Harmonic Order	Forward	Reverse
Fundamental	x	
2		x
4	x	
5		x
7	x	
8		x
10	x	
11		x
13	x	
14		x
16	x	
17		x
19	x	
20		x
22	x	
23		x
25	x	
26		x
28	x	
29		x
31	x	

Note: The pattern is repeated for higher order harmonics.

single-phase loads between the phase and neutral, all positive and negative sequence harmonics will cancel out leaving only the zero sequence harmonics.

In an unbalanced three-phase system, serving single-phase load, the neutral carries zero sequence and the residual unbalance of positive and negative sequence currents. Even harmonics are absent in the line because of phase symmetry (Chapter 2), and unsymmetrical waveforms will add even harmonics to the phase conductors, for example, half-controlled three-phase bridge circuit discussed in Chapter 4.

1.5.1 Sequence Impedances of Power System Components

Positive, negative, and zero sequence impedances vary over large limits, depending on the power system equipment. For example, for transformers, positive and negative sequence impedances may be considered equal, but zero sequence impedance can be infinite depending on transformer winding connections and grounding. The zero sequence impedance of transmission lines can be two to three times that of the positive or negative sequence impedance. Even for fundamental frequency current flow, the accurate modeling of sequence impedances is important and the sequence impedances to harmonics must be modeled (Chapter 12).

1.6   HARMONIC INDICES

1.6.1   Harmonic Factor

An index of merit has been defined as a harmonic distortion factor [5] (harmonic factor). It is the ratio of the root mean square of the harmonic content to the root mean square value of the fundamental quantity, expressed as a percentage of the fundamental:

$$DF = \sqrt{\frac{\sum \text{of squares of amplitudes of all harmonics}}{\text{Square of the amplitude of the fundamental}}} \times 100\% \qquad (1.11)$$

The most commonly used index, total harmonic distortion (THD), which is in common use is the same as DF.

1.6.2   Equations for Common Harmonic Indices

We can write the following equations.

RMS voltage in presence of harmonics can be written as

$$V_{rms} = \sqrt{\sum_{h=1}^{h=\infty} V_{h,rms}^2} \qquad (1.12)$$

And similarly, the expression for the current is

$$I_{rms} = \sqrt{\sum_{h=1}^{h=\infty} I_{h,rms}^2} \qquad (1.13)$$

The total distortion factor for the voltage is

$$THD_V = \frac{\sqrt{\sum_{h=2}^{h=\infty} V_{h,rms}^2}}{V_{f,rms}} \qquad (1.14)$$

where $V_{f,rms}$ is the fundamental frequency voltage. This can be written as

$$THD_V = \sqrt{\left(\frac{V_{rms}}{V_{f,rms}}\right)^2 - 1} \qquad (1.15)$$

or

$$V_{rms} = V_{f,rms}\sqrt{1 + THD_V^2} \qquad (1.16)$$

Similarly,

$$\text{THD}_I = \frac{\sqrt{\sum_{h=2}^{h=\infty} I_{h,\text{rms}}^2}}{I_{f,\text{rms}}} = \sqrt{\left(\frac{I_{\text{rms}}}{I_{f,\text{rms}}}\right)^2 - 1} \tag{1.17}$$

$$I_{\text{rms}} = I_{f,\text{rms}} \sqrt{1 + \text{THD}_I^2} \tag{1.18}$$

where $I_{f,\text{rms}}$ is the fundamental frequency current.

The total demand distortion (TDD) is defined as

$$\text{TDD} = \frac{\sqrt{\sum_{h=2}^{h=\infty} I_h^2}}{I_L} \tag{1.19}$$

where I_L is the load demand current.

The partial weighted harmonic distortion (PWHD) of current is defined as

$$\text{PWHD}_I = \frac{\sqrt{\sum_{h=14}^{h=40} h I_h^2}}{I_{f,\text{rms}}} \tag{1.20}$$

Similar expression is applicable for the voltage. The PWHD evaluates influence of current or voltage harmonics of higher order. The sum parameters are calculated with single harmonic current components I_h.

1.6.3 Telephone Influence Factor

Harmonics generate telephone Influence through inductive coupling. The telephone influence factor (TIF) for a voltage or current wave in an electrical supply circuit is the ratio of the square root of the sum of the squares of the weighted root mean square values of all the sine wave components (including AC waves both fundamental and harmonic) to the root mean square value (unweighted) of the entire wave:

$$\text{TIF} = \frac{\sqrt{\sum_f W_f^2 I_f^2}}{I_{\text{rms}}} \tag{1.21}$$

where I_f is the single frequency rms current at frequency f, W_f is the single frequency TIF weighting at frequency f. The voltage can be substituted for current. This definition may not be so explicit, see example in Chapter 8 for calculation. A similar expression can be written for voltage.

IT product is the inductive influence expressed in terms of the product of its root mean square magnitude I in amperes times its TIF.

$$IT = TIF * I_{rms} = \sqrt{\sum (W_f I_f)^2} \qquad (1.22)$$

kVT product is the inductive influence expressed in terms of the product of its root mean square magnitude in kV times its TIF.

$$kVT = TIF * kV_{rms} = \sqrt{\sum (W_f V_f)^2} \qquad (1.23)$$

The telephone weighting factor that reflects the present C message weighting and the coupling normalized to 1 kHz is given by:

$$W_f = 5P_f f \qquad (1.24)$$

where $P_f = C$ message weighting at frequency f under consideration. See Section 8.12 for further details.

1.7 POWER FACTOR, DISTORTION FACTOR, AND TOTAL POWER FACTOR

For sinusoidal voltages and currents, the power factor is defined as kW/kVA and the power factor angle ϕ is

$$\varphi = \cos^{-1} \frac{kW}{kVA} = \tan^{-1} \frac{kvar}{kW} \qquad (1.25)$$

The power factor in presence of harmonics comprises two components: displacement and distortion. The effect of the two is combined in *total power factor*. The displacement component is the ratio of active power of the fundamental wave in watts to apparent power of fundamental wave in volt-amperes. This is the power factor as seen by the watt-hour and var-hour meters. The distortion component is the part that is associated with harmonic voltages and currents.

$$PF_t = PF_f \times PF_{distortion} \qquad (1.26)$$

At fundamental frequency the displacement power factor will be equal to the total power factor, as the displacement power factor does not include kVA due to harmonics, while the total power factor does include it. For harmonic generating loads, the total power factor will always be less than the displacement power factor.

Continuing with the relation between power factor and displacement factor, the power factor of a converter with DC-link reactor is given by the expression from IEEE 519, Ref. [5]:

$$\text{Total PF} = \frac{q}{\pi} \sin\left(\frac{\pi}{q}\right) \qquad (1.27)$$

where q is the number of converter pulses and π/q is the angle in radians (see Chapter 4). This ignores commutation overlap and no-phase overlap, and neglects transformer magnetizing current. For a six-pulse converter, the *maximum* power factor is $3/\pi = 0.955$. A 12-pulse converter has a theoretical *maximum* power factor of 0.988. The power factor drops drastically with the increase in firing angle.

Note that the power factor is a function of the drive topology, for example, with pulse width modulation, the input power factor is dependent on the type of converter only and the motor power factor is compensated by a capacitor in the DC link.

In the case of sinusoidal voltage and current, the following relationship holds

$$S^2 = P^2 + Q^2 \tag{1.28}$$

where P is the active power, Q is the reactive volt-ampere, and S is the volt-ampere. This relationship has been amply explored in load flow programs:

$$S = V_f I_f, \quad Q = V_f I_f \ \sin(\theta_f - \delta_f), \quad P = V_f I_f \ \cos(\theta_f - \delta_f), \quad \text{and} \ \ PF = P/S \tag{1.29}$$

$\theta_f - \delta_f =$ phase angle between fundamental voltage and fundamental current.

In the case of nonlinear load or when the source has nonsinusoidal waveform, the active power P can be defined as

$$P = \sum_{h=1}^{h=\infty} V_h I_h \cos(\theta_h - \delta_h) \tag{1.30}$$

Q can be written as

$$Q = \sum_{h=1}^{h=\infty} V_h I_h \sin(\theta_h - \delta_h) \tag{1.31}$$

V_h and I_h are in rms values, and the apparent power can be defined as

$$S = \sqrt{P^2 + Q^2 + D^2} \tag{1.32}$$

where D is the distortion power. Consider D^2 up to the third harmonic:

$$
\begin{aligned}
D^2 = {}& (V_0^2 + V_1^2 + V_2^2 + V_3^2)(I_0^2 + I_1^2 + I_2^2 + I_3^2) \\
& - (V_0 I_0 + V_1 I_1 \ \cos\theta_1 + V_2 I_2 \ \cos\theta_2 + V_3 I_3 \ \cos\theta_3)^2 \\
& - (V_1 I_1 \ \sin\theta_1 + V_2 I_2 \ \sin\theta_2 + V_3 I_3 \ \sin\theta_3)^2
\end{aligned} \tag{1.33}
$$

An expression for distortion power factor can be arrived from current and voltage harmonic distortion factors. From the definition of these factors, rms harmonic voltages and currents can be written as

$$V_{\text{rms}(h)} = V_f \sqrt{1 + \left(\frac{\text{THD}_V}{100}\right)^2} \tag{1.34}$$

$$I_{\text{rms}(h)} = I_f \sqrt{1 + \left(\frac{\text{THD}_I}{100}\right)^2} \tag{1.35}$$

Therefore, the total power factor is

$$\text{PF}_{\text{tot}} = \frac{P}{V_f I_f \sqrt{1 + \left(\frac{\text{THD}_V}{100}\right)^2} \sqrt{1 + \left(\frac{\text{THD}_I}{100}\right)^2}} \tag{1.36}$$

Neglecting the power contributed by harmonics and also voltage distortion, as it is generally small, that is,

$$\text{THD}_V \cong 0 \tag{1.37}$$

$$\text{PF}_{\text{tot}} = \cos(\theta_f - \delta_f) \cdot \frac{1}{\sqrt{1 + \left(\frac{\text{THD}_I}{100}\right)^2}}$$

$$= \text{PF}_{\text{displacement}} \text{PF}_{\text{distortion}} \tag{1.38}$$

The total power factor is the product of displacement power factor (which is the same as the fundamental power factor) and is multiplied by the distortion power factor as defined earlier.

The discussion is continued in Chapter 4. The modern trends in converter technology are to compensate for line harmonics and improve power factor to approximately unity simultaneously (Chapter 6).

1.8 POWER THEORIES

A number of power theories exist to explain the active, reactive, and instantaneous power relations in presence of harmonics, each fraught with some controversies:

1. Fryze theory in time domain
2. Shepherd and Zakikhani theory in frequency domain
3. Czarnecki power theory in frequency domain
4. Nabe and Akagi instantaneous power theory

   See references [12–16].

1.8.1 Single-Phase Circuits: Sinusoidal

The instantaneous power is

$$p = vi = 2VI \sin \omega t \sin(\omega t - \theta) = p_a + p_q \tag{1.39}$$

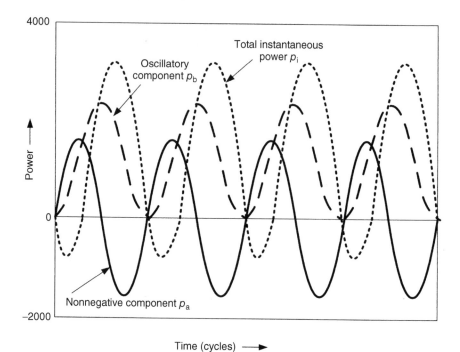

Figure 1.4   The waveform of separated components of instantaneous power in a single-phase circuit, with linear resistive-inductive load.

The active power also called real power is the average value of instantaneous power measured over a certain time period, say, τ to $\tau + kT$

We will denote instantaneous values in lowercase (v and i in (1.39) are in peak values).

$$p_a = VI \cos \theta [1 - \cos(2\omega t)] = P[1 - \cos(2\omega t]$$

$$p_q = -VI \sin \theta \ \sin(2\omega t) = -Q \ \sin(2\omega t) \tag{1.40}$$

The energy flows unidirectional from source to load $p_a \geq 0$. The instantaneous active power has two terms, active or real power and the intrinsic power $-P \cos 2\omega t$, which is always present when energy is transferred from source to load. If load is inductive $Q > 0$, and if load is capacitive $Q < 0$.

Figure 1.4 illustrates the instantaneous power components in single-phase circuits: the nonnegative component p_a, the oscillatory component p_b, and total instantaneous power p_i are shown.

1.8.2   Single-Phase Circuits: Nonsinusoidal

We can write

$$v = v_1 + v_H$$

$$i = i_1 + i_H$$

$$v_H = V_0 + \sqrt{2} \sum_{h \neq 1} V_h \sin(h\omega t - \alpha_h)$$

$$i_H = I_0 + \sqrt{2} \sum_{h \neq 1} I_h \sin(h\omega t - \beta_h) \tag{1.41}$$

v_1 and i_1 are power frequency components and v_H and i_H are other components.
   The active power (rms value) is

$$p_a = V_0 I_0 + \sum_h V_h I_h \cos \theta_h [1 - \cos(2h\omega t - 2\alpha_h)] \tag{1.42}$$

It has two terms: $P_h = V_h I_h \cos \theta_h$ and the intrinsic harmonic power $-P_h \cos(2h\omega t - 2\alpha_h)$, which does not contribute to the net transfer of energy or additional power loss in the conductors.
   Also, fundamental active power is

$$P_1 = V_1 I_1 \cos \theta_1 \tag{1.43}$$

And harmonic active power is

$$P_H = V_0 I_0 + \sum_{h \neq 1} V_h I_h \cos \theta_h = P - P_1 \tag{1.44}$$

P_q does not represent a net transfer of energy, its average value is nil. The current related to these nonactive components causes additional power loss in the conductors.
   The apparent power is

$$S^2 = (V_1^2 + V_H^2)(I_1^2 + I_H^2) = S_1^2 + S_N^2 \tag{1.45}$$

where

$$S_N^2 = (V_1 I_H)^2 + (V_H I_1)^2 + (V_H I_H)^2$$

$$= D_1^2 + D_V^2 + S_H^2 \tag{1.46}$$

where

$$D_1 = \text{current distortion power (var)} = S_1(\text{THD}_I)$$

$$D_V = \text{voltage distortion power (var)} = S_1(\text{THD}_V)$$

$$S_H = \text{harmonic apparent power (VA)} = V_H I_H$$

$$= S_1(\text{THD}_I)(\text{THD}_V) = \sqrt{P_H^2 + D_H^2}$$

where D_H is the harmonic distortion power.

As THD_V is $\ll \text{THD}_I$,

$$S_N = S_1(\text{THD}_I)$$

The fundamental power factor is

$$\text{PF}_1 = \cos\theta_1 = \frac{P_1}{S_1} \tag{1.47}$$

It is also called the displacement power factor.

And

$$\text{PF} = P/S = \frac{[1 + (P_H/P_1)]\text{PF}_1}{\sqrt{1 + \text{THD}_I^2 + \text{THD}_V^2 + (\text{THD}_I\text{THD}_V)^2}} \approx \frac{1}{\sqrt{1 + \text{THD}_I^2}}\text{PF}_1 \tag{1.48}$$

$$D_I > D_V > S_H > P_H$$

Equation (1.48) is the same as Eq. (1.38).

1.8.3   Three-Phase Systems

We can consider

- Balanced three-phase voltages and currents
- Asymmetrical voltages or load currents
- Nonlinear loads

Figure 1.5(a) shows balanced three-phase voltages and currents and balanced resistive load, and Fig. 1.5(b) depicts the instantaneous power in Fig. 1.5(a). The summation of phase instantaneous active powers in three phases is constant. Thus, the concepts arrived at in single-phase circuits cannot be applied. We examined that in single-phase circuits the active power has an intrinsic power component.

In three-phase circuits, it is impossible to separate reactive power on the basis of instantaneous power. Reactive power interpretation of single-phase circuits cannot be applied.

Figure 1.6 shows waveforms of voltages and currents in three-phase circuits with unbalanced resistive load. Now, the instantaneous active power is no longer constant. Considering three-phase circuit as three single-phase circuits leads to major misinterpretation of power phenomena.

Figure 1.7 depicts the symmetrical *nonlinear* load current and symmetrical waveforms of the supply voltage. Again the instantaneous active power is no longer constant. The individual instantaneous active powers in phases are shown in Fig. 1.8.

The extension of concept of apparent power in three-phase circuits has led to

Arithmetic apparent power:

$$V_a I_a + V_b I_b + V_c I_c = S_A \tag{1.49}$$

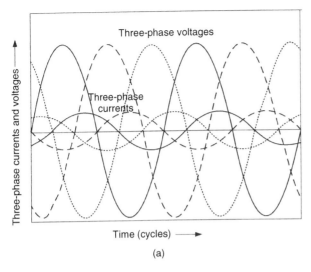

(a)

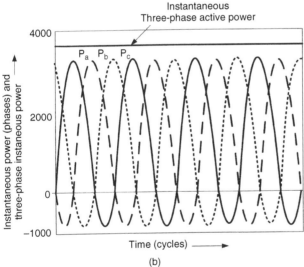

(b)

Figure 1.5   (a) Balanced three-phase voltages and currents in a three-phase system, (b) phase instantaneous powers, p_a, p_b, and p_c and total instantaneous power.

V_a, V_b, ... I_a, I_b .... in rms values.

Geometric apparent power:

$$S_G = \sqrt{P^2 + Q^2} \tag{1.50}$$

where three-phase active and reactive powers P and Q are

$$I_{eH} = \sqrt{\frac{I_{aH}^2 + I_{bH}^2 + I_{cH}^2}{3}} = \sqrt{I_e^2 - I_{e1}^2} \tag{1.55}$$

Similar expressions can be written for voltages.

Resolution of S_e is implemented as

$$S_e^2 = S_{e1}^2 + S_{eN}^2 \tag{1.56}$$

S_{e1} is the fundamental apparent power and S_{eN} is nonfundamental apparent power.

$$S_{e1} = 3V_{e1}I_{e1}$$

$$S_{eN}^2 = D_{eI}^2 + D_{eV}^2 + S_{eH}^2 = 3V_{e1}I_{eH} + 3V_{eH}I_{e1} + 3V_{eH}I_{eH} \tag{1.57}$$

$$S_{eN} = \sqrt{\text{THD}_{eI}^2 + \text{THD}_{eV}^2 + (\text{THD}_{eI}\text{THD}_{eV})^2} \tag{1.58}$$

where

$$D_{eI} = S_{e1}(\text{THD}_{eI}) \quad D_{eV} = S_{e1}(\text{THD}_{eV}) \quad D_{eH} = S_{e1}(\text{THD}_{eV})(\text{THD}_{eI}) \tag{1.59}$$

are the components of nonfundamental apparent power.

The load unbalance can be evaluated using unbalance power:

$$S_{U1} = \sqrt{S_{e1}^2 - (S_1^+)^2} \tag{1.60}$$

where

$$S_1^+ = \sqrt{(P_1^+)^2 + (Q_1^+)^2} \tag{1.61}$$

Here,

$$P_1^+ = 3V_1^+ I_1^+ \cos\theta_1^+$$

$$Q_1^+ = 3V_1^+ I_1^+ \sin\theta_1^+$$

The fundamental positive sequence power factor is

$$PF_1^+ = \frac{P_1^+}{S_1^+} \tag{1.62}$$

It plays the same role as the fundamental power factor has in nonsinusoidal single-phase circuits.

The combined power factor is

$$PF = \frac{P}{S_e} \tag{1.63}$$

Table 1.3 shows these relations.

TABLE 1.4   Active Powers, Example 1.1

P_1 (W)	P_3 (W)	P_5 (W)	P_7 (W)	P (W)	P_H (W)
8660.00	−13.94	−11.78	−1.74	8632.54	−27.46

Source: Ref. [11].

TABLE 1.5   Reactive Powers, Example 1.1

Q_1 (var)	Q_3 (var)	Q_5 (var)	Q_7 (var)
5000.00	159.39	−224.69	49.97

Source: Ref. [11].

TABLE 1.6   Distortion Powers and Their Components, Example 1.1

D_{13} (var)	D_{15} (var)	D_{17} (var)	D_I (var)
2000.00	1500.00	1000.00	2692.58
D_{31} (var)	D_{51} (var)	D_{71} (var)	D_V (var)
800.00	1500.00	500.00	1772.00

Source: Ref. [11].

$$\Delta P_B = \frac{r}{V^2}(4984.67)^2 = 431.37\,\text{W}$$

This is incorrect. It should be calculated as

$$\Delta P = \frac{r}{V^2}(Q_1^2 + Q_3^2 + Q_5^2 + Q_7^2) = 435.39\,\text{W}$$

The cross products that produce distortion powers are in Table 1.6, and the cross products that belong to *harmonic* apparent power are in Table 1.7.

The system has $V = 101.56$ V, $I = 103.56$A, $\text{THD}_V = 0.177$, $\text{THD}_I = 0.269$, fundamental power factor $\text{PF}_1 = 0.866$, $\text{PF} = 0.821$.

The power components are shown in Fig. 1.9.

1.8.5 Instantaneous Power Theory

The Nabe–Akagi instantaneous reactive power p–q theory is based on Clark's component transformations [10] and provides power properties in three-phase circuits. Figure 1.10 shows the transformation of $a - b - c$ coordinates into $\alpha - \beta - 0$ coordinates. The description of power properties of the electrical circuits using instantaneous voltage and current values without the use of Fourier series generates interest in this theory, used for switching compensators and active filter controls.

The instantaneous power method calculates the desired current so that the instantaneous active power and reactive power in a three-phase system are kept constant, that is, the active filter compensates for variation in instantaneous power [17].

By linear transformation, the phase voltages e_a, e_b, e_c and load currents i_a, i_b, i_c are transformed into an $\alpha - \beta$ (two-phase) coordinate system:

$$\begin{vmatrix} e_\alpha \\ e_\beta \end{vmatrix} = \sqrt{\frac{2}{3}} \begin{vmatrix} 1 & -\frac{1}{2} & -\frac{1}{2} \\ 0 & \frac{\sqrt{3}}{2} & -\frac{\sqrt{3}}{2} \end{vmatrix} \begin{vmatrix} e_a \\ e_b \\ e_c \end{vmatrix} \tag{1.64}$$

and

$$\begin{vmatrix} i_\alpha \\ i_\beta \end{vmatrix} = \sqrt{\frac{2}{3}} \begin{vmatrix} 1 & -\frac{1}{2} & -\frac{1}{2} \\ 0 & \frac{\sqrt{3}}{2} & -\frac{\sqrt{3}}{2} \end{vmatrix} \begin{vmatrix} i_a \\ i_b \\ i_c \end{vmatrix} \tag{1.65}$$

The instantaneous real power ρ and the instantaneous imaginary power q are defined as

$$\begin{vmatrix} p \\ q \end{vmatrix} = \begin{vmatrix} e_\alpha & e_\beta \\ -e_\beta & e_\alpha \end{vmatrix} \begin{vmatrix} i_\alpha \\ i_\beta \end{vmatrix} \tag{1.66}$$

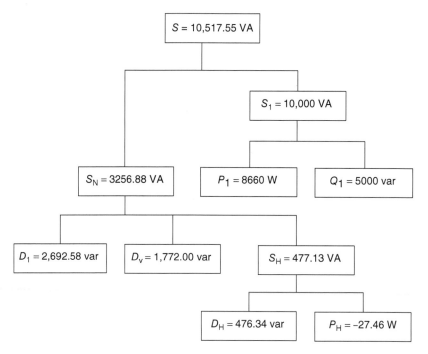

Figure 1.9   Calculations tree of various power components, Example 1.1.

TABLE 1.7   Distortion Harmonic Powers, Example 1.1

D_{35} (var)	D_{37} (var)	D_{53} (var)	D_{57} (var)	D_{73} (var)	D_{75} (var)
120.00	80.00	300.00	150.00	100.00	75.00

Source: Ref. [11].

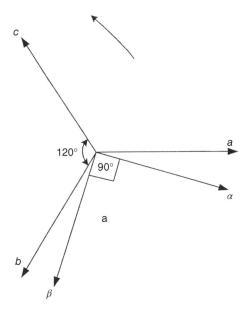

Figure 1.10   Transformation of $a - b - c$ coordinates into $\alpha - \beta - 0$ coordinates, $p - q$ theory.

Here, p and q are not conventional watts and vars. The p and q are defined by the instantaneous voltage in one phase and the instantaneous current in the other phase:

$$p = e_\alpha i_\alpha + e_\beta i_\beta = e_a i_a + e_b i_b + e_c i_c \tag{1.67}$$

To define instantaneous reactive power, the space vector of imaginary power is defined as

$$q = e_\alpha i_\beta + e_\beta i_\alpha$$

$$= \frac{1}{\sqrt{3}}[i_a(e_c - e_b) + i_b(e_a - e_c) + i_c(e_b - e_a)] \tag{1.68}$$

Equation (1.66) can be written as

$$\begin{vmatrix} i_\alpha \\ i_\beta \end{vmatrix} = \begin{vmatrix} e_\alpha & e_\beta \\ -e_\beta & e_\alpha \end{vmatrix}^{-1} \begin{vmatrix} p \\ q \end{vmatrix} \tag{1.69}$$

These are divided into two kinds of currents:

$$\begin{vmatrix} i_\alpha \\ i_\beta \end{vmatrix} = \begin{vmatrix} e_\alpha & e_\beta \\ -e_\beta & e_\alpha \end{vmatrix}^{-1} \begin{vmatrix} p \\ 0 \end{vmatrix} + \begin{vmatrix} e_\alpha & e_\beta \\ -e_\beta & e_\alpha \end{vmatrix}^{-1} \begin{vmatrix} 0 \\ q \end{vmatrix} \tag{1.70}$$

This can be written as

$$\begin{vmatrix} i_\alpha \\ i_\beta \end{vmatrix} = \begin{vmatrix} i_{\alpha p} \\ i_{\beta p} \end{vmatrix} + \begin{vmatrix} i_{\alpha q} \\ i_{\beta q} \end{vmatrix} \tag{1.71}$$

where $i_{\alpha p}$ is the α-axis instantaneous active current:

$$i_{\alpha p} = \frac{e_\alpha}{e_\alpha^2 + e_\beta^2} p \tag{1.72}$$

$i_{\alpha q}$ is the α-axis instantaneous reactive current:

$$i_{\alpha q} = \frac{-e_\beta}{e_\alpha^2 + e_\beta^2} q \tag{1.73}$$

$i_{\beta p}$ is the β-axis instantaneous active current:

$$i_{\beta p} = \frac{e_\alpha}{e_\alpha^2 + e_\beta^2} p \tag{1.74}$$

and $i_{\beta q}$ is the β-axis instantaneous reactive current:

$$i_{\beta q} = \frac{e_\alpha}{e_\alpha^2 + e_\beta^2} q \tag{1.75}$$

The following equations exist:

$$p = e_\alpha i_{\alpha P} + e_\beta i_{\beta P} \equiv P_{\alpha P} + P_{\beta P}$$
$$0 = e_\alpha i_{\alpha q} + e_\beta i_{\beta q} \equiv P_{\alpha q} + P_{\beta q} \tag{1.76}$$

where the α-axis instantaneous active and reactive powers are

$$P_{\alpha p} = \frac{e_\alpha^2}{e_\alpha^2 + e_\beta^2} p \quad P_{\alpha q} = \frac{-e_\alpha e_\beta}{e_\alpha^2 + e_\beta^2} q \tag{1.77}$$

The β-axis instantaneous active power and reactive power are

$$P_{\beta q} = \frac{e_\beta^2}{e_\alpha^2 + e_\beta^2} p \quad P_{\beta q} = \frac{e_\alpha e_\beta}{e_\alpha^2 + e_\beta^2} q \tag{1.78}$$

The sum of the instantaneous active powers in two axes coincides with the instantaneous real power in the three-phase circuit. The instantaneous reactive powers $P_{\alpha q}$ and $P_{\beta q}$ cancel each other and make no contribution to the instantaneous power flow from the source to the load.

Consider instantaneous power flow in a three-phase cycloconverter. The instantaneous reactive power on the source side is the instantaneous reactive power circulating between source and cycloconverter while the instantaneous reactive power on the output side is the instantaneous reactive power between the cycloconverter and the load. Therefore, there is no relationship between the instantaneous reactive powers on the input and output sides, and the instantaneous imaginary power on the input side is not equal to the instantaneous imaginary power on the output side. However, assuming zero active power loss in the converter, the instantaneous real power on the input side is equal to the real output power. An application to active filters is discussed in Chapter 6.

The author in Ref. [15] critiques this theory that it suggests an erroneous interpretation of three-phase power circuits. According to the theory, instantaneous imaginary current can occur in the current of a load with zero reactive power, Also instantaneous active current may occur in the current of a load with zero active power.

1.9   AMPLIFICATION AND ATTENUATION OF HARMONICS

Harmonics originating from their source are propagated in the power systems and their impact can be present at a distance [18]. In this process, the harmonics can be either amplified or mitigated. Capacitor banks in the power system are a major source of harmonic amplifications and waveform distortions. Many different types of harmonic sources may be dispersed throughout the system, and the current and voltage distortions due to these become of concern. Utilities must maintain a certain voltage quality at the consumer premises and, in turn, the harmonics injected into the power systems by a consumer must be controlled and limited. The nature of power system: industrial plant distributions, commercial distribution systems, and utility distribution or transmission systems are important in this aspect. An analysis requires correct estimation of the harmonic generation at a certain point in the power systems, modeling of system components and harmonics themselves for accurate results, for example, constant current injection models for all types of harmonic generation may not be accurate. Based on the accurate harmonic analysis, provisions of active harmonic mitigation strategies at the source of harmonic generation can be applied to limit the harmonics. Passive filters are another important option, especially in large Mvar ratings. These subjects are covered in this book.

New Power Conversion Techniques

Advances in power electronics have resulted in techniques for improving the current wave shape and power factor simultaneously, minimizing the filter requirements [19]. In general, these systems use high-frequency switching to achieve greater flexibility in power conversion and can reduce the lower order harmonics also. Distortion is created at high-frequency switching, which is generally above 20 kHz, and the distortion cannot penetrate into the system (see Chapter 6).

Some publications (books only) on harmonics are separately listed in References. Also some important ANSI/IEEE standards, though referenced appropriately in the rest of the book, are listed.

REFERENCES

1. IEEE Working Group on Power System Harmonics. "Power system harmonics: an overview," IEEE Transactions on Power Apparatus and Systems PAS, vol. 102, pp. 2455–2459, 1983.
2. E. L. Owen, "A history of harmonics in power systems," IEEE Industry Application Magazine, vol. 4, no. 1, pp. 6–12, 1998.
3. M. E. Baran and N. R. Mahajan, "DC distribution for industrial systems: opportunities and challenges," IEEE Transactions on Industry Applications, vol. 39, no. 6, pp. 1596–1601, 2003.
4. J. G. Ciezki and R. W. Ashton, "Selection and stability issues associated with a navy shipboard DC zonal electrical distribution system," IEEE Transactions on Power Delivery, vol. 15, pp. 665–669, 2000.
5. IEEE Standard 519, IEEE recommended practices and requirements for harmonic control in power systems, 1992.
6. ANSI/IEEE Standard 446, IEEE recommended practice for emergency and standby power systems for industrial and commercial applications, 1987.
7. G. O. Calabrase. Symmetrical Components Applied to Electrical Power Networks, Ronald Press Group, New York, 1959.
8. J. L. Blackburn. Symmetrical Components for Power Systems Engineering, Marcel Dekker, New York, 1993.
9. Westinghouse. Westinghouse Transmission and Distribution Handbook, Fourth Edition, Pittsburgh, 1964.
10. J. C. Das. Power System Analysis, Short Circuit Load Flow and Harmonics, 2nd Edition, CRC Press, 2012.
11. L. S. Czarnecki. "What is wrong with Budenu's concept of reactive power and distortion power and why it should be abandoned," IEEE Transactions on Instrumentation and Measurement, vol. IM-36, no. 3, 1987.
12. IEEE Working Group on Non-Sinusoidal Situations. "Practical definitions for powers in systems with non-sinusoidal waveforms and unbalanced loads," IEEE Transactions on Power Delivery, vol. 11, no. 1, pp. 79–101, 1996.
13. IEEE Std. 1459. IEEE standard definitions for the measurement of electrical power quantities under sinusoidal, nonsinusoidal, balanced or unbalanced conditions, 2010.
14. N. L. Kusters and W. J. M. Moore. "On the definition of reactive power under nonsinusoidal conditions," IEEE Transactions on Power Applications, vol. PAS-99, pp. 1845–1854, 1980.
15. L. S. Czarnecki. Energy Flow and Power Phenomena in Electrical Circuits: Illusions and Reality. Springer Verlag, Electrical Engineering, 2000.
16. S. W. Zakikhani. "Suggested definition of reactive power for nonsinusoidal systems," IEE Proceedings, no. 119, pp. 1361–1362, 1972.
17. H. Akagi and A. Nabe. "The p-q theory in three-phase systems under non-sinusoidal conditions," ETEP, vol. 3, pp. 27–30, 1993.
18. A. E. Emanual. "On the assessment of harmonic pollution," IEEE Transactions on Power Delivery, vol. 10, no. 3, pp. 1693–1698, 1995.
19. F. L. Luo and H. Ye. Power Electronics, CRC Press, Boca Raton, FL, 2010.

Books on Power System Harmonics

20. E. Acha and M. Madrigal. Power System Harmonics: Computer Modeling and Analysis, John Wiley & Sons, 2001
21. J. Arrillaga, B. C. Smith, N. R. Watson and A. R. Wood. Power System Harmonic Analysis, John Wiley & Sons, 2000
22. J. Arrillaga and N. R. Watson. Power System Harmonics, 2nd Edition, John Wiley & Sons, 2003.
23. F. C. De La Rosa, Harmonics and Power Systems, CRC Press, 2006.
24. G. J. Wakileh. Power System Harmonics: Fundamentals, Analysis and Filter Design, Springer, 2001.
25. J. C. Das. Power System Analysis, Short Circuit Load Flow and Harmonics, 2nd Edition, CRC Press, 2012.

26. S. M. Ismail, S. F. Mekhamer and A. Y. Abdelaziz. Power System Harmonics in Industrial Electrical Systems-Techno-Economical Assessment, Lambert Academic Publishing, 2013.

27. A. Fadnis, Harmonics in Power Systems: Effects of Power Switching Devices in Power Systems, Lambert Academic Publishing, 2012.

28. A. Nassif, Harmonics in Power Systems-Modeling, Measurement and Mitigation, CRC Press, 2010.

29. E.W. Kimbark. Direct Current Transmission, Vol. 1, John Wiley & Sons, New York, 1971.

Books on Power Quality

30. J. Arrillaga, S. Chen and N. R. Watson, Power Quality Assessment, John Wiley & Sons, 2000.

31. A. Baggini, Handbook of Power Quality, John Wiley & Sons, 2008

32. M. H. J. Bollen, Understanding Power Quality Problems, IEEE, 2000.

33. B. Kennedy, Power Quality Primer, McGraw-Hill, 2000.

34. R. C. Dugan, M. F. McGranaghan and H. W. Beaty, Electrical Power Systems Quality, McGraw-Hill, 1976.

35. E.F. Fuchs and M.A.S. Masoum, Power Quality in Power Systems and Electrical Machines, Elsevier, 2008.

36. A. Ghosh and G. Ledwich, Power Quality Enhancement Using Custom Power Devices, Kulwar Academic Publishers, Norwell, MA, 2002.

37. G. T. Heydt, Electrical Power Quality, 2nd Edition, Stars in a Circle Publications, 1991.

38. A. Kusko, and M. T. Thompson, Power Quality in Electrical Systems, McGraw-Hill Professional, 2007.

39. W. Mielczarski, G. J. Anders, M. F. Conlon, W. B. Lawrence, H. Khalsa, and G. Michalik, Quality of Electricity Supply and Management of Network Losses, Puma Press, Melbourne, 1997.

40. A. Moreno-Muoz (Ed.) Power Quality Mitigation Technologies in a Distributed Environment, Springer, 2007.

41. C. Sankran, Power Quality, CRC Press, 2002.

Major IEEE Standards Related to Harmonics

42. IEEE Standard 519, IEEE recommended practices and requirements for harmonic control in power systems, 1992.

43. IEEE Standard C37.99, IEEE Guide for Protection of Shunt Capacitor Banks, 2000

44. IEEE P519.1/D9a. Draft for applying harmonic limits on power systems, 2004.

45. ANSI/IEEE Standard C37.99, Guide for protection of shunt capacitor banks, 2000.

46. IEEE P1036/D13a. Draft Guide for Application of Shunt Power Capacitors, 2006.

47. IEEE Standard C57.110. IEEE recommended practice for establishing liquid-filled and dry-type power and distribution transformer capability when supplying nonsinusoidal load currents, 2008.

48. IEEE Standard 18. IEEE standard for shunt power capacitors, 2002.

49. IEEE Standard 1531, IEEE guide for application and specifications of harmonic filters, 2003.

CHAPTER 2

FOURIER ANALYSIS

The French mathematician J. B. J. Fourier (1758–1830) showed that arbitrary peri-odic functions could be represented by an infinite series of sinusoids of harmonically related frequencies. This was related to heat flow as the electrical applications were not developed at that time. We first define periodic functions.

2.1 PERIODIC FUNCTIONS

A function is said to be periodic if it is defined for all real values of t and if there is a positive number T such that

$$f(t) = f(t + T) = f(t + 2T) = f(t + nT) \tag{2.1}$$

then T is called the period of the function.

If k is any integer and $f(t + kT) = f(t)$ for all values of t and if two functions $f_1(t)$ and $f_2(t)$ have the same period T, then the function $f_3(t) = af_1(t) + bf_2(t)$, where a and b are constants, also has the same period T. Figure 2.1 shows periodic functions.

The functions

$$f_1(t) = \cos \frac{2\pi n}{T} t = \cos n\omega_0 t$$

$$f_2(t) = \sin \frac{2\pi n}{T} t = \sin n\omega_0 t \tag{2.2}$$

are of special interest. Each frequency of the sinusoids $n\omega_0$ is said to be of nth har-monic of the fundamental frequency ω_0, and each of these frequencies is related to period t.

2.2 ORTHOGONAL FUNCTIONS

Two functions $f_1(t)$ and $f_2(t)$ are orthogonal over the interval (T_1, T_2) if

$$\int_{T_1}^{T_2} f_1(t) f_2(t) = 0 \tag{2.3}$$

Power System Harmonics and Passive Filter Designs, First Edition. J.C. Das.
© 2015 The Institute of Electrical and Electronics Engineers, Inc. Published 2015 by John Wiley & Sons, Inc.

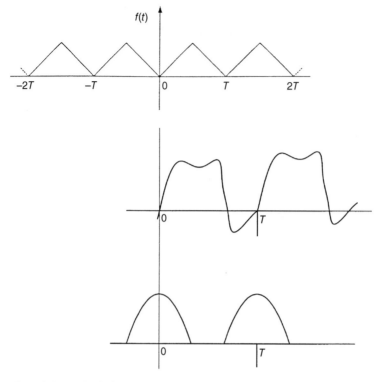

Figure 2.1   Periodic functions.

Figure 2.2 shows the orthogonal functions over a period T. Observe that

$$\int_0^T \sin\, m\omega_0 t\ dt = 0 \quad \text{all}\ m$$

$$\int_0^T \cos\, n\omega_0 t\ dt = 0 \quad \text{all}\ n \neq 0 \tag{2.4}$$

The average value of a sinusoid over m or n complete cycles is zero; therefore, the following three cross products are also zero.

$$\int_0^T \sin m\omega_0 t\, dt.\cos n\omega_0 t\, dt = 0 \quad \text{all}\ m, n$$

$$\int_0^T \sin m\omega_0 t\, dt.\sin n\omega_0 t\, dt = 0 \quad m \neq n$$

$$\int_0^T \cos m\omega_0 t\, dt.\cos n\omega_0 t\, dt = 0 \quad m \neq n \tag{2.5}$$

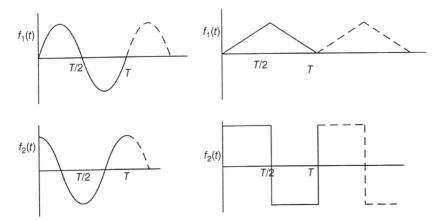

Figure 2.2   Orthogonal functions.

Nonzero values occur when $m = n$:

$$\int_0^T \sin^2 m\omega_0 t \, dt = T/2 \quad \text{all } m$$

$$\int_0^T \cos^2 m\omega_0 t \, dt = T/2 \quad \text{all } n \tag{2.6}$$

2.3   FOURIER SERIES AND COEFFICIENTS

A periodic function can be expanded in a Fourier series. The series has the expression:

$$f(t) = a_0 + \sum_{n=1}^{\infty} \left(a_n \cos \left(\frac{2\pi nt}{T} \right) + b_n \sin \left(\frac{2\pi nt}{T} \right) \right) \tag{2.7}$$

where a_0 is the average value of function $f(t)$. It is also called the DC component, and a_n and b_n are called the coefficients of the series. A series such as Eq. (2.7) is called a trigonometric Fourier series. The Fourier series of a periodic function is the sum of sinusoidal components of different frequencies. The term $2\pi/T$ can be written as ω. The nth term $n\omega$ is then called the nth harmonic and $n = 1$ gives the fundamental; $a_0, a_n,$ and b_n are calculated as follows:

$$a_0 = \frac{1}{T} \int_{-T/2}^{T/2} f(t) dt \tag{2.8}$$

$$a_n = \frac{2}{T} \int_{-T/2}^{T/2} \cos \left(\frac{2\pi nt}{T} \right) dt \quad \text{for } n = 1, 2, \dots, \infty \tag{2.9}$$

$$b_n = \frac{2}{T} \int_{-T/2}^{T/2} \sin\left(\frac{2\pi nt}{T}\right) dt \qquad \text{for } n = 1, 2, \ldots, \infty \qquad (2.10)$$

These equations can be written in terms of angular frequency:

$$a_0 = \frac{1}{2\pi} \int_{-\pi}^{\pi} f(x)\, \omega t d\omega t \qquad (2.11)$$

$$a_n = \frac{1}{\pi} \int_{-\pi}^{\pi} f(x)\, \omega t \, \cos(n\omega t) \, d\omega t \qquad (2.12)$$

$$b_n = \frac{1}{\pi} \int_{-\pi}^{\pi} f(x)\, \omega t \sin(n\omega t) \, d\omega t \qquad (2.13)$$

This gives

$$x(t) = a_0 + \sum_{n=1}^{\infty} [a_n \cos(n\omega t) + b_n \sin(n\omega t)] \qquad (2.14)$$

We can write

$$a_n \cos n\omega t + b_n \sin \omega t = [a_n^2 + b_n^2]^{1/2}[\sin \phi_n \cos n\omega t + \cos \phi_n \sin n\omega t]$$
$$= [a_n^2 + b_n^2]^{1/2} \sin(n\omega t + \phi_n) \qquad (2.15)$$

where

$$\phi_n = \tan^{-1} \frac{a_n}{b_n}$$

The coefficients can be written in terms of two separate integrals:

$$a_n = \frac{2}{T} \int_0^{T/2} x(t) \cos\left(\frac{2\pi nt}{T}\right) dt + \frac{2}{T} \int_{-T/2}^{0} x(t) \cos\left(\frac{2\pi nt}{T}\right) dt$$

$$b_n = \frac{2}{T} \int_0^{T/2} x(t) \sin\left(\frac{2\pi nt}{T}\right) dt + \frac{2}{T} \int_{-T/2}^{0} x(t) \sin\left(\frac{2\pi nt}{T}\right) dt \qquad (2.16)$$

Example 2.1:   Find the Fourier series of a periodic function of period 1 defined by

$$f(x) = 1/2 + x, \quad -1/2 < x \le 0$$
$$= 1/2 - x, \quad 0 < x < 1/2$$

When the period of the function is not 2π, it is converted to length 2π, and the independent variable is also changed proportionally. Say, if the function is defined in interval $(-t, t)$, then 2π is interval for the variable $= \pi x/t$, so put $z = \pi x/t$ or $x = zt/\pi$. The function $f(x)$ of $2t$ is transformed to function $f(tz/\pi)$ or $F(z)$ of 2π. Let

$$f(x) = \frac{a_0}{2} + a_1 \cos \frac{\pi x}{t} + a_2 \cos \frac{2\pi x}{t} + \ldots b_1 \sin \frac{\pi x}{t} + a_2 \sin \frac{2\pi x}{t} + \ldots$$

$$2t = 1$$

By definition,

$$a_0 = \frac{1}{1/2} \int_{-1/2}^{0} \left(\frac{1}{2} + x \right) dx + \frac{1}{1/2} \int_{0}^{1/2} \left(\frac{1}{2} - x \right) dx = 1/2$$

$$a_n = \frac{1}{t} \int_{-t}^{t} f(x) \cos \frac{n\pi x}{t} \, dx$$

$$= \frac{1}{1/2} \int_{-1/2}^{0} \left(\frac{1}{2} + x \right) \cos \frac{n\pi x}{1/2} dx + \int_{0}^{1/2} \left(\frac{1}{2} - x \right) \cos \frac{n\pi x}{1/2} dx$$

$$= 2 \left[\left(\frac{1}{2} + x \right) \frac{\sin 2n\pi x}{2n\pi} - (1) \left(\frac{\cos 2n\pi x}{4n^2\pi^2} \right) \right]_{-1/2}^{0}$$

$$+ 2 \left[\left(\frac{1}{2} - x \right) \frac{\sin 2n\pi x}{2n\pi} - (-1) \left(\frac{-\cos 2n\pi x}{4n^2\pi^2} \right) \right]_{0}^{1/2}$$

$$= \frac{2}{n^2\pi^2} \quad \text{for} \quad n = \text{odd}$$

$$= 0 \quad \text{for} \quad n = \text{even}$$

$$b_n = \frac{1}{t} \int_{-t}^{t} f(x) \sin \frac{n\pi x}{t} \, dx$$

$$= \frac{1}{1/2} \int_{-1/2}^{0} \left(\frac{1}{2} + x \right) \sin \frac{n\pi x}{1/2} dx + \int_{0}^{1/2} \left(\frac{1}{2} - x \right) \sin \frac{n\pi x}{1/2} dx$$

$$= 2 \left[\left(\frac{1}{2} + x \right) \frac{-\cos 2n\pi x}{2n\pi} - (1) \left(-\frac{\sin 2n\pi x}{4n^2\pi^2} \right) \right]_{-1/2}^{0}$$

$$+ 2 \left[\left(\frac{1}{2} - x \right) \frac{-\cos 2n\pi x}{2n\pi} - (-1) \left(\frac{-\sin 2n\pi x}{4n^2\pi^2} \right) \right]_{0}^{1/2} = 0$$

Substituting the values

$$f(x) = \frac{1}{4} + \frac{2}{\pi^2} \left[\frac{\cos 2\pi x}{1^2} + \frac{\cos 6\pi x}{3^2} + \frac{\cos 10\pi x}{5^2} - \dots \right]$$

2.4   ODD SYMMETRY

A function $f(x)$ is said to be an odd or skew symmetric function, if

$$f(-x) = -f(x) \tag{2.17}$$

The area under the curve from $-T/2$ to $T/2$ is zero. This implies that

$$a_0 = 0, a_n = 0 \tag{2.18}$$

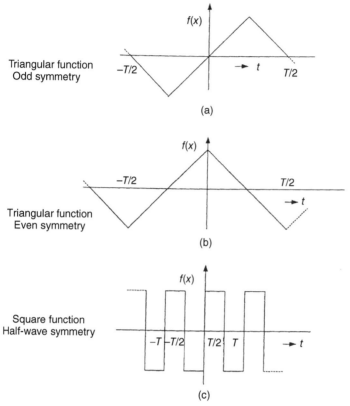

Triangular function
Odd symmetry

(a)

Triangular function
Even symmetry

(b)

Square function
Half-wave symmetry

(c)

Figure 2.3   (a) Triangular function with odd symmetry, (b) triangular function with even symmetry, and (c) square function with half-wave symmetry.

$$b_n = \frac{4}{T} \int_0^{T/2} f(t) \sin\left(\frac{2\pi nt}{T}\right) dt \qquad (2.19)$$

Figure 2.3(a) shows a triangular function, having odd symmetry, the Fourier series contains only sine terms.

2.5   EVEN SYMMETRY

A function $f(x)$ is even symmetric, if

$$f(-x) = f(x) \qquad (2.20)$$

The graph of such a function is symmetric with respect to the y-axis. The y-axis is a mirror reflection of the curve.

$$a_0 = 0, b_n = 0 \qquad (2.21)$$

$$a_n = \frac{4}{T} \int_0^{T/2} f(t) \cos\left(\frac{2\pi nt}{T}\right) dt \qquad (2.22)$$

Figure 2.3(b) shows a triangular function with even symmetry. The Fourier series contains only cosine terms. Note that the odd and even symmetry has been obtained with the triangular function by shifting the origin.

2.6   HALF-WAVE SYMMETRY

A function is said to have half-wave symmetry if

$$f(x) = -f(x + T/2) \qquad (2.23)$$

Figure 2.3(c) shows that a square-wave function has half-wave symmetry, with respect to the period $-T/2$. The negative half-wave is the mirror image of the positive half, but phase shifted by $T/2$ (or π radians). Due to half-wave symmetry, the average value is zero. The function contains only odd harmonics.

If n is odd, then

$$a_n = \frac{4}{T} \int_0^{T/2} x(t) \cos\left(\frac{2\pi nt}{T}\right) dt \qquad (2.24)$$

and $a_n = 0$ for $n = $ even.

$$b_n = \frac{4}{T} \int_0^{T/2} x(t) \sin\left(\frac{2\pi nt}{T}\right) dt \qquad (2.25)$$

for $n = $ odd, and it is zero for $n = $ even.

Example 2.2:   Calculate the Fourier series for an input current to a six-pulse converter, with a firing angle of α.

Then, as the wave is symmetrical, DC component is zero.

The waveform pattern with firing angle α is shown in Fig. 2.4.

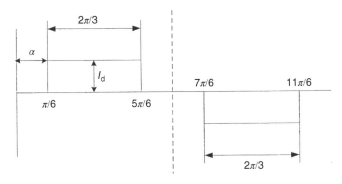

Figure 2.4   Waveform for Example 2.2.

The Fourier series of the input current is

$$\sum_{n=1}^{\infty}(a_n \cos n\omega t + b_n \sin n\omega t)$$

$$a_n = \frac{1}{\pi}\left[\int_{\pi/6+\alpha}^{5\pi/6+\alpha} I_d \cos n\omega t \ d(\omega t) - \int_{7\pi/6+\alpha}^{11\pi/6+\alpha} I_d \cos n\omega t \ d(\omega t)\right]$$

$$= -\frac{4I_d}{n\pi}\sin\frac{n\pi}{3}\sin n\alpha, \quad \text{for} \quad n = 1,3,5, \ldots$$

$$= 0, \quad \text{for } n = 2,6, \ldots$$

$$b_n = \frac{1}{\pi}\left[\int_{\pi/6+\alpha}^{5\pi/6+\alpha} I_d \sin n\omega t \ d(\omega t) - \int_{7\pi/6+\alpha}^{11\pi/6+\alpha} I_d \sin n\omega t \ d(\omega t)\right]$$

$$= \frac{4I_d}{n\pi}\sin\frac{n\pi}{3}\cos n\alpha \quad \text{for} \quad n = 1,3,5..$$

$$= 0, \text{ for } \quad n = \text{even}$$

We can write the Fourier series as

$$i = \sum_{n=1,2,\ldots}^{\infty} \sqrt{2}I_n \sin(n\omega t + \phi_n)$$

where i is the instantaneous current and

$$\phi_n = \tan^{-1}\frac{a_n}{b_n} = -n\alpha$$

Rms value of nth harmonic is

$$I_{n,\text{rms}} = \frac{1}{\sqrt{2}}(a_n^2 + b_n^2)^{1/2}$$

$$= \frac{2\sqrt{2}I_d}{n\pi}\sin\frac{n\pi}{3}$$

The fundamental rms current is

$$I_1 = \frac{\sqrt{6}}{\pi}I_d = 0.7797I_d$$

Example 2.3:   A single-phase full bridge supplies a motor load. Assuming that the motor DC current is ripple free, determine the input current (using Fourier analysis), harmonic factor, distortion factor, and power factor for an ignition delay angle of α.

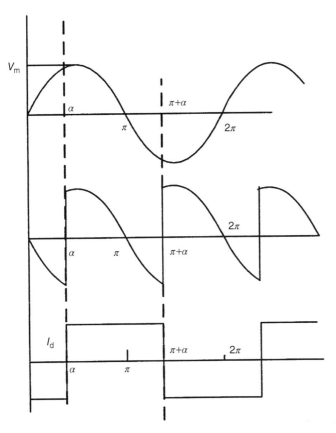

Figure 2.5   Waveforms of fully controlled single-phase bridge (Example 2.3).

The waveform of full-wave single-phase bridge rectifier is shown in Fig. 2.5. The average value of DC voltage is

$$V_{DC} = \int_{\alpha}^{\pi+\alpha} V_m \sin \omega t \ d(\omega t)$$

$$= \frac{2V_m}{\pi} \cos \alpha$$

It can be controlled by change of conduction angle α.

From Fig. 2.5, the instantaneous input current can be expressed in the Fourier series as

$$I_{input} = I_{DC} + \sum_{n=1,2,\,...}^{\infty} (a_n \cos n\omega t + b_n \sin n\omega t)$$

$$I_{DC} = \frac{1}{2\pi} \int_{\alpha}^{2\pi+\alpha} i(t)d(\omega t) = \frac{1}{\pi} \left[\int_{\alpha}^{\pi+\alpha} I_a d(\omega t) + \int_{\pi+\alpha}^{2\pi+\alpha} I_a d(\omega t) \right] = 0$$

Also

$$a_n = \frac{1}{\pi} \int_{\alpha}^{2\pi+\alpha} i(t) \cos n\omega t \ d(\omega t)$$

$$= -\frac{4I_a}{n\pi} \sin n\alpha \quad \text{for } n = 1, 3, 5$$

$$= 0 \quad \text{for } n = 2, 4, \ldots$$

$$b_n = \frac{1}{\pi} \int_{\alpha}^{2\pi+\alpha} i(t) \sin n\omega t \ d(\omega t)$$

$$= \frac{4I_a}{n\pi} \cos n\alpha \quad \text{for } n = 1, 3, 5$$

$$= 0 \quad \text{for } n = 2, 4, \ldots$$

We can write the instantaneous input current as

$$i_{\text{input}} = \sum_{n=1,2,\ldots}^{\infty} \sqrt{2} I_n \ \sin(\omega t + \phi_n)$$

where

$$\phi_n = \tan^{-1} \left(\frac{a_n}{b_n} \right) = -n\alpha$$

$\phi_n = -n\alpha$ is the displacement angle of the nth harmonic current. The rms value of the nth harmonic input current is

$$I_n = \frac{1}{\sqrt{2}} (a_n^2 + b_n^2)^{1/2} = \frac{2\sqrt{2}}{n\pi} I_a$$

The rms value of the fundamental current is

$$I_1 = \frac{2\sqrt{2}}{\pi} I_d$$

Thus, the rms value of the input current is

$$I_{\text{rms}} = \left(\sum_{n=1}^{\infty} I_n^2 \right)^{1/2}$$

The harmonic factor is

$$\text{HF} = \left[\left(\frac{I_{\text{rms}}}{I_1} \right)^2 - 1 \right]^{1/2} = 0.4834$$

The displacement factor is

$$DF = \cos \phi_1 = \cos(-\alpha)$$

The power factor is

$$PF = \frac{V_{rms}I_1}{V_{rms}I_{rms}} \cos \phi_1 = \frac{2\sqrt{2}}{\pi} \cos \alpha$$

2.7   HARMONIC SPECTRUM

The Fourier series of a square-wave function is

$$f(t) = \frac{4k}{\pi} \left(\frac{\sin \omega t}{1} + \frac{\sin 3\omega t}{3} + \frac{\sin 5\omega t}{5} + \cdots \right) \qquad (2.26)$$

where k is the amplitude of the function. The magnitude of the nth harmonic is $1/n$, when the fundamental is expressed as one per unit.

The construction of a square wave from the component harmonics is shown in Fig. 2.6(a), and the plotting of harmonics as a percentage of the magnitude of the fundamental gives the harmonic spectrum of Fig. 2.6(b). A harmonic spectrum indicates the relative magnitude of the harmonics with respect to the fundamental and is not indicative of the sign (positive or negative) of the harmonic nor its phase angle.

2.8   COMPLEX FORM OF FOURIER SERIES

A vector with amplitude A and phase angle θ with respect to a reference can be resolved into two oppositely rotating vectors of half the magnitude so that

$$|A| \cos \theta = |A/2|e^{j\theta} + |A/2|e^{-j\theta} \qquad (2.27)$$

Thus,

$$a_n \cos n\omega t + b_n \sin n\omega t \qquad (2.28)$$

can be substituted by

$$\cos(n\omega t) = \frac{e^{jn\omega t} + e^{-jn\omega t}}{2} \qquad (2.29)$$

$$\sin(n\omega t) = \frac{e^{jn\omega t} - e^{-jn\omega t}}{2j} \qquad (2.30)$$

Thus,

$$x(t) = \frac{a_0}{2} + \frac{1}{2} \sum_{n=1}^{n=\infty} (a_n - jb_n)e^{jn\omega t} + \frac{1}{2} \sum_{n=1}^{n=\infty} (a_n - jb_n)e^{-jn\omega t} \qquad (2.31)$$

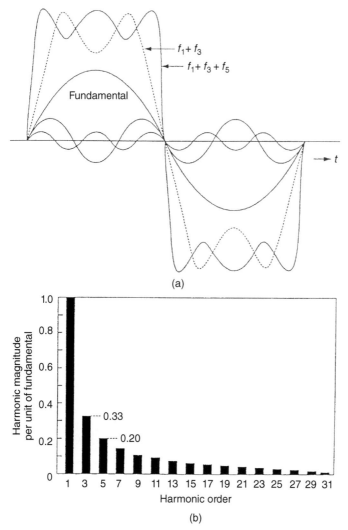

(a)

(b)

Figure 2.6 (a) Construction of a square wave from its harmonic components and (b) harmonic spectrum.

We introduce negative values of n in the coefficients, that is,

$$a_{-n} = \frac{2}{T}\int_{-T/2}^{T/2} x(t)\cos(-n\omega t)dt = \frac{2}{T}\int_{-T/2}^{T/2} x(t)\cos(n\omega t)dt = a_n \quad n = 1, 2, 3, \ldots$$

(2.32)

$$b_{-n} = \frac{2}{T}\int_{-T/2}^{T/2} x(t)\sin(-n\omega t)dt = -\frac{2}{T}\int_{-T/2}^{T/2} x(t)\sin(n\omega t)dt = -b_n \quad n = 1, 2, 3, \ldots$$

(2.33)

Hence,

$$\sum_{n=1}^{\infty} a_n e^{-jn\omega t} = \sum_{n=-1}^{\infty} a_n e^{jn\omega t} \qquad (2.34)$$

and

$$\sum_{n=1}^{\infty} jb_n e^{-jn\omega t} = \sum_{n=-1}^{\infty} jb_n e^{jn\omega t} \qquad (2.35)$$

Therefore, substituting in Eq. (2.31), we obtain

$$x(t) = \frac{a_0}{2} + \frac{1}{2} \sum_{n=-\infty}^{\infty} (a_n - jb_n) e^{jn\omega t} = \sum_{n=-\infty}^{\infty} c_n e^{jn\omega t} \qquad (2.36)$$

This is the expression for a Fourier series expressed in exponential form, which is the preferred approach for analysis. The coefficient c_n is complex and is given by

$$c_n = \frac{1}{2}(a_n - jb_n) = \frac{1}{T} \int_{-T/2}^{T/2} x(t) e^{-jn\omega t} dt \quad n = 0, \pm 1, \pm 2, \ldots \qquad (2.37)$$

2.9   FOURIER TRANSFORM

Fourier analysis of a continuous periodic signal in the time domain gives a series of discrete frequency components in the frequency domain. The Fourier integral is defined by the expression:

$$X(f) = \int_{\infty}^{-\infty} x(t) e^{-j2\pi ft} dt \qquad (2.38)$$

If the integral exists for every value of parameter f (frequency), then this equation describes the Fourier transform. The Fourier transform is a complex quantity:

$$X(f) = R(f) + jI(f) = |X(f)| e^{j\phi(f)} \qquad (2.39)$$

where $R(f)$ is the real part of the Fourier transform and $I(f)$ is the imaginary part of the Fourier transform. The amplitude or *Fourier spectrum of x(t)* is given by

$$|X(f)| = \sqrt{R^2(f) + I^2(f)} \qquad (2.40)$$

The phase angle of the Fourier transform is given by

$$\phi(f) = \tan^{-1} \frac{I(f)}{R(f)} \qquad (2.41)$$

The inverse Fourier transform or the backward Fourier transform is defined as

$$x(t) = \int_{-\infty}^{\infty} X(f)e^{j2\pi ft} df \qquad (2.42)$$

Inverse transformation allows determination of a function in time domain from its Fourier transform. Equations (2.38) and (2.42) are a Fourier transform pair, and the relationship can be indicated by

$$x(t) \leftrightarrow X(f) \qquad (2.43)$$

Fourier transform pair is also written as

$$X(w) = a_1 \int_{-\infty}^{\infty} x(t)e^{-j\omega t} dt$$

$$x(t) = a_2 \int_{-\infty}^{\infty} X(\omega)e^{j\omega t} d\omega$$

where a_1 and a_2 can take different values depending on the user, some take $a_1 = 1$ and $a_2 = 1/2\pi$, or set $a_1 = 1/2\pi$ and $a_2 = 1$ or $a_1 = a_2 = 1/\sqrt{2\pi}$. The requirement is that $a_1 \times a_2 = 1/2\pi$. In most texts, it is defined as

$$X(w) = \int_{-\infty}^{\infty} x(t)e^{-j\omega t} dt$$

$$x(t) = \frac{1}{2\pi} \int_{-\infty}^{\infty} X(\omega)e^{j\omega t} d\omega$$

However, definitions in equations (2.38) and (2.42) are consistent with Laplace transform.

Example 2.4: Consider a function defined as

$$x(t) = \beta e^{-\alpha t} \, t > 0$$

$$= 0 \; t < 0 \qquad (2.44)$$

It is required to write its forward Fourier transform.
From Eq. (2.38),

$$X(f) = \int_{0}^{\infty} \beta e^{-\alpha t} e^{-j2\pi ft} dt$$

$$= \frac{-\beta}{\alpha + j2\pi f} e^{-(\alpha + j2\pi f)t} \Big|_{0}^{\infty}$$

$$= \frac{\beta}{\alpha + j2\pi f} = \frac{\beta\alpha}{\alpha^2 + (2\pi f)^2} - j\frac{2\pi f\beta}{\alpha^2 + (2\pi f)^2}$$

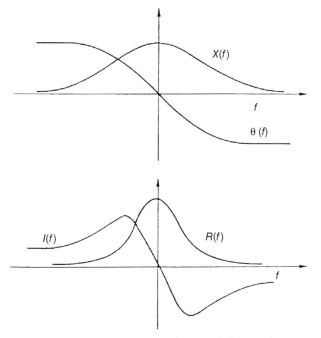

Figure 2.7   Real, imaginary, magnitude, and phase angle representations of the Fourier transform (Example 2.5).

$$R(f) = \frac{\beta\alpha}{\alpha^2 + (2\pi f)^2}$$

$$I(f) = -j\frac{2\pi f\beta}{\alpha^2 + (2\pi f)^2}$$

Thus, $X(f)$ is

$$\frac{\beta}{\sqrt{\alpha^2 + (2\pi f)^2}}e^{j\,\tan^{-1}[-2\pi f/\alpha]} \qquad (2.45)$$

This is plotted in Fig. 2.7.

Example 2.5:   Convert the function arrived at in Example 2.4 to $x(t)$.
    The inverse Fourier transform is

$$x(t) = \int_{-\infty}^{\infty} X(f)e^{j2\pi ft}df$$

$$= \int_{-\infty}^{\infty}\left[\frac{\beta\alpha}{\alpha^2 + (2\pi f)^2} - j\frac{2\pi f\beta}{\alpha^2 + (2\pi f)^2}\right]e^{j2\pi ft}df$$

$$= \int_{-\infty}^{\infty} \left[\frac{\beta\alpha \cos(2\pi ft)}{\alpha^2 + (2\pi f)^2} + \frac{2\pi f\beta \sin(2\pi ft)}{\alpha^2 + (2\pi f)^2} \right] df$$

$$+ j \int_{-\infty}^{\infty} \left[\frac{\beta\alpha \sin(2\pi ft)}{\alpha^2 + (2\pi f)^2} + \frac{2\pi f\beta \cos(2\pi ft)}{\alpha^2 + (2\pi f)^2} \right] df$$

The imaginary term is zero, as it is an odd function.

This can be written as

$$x(t) = \frac{\beta\alpha}{(2\pi)^2} \int_{-\infty}^{\infty} \frac{\cos(2\pi tf)}{(\alpha/2\pi)^2 + f^2} df + \frac{2\pi\beta}{(2\pi)^2} \int_{-\infty}^{\infty} \frac{f \sin(2\pi tf)}{(\alpha/2\pi)^2 + f^2} df$$

As

$$\int_{-\infty}^{\infty} \frac{\cos \alpha x}{b^2 + x^2} dx = \frac{\pi}{b} e^{-ab}$$

and

$$\int_{-\infty}^{\infty} \frac{x \sin \alpha x}{b^2 + x^2} dx = \pi e^{-ab}$$

$x(t)$ becomes

$$x(t) = \frac{\beta\alpha}{(2\pi)^2} \left[\frac{\pi}{\alpha/2\pi} e^{-(2\pi t)(\alpha/2\pi)} \right] + \frac{2\pi\beta}{(2\pi)^2} [\pi e^{-(2\pi t)(\alpha/2\pi)}]$$

$$= \frac{\beta}{2} e^{-\alpha t} + \frac{\beta}{2} e^{-\alpha t} = \beta e^{-\alpha t} \; t > 0$$

that is,

$$\beta e^{-\alpha t} t > 0 \leftrightarrow \frac{\beta}{\alpha + j2\pi f} \tag{2.46}$$

Example 2.6: Consider a function defined by

$$x(t) = K; \text{ for } |t| \leq T/2$$

$$= 0; \text{ for } |t| > T/2 \tag{2.47}$$

It is a bandwidth limited rectangular function (Fig. 2.8(a)); the Fourier transform is

$$X(f) = \int_{-T/2}^{T/2} K e^{-j2\pi fT} dt = KT \left[\frac{\sin(\pi fT)}{\pi fT} \right] \tag{2.48}$$

The term in parentheses in Eq. (2.48) is called the *sinc function*. The function has zero value at points $f = n/T$. Figure 2.8(b) shows zeros and side lobes.

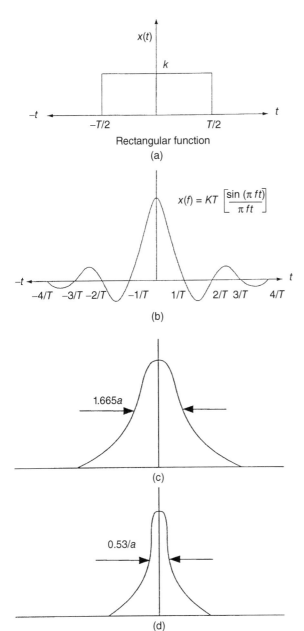

Rectangular function
(a)

$$x(f) = KT \left[\frac{\sin (\pi ft)}{\pi ft} \right]$$

(b)

(c)

(d)

Figure 2.8   (a) Bandwidth limited rectangular function, (b) the sinc function showing side lobes, and (c) and (d) a Gaussian function with its transform.

2.9.1   Fourier Transform of Some Common Functions

Gaussian Function   Consider the function:

$$x(t) = e^{-x^2/a^2} \qquad (2.49)$$

where a is the width parameter. The value of $x(t) = 1/2$ when $(x/a)^2 = \log_e 2$ or $x = \pm 0.9325a$, so that the full width at half maximum (FWHM) = 1.655a. It is shown in Fig. 2.8(c).

$$X(f) = \int_{-\infty}^{\infty} e^{-x^2/a^2} e^{-j2\pi ft} dx$$

$$= a\sqrt{\pi} e^{-\pi^2 a^2 f^2}$$

The Fourier transform is another Gaussian function with width $1/(\pi a)$.

Note that the original function has a width of $1.665a$ at half maximum. The Fourier transform has a narrower width (Fig. 2.8(d)).

Some Common Transforms Figure 2.9 (a–j) shows graphically the Fourier transforms of some common functions.

The following transforms exist:

(a) Fourier transformer of an impulse function:

$$x(t) = K\delta(t)$$

$$X(f) = K \tag{2.50}$$

This means that the Fourier transform of a delta function is unity.

$$\delta(t) \leftrightarrow 1 \tag{2.51}$$

For a pair of delta functions, equally placed on either side of the origin, the Fourier transform is a cosine wave:

$$\delta(t - a)\delta(t - a) = e^{j2\pi fa} + e^{-j2\pi fa}$$

$$= 2 \cos(2\pi fa) \tag{2.52}$$

(b) Fourier transform of a constant amplitude waveform:

$$x(t) = K$$

$$X(f) = K\delta(f) \tag{2.53}$$

(c) Fourier transform of a pulse waveform:

$$x(t) = A \quad |t| < T_0$$

$$= (A/2) \quad |t| = T_0$$

$$= 0 \quad |t| > T_0$$

$$X(f) = 2AT_0 \frac{\sin(2\pi T_0 f)}{2\pi T_0 f} \tag{2.54}$$

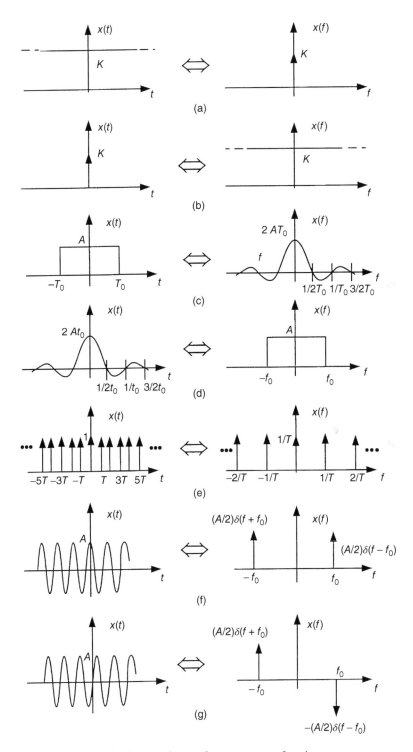

Figure 2.9   (a–j) Fourier transforms of some common functions.

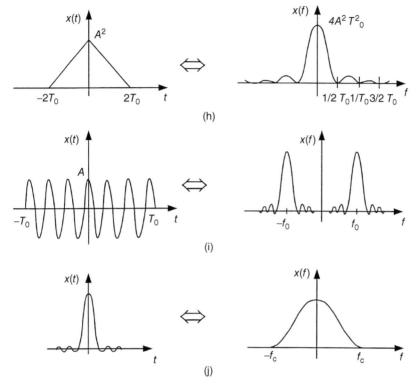

Figure 2.9   (*Continued*)

(d)   This represents situation in reverse.

(e)   The Fourier transform of sequence of equal distance pulses is another sequence of equal distance pulses.

$$x(t) = \sum_{n=-\infty}^{\infty} \delta(t - nT)$$

$$X(f) = \frac{1}{T} \sum_{n=-\infty}^{\infty} \delta\left(f - \frac{n}{T}\right) \qquad (2.55)$$

(f), (g)   Fourier transform of periodic functions

$$x(t) = A \cos(2\pi f_0 t)$$

$$X(f) = \frac{A}{2}\delta(f - f_0) + \frac{A}{2}\delta(f + f_0)$$

$$x(t) = A \sin(2\pi f_0 t)$$

$$X(f) = -j\frac{A}{2}\delta(f - f_0) + j\frac{A}{2}\delta(f + f_0) \qquad (2.56)$$

(h) Fourier transform of triangular function:

$$x(t) = A^2 - \frac{A^2}{2T_0}$$

$$= 0 \quad |t| < 2T_0$$

$$= 0 \quad |t| > 2T_0$$

$$X(f) = A^2 \frac{\sin^2(2\pi T_0 f)}{(\pi f)^2} \qquad (2.57)$$

(i) Fourier transform of

$$x(t) = A\,\cos(2\pi f_0 t) \quad |t| < T_0$$

$$= 0 \qquad\qquad |t| > T_0$$

$$X(f) = A^2 T_0 \left[\frac{\sin(2\pi T_0 f)}{2\pi T_0 f}(f + f_0) + \frac{\sin(2\pi T_0 f)}{2\pi T_0 f}(f - f_0) \right] \qquad (2.58)$$

(j) Fourier transform of

$$x(t) = \frac{1}{2}q(t) + \frac{1}{4}q\left(t + \frac{1}{2f_c}\right) + \frac{1}{4}q\left(t - \frac{1}{2f_c}\right)$$

where

$$q(t) = \frac{\sin(2\pi f_c t)}{\pi t}$$

$$X(f) = \frac{1}{2} + \frac{1}{2}\cos\left(\frac{\pi f}{f_c}\right) \quad |f| \leq f_c$$

$$= 0 \qquad\qquad |f| > f_c \qquad (2.59)$$

(k) Fourier transform of Dirac comb. A Dirac comb is a set of equally spaced δ functions, usually denoted by Cyrillic letter III

$$\text{III}_a(t) = \sum_{n=-\infty}^{\infty} \delta(t - na) \qquad (2.60)$$

The Fourier transform is another Dirac comb:

$$\text{III}_a(t) \Leftrightarrow \frac{1}{a}\text{III}_{1/a}(f) \qquad (2.61)$$

2.10 DIRICHLET CONDITIONS

Fourier transforms cannot be applied to all functions. The Dirichlet conditions are

- Functions $X(f)$ and $f(t)$ are square integrable:

$$\int_{-\infty}^{\infty} [X(f)]^2 dx \quad X(f) \to 0 \;\; as \;\; |X| \to \infty \tag{2.62}$$

This implies that the function is finite. A function shown in Fig. 2.10(a) or (b) does not meet this criterion

- $X(f)$ and $x(t)$ are single valued. The function shown in Fig. 2.10(a) does not meet this criterion. There are three values at point A.

- $X(f)$ and $x(t)$ are piecewise continuous. The functions can be broken into separate pieces, so that these can be isolated discontinuous, any number of times.

- Functions $X(f)$ and $x(t)$ have upper and lower bonds. This is the condition that is *sufficient* but not proved to be *necessary*.

Mostly the functions do behave so that Dirichlet conditions are fulfilled. Consider the so-called "sign" function shown in Fig. 2.11(a) and defined as

$$\mathrm{sgn}(t) = -1 \quad -\infty < t < 0$$

$$= +1 \quad 0 < t < \infty \tag{2.63}$$

Divide by 2 and add 1/2 to give a Heavyside step of unit height.

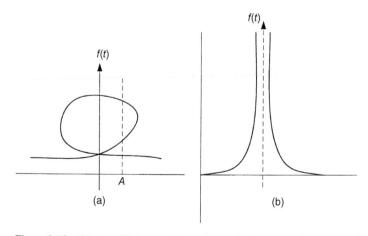

Figure 2.10   (a) A multiple valued function and (b) a discontinuous function.

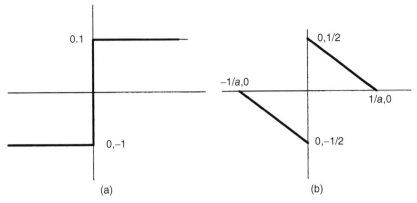

Figure 2.11   (a) A sgn function and (b) representation of a Heavyside step function by two functions that obey Dirichlet constraints.

The function sign(t)$/2$ does not obey Dirichlet conditions but can be approximated by considering it as a limiting case of a pair of ramp functions (Fig. 2.11(b))

$$x(t) = \lim_{a\to 0}\frac{-(at+1)}{2} \quad -1/a < x < 0$$

$$= \lim_{a\to 0}\frac{(1-at)}{2} \quad 0 < x < 1/a \tag{2.64}$$

A unit step function $u(t)$ can be written in terms of sign function:

$$u(t) = \frac{1}{2} + \frac{1}{2}\text{sgn}(t)$$

Its Fourier transform is

$$\pi\delta(\omega) + \frac{1}{j\omega}$$

Some relations where Dirichlet conditions are applicable are

$$X_1(f) + X_2(f) \leftrightarrow x_1(t) + x_2(t)$$

$$X(f+a) \leftrightarrow x(t)e^{j2\pi fa}$$

$$X(f-a) \leftrightarrow x(t)e^{-j2\pi fa} \tag{2.65}$$

Note that if $X(f)$ is a delta function, then

$$\delta(X+a) \leftrightarrow x(t)e^{j2\pi fa}$$

$$\delta(X-a) \leftrightarrow x(t)e^{-j2\pi fa} \tag{2.66}$$

2.11 POWER SPECTRUM OF A FUNCTION

The notion of power spectrum is important in electrical engineering. Consider that the voltage at a point varies with time denoted by $V(t)$. Let $X(f)$ be the Fourier transform of $V(t)$, which can even be negative. Then the power per unit frequency interval being transmitted is proportional to

$$X(f)X(f)^* \tag{2.67}$$

The superscript "*" describes a conjugate. The constant of proportionality depends on load impedance. The function

$$X(f)X(f)^* = |X(f)|^2 \tag{2.68}$$

is called the power spectrum or the spectral power density (SPD) of $V(t)$.

Using equation (2.32), P can be written as

$$P = \frac{1}{T} \int_{-T/2}^{T/2} x^2(t)dt = \frac{1}{T} \int_{-T/2}^{T/2} x(t)(c_n e^{jn\omega t})dt \tag{2.69}$$

Interchanging operation of summation and integration:

$$P = \frac{1}{T} e^{jn\omega t} c_n \int_{-T/2}^{T/2} x(t)(e^{jn\omega t})dt$$

$$= \sum_{n=-\infty}^{\infty} c_n c_{-n}$$

As

$$c_n^* = c_{-n}$$

$$P = \sum_{n=-\infty}^{\infty} |c_n|^2 = |c_0|^2 + 2\sum_{n=1}^{\infty} |F_n|^2 \tag{2.70}$$

This is *Parseval's theorem as applied to exponential Fourier series*. Power in a periodic signal is sum of component powers in exponential Fourier series.

$|c_n|^2$ plotted as a function of $n\omega$ is called power spectrum of $x(t)$

Example 2.7:   Fourier series is required for a function of periodic pulse train shown in Fig. 2.12. This can be called a Dirac comb.

From Eq. (2.32),

$$c_n = \frac{1}{T} \int_{-T/2}^{T/2} x(t)e^{-jn\omega t}dt = \frac{Ad}{T} \sin c\left(\frac{n\pi d}{T}\right)$$

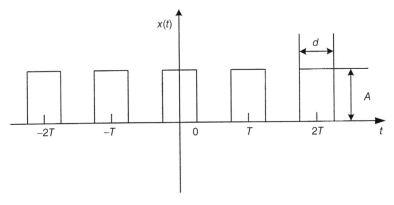

Figure 2.12   A periodic pulse train, Dirac comb.

If $A = 1$, $d = 1/16$, and $T = 1/4$, then

$$c_n = \frac{1}{4} \sin c \left(\frac{n\pi}{4} \right)$$

Thus, Fourier series is given by

$$\sum_{n=-\infty}^{\infty} c_n e^{jn\omega t}$$

The Fourier transform of $x(t)$ is

$$\frac{2\pi A d}{T} \sum_{n=-\infty}^{\infty} \sin c \left(\frac{n\pi d}{T} \right) \delta(\omega - n\omega_0) \quad \omega_0 = \frac{2\pi}{T}$$

The spectrum has first zero crossing at $n = 4$. The power within first zero crossing is

$$P_{n=4} = |c_0|^2 + 2\{|c_1|^2 + |c_2|^2 + |c_3|^2\}$$

$$= \left(\frac{1}{4} \right)^2 + \frac{2}{4^2} \left[\sin c^2 \left(\frac{\pi}{4} \right) + \sin c^2 \left(\frac{\pi}{2} \right) + \sin c^2 \left(\frac{3\pi}{4} \right) \right]$$

$$= \frac{1}{16} + \frac{1}{8}(0.811 + 0.405 + 0.090) = 0.226$$

The total power of the $x(t)$ is

$$P = \frac{1}{T} \int_{-T/2}^{T/2} x^2(t) dt = \frac{1}{4} \int_{-1/32}^{1/32} 1 dt = 0.25$$

2.12   CONVOLUTION

2.12.1   Time Convolution

If

$$x_1(t) \leftrightarrow X_1(\omega)$$

$$x_2(t) \leftrightarrow X_2(\omega) \tag{2.71}$$

then

$$x_1(t) * x_2(t) \leftrightarrow X_1(\omega)X_2(\omega) \tag{2.72}$$

This signifies that convolution in the time domain is multiplication in the frequency domain. Convolution is generally carried out in the frequency domain.

2.12.2   Frequency Convolution

$$x_1(t)x_2(t) \leftrightarrow \frac{1}{2\pi}X_1(\omega) * X_2(\omega) \tag{2.73}$$

Thus, the convolution operation in one domain is transformed to a product operation in the other domain. This has led to the use of transform method, though the time domain is becoming more attractive, for dealing with large dimensional systems. The use of block diagrams and signal flow graphs in the transform domain treats the convolution as an algebraic operator.

The distributive rule:

$$X_1(f) * [X_2(f) + X_3(f)] = X_1(f) * X_2(f) + X_1(f) * X_3(f) \tag{2.74}$$

The commutative rule:

$$X_1(f) * X_2(f) = X_2(f) * X_1(f) \tag{2.75}$$

The associative rule

$$X_1(f) * [X_2(f) * X_3(f)] = [X_1(f) * X_2(f)] * X_3(f) \tag{2.76}$$

Convolution of three functions:

$$X_1(f) * X_2(f) * X_3(f) = \int_{-\infty}^{\infty} \int_{-\infty}^{\infty} X_1(f - f')X_2(f' - f'')df'df'' \tag{2.77}$$

The shift theorem is

$$X(f - a) = X(f) * \delta(t - a) \leftrightarrow x(t)e^{-j2\pi fa} \tag{2.78}$$

Convolution of a pair of δ functions with another function:

$$[\delta(t - a) + \delta(t + a)] * F(X) \leftrightarrow 2\ \cos(2\pi fa) \cdot x(t) \qquad (2.79)$$

The convolution of two Gaussian functions is

$$e^{-x^2/a} * e^{-x^2/b} \leftrightarrow ab\pi e^{-\pi^2 f^2(a^2 + b^2)} \qquad (2.80)$$

Some relations that can be applied to convolution are

$$[A(f) * B(f)] \cdot [C(f) * D(f)] \leftrightarrow [a(t) \cdot b(t)] * [c(t) \cdot d(t)] \qquad (2.81)$$

Note that
$$[A(f) * B(f)] \cdot C(f) \neq A(f) * [B(f) \cdot C(f)] \qquad (2.82)$$

$$[A(f) * B(f) + C(f) \cdot D(f)] \cdot E(f) \leftrightarrow [a(t) \cdot b(t) + c(t) * d(t)] * e(t) \qquad (2.83)$$

2.12.3 The Convolution Derivative Theorem

The derivative theorem is
$$\frac{dX}{df} \leftrightarrow -j2\pi f x(t) \qquad (2.84)$$

Therefore,

$$\frac{d}{df}[X_1(f) * X_2(f)] \leftrightarrow X_1(f) * \frac{dX_2(f)}{df} = \frac{dX_1(f)}{df} * X_2(f) \qquad (2.85)$$

Table 2.1 summarizes some properties of Fourier transform, and Table 2.2 gives some useful transform pairs.

2.12.4 Parseval's Theorem

We defined Parseval's theorem in connection with exponential Fourier series. This is also called Rayleigh theorem or simply the power theorem.

$$\int_{-\infty}^{\infty} X_1(f) X_2^*(f) df = \int_{-\infty}^{\infty} f_1(t) f_2^*(t) dt \qquad (2.86)$$

2.13 SAMPLED WAVEFORM: DISCRETE FOURIER TRANSFORM

The sampling theorem states that if the Fourier transform of a function $x(t)$ is zero for all frequencies greater than a certain frequency f_c, then the continuous function $x(t)$ can be uniquely determined by a knowledge of the sampled values. The constraint is that $x(t)$ is zero for frequencies greater than f_c, that is, the function is band limited at

TABLE 2.1 Properties of Fourier Transform

Property	Formulation
Linearity	$a_1 x_1(t) + a_2 x_2(t) \Leftrightarrow a_1 X_1(\omega) + a_2 X_2(\omega)$
Transformation	$x(t) \Leftrightarrow X(\omega)$
Symmetry	$X(t) \Leftrightarrow 2\pi x(-\omega)$
Scaling	$x(at) \Leftrightarrow (1/|a|)X(\omega/a)$
Delay	$x(t - t_0) \Leftrightarrow e^{-j2\pi f t_0} X(\omega)$
Modulation	$e^{-j2\pi f_0 t} x(t) \Leftrightarrow -X(\omega - \omega_0)$
Time convolution	$x_1(t) * x_2(t) \Leftrightarrow X_1(\omega) X_2(\omega)$
Frequency convolution	$x_1(t) x_2(t) \Leftrightarrow (1/2\pi) X_1(\omega) * X_2(\omega)$
Time differentiation	$\dfrac{d^n}{dt^n} x(t) \Leftrightarrow (j\omega)^n X(\omega)$
Time integration	$\displaystyle\int_{-\infty}^{t} x(t)dt \Leftrightarrow \dfrac{X(\omega)}{j\omega} + \pi X(0)\delta(\omega)$
Frequency differentiation	$-jt x(t) \Leftrightarrow \dfrac{dX(\omega)}{d\omega}$
Frequency integration	$\dfrac{x(t)}{-jt} \Leftrightarrow \displaystyle\int X(\omega)d\omega$

TABLE 2.2 Some Useful Transforms

$x(t)$	$X(\omega)$
$e^{-at}u(t)$	$\dfrac{1}{a + j\omega}$
$te^{-at}u(t)$	$\dfrac{1}{(a + j\omega)^2}$
$\dfrac{t^{n-1}}{(n - 1)!} e^{-at}u(t)$	$\dfrac{1}{(a + j\omega)^n}$
$\dfrac{\omega_0}{2\pi} \sin c\left(\dfrac{\omega_0 t}{2}\right)$	$1, \quad |\omega| < \omega_0/2$ $= 0 \quad \text{otherwise}$
$e^{-a|t|}$	$\dfrac{2a}{a^2 + \omega^2}$
$\dfrac{1}{a^2 + t^2}$	$\dfrac{\pi}{2} e^{-a|\omega|}$
$e^{-at} \sin \omega_0 t u(t)$	$\dfrac{\omega_0}{(a + j\omega)^2 + \omega_0^2}$
$e^{-at} \cos \omega_0 t u(t)$	$\dfrac{a + j\omega}{(a + j\omega)^2 + \omega_0^2}$
$\sin \omega_0 t$	$j\pi[\delta(\omega + \omega_0) - \delta(\omega - \omega_0)]$
$\cos \omega_0 t$	$\pi[\delta(\omega - \omega_0) + \delta(\omega - \omega_0)]$
$\displaystyle\sum_{n=-\infty}^{\infty} c_n e^{jn\omega_0 t}$	$2\pi \displaystyle\sum_{n=-\infty}^{\infty} c_n \delta(\omega - n\omega_0)$

Figure 2.13   High frequency impersonating a low frequency to illustrate aliasing.

frequency f_c. The second constraint is that the sampling spacing must be chosen so that

$$T = 1/(2f_c) \tag{2.87}$$

The frequency $1/T = 2f_c$ is known as the *Nyquist sampling rate.*

Aliasing means that the high-frequency components of a time function can impersonate a low frequency if the sampling rate is low. Figure 2.13 shows a high frequency as well as a low frequency that share identical sampling points. Here, a high frequency is impersonating a low frequency for the same sampling points.

The sampling rate must be high enough for the highest frequency to be sampled at least twice per cycle, $T = 1/(2f_c)$. An input signal $x(t)$ will be represented correctly if this condition is met. The Nyquist frequency is also called *folding frequency.*

Often the functions are recorded as sampled data in the time domain, the sampling being done at a certain frequency. The Fourier transform is represented by the summation of discrete signals where each sample is multiplied by

$$e^{-j2\pi fnt_1} \tag{2.88}$$

that is,

$$X(f) = \sum_{n=-\infty}^{\infty} x(nt_1)e^{-j2\pi fnt_1} \tag{2.89}$$

Figure 2.14 illustrates sampled time domain function and frequency spectrum for a discrete time domain function.

When the frequency domain spectrums as well as the time domain function are sampled functions, the Fourier transform pair is made of discrete components:

$$X(f_k) = \frac{1}{N} \sum_{n=0}^{N-1} x(t_n)e^{-j2\pi kn/N} \tag{2.90}$$

$$X(t_n) = \sum_{k=0}^{N-1} X(f_k)e^{j2\pi kn/N} \tag{2.91}$$

Figure 2.15(a) and (b) shows discrete time and frequency functions. *The discrete Fourier transform* approximates the continuous Fourier transform.

However, errors can occur in the approximations involved. Consider a cosine function $x(t)$ and its continuous Fourier transform $X(f)$, which consists of two impulse functions that are symmetric about zero frequency (Fig. 2.16(a)).

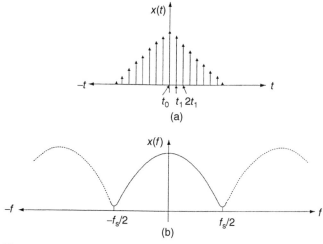

Figure 2.14 (a) Sampled time domain function and (b) frequency spectrum for the time domain function.

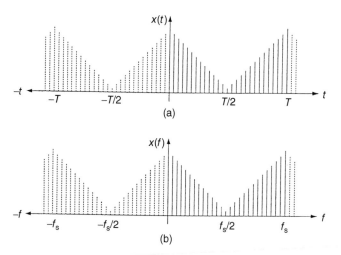

Figure 2.15 (a) and (b) discrete time and frequency domain functions.

The finite portion of $x(t)$, which can be viewed through a unity amplitude window $w(t)$, and its Fourier transform $W(f)$, which has side lobes, are shown in Fig. 2.16(b).

Figure 2.16(c) shows that the corresponding convolution of two frequency signals results in blurring of $X(f)$ into two $\sin x/x = \sin c(x)$ shaped pulses. Thus, the estimate of $X(f)$ is fairly corrupted.

The sampling of $x(t)$ is performed by multiplying with $c(t)$ (Fig. 2.16(d)); the resulting frequency domain function is shown in Fig. 2.16(e).

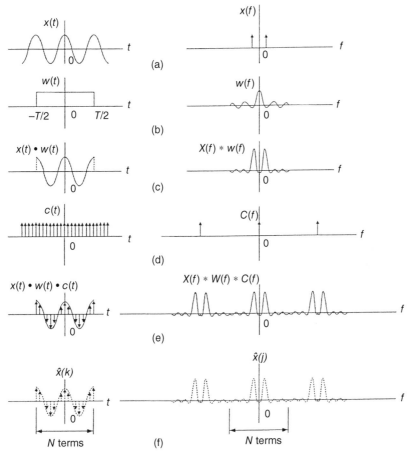

Figure 2.16 Fourier coefficients of the discrete transform viewed as corrupted estimate for the continuous Fourier transform: (a) $x(t)$ and Fourier transform $X(f)$, (b) unit amplitude window $w(t)$ and $W(f)$, (c) convolution of $x(t)$ and $w(f)$, (d) discrete sampling function, (e) convolution $x(t)$, $w(t)$, and $c(t)$, and (f) discrete bandwidth limited function based on (e). Source: Ref. [1].

The continuous frequency domain function shown in Fig. 2.16(e) can be made discrete if the time function is treated as one period of a periodic function. This forces both the time domain and frequency domain functions to be infinite in extent, periodic and discrete (Fig. 2.16(f)). *The discrete Fourier transform* is reversible mapping of N terms of the time function into N terms of the frequency function. Some problems are outlined later.

2.13.1 Leakage

Leakage is inherent in the Fourier analysis of any finite record of data. The record of the data is obtained by looking at the function for a period T and neglecting everything

Figure 2.17   An extended data window. Source: [B1].

that happens before and after this period. The function may not be localized on the frequency axis and has side lobes (Fig. 2.8(b)). The objective is to localize the contribution of a given frequency by reducing the leakage through these side lobes. The usual approach is to apply a data window in the time domain, which has lower side lobes in the frequency domain, as compared to a rectangular data window. An extended cosine bell data window, called Tukey's interim data window, is shown in Fig. 2.17. A raised cosine wave is applied to the first and last 10% of the data, and a weight of unity is applied for the middle 90% of the data. A number of other types of windows that give more rapidly decreasing side lobes have been described in the literature. Some of the window types are as follows:

- Rectangular
- Triangular
- Cosine squared (hanning)
- Hamming
- Gaussian
- Dolph–Chebyshev

For periodic functions, the rectangular window results in zero spectral leakage and high spectral resolution. The rectangular window spans exactly one period, the zeros in the spectrum of the window coincide with all harmonics except one. This results in no spectral leakage under ideal conditions.

A window function often incorporated in spectrum analyzers is Hanning window.

$$W(t) = 0.5 - 0.5 \, \cos \frac{2\pi t}{T}, \quad \text{for} \quad -0.5T < t < 0.5T \tag{2.92}$$

The function is easily generated from sinusoidal signals. The main lobe noise bandwidth is greater than that in a rectangular window. The highest side lobe is at $-32\,\text{dB}$ and side fall-off rate is $-60\,\text{dB}$ (see Fig. 2.18(a) and (b) for comparison of rectangular and Hanning windows).

The *Hamming* window function is

$$W(t) = 0.54 - 0.46 \, \cos \frac{2\pi t}{T}, \quad \text{for} \quad -0.5T < t < 0.5T \tag{2.93}$$

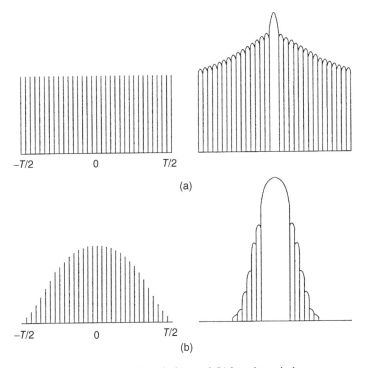

Figure 2.18   (a) Rectangular window and (b) hanning window.

2.13.2   Picket Fence Effect

An analogy between the output of fast Fourier transform (FFT) algorithm and a bank of band-pass filters is shown in Fig. 2.19. Each Fourier coefficient ideally acts as a filter having a rectangular response in the frequency domain. In practice, the response is of the type with side lobes. In Fig. 2.19, main lobes only have been plotted to represent output of FFT. The width of each lobe is proportional to the original record length.

When the signal being viewed is not one of these orthogonal frequencies, the picket pence effect becomes evident. The picket fence effect can reduce the amplitude of the signal in the spectral windows, when the signal being analyzed falls in between the orthogonal frequencies, say between the third and fourth harmonics. The signal will be experienced by both the third and fourth harmonic spectral windows, and in the worst case halfway between the computed harmonics. The signal is then reduced to 0.637 in both the spectral windows. Squaring this number, the peak power is reduced to 0.406.

By analyzing the data with a set of samples that are identically zero, the FFT algorithm can compute a set of coefficients with terms lying in between the original harmonics. As the width of the window is related solely to the record length, the width of these new spectral windows remains unchanged: that means a considerable overlap.

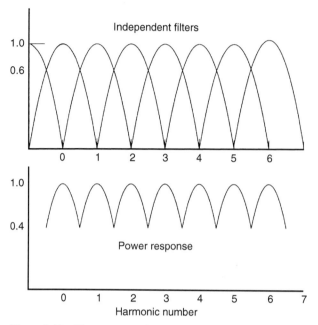

Figure 2.19   The response of the discrete Fourier transform Fourier coefficients viewed as a set of band-pass filters. Source: Ref. [1].

Consider that the original series is represented by $g(k)$ for $k = 0, 1, \ldots N - 1$, then the new series can be represented by

$$\widehat{g}(k) = g(k) \quad \text{for} \quad 0 \le k < N$$

$$\widehat{g}(k) = 0 \qquad \text{for} \quad N \le k < 2N \tag{2.94}$$

This is called *zero padding* (Fig. 2.20). The ripple in the power spectrum is reduced from 60% to 20%.

2.14   FAST FOURIER TRANSFORM

The FFT is simply an algorithm that can compute the discrete Fourier transform more rapidly than any other available algorithm.

Define

$$W = e^{-j2\pi/N} \tag{2.95}$$

The frequency domain representation of the waveform is

$$X(f_k) = \frac{1}{N} \sum_{n=0}^{N=1} x(t_n) W^{kn} \tag{2.96}$$

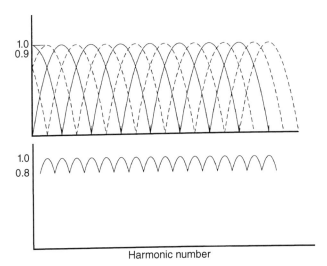

Figure 2.20   Reduction of the picket fence effect by computing redundant overlapping sets of Fourier coefficients. Source: Ref. [1].

The equation can be written in matrix form:

$$\begin{vmatrix} X(f_0) \\ X(f_1) \\ \cdot \\ X(f_k) \\ \cdot \\ X(f_{N-1}) \end{vmatrix} = \frac{1}{N} \begin{vmatrix} 1 & 1 & \cdot & 1 & \cdot & 1 \\ 1 & W & \cdot & W^k & \cdot & W^{N-1} \\ \cdot & \cdot & & \cdot & & \cdot \\ 1 & W^k & \cdot & W^{k2} & \cdot & W^{k(N-1)} \\ \cdot & \cdot & & \cdot & & \cdot \\ 1 & W^{N-1} & \cdot & W^{(N-1)k} & \cdot & W^{(N-1)^2} \end{vmatrix} \begin{vmatrix} x(t_0) \\ x(t_1) \\ \cdot \\ x(t_n) \\ \cdot \\ x(t_{N-1}) \end{vmatrix} \tag{2.97}$$

or in a condensed form:

$$[\overline{X}(f_k)] = \frac{1}{N}[\overline{W}^{-kn}][\overline{x}(t_n)] \tag{2.98}$$

where $[\overline{X}(f_k)]$ is a vector representing N components of the function in the frequency domain, while $[\overline{x}(t_n)]$ is a vector representing N samples in the time domain. Calculation of N frequency components from N time samples, therefore, requires a total of $N \times TN$ multiplications.

For $N = 4$:

$$\begin{vmatrix} X(0) \\ X(1) \\ X(2) \\ X(3) \end{vmatrix} = \begin{vmatrix} 1 & 1 & 1 & 1 \\ 1 & W^1 & W^2 & W^3 \\ 1 & W^2 & W^4 & W^6 \\ 1 & W^3 & W^6 & W^9 \end{vmatrix} \begin{vmatrix} x(0) \\ x(1) \\ x(2) \\ x(3) \end{vmatrix} \tag{2.99}$$

However, each element in matrix $[\overline{W}^{-kn}]$ represents a unit vector with clockwise rotation of $2n/N$, $(n = 0, 1, 2, \ldots, N-1)$. Thus, for $N = 4$ (i.e., four sample points), $2\pi/N = 90°$. Thus,

$$W^0 = 1 \tag{2.100}$$

$$W^1 = \cos \pi/2 - j \sin \pi/2 = -j \tag{2.101}$$

$$W^2 = \cos \pi - j \sin \pi = -1 \tag{2.102}$$

$$W^3 = \cos 3\pi/2 - j \sin 3\pi/2 = j \tag{2.103}$$

$$W^4 = W^0 \tag{2.104}$$

$$W^6 = W^2 \tag{2.105}$$

Hence, the matrix can be written in the form:

$$
\begin{vmatrix} X(0) \\ X(1) \\ X(2) \\ X(3) \end{vmatrix} =
\begin{vmatrix} 1 & 1 & 1 & 1 \\ 1 & W^1 & W^2 & W^3 \\ 1 & W^2 & W^0 & W^2 \\ 1 & W^3 & W^2 & W^1 \end{vmatrix}
\begin{vmatrix} x_0(0) \\ x_0(1) \\ x_0(2) \\ x_0(3) \end{vmatrix} \tag{2.106}
$$

This can be factorized into

$$
\begin{vmatrix} X(0) \\ X(2) \\ X(1) \\ X(3) \end{vmatrix} =
\begin{vmatrix} 1 & W^0 & 0 & 0 \\ 1 & W^2 & 0 & 0 \\ 0 & 0 & 1 & W^1 \\ 0 & 0 & 1 & W^3 \end{vmatrix}
\begin{vmatrix} 1 & 0 & W^0 & 0 \\ 0 & 1 & 0 & W^0 \\ 1 & 0 & W^2 & 0 \\ 0 & 1 & 0 & W^2 \end{vmatrix}
\begin{vmatrix} x_0(0) \\ x_0(1) \\ x_0(2) \\ x_0(3) \end{vmatrix} \tag{2.107}
$$

Equation (2.106) yields square matrix in Eq. (2.107) except that rows 1 and 2 in first column vector have been interchanged.

First let,

$$
\begin{vmatrix} x_1(0) \\ x_1(1) \\ x_1(2) \\ x_1(3) \end{vmatrix} =
\begin{vmatrix} 1 & 0 & W^0 & 0 \\ 0 & 1 & 0 & W^0 \\ 1 & 0 & W^2 & 0 \\ 0 & 1 & 0 & W^2 \end{vmatrix}
\begin{vmatrix} x_0(0) \\ x_0(1) \\ x_0(2) \\ x_0(3) \end{vmatrix} \tag{2.108}
$$

The column vector on the left is equal to the product of second matrix and last column vector in Eq. (2.107).

Element $x_1(0)$ is computed with one complex multiplication and one complex addition:

$$x_1(0) = x_0(0) + W^0 x_0(2) \tag{2.109}$$

Element $x_1(1)$ is also calculated by one complex multiplication and addition. One complex addition is required to calculate $x_1(2)$:

$$x_1(2) = x_0(0) + W^2 x_0(2) = x_0(0) - W^0 x_0(2) \tag{2.110}$$

because
$W^0 = -W^2$ and $W^0 x_0(2)$ is already computed in Eq. (2.109).

Then Eq. (2.107) is

$$\begin{vmatrix} X(0) \\ X(2) \\ X(1) \\ X(3) \end{vmatrix} = \begin{vmatrix} x_2(0) \\ x_2(1) \\ x_2(2) \\ x_2(3) \end{vmatrix} = \begin{vmatrix} 1 & 0 & W^0 & 0 \\ 0 & 1 & 0 & W^0 \\ 1 & 0 & W^2 & 0 \\ 0 & 1 & 0 & W^3 \end{vmatrix} \begin{vmatrix} x_1(0) \\ x_1(1) \\ x_1(2) \\ x_1(3) \end{vmatrix} \tag{2.111}$$

The term $x_2(0)$ is determined by one complex multiplication and addition:

$$x_2(0) = x_1(0) + W^0 x_1(1) \tag{2.112}$$

$x_1(3)$ is computed by one complex addition and no multiplication.

Computation requires four complex multiplications and eight complex additions. Computation of Eq. (2.99) requires 16 complex multiplications and 12 complex additions. The computations are reduced. In general, the direct method requires N^2 multiplications and $N(N - 1)$ complex additions.

For $N = 2^\gamma$, the FFT algorithm is simply factoring $N \times N$ matrix into γ matrices, each of dimensions $N \times N$. These have the properties of minimizing the number of complex multiplications and additions (Fig. 2.21).

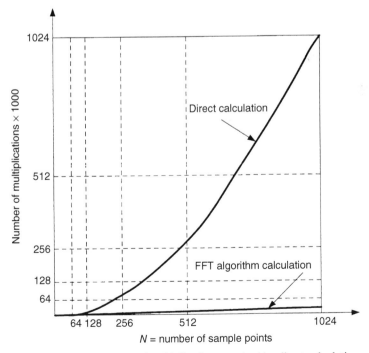

Figure 2.21   Comparison of multiplications required by direct calculations and FFT algorithm.

The matrix factoring does introduce the discrepancy that instead of

$$X(n) = \begin{vmatrix} X(0) \\ X(1) \\ X(2) \\ X(3) \end{vmatrix} \tag{2.113}$$

it yields

$$\overline{X}(n) = \begin{vmatrix} X(0) \\ X(2) \\ X(1) \\ X(3) \end{vmatrix} \tag{2.114}$$

See Eqs. (2.106) and (2.107). This can be easily rectified in matrix manipulation by unscrambling.

The $\overline{X}(n)$ can be rewritten by replacing n with its binary equivalent:

$$\overline{X}(n) = \begin{vmatrix} X(0) \\ X(2) \\ X(1) \\ X(3) \end{vmatrix} = \begin{vmatrix} X(00) \\ X(10) \\ X(01) \\ X(11) \end{vmatrix} \tag{2.115}$$

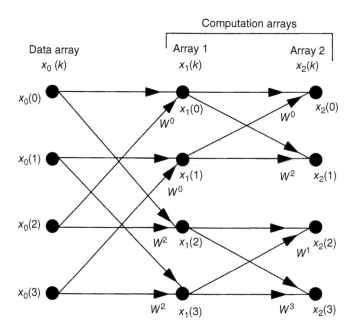

Figure 2.22   FFT signal flow graph $N = 4$.

If the bits are flipped over, then

$$\overline{X}(n) = \begin{vmatrix} X(00) \\ X(10) \\ X(01) \\ X(11) \end{vmatrix} \quad \text{flips to} \quad \begin{vmatrix} X(00) \\ X(01) \\ X(10) \\ X(11) \end{vmatrix} = X(n) \tag{2.116}$$

2.14.1 Signal Flow Graph

Equation (2.111) can be converted into a signal flow graph as shown in Fig. 2.22. The data vector or array $x_0(k)$ is represented by a vertical column on the left of the graph. The second vertical array of nodes is vector $x_1(k)$ and the next vector is $x_2(k)$. There will be γ computation arrays where $N = 2^\gamma$. Each node is entered by two solid lines representing transmission paths from previous nodes. A path transmits or brings a quantity from a node in one array, multiplies the quantity by W^p, and inputs into the node in the next array. Absence of factor W^p implies that it is equal to 1. Signal flow graph is a concise method of representing the computations required in the factored matrix FFT algorithm.

Ref. [2–19] provide further reading.

REFERENCES

1. G. D. Bergland. "A guided tour of the fast fourier transform". IEEE Spectrum, pp. 41–52, July 1969.
2. J. F. James, A student's Guide to Fourier Transforms, 3rd Edition, Cambridge University Press, UK, 2012.
3. I. N. Sneddon, Fourier Transforms, Dover Publications Inc. New York, 1995.
4. H. F. Davis, Fourier Series and Orthogonal Functions, Dover Publications, New York, 1963.
5. R. Roswell, Fourier Transform and Its Applications, McGraw-Hill, New York, 1966.
6. R.N. Bracewell, Fourier Transform and Its Applications, 2nd Edition, McGraw-Hill, New York, 1878.
7. E. M. Stein, R. R. Shakarchi, Fourier Analysis (Princeton Lectures in Analysis), Princeton University Press, NJ, 2003.
8. B. Gold, C. M. Rader, Digital Processing of Signals, McGraw Hill, New York, 1969.
9. A. V. Oppenheim, R. W. Schafer, and T. G. Stockham, "Nonlinear filtering of multiplied and convoluted signals," IEEE Transactions on Audio and Electroacoustics, vol. AU-16, pp. 437–465, 1968.
10. P. I. Richards, "Computing reliable power spectra," IEEE Spectrum, vol. 4, pp. 83–90, 1967.
11. J.W. Cooley, P.A.W Lewis, and P.D. Welch, "Application of fast Fourier transform to computation of Fourier Integral, Fourier series and convolution integrals," IEEE Transactions on Audio and Electroacoustics, vol. AU-15, pp. 79–84, 1967.
12. H. D. Helms, "Fast Fourier transform method for calculating difference equations and simulating filters," IEEE Transactions on Audio and Electroacoustics, vol. AU-15, pp. 85–90, 1967.
13. G. D. Bergland, "A fast Fourier transform algorithm using base 8 iterations," Mathematics of Computation, vol.22, pp. 275–279, 1968.
14. J. W. Cooley, Harmonic analysis complex Fourier series, SHARE Doc. 3425, 1966.
15. R. C. Singleton, "On computing the fast Fourier transform," Communication of the ACM, vol. 10, pp. 647–654, 1967.
16. R. Yavne, An economical method for calculating the discrete Fourier transform, Fall Joint Computer Conference, IFIPS Proceedings on vol. 33, pp. 115–125, Spartan Books, Washington DC.
17. J. Arsac, Fourier Transform, Prentice Hall, Englewood Cliffs, NJ, 1966.
18. G. D. Bergland, "A fast Fourier algorithm for real value series," Numerical Analysis, vol. 11, no. 10, pp. 703–710, 1968.
19. F. F. Kuo, Network Analysis and Synthesis, John Wiley and Sons, New York, 1966.

CHAPTER 3

HARMONIC GENERATION-1

Harmonics are generated not only by nonlinear electronically controlled loads, but also by the conventional power equipment, such as transformers, motors and generators. The saturation under normal operation and switching transients will give rise to harmonics.

In Chapter 1 we stated that harmonics *are one of the main concerns* for the power quality problems. The major power quality problems are summarized in Table 3.1. Power quality implies a wide variety of electromagnetic phenomena ranging from a few nanoseconds (e.g., lightning strokes) to steady state disturbances. The IEEE SCC2, (Standard Coordinating Committee 22) has led the main efforts of coordinating power quality standards in the United States. There are some compatibility issues between IEC and US terminology and the committee liaisons with IEC and CIGRE. Table 3.1 shows categorization of electromagnetic phenomena. Table 10.1 (Chapter 10) provides listing of major IEC standards on electromagnetic compatibility. This book concentrates on item 1.0 and flicker of item 2.0 in Table 3.1, related in general to harmonics. The other power quality issues are not discussed.

3.1 HARMONICS IN TRANSFORMERS

Harmonics in transformers originate as a result of saturation (which can occur due to residual or trapped flux after a severe fault), switching, high-flux densities (which can occur due to high voltages and which in turn give rise to saturation), winding connections and grounding; for example phenomena of oscillating neutrals in ungrounded wye-wye connected transformers, discussed further.

3.1.1 Linear Model of a Two-Winding Transformer

A two-winding transformer model is shown in Fig. 3.1(a). This figure shows the equivalent circuit of a transformer supplying load connected to its secondary windings (a single-phase two winding transformer is shown for simplicity). The load current is I_2 at a terminal voltage V_2 at a lagging power factor angle ϕ_2. Exciting the primary winding with voltage V_1 produces changing flux linkages. Though the coils in a transformer are tightly coupled by interleaving the windings and are wound

---

Power System Harmonics and Passive Filter Designs, First Edition. J.C. Das.
© 2015 The Institute of Electrical and Electronics Engineers, Inc. Published 2015 by John Wiley & Sons, Inc.

TABLE 3.1   Major Power Quality Problems

No	Category	Spectral Content	Typical Duration	Typical Voltage Magnitude
1	*Waveform Distortion*			
	DC offset	0–5000Hz	Steady state	0–0.25%
	Harmonics	0–6kHz	Steady state	0–30%
	Interharmonics	Broad band	Steady state	0–5%
	Notching		Steady state	
	Noise-Common mode and normal mode		Steady state	
2	Voltage fluctuations		Intermittent	01–7%
	Flicker		Steady state	Pst and Plt
	Voltage unbalance			0.5–4%
3	*Transients-Impulsive*			
	Nanosecond	5 ns rise	<50 ns	
	Microsecond	1 μs rise	50 ns–1 ms	
	Millisecond	0.1 ms rise	>1 ms	
4	*Transients-Oscillatory*			
	Low frequency	<5 kHz	0.3–50 ms	0–4 pu
	Medium frequency	5–500 kHz	20 μs	0–8 pu
	High frequency	0.5% MHz	5 μs	0–4 pu
5	*Short-duration variations*			
	a. *Instantaneous*			
	Interruptions		0.5–30 cycle	<0.1 pu
	Sags (dips)		0.5–30 cycle	0.1–0.9 pu
	Swell		0.5–30 cycle	1.1–1.8 pu
	b. *Momentary*			
	Interruptions		30 cycles–3 s	<0.1 pu
	Sags		30 cycles–3 s	0.1–0.9 pu
	Swell		30 cycles–3 s	1.1–1.4 pu
	c. *Temporary*			
	Interruptions		3 s–1 min	<0.1 pu
	Sags		3 s–1 min	0.1–0.9 pu
	Swell		3 s–1 min	1.1–1.2 pu
6	*Long Duration variations*			
	Sustained interruption		●1min	0.0 pu
	Undervoltage		●1 min	0.8-0.9 pu
	Overvoltage		●1 min	1.1-1.2 pu
7	*Frequency variations*		<10 s	

on a magnetic material of high permeability, all the flux produced by primary windings does not link with the secondary windings. The winding leakage flux gives rise to leakage reactances. Φ_m is the main or mutual flux, assumed to be constant. In the secondary winding, the ideal transformer produces an EMF E_2 due to mutual flux linkages and in the primary windings induced EMF, E_1 opposes the EMF induced in the secondary windings, E_2. There has to be a primary magnetizing current even

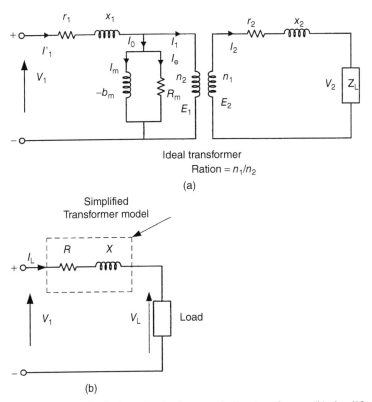

Figure 3.1   (a) Equivalent circuit of a two winding transformer, (b) simplified equivalent circuit.

at no load, in time- phase with its associated flux, to excite the core. The pulsation of flux in the core produces losses. Considering that the no-load current is sinusoidal (which is not true under magnetic saturation), it must have a core loss component due to hysteresis and eddy currents:

$$I_0 = \sqrt{I_m^2 + I_e^2}$$ (3.1)

where I_m is the magnetizing current, I_e is the core loss component of the current, and I_0 is the no-load current; I_m and I_e are in phase quadrature. The generated EMF because of flux Φ_m is given by

$$E_2 = 4.44 f n_2 \Phi_m$$ (3.2)

where E_2 is in volts when Φ_m is in Wb/m^2, n_2 is the number of secondary turns, and f is the frequency. As primary ampère turns must be equal to the secondary ampère turns, that is, $E_1 I_1 = E_2 I_2$, we can write:

$$\frac{E_1}{E_2} = \frac{n_1}{n_2} = n \text{ and}$$

$$\frac{I_1}{I_2} \approx \frac{n_2}{n_1} = \frac{1}{n} \tag{3.3}$$

The current relation holds because the no-load current is small. The terminal relations can now be derived. On the primary side, the current is compounded to consider the no-load component of the current, the primary voltage is equal to $- E_1$ (to neutralize the EMF of induction) and $I_1' r_1$ and $I_1' x_1$ are resistive and inductive (lagging power factor) drop in the primary windings. On the secondary side the terminal voltage is given by the induced EMF E_2 less $I_2 r_2$ and $I_2 x_2$ voltage drops in the secondary windings. The phasor diagram corresponding to Fig. 3.1(a) is in Fig. 3.2(a). See Refs. [1–3].

By pulling the secondary parameters of the transformer to the primary side multiplied by the square of the turns-ratio, and ignoring the magnetization and eddy current branches a very simplified model results, as shown in Fig. 3.1(b). Its phasor diagram is in Fig. 3.2(b). The percentage impedance specified by the manufacturer and X/R ratio is based on this model. This model is good for linear fundamental

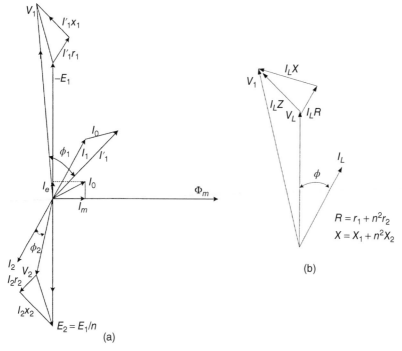

Figure 3.2   (a) Phasor diagram corresponding to the equivalent circuit in Fig. 3.1(a), (b) phasor diagram corresponding to Fig. 3.1(b).

frequency load flow and cannot be applied to harmonic analysis. The appropriate transformer models are discussed in Chapter 12.

The expression for hysteresis loss is

$$P_h = K_h f B_m^s \tag{3.4}$$

where K_h is a constant and s is the Steinmetz exponent, which varies from 1.5 to 2.5, depending on the core material; generally, it is equal to 1.6.

The eddy current loss is

$$P_e = K_e f^2 B_m^2 \tag{3.5}$$

where K_e is a constant. Eddy current loss occurs in core laminations, conductors, tanks, and clamping plates. The core loss is the sum of the eddy current and hysteresis loss. In Fig. 3.2(a), the primary power factor angle ϕ_1 is $> \phi_2$.

3.1.2 B-H Curve and Peaky Magnetizing Current

For economy in design and manufacture, transformers are operated close to the knee point of saturation characteristics of magnetic materials. Figure 3.3 shows a $B - H$ curve and the magnetizing current waveform.

Referring to Fig. 3.3, the magnetic current variation can be plotted as illustrated. An ascending flux density corresponding to point P on hysteresis loop and point Q on the sinusoidal wave requires x times the magnetizing current. If a number of such points are plotted, the resulting current curve deviates considerably from the sinusoidal. Also see Ref. [4].

A sinusoidal flux wave, required by sinusoidal applied voltage, demands a magnetizing current with a harmonic content. Conversely, with a sinusoidal magnetizing

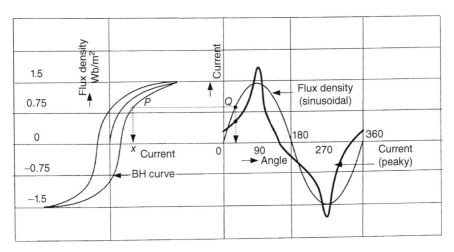

Figure 3.3   Transformer B-H curve and derivation of peaky magnetizing current.

current, the induced EMF is peaky and the flux wave is flat topped. The peaky current wave has odd harmonics: a flat-topped flux density waveform can be written as:

$$B = B_1 \sin \omega t + B_3 \sin 3\omega t + B_5 \sin 5\omega t + \ ... \tag{3.6}$$

This gives :

$$e = \omega A_c (B_1 \cos \omega t + B_3 \cos 3\omega t + B_5 \cos 5\omega t + \ ... \ ) \quad \text{Volts per turn} \tag{3.7}$$

where A_c is the area over which the flux density acts.

In a system of three-phase balanced voltages, the triplen harmonic voltages are cophasial:

In phase a:

$$v_3 \sin(3\omega t + \alpha_3) \tag{3.8}$$

In phase b:

$$v_3 \sin[3(\omega t - 120^o) + \alpha_3] = v_3 \sin(3\omega t + \alpha_3) \tag{3.9}$$

In phase c:

$$v_3 \sin[3(\omega t - 240^o) + \alpha_3] = v_3 \sin(3\omega t + \alpha_3) \tag{3.10}$$

The third harmonics and all triplen harmonics are in time phase with each other.

3.1.3   Effect of Transformer Construction and Winding Connections

If the impedance to the third harmonic is negligible, only a very small third harmonic EMF is required to circulate a magnetizing current additive to the fundamental frequency, so as to maintain a sinusoidal flux. The zero sequence impedance of the transformers varies over large values depending on winding connections and construction, shell or core type, three limbs or five limbs.

Single-phase Transformers   With sinusoidal supply voltage and flux, supply voltage itself cannot give rise to third harmonic current. A triplen frequency EMF must be generated in the transformer by a triplen frequency flux. This EMF must circulate third harmonic current through transformer, through supply system on the primary side and through transformer and load on the secondary side. If the supply source impedance is negligible compared to third harmonic impedance of the transformer, then the third harmonic EMF will be balanced with the corresponding impedance drop and little third harmonic voltage will appear on the lines. These remarks are applicable to other harmonics also, that is, if the impedance to a certain harmonic is low, the harmonic voltage developed across that impedance will also be low.

Three-phase banks of single-phase transformers In three phase banks of single phase transformers, the magnetic circuits are not linked and each core must produce the flux demanded by the connections.

In a delta-delta connected three-phase transformers, harmonics of the order 5th, 7th, 11th, ... produce voltages displaced by 120° mutually, while triplen harmonics are cophasial. The delta connection of phases forms a closed path for the triplen harmonics, and will circulate corresponding harmonic currents in the delta, none appearing in the lines. Wye–delta and delta–wye connections substantiality operate in the same manner.

In a wye–wye connection without neutrals, as all the triplen harmonics are directed either inwards or outwards, these cancel between the lines, no third harmoni currents flow, and the flux wave in the transformer is flat topped. The effect on a wye-connected neutral is to make it oscillate at three times the fundamental frequency, giving rise to distortion of the phase voltages, Fig. 3.4. This is called phenomenon of *oscillating neutral*. Practically, ungrounded wye-wye transformers are not used. Even when both the primary and secondary windings are grounded, tertiary delta-connected windings are included in wye–wye connected transformers for neutral stabilization

In wye–wye connection with grounded neutrals, the neutral will carry three times the third harmonic currents that are all in phase.

A tertiary delta connected winding will suppress the third harmonic currents. Often the tertiary windings can be designed to serve loads at a different voltage.

Three-phase transformers We can distinguish between core type construction and shell type construction, which have distinct properties with respect to the flow of zero sequence currents. Figure 3.5 illustrates the basic construction of core and shell type transformers.

In a core type transformer (Fig. 3.5(a)) the flux in a core must return through the other two core limbs, that is, the magnetic circuit of a phase is completed through the other two phases in parallel. The magnetizing current that is required for the core and part of the yoke and will be different in each phase, though this difference is not significant, as yoke reluctance is only a small percentage of the total core reluctance. However, the zero sequence or triplen harmonics will be all directed in one direction, in each core leg. The return path lies through the insulating medium and transformer

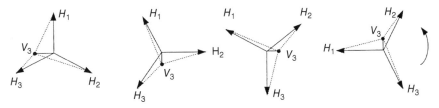

Figure 3.4   Phenomenon of oscillating neutrals in wye–wye connected ungrounded three-phase transformers. Source: Ref. [1].

tank (Fig.3.5(a)). Sometimes five-limb construction, (Fig. 3.5(c)) is adopted to provide return path for zero sequence harmonics.

Shell form is more rugged and expensive and generally used for large-sized
transformers. In three-single phase transformers or shell type construction the magnetic circuits are complete in themselves and do not interact (Fig. 3.5(b)).

3.1.4 Control of Harmonics in Core Type Transformers

A three-limb core type transformer under some saturation will have a magnetizing
current (waveform as shown in Fig. 3.5(d)). It has a strong fifth harmonic, the third is
precluded by transformer connections and further the flux is forced to be nearly sinusoidal by high reluctance offered to third harmonics (Fig. 3.5(a)). If the transformer
is provided with a five limb core construction (Fig. 3.5(c)), the path for the third harmonic s is provided by the end limbs and the flux wave flattens. Figure 3.5(e) shows
that the fifth harmonic has now a reversed sign. If the end limbs are so designed that
these have intermediate reluctance, the fifth harmonic can be eliminated (Fig. 3.5(f)).

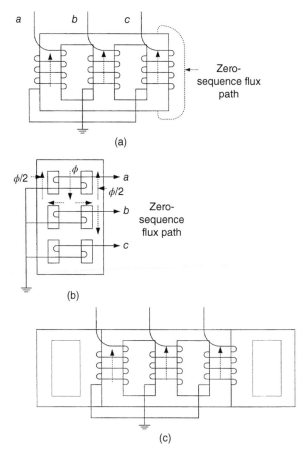

Figure 3.5
(a) Construction of core type transformer, the zero sequence flux path, (b) construction of a shell type transformer, the zero sequence path, (c) a five limb core type transformer for cancellation of harmonics, (d) magnetizing current core type transformer, (e) magnetizing current core type 5-limb transformer, (f) magnetizing current with optimum design of the yoke reluctance five-limb transformer. Figure 3.5 (d), (e) and (f) are applicable to only core type wye−wye connected transformers, (magnetizing currents in steady state).

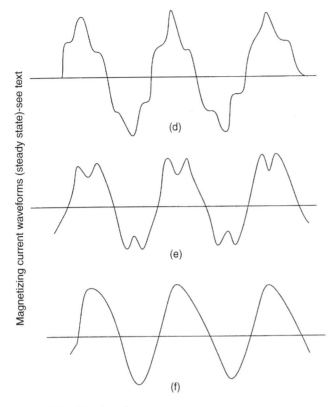

Figure 3.5   (*Continued*)

This method is applicable to wye–wye connected and insulated neutral connection or wye/zigzag transformers and not to delta connection.

If a wye–wye transformer and a delta-delta connected transformer are paralleled, the combination will suppress fifth and seventh harmonic in line connections, as these are in phase-opposition in the two transformers. Note that here we are not discussing the energization inrush currents.

It can be said that power transformers generate very low levels of harmonic currents in steady-state operation, and the harmonics are controlled by design and transformer winding connections. *The fifth and seventh harmonics may be less than 0.1% of the transformer full-load current and third is eliminated by transformer connections or five-limb core design.*

3.2   ENERGIZATION OF A TRANSFORMER

Energizing a power transformer does generate a high order of harmonics including a DC component. When a transformer is switched with its secondary circuit open, it acts like a reactor and at every instant the voltage applied must be balanced by the EMF induced by the flux generated in the core by the magnetizing current and by small voltage drops in resistance and leakage reactance components. Figure 3.6(a)–(c) shows three conditions of energizing of a power transformer: (a) the switch closed at the peak value of the voltage, and the core is initially demagnetized (b) the switch closed at zero value of the voltage, and the core is demagnetized; and (c) energizing with some residual trapped flux in the magnetic core due to retentivity of the magnetic materials. The EMF must be immediately established and it requires a flux having maximum rate of change. When the switch is closed at the peak value of the voltage, the flux and EMF wave assume the normal relations for an inductive circuit (Fig. 3.6(a)). In a three-phase circuit, as the voltages are displaced by 120 electrical degrees, at best the switch will be closed at the maximum voltage in one phase. Note that there is no control over the instant at which the switch is closed. Therefore, practically the conditions shown in Fig. 3.6(b) will apply. In the real world situation the transformer cores do retain a residual flux, which can vary depending on the conditions prior to switching—say, energization after a short-circuit will trap much residual flux. Figure 3.6(d) shows the hysteresis loop under these three conditions and Fig. 3.6(e) shows the waveform of magnetizing inrush current, which resembles a rectified current and its peak value may reach 8–15 times the transformer full-load current - mainly depending on the transformer size. The asymmetrical loss due to conductor and core heating rapidly reduces the flux wave to symmetry about the time axis and typically the inrush currents last for a short duration of about 0.1–0.2 sec. In presence of capacitors the magnitude of inrush current and its time duration can increase, see Chapter 11.

3.2.1   DC Core Saturation of Transformers

The transformer core may contain a DC flux, for example, if a transformer is serving a three-phase converter load with unbalanced firing angles or under geomagnetic transients, Section 3.3.4. Under this condition the excitation current will contain a DC component of the current and both odd and even harmonics will be present, Typical harmonics generated by the transformer inrush current are shown in Fig. 3.7, which shows harmonics with DC saturation.

The DC component can also occur due to unbalance flux in the core, that is, one phase may trap a residual flux higher than the other two. This can occur after an unsymmetrical fault clearance and subsequent switching of the transformer. The generated fundamental frequency EMF is given by

$$V = 4.44 \, f T_{ph} B_m A_c \tag{3.11}$$

where T_{ph} is the number of turns in a phase, B_m is the flux density (consisting of fundamental and higher order harmonics), and A_c is the area of core. In order to maintain flux density, the ratio V/f should remain equal to 1. Consider that the rated voltage rises and frequency does not rise, which signifies increased flux. Thus, the factor V/f is a measure of the overexcitation, though, practically, these currents do not normally cause a wave distortion of any significance. Exciting currents increase rapidly with voltage, and transformer standards specify that transformer must be capable of

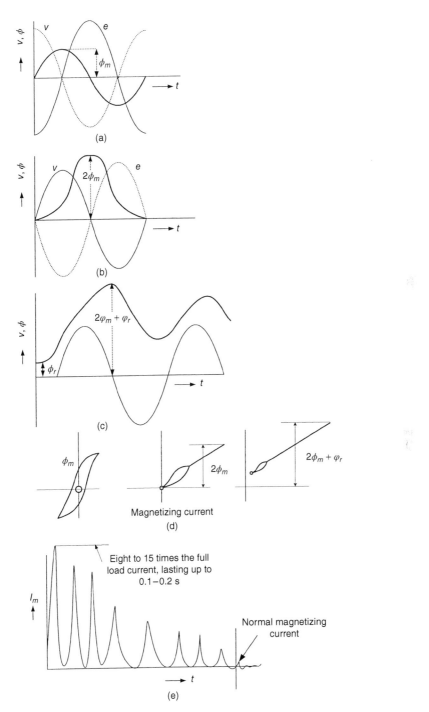

Figure 3.6   (a–e) related to the switching inrush current transients of transformers, see text.

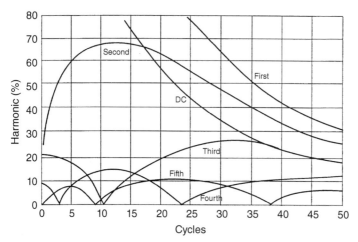

Figure 3.7   Harmonic components of the inrush current of a transformer with DC saturation. Source: Westinghouse Training Center, Course No. C/E 57, Power System Harmonics.

application of 110% voltage without overheating. Under certain system upset conditions, the transformers may be subjected to even higher voltages and overexcitation.

Example 3.1:   The switching current transients of a 10 MVA , delta–wye connected transformer, wye windings resistance grounded are simulated using EMTP. The system configuration is shown in Fig. 3.8(a), which shows the source (13.8 kV) positive and zero sequence impedance to which the transformer is connected. Figure 3.8(b) shows the inrush current transients in the three phases. The maximum peak inrush current in phase B is 600A, the 10 MVA transformer load current at 138 kV is 41.8A. Thus, the inrush current is approximately 10 times the full load current.

3.2.2   Sympathetic Inrush Current

Consider that two three-phase transformers "A" and "B" are connected on the primary side on the same bus. The transformer A serves its rated load, while the transformer B is off-line. When the transformer B is switched in, the inrush current is not only limited to transformer B, *but is also reflected in the line current of transformer A*. This phenomenon is called sympathetic inrush current of the transformer. See Ref. [5] for further details.

3.3   DELTA WINDINGS OF THREE-PHASE TRANSFORMERS

Consider a delta primary and wye connected grounded secondary windings transformer with nonlinear loads on the secondary wye-connected winding. These loads

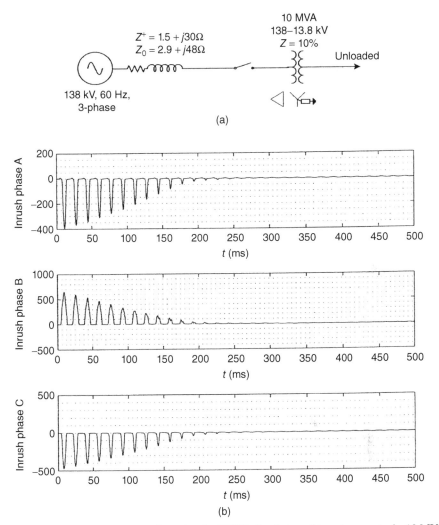

Figure 3.8   (a) A system configuration for EMTP simulation of inrush current of a 10 MVA transformer and (b) switching inrush currents in three-phases (Example 3.1).

may have some noncharacteristics triplen harmonics. No triplen harmonics will appear on the power lines serving the delta windings, because delta-windings are a sink to the triplen harmonics. Thus, pattern of harmonics on the primary delta windings lines will be different from that on the secondary side. Practically, all commercially available harmonic analyses programs eliminate triplen harmonics on the power lines feeding the delta windings and also consider the phase shifts in the transformer windings to alter the phase angle of harmonics on the primary side of the transformer.

3.3.1 Phase Shift in Three-Phase Transformers Winding Connections

The angular displacement of a poly-phase transformer is the time angle expressed in degrees between the line-to-neutral voltage of the reference identified terminal and the line-to-neutral voltage of the corresponding identified low-voltage terminal. For transformers manufactured according to the ANSI/IEEE standard [3], the *low-voltage side, whether in wye or delta* connection, has a phase shift of 30° lagging with respect to the high-voltage side of phase-to-neutral voltage vectors. Figure 3.9 shows ANSI/IEEE [3] transformer connections and a phasor diagram of the delta side and wye side voltages. *These relations and phase displacements are applicable to positive sequence voltages.*

The International Electrotechnical Commission (IEC) allocates vector groups, giving the type of phase connection and the *angle of advance* turned though in passing from the vector representing the high-voltage side EMF to that representing the low-voltage side EMF at the corresponding terminals. The angle is indicated much like the hour needle of a clock, the high-voltage vector being at 12 o'clock (zero) and the corresponding low-voltage vector being represented by the hour hand. The total rotation corresponding to hour hand of the clock is 360°. Thus, Dy11 and Yd11 symbols specify 30° lead (11 being the hour hand of the clock) and Dy1 and Yd1 signify 30° lag (1 being the hour hand of the clock). See Ref. [2,6] for more details.

$$I_H = I_L < 30^o \quad h = 3n + 1 = 1, \ 4, \ 7, \ 10, \ 13 \ ...$$

$$= I_L < -30^o \quad h = 3n - 1 = 2, \ 5, \ 8, \ 11, \ 14 \ ...$$

All commercial harmonic analysis programs apply this shift based on input transformer connection. The phase shifts in transformer windings are shown in Fig. 3.10 from Ref. [7].

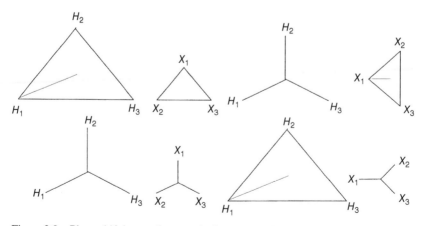

Figure 3.9   Phase shift in transformer windings, according to ANSI/IEEE standards. Source: Ref. [3].

3.3 DELTA WINDINGS OF THREE-PHASE TRANSFORMERS**   **85**

3.3.2 Phase Shift for Negative Sequence Components

If a voltage of negative phase sequence is applied to a delta–wye connected transformer, the phase angle displacement will be equal to the positive sequence phasors, but in the opposite direction. Therefore, when the positive sequence currents and voltages on one side lead the positive sequence current and voltages to the other side by 30°, the corresponding negative sequence currents and voltages will lag by 30°.

Transformer Type	Wdg #	Connection	Voltage Phasors	Phase Shift
D/d300°	1	Delta		300° lag
	2	Delta 300° lag		0°
D/y30°	1	Delta		0°
	2	Wye (gnd 1/2) 30° lag		330° lag
D/y150°	1	Delta		0°
	2	Wye (gnd 1/2) 150° lag	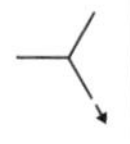	210° lag
D/y210°	1	Delta		0°
	2	Wye (gnd 1/2) 210° lag	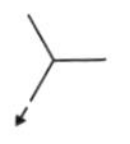	150° lag
D/y330°	1	Delta	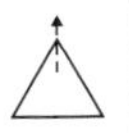	0°
	2	Wye (gnd 1/2) 330° lag	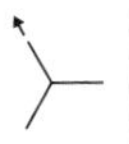	30° lag
Y/z30°	1	Wye (gnd 1/2)		30° lag
	2	Zig-zag (gnd 2/3) 30° lag		0°

Transformer Type	Wdg #	Connection	Voltage Phasors	Phase Shift
Y/z150°	1	Wye (gnd 1/2)	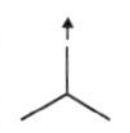	150° lag
	2	Zig-zag (gnd 2/3) 150° lag	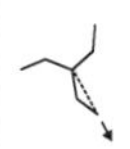	0°
Y/z210°	1	Wye (gnd 1/2)	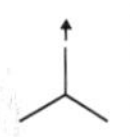	210° lag
	2	Zig-zag (gnd 2/3) 210° lag	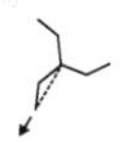	0°
Y/z330°	1	Wye (gnd 1/2)		330° lag
	2	Zig-zag (gnd 2/3) 330° lag	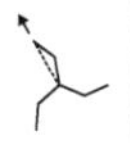	0°
D/z0°	1	Delta	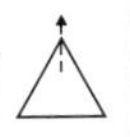	0°
	2	Zig-zag (gnd 1/2) 0° lag	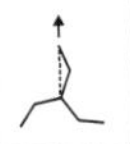	0°
D/z60°	1	Delta	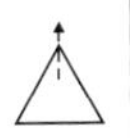	60° lag
	2	Zig-zag (gnd 1/2) 60° lag		0°
D/z120°	1	Delta		120° lag
	2	Zig-zag (gnd 1/2) 120° lag	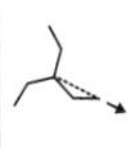	0°

(a)

Figure 3.10   (a) and (b) Phase shift in three-phase two-winding and three winding transformers.

Transformer Type	Wdg #	Connection	Voltage Phasors	Phase Shift
Y/y180°/d150°	1	Wye (gnd 1/2)		150° lag
	2	Wye (gnd 2/3) 180° lag		330° lag
	3	Delta 150° lag		0°
Y/y180°/d210°	1	Wye (gnd 1/2)		210° lag
	2	Wye (gnd 2/3) 180° lag		30° lag
	3	Delta 210° lag		0°
Y/y180°/d330°	1	Wye (gnd 1/2)		330° lag
	2	Wye (gnd 2/3) 180° lag		150° lag
	3	Delta 330° lag		0°
Y/d30°/y0°	1	Wye (gnd 1/2)		30° lag
	2	Delta 30° lag		0°
	3	Wye (gnd 2/3) 0°		30° lag

Transformer Type	Wdg #	Connection	Voltage Phasors	Phase Shift
Y/d30°/y180°	1	Wye (gnd 1/2)		30° lag
	2	Delta 30° lag		0°
	3	Wye (gnd 1/2) 180° lag		210° lag
Y/d30°/d30°	1	Wye (gnd 1/2)		30° lag
	2	Delta 30° lag		0°
	3	Delta 30° lag		0°
Y/d30°/d150°	1	Wye (gnd 1/2)		30° lag
	2	Delta 30° lag		0°
	3	Delta 150° lag		240° lag
Y/d30°/d210°	1	Wye (gnd 1/2)		30° lag
	2	Delta 30° lag		0°
	3	Delta 210° lag		180° lag

(b)

Figure 3.10 (*Continued*)

If the positive sequence voltages and currents on one side lag the positive sequence voltages, then the negative sequence voltages and currents will lead by 30°.

Example 3.2: Consider a balanced three-phase delta load connected across an unbalanced three-phase supply system (as shown in Fig. 3.11). The currents in lines *a* and *b* are given and the currents in the delta-connected load and also the symmetrical components of line and delta currents are required to be calculated. From these

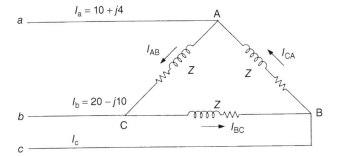

Figure 3.11   An unbalanced delta connected load (Example 3.2).

calculations, the phase shifts of positive and negative sequence components in delta windings and line currents can be established.

$$I_a = 10 + j4$$

$$I_b = 20 - j10$$

The line current in phase c is given by

$$I_c = -(I_a + I_b)$$

$$= -30 + j6.0 \text{A}$$

The currents in delta windings are

$$I_{AB} = \frac{1}{3}(I_a - I_b) = -3.33 + j4.67 = 5.735 < 144.51^\circ \text{A}$$

$$I_{BC} = \frac{1}{3}(I_b - I_c) = 16.67 - j5.33 - 17.50 < -17.7^\circ \text{A}$$

$$I_{CA} = \frac{1}{3}(I_c - I_a) = -13.33 + j0.67 = 13.34 < 177.12^\circ \text{A}$$

Calculate the sequence component of the currents I_{AB}:

$$\begin{vmatrix} I_{AB0} \\ I_{AB1} \\ I_{AB2} \end{vmatrix} = \frac{1}{3} \begin{vmatrix} 1 & 1 & 1 \\ 1 & a & a^2 \\ 1 & a^2 & a \end{vmatrix} \begin{vmatrix} I_{AB} \\ I_{BC} \\ I_{CA} \end{vmatrix} = \frac{1}{3} \begin{vmatrix} 1 & 1 & 1 \\ 1 & a & a^2 \\ 1 & a^2 & a \end{vmatrix} \begin{vmatrix} 5.735 < 144.51^\circ \\ 17.50 < -17.7^\circ \\ 13.34 < 177.12^\circ \end{vmatrix}$$

This calculation gives

$$I_{AB1} = 9.43 < 89.57^\circ \text{A}$$
$$I_{AB2} = 7.181 < 241.76^\circ \text{A}$$
$$I_{AB0} = 0 \text{A}$$

As stated in Chapter 1, theory of symmetrical components is not explained in this book. See references quoted in Chapter 1.

Calculate the sequence component of current I_a. This calculation gives

$$I_{a1} = 16.33 < 59.57° \, A$$

$$I_{a2} = 12.437 < 271.76° \, A$$

$$I_{a0} = 0A$$

This shows that the positive sequence current in the delta winding is $1/\sqrt{3}$ times the line positive sequence current, and the phase displacement is $+30°$, that is,

$$I_{AB1} = 9.43 < 89.57° = \frac{I_{a1}}{\sqrt{3}} < 30° = \frac{16.33}{\sqrt{3}} < (59.57° + 30°)A$$

The negative sequence current in the delta winding is $1/\sqrt{3}$ times the line negative sequence current, and the phase displacement is $-30°$, that is,

$$I_{AB2} = 7.181 < 241.76° = \frac{I_{a2}}{\sqrt{3}} < -30° = \frac{12.437}{\sqrt{3}} < (271.76° - 30°)A$$

This example illustrates that the negative sequence currents and voltages undergo a phase shift which is the reverse of the positive sequence currents and voltages.

Example 3.3: Consider a harmonic spectrum as shown in Table 3.2, which is applied to the secondary of a delta-primary, wye secondary (grounded) transformer of 2 MVA connected to a 13.8 kV system, three-phase short-circuit MVA = 750, $X/R = 10$. The harmonic current waveforms on the primary and secondary side of the transformer are plotted in Fig. 3.12. These differ because of shift in the positive and negative sequence harmonics (Chapter 1) passing through the transformer to the 13.8 kV source.

Example 3.4: Draw the zero sequence network of a wye–wye transformer with tertiary delta, both the wye-neutrals are solidly grounded and also show the flow of third harmonic currents.

The zero sequence impedance circuit is drawn as shown in Fig. 3.13(a) and the flow of zero sequence currents is illustrated in Fig. 3.13 (b). See Ref. [8,9] for detail description of sequence networks of transformers and also Chapter 12.

TABLE 3.2 Harmonic Injection at the Secondary of the Transformer

h	5	7	11	13	17	19	23	25	29	31
Mag	17	12	7	5	2.8	1.5	0.50	0.40	0.30	0.20

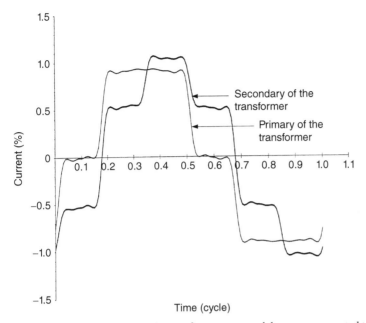

Figure 3.12   Simulated waveforms of currents on a delta–wye connected transformer on the primary and secondary sides (Example 3.3).

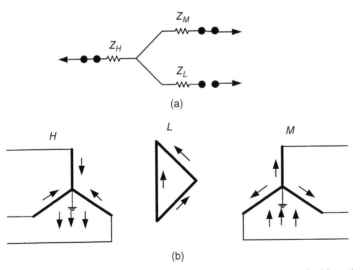

Figure 3.13   (a) Zero sequence circuit of a wye–wye connected with tertiary delta transformer; both wye windings solidly grounded, (b) flow of zero sequence currents in the windings and lines (Example 3.4).

3.3.3 Distortion due to Saturation

Saturation of the transformers can cause serious distortions. Overfluxing can occur due to higher voltages and factor V/f is a measure of overfluxing. ANSI/ IEEE device 24, V/f relay is commonly used to take the transformer off-line due to overfluxing. Generally the transformers are tripped if V/f exceeds 1.10 for a short duration. Non linear modeling of transformers is discussed in Ref. [2,10] (also see Chapter 12).

Example 3.5: The transformer of Example 3.1 is subjected to overvoltage and some initial trapped flux. It is loaded to 7 MVA. The distorted primary currents due to saturation in three phases are shown in Fig. 3.14, which is the result of EMTP simulation. This simulation shows that (1) the harmonic currents increase considerably and the inrush transients decay much slowly lasting for a considerable period of time. In Fig. 3.14, the initial inrush period of 1 sec is not shown.

3.3.4 Geomagnetically Induced Currents

Geomagnetically induced currents (GIC) flow in earth surface due to solar magnetic disturbance (SMD) and these are typically of 0.001 to 0.1Hz and can reach peak values of 200A. These can enter transformer windings through grounded neutrals (Fig 3.15), and bias the transformer core to 1/2 cycle saturation. As a result the transformer magnetizing current is greatly increased. Harmonics increase and these could cause reactive power consumption, capacitor overload and false operation of protective relays; etc.

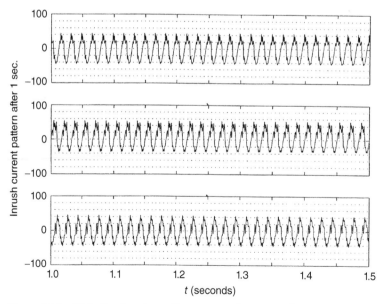

Figure 3.14   Distortion in primary line currents due to transformer saturation, EMTP simulation (Example 3.5).

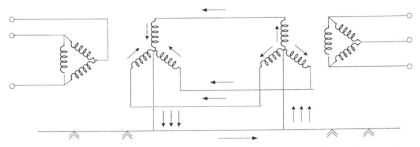

Figure 3.15   GIC entering the grounded neutrals of wye-connected transformers. Source: Ref. [10].

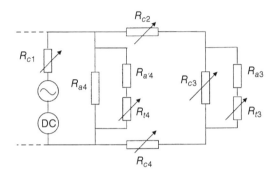

Figure 3.16   Transformer model for GIC simulation. Source: Ref. [10].

A GIC model is shown in Fig 3.16, as developed by authors of Ref. [11]. This saturation magnetic circuit model of a single phase shell form of transformer, valid for GIC levels was developed based on 3D finite element analysis (FEM) results. This model is able to simulate not only the four linear and knee region equations, but also the heavy saturated region and the so called "air-core" region; four major flux paths are included. All R elements represent reluctances in different branches. Subscripts c, a, and t stand for core, air, and tank respectively and 1, 2, 3 and 4 represent major branches of flux paths. Branch 1 represents sum of core and air fluxes within the excitation windings, branch 2 represents flux path in yoke, branch 3 represents sum of fluxes entering the side leg, part of which leaves the side leg and enters the tank. Branch 4 represents flux leaving the tank from the center leg. An iterative program is used to solve the circuit of Fig. 3.16 so that nonlinearity is considered.

Reference [12] details harmonic interaction in the presence of GIC for the Quebec–New England phase II HVDC Transmission. Significant amplitudes of GIC could flow in the neutrals of transformers in an isolated network, and induce large harmonic distortions. This reference describes 735 kV and 315 kV AC systems serving Radisson 315 kV converters ± 450 KV DC. There is a reflection of DC side impedances to AC side due to nonlinear switching action of the converters.

The converter acts like a modulator of DC side oscillations when transforming them to AC side:

$f_{DC} + f$ Positive sequence

$f_{DC} - f$ Negative sequence

Thus, a ripple at 6 times the fundamental frequency will be transferred as fifth and seventh harmonics on the AC side. Negative sequence AC side voltage oscillation at a given frequency will see the DC side impedance at that frequency plus 60 Hz, with converter transformer turns ratio and other scaling as appropriate. Similarly for positive sequence AC side frequency will see DC side impedance at that frequency minus 60 Hz.

3.4 HARMONICS IN ROTATING MACHINE WINDINGS

The armature windings of rotating machines (motors or generators) consist of phase coils that span approximately a pole-pitch. A phase winding consists of a number of coils connected in series, and the EMF generated in these coils is time displaced in phase by a certain angle. The air gap in the machines is bounded on either side by iron surfaces and provided with slots and duct openings and is skewed (for sinusoidal voltage generation). Simple methods of estimating the reluctance of the gap to carry a certain flux across the gap are not applicable.

Figure 3.17(a) illustrates the outline of the half-pole arc and the flux density. Assume that the flux lines leaving and entering the iron are perpendicular. If flux density is assumed to be 100% at the gap length l_g, then at other points it can be assumed as $100 l_g / l$ where l is the length of the flux lines elsewhere along the pole surface. The flux density is not sinusoidal (Fig. 3.17(b)) and it can be resolved into harmonic components with Fourier analysis. Figure 3.17(c) shows resolution into third, fifth and seventh harmonics.

The harmonic EMF in the phase windings is affected by:

- The harmonic flux components
- The phase spread and fractional slotting (when the number of slots per pole per phase are not an integer number)
- The coil span
- The winding connections and
- Tooth ripples

The three phase windings in the rotating machines are designed to reduce the harmonics by proper chording, slotting, fractional slot per pole, skewing, and so on, which are not discussed in detail [13]. The wide spread of 120° eliminates triplen harmonics, because the effective spread at third harmonic becomes $3 \times 120° = 360°$. When a coil is pitched short or long by π/h (or $3\pi/h$, $5\pi/h \dots$), then no harmonic of the order h will appear in the coil EMF. Generally the third harmonics are eliminated by wye or delta connection of the phases and the coil spans are selected to reduce as much as possible fifth and seventh harmonics.

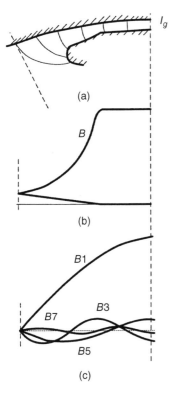

Figure 3.17   (a) Magnetic lines of force in the polar gap, (b) flux density distribution, and (c) resolution into harmonics.

Coil span factor

Short-pitching or chording of the windings reduces the fundamental frequency EMF. Referring to Fig. 3.18, the chording factor for the fundamental is:

$$k_{ef} = \cos \frac{\varepsilon}{2} \qquad (3.12)$$

For a coil span of 2/3rd of a pole pitch, $\varepsilon = \pi/3$, and $k_{ef} = 0.866$, that is, fundamental frequenmcy EMF is reduced by this factor.

For the h-th harmonic it becomes:

$$k_{eh} = \cos \frac{h\varepsilon}{2} \qquad (3.13)$$

Table 3.3 shows the harmonic generation for a pitch factor of 83.3% ($\varepsilon = \pi/6$). The factors for fifth and seventh harmonics are small and third and nineth harmonics are eliminated due to winding connections.

Fractional slot windings and skewing are some other means to reduce harmonics.

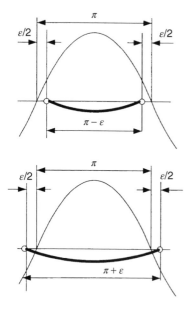

Figure 3.18 Chorded coil-spans for reduction of harmonics in machine windings.

TABLE 3.3 Selecting a Coil Span Factor to Reduce Harmonics in Armature Windings

Order of harmonic	Fundamental	3	5	7	9
Coil span factor	0.966	0.707	0.259	0.259	0.707

3.4.1 EMF of the Windings

The EMF equation can be written as:

$$E_{\text{phf}} = 4.44K_{\text{wf}}fT_{\text{ph}}\phi_f \tag{3.14}$$

where ϕ_f is the fundamental frequency flux, K_{wf} is the fundamental frequency winding factor, T_{ph} are the turns per phase, f is the fundamental frequency and E_{phf} is the fundamental frequency EMF. A similar expression for harmonic flux is:

$$E_{\text{phh}} = 4.44K_{\text{wh}}f_hT_{\text{ph}}\phi_h \tag{3.15}$$

3.4.2 Distribution Factor

Consider m coils per phase group. Then the generation of EMF in each coil, e_a, e_b, e_c is at an angle, say ϕ. The distribution factor for fundamental frequency becomes:

$$k_{\text{mf}} = \frac{\sin(1/2)\sigma}{g'\sin(1/2)(\sigma/g')} \tag{3.16}$$

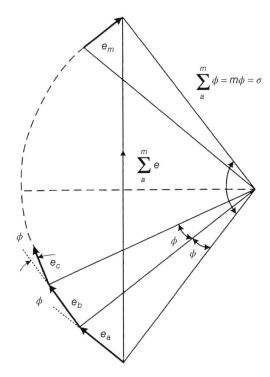

$$\sum_{a}^{m} \phi = m\phi = \sigma$$

Figure 3.19   Phase EMF of stator windings.

where g' = number of slots per pole per phase and σ is the phase spread (Fig. 3.19). For a three-phase winding 2 pole pitches with $\sigma = 60°$ and $g' = 2.5$ (total number of slots = 15), effective angular displacement of coils = 12°, and $k_{mf} = 0.956$.

A similar expression for the distribution factor for harmonics can be written as

$$k_{mh} = \frac{\sin(1/2)h\sigma}{g' \sin(1/2)(h\sigma/g')} \tag{3.17}$$

For $g' > 5$, a uniform distribution can be assumed for which

$$k_{mf} = \frac{\sin(1/2)\sigma}{(1/2)\sigma} \tag{3.18}$$

$$k_{mh} = \frac{\sin(1/2)h\sigma}{(1/2)h\sigma} \tag{3.19}$$

Table 3.4 shows k_{mh} for phase spreads. A zero value indicates that vector closes on itself leaving no resultant. The wide spread of 120° eliminates all triplen harmonics.

TABLE 3.4   Values of k_{mn}

Number of Phases	Phase spread, δ	k_{m1}	K_{m3}	K_{m5}	K_{m7}	K_{m9}
3	60°	0.955	0.637	0.191	−0.135	−0.212
3 or 1	120°	0.827	0	−0.165	0.118	0

3.4.3   Armature Reaction

When the machine is loaded the effect of armature reaction due to flow of currents is the determining factor for the resultant flux. It rotates in synchronism with the change of phase current and is not of constant value. Figure 3.20 shows that armature reaction varies between a pointed and flat-topped trapezium for a phase spread of $\pi/3$. Fourier analysis of the pointed waveform in Fig. 3.20 gives

$$F = \frac{4}{\pi} F_m \cos \omega t \left[\sum_{h=1}^{h=\infty} \frac{1}{h} \left(\frac{\sin(1/2)h\sigma}{(1/2)h\sigma} \right) \sin\ (hx) \right] \qquad (3.20)$$

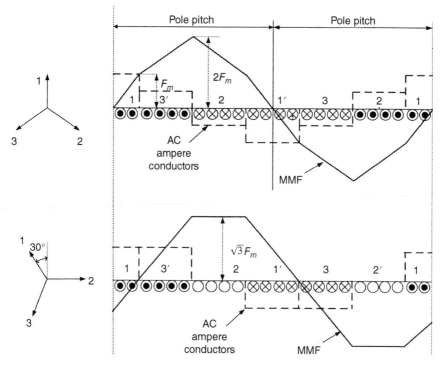

Figure 3.20   Armature reaction of three-phase windings.

The MMFs of three phases will be given by considering the time displacement of currents and space displacement of axes as:

$$F_{p-1} = \frac{4}{\pi} F_m \cos \omega t \left[\sum_{h=1}^{h=\infty} \frac{1}{h} k_{\mathrm{mh}} \sin (hx) \right] \tag{3.21}$$

$$F_{p-2} = \frac{4}{\pi} F_m \left(\cos \omega t - \frac{2\pi}{3} \left[\sum_{h=1}^{h=\infty} \frac{1}{h} k_{\mathrm{mh}} \sin[\, h\left(x - \frac{2\pi}{3} \right) \right] \tag{3.22}$$

$$F_{p-3} = \frac{4}{\pi} F_m \left(\cos \omega t + \frac{2\pi}{3} \left[\sum_{h=1}^{h=\infty} \frac{1}{h} k_{\mathrm{mh}} \sin[\, h\left(x + \frac{2\pi}{3} \right) \right] \tag{3.23}$$

This gives:

$$F_t = \frac{6}{\pi} F_m \left[F_{mi} \sin (x - \omega t) + \frac{1}{5} k_{m5} \sin(5x - \omega t) - \frac{1}{7} k_{m7} \sin(7x - \omega t) + \cdots \right] \tag{3.24}$$

where k_{m5} and k_{m6} are harmonic winding factors.

The MMF has a constant fundamental, and harmonics of the order of 5, 7, 11, 13 ... or $6m \pm 1$, where m is any positive integer. The third harmonic and its multiples (triplen harmonics) are absent; though in practice some triplen harmonics are produced. The harmonic flux components are affected by phase spread, fractional slotting, and coil span. The pointed curve is obtained when $\sigma = 60°$ and $\omega t = 0$ and from Eq. (3.24), it has a peak value of:

$$F_{1,\mathrm{peak}} = \frac{18}{\pi^2} F_m \left[1 + \frac{1}{25} + \frac{1}{49} + \cdots \right] \approx 2F_m \tag{3.25}$$

The flat topped curve is obtained when $\omega t = \pi/6$ and has the maximum amplitude:

$$F_{1,\mathrm{peak}} = \frac{18}{\pi^2} F_m \frac{\sqrt{3}}{2} \left[1 + \frac{1}{25} + \frac{1}{49} + \cdots \right] \approx \sqrt{3} F_m \tag{3.26}$$

The harmonics are generally of small magnitude.

3.5   COGGING AND CRAWLING OF INDUCTION MOTORS

Parasitic magnetic fields are produced in an induction motor due to harmonics in the MMF originating from:

- Windings
- Certain combination of rotor and stator slotting
- Saturation

- Air gap irregularity
- Unbalance and harmonics in the supply system voltage.

The harmonics move with a speed reciprocal to their order, either with or against the fundamental. Harmonics of the order of $6m + 1$ move in the same direction as the fundamental magnetic field while those of $6m - 1$ move in the opposite direction.

3.5.1 Harmonic Induction Torques

The harmonics can be considered to produce, by an additional set of rotating poles, rotor EMF's, currents, and harmonic torques akin to the fundamental frequency at synchronous speeds depending on the order of the harmonics. Then, the resultant speed torque curve will be a combination of the fundamental and harmonic torques. This produces a saddle in the torque speed characteristics and the motor can crawl at the lower speed of 1/7th of the fundamental, Fig. 3.21(a). This torque speed curve is called *harmonic induction torque curve*.

This harmonic torque can be augmented by stator and rotor slotting. In n-phase winding, with g' slots per pole per phase, EMF distribution factors of the harmonics are:

$$h = 6Ag' \pm 1 \tag{3.27}$$

where A is any integer, 0, 1, 2, 3 ... .

The harmonics of the order $6Ag' + 1$ rotate in the same direction as the fundamental, while those of order $6Ag' - 1$ rotate in the opposite direction.

A four-pole motor with 36 slots, $g' = 3$ slots per pole per phase, will give rise to 17th and 19th harmonic torque saddles, observable at $+1/19$ and $-1/17$ speed, similar to the saddles shown in Fig. 3.21(a).

Consider 24 slots in the stator of a four-pole machine. Then $g' = 2$ and 11th and 13th harmonics will be produced strongly. The harmonic induction torque thus produced can be augmented by the rotor slotting. For a rotor with 44 slots, 11th harmonic has 44 half waves each corresponding to a rotor bar in a squirrel cage induction motor. This will accentuate 11th harmonic torque and produce strong vibrations.

If the number of stator slots is equal to the number of rotor slots, the motor may not start at all, a phenomenon called *cogging*.

The phenomena will be more pronounced in squirrel cage induction motors as compared to wound rotor motors, as the effect of harmonics can be reduced by coil pitch, Section 3.4. In the cage induction motor design, S_2 (number of slots in the rotor) should not exceed S_1 (number of slots in the stator) by more than 50–60%, otherwise there will be some tendency towards saddle harmonic torques.

3.5.2 Harmonic Synchronous Torques

Consider that the fifth and seventh harmonics are present in the gap of a three-phase induction motor. With this harmonic content and with certain combination of the

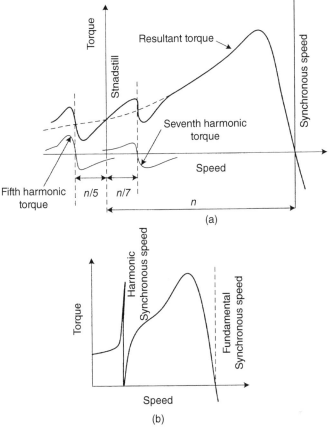

Figure 3.21   (a) Harmonic induction torques and (b) synchronous torques in an induction machine.

stator and rotor slots, it is possible to get a stator and rotor harmonic torque producing a *harmonic synchronizing* torque as in a synchronous motor. There will be a tendency to develop sharp synchronizing torque at some lower synchronous speed, (Fig. 3.21(b)). The motor may crawl at a lower speed.

The rotor slotting will produce harmonics of the order of:

$$h = \frac{S_2}{p} \pm 1 \qquad (3.28)$$

where S_2 is the number of rotor slots. Here, the plus sign means rotation with the machine. Consider a four-pole (p = number of pair of opoles = 2) motor with $S_1 = 24$ and with $S_2 = 28$. The stator produces reversed 11th harmonic (reverse going) and 13th harmonic (forward going). The rotor develops a reversed 13th and forward 15th harmonic. The 13th harmonic is produced by both stator and rotor, but is of opposite rotation. The synchronous speed of the 13th harmonic is 1/13 of the fundamental

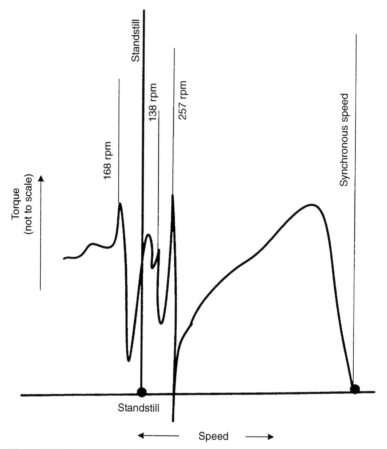

Figure 3.22 Torque speed curve of four-pole, 60 Hz motor, considering harmonic synchronous torques.

synchronous speed. Relative to rotor it becomes:

$$-\frac{(n_s - n_r)}{13} \tag{3.29}$$

where n_s is the synchronous speed and n_r is the rotor speed. The rotor, therefore rotates its own 13th harmonic at a speed of

$$-\frac{(n_s - n_r)}{13} + n_r \tag{3.30}$$

relative to the stator. The stator and rotor 13th harmonic fall into step when:

$$+\frac{n_s}{13} = -\frac{(n_s - n_r)}{13} + n_r \tag{3.31}$$

TABLE 3.5   Typical Synchronous Torques 4-pole Cage Induction Motors

Stator Slots	Rotor Slots	Stator Harmonics		Rotor Harmonics	
S1	S2	Negative	Positive	Negative	Positive
24	20	−11	+13	−9	+11
24	28	−11	+13	−13	+15
36	32	−17	+19	−15	+17
36	40	−17	+19	−19	+21
48	44	−23	+25	−21	+23

This gives $n_r = n_s/7$, i.e., torque discontinuity is produced not by 7th harmonic but by 13th harmonic in the stator and rotor rotating in opposite directions. The torque speed curve is shown in Fig.3.22.

- The synchronous torque at $1800/7 = 257$ rpm
- Induction torque due 13th stator harmonic = 138 rpm
- Induction torque due to reversed 11th harmonic = 164 rpm

Typical synchronous torques in 4-pole cage induction motors are listed in Table 3.5. If $S_1 = S_2$, the same order harmonics will be strongly produced, and each pair of harmonics will produce a synchronizing torque, and the rotor may remain at standstill (cogging), unless the fundamental frequency torque is large enough to start the motor.

The harmonic torques are avoided in the design of induction machines by proper selection of the rotor and stator slotting and winding designs

3.5.3   Tooth Ripples in Electrical Machines

Tooth ripples in electrical machinery are produced by slotting as these affect air-gap permeance and give rise to harmonics. Figure 3.23 shows ripples in the air-gap flux distribution (exaggerated) because of variation in gap permeance. The frequency of flux pulsations correspond to the rate at which slots cross the pole face, i.e., it is given by $2gf$, where g is the number of slots per pole and f is the system frequency. *The ripples do not move with respect to the conductors but pass over the flux distribution curve.* This stationary pulsation may be regarded as two waves of fundamental space distribution rotating at angular velocity $2g\omega$ in forward and backward directions. The component fields will have velocities of $(2g \pm 1)\omega$ relative to the armature winding and will generate harmonic EMFs of frequencies $(2g \pm 1)f$ cycles per second. However, this is not the main source of tooth ripples. Since the ripples do not move with respect to conductors, these cannot generate an EMF of pulsation. With respect to the rotor the flux waves have a relative velocity of $2g\omega$ and generate EMFs of $2gf$ frequency. Such currents superimpose an MMF variation of $2gf$ on the resultant pole MMF. These can be again resolved into forward and backward moving components with respect to the *rotor*, and $(2g \pm 1)\omega$ with respect to the *stator*. Thus, stator EMFs

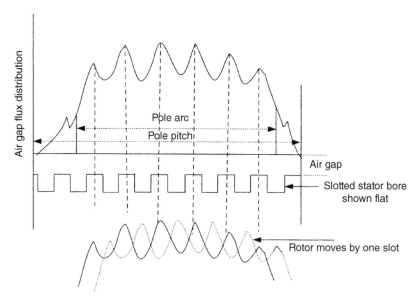

Figure 3.23   Tooth ripples in electrical machines, air gap flux distribution.

at frequencies $(2g \pm 1)f$ are generated, which are the principal tooth ripples. Ref. [13–20] may be seen for electrical machines.

3.6   SYNCHRONOUS GENERATORS

3.6.1   Voltage Waveform

The synchronous generators produce almost sinusoidal voltages. The terminal voltage wave of synchronous generators must meet the requirements of NEMA and IEEE Standard 115 [21] which states that the deviation factor of the open line-to-line terminal voltage of the generator shall not exceed 0.1.

Figure 3.24 shows a plot of a hypothetical generated wave, superimposed on a sinusoid, and the deviation factor is defined as

$$F_{\text{DEV}} = \frac{\Delta E}{E_{\text{OM}}} \tag{3.32}$$

where E_{OM} is calculated from a number of samples of instantaneous values:

$$E_{\text{OM}} = \sqrt{\frac{2}{J} \sum_{j=1}^{J} E_j^2} \tag{3.33}$$

The deviation from a sinusoid is very small.

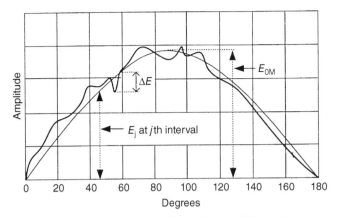

Figure 3.24   Measurement of deviation factor of the synchronous generator generated voltage. Source: Ref. [20].

Though the generators produce little harmonics by themselves these are sensitive to harmonic loading. This is so because synchronous generators have limited negative sequence capability, Chapter 8.

3.6.2   Third Harmonic Voltages and Currents

Generator neutrals have predominant third harmonic voltages. In a wye-connected generator, with the neutral grounded through high impedance, the third harmonic voltage increases toward the neutral, while the fundamental frequency voltage decreases. The third harmonic voltages at line and neutral can vary considerably with load.

Figure 3.25 shows the fundamental frequency and third harmonic voltage distribution, typical of synchronous generators. Figure 3.25(a) shows that the fundamental frequency voltage decreases linearly to the neutral point, whereas 3.24(b) shows that the third harmonic distribution crosses somewhere in the middle of the windings under normal operating condition. For a ground fault in the stator windings towards the neutral, the third harmonic voltage distribution is shown in Fig. 3.25(c); it increases at the line terminals and decreases at neutral. When the neutrals of the generators are grounded through high resistance, current limited to no more than 10A, (equal to the stray capacitance current of the system; generally in system configuration where the generator and step up transformer are directly connected to a utility HV system) this fundamental and third harmonic voltage profile is used to provide 100% stator winding protection against ground faults [22,23].

Normally, the generator winding wye neutrals are not solidly grounded. Sometimes generators of smaller ratings may be solidly grounded. When such solidly grounded generators are paralleled on the same bus; large third harmonic currents can circulate in the generator windings. Thermal overloads can occur. It is a good practice not to solidly ground *any* generator. A resistance introduced in the generator neutral limits these third harmonic currents. A third harmonic current of 20A was

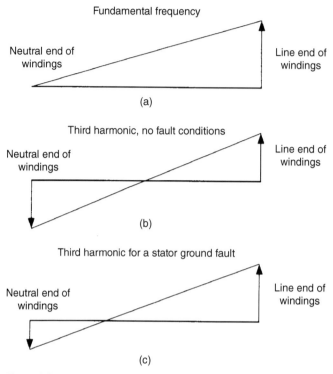

Figure 3.25   Fundamental and third harmonic voltage distributions from line terminal to neutral in synchronous generator stator windings. Source: Ref. [22].

measured in a 13.8 kV 40 MVA generator grounded through 400A resistor. For solidly grounded bus connected generators it can approach 60% or more of the generator full load current.

3.7   SATURATION OF CURRENT TRANSFORMERS

Saturation of current transformers under fault conditions produces harmonics in the secondary circuits. Accuracy classification of current transformers is designated by one letter, C or T, depending on current transformer construction [24]. Classification C covers bushing type transformers with uniformly distributed windings, and the leakage flux has a negligible effect on the ratio within the defined limits. A transformer with relaying accuracy class C200 means that the percentage ratio correction will not exceed 10% at any current from 1−20 times the rated secondary current at a standard burden of 2.0 ohms, which will generate 200 V, Ref. [24]. The secondary voltage as given by maximum fault current reflected on the secondary side multiplied by connected burden $(R + jX)$ should not exceed the assigned C accuracy class. This is very preliminary CT selection calculation for steady state. Avoiding saturation under transient conditions is important; as a relaying class CT must reproduce the high and

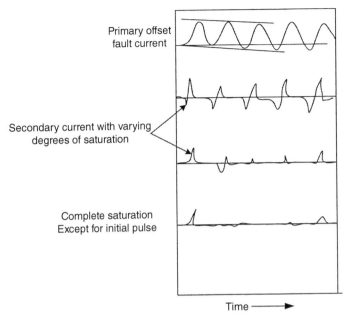

Time ⟶

Figure 3.26   Progressive saturation of a CT secondary current on an asymmetrical fault giving rise to harmonics. Source: Ref. [24].

asymmetrical fault currents accurately. These aspects of proper CT application are not discussed. When current transformers are improperly applied saturation can occur, as shown in Fig. 3.26, Ref. [25]. A completely saturated CT does not produce a current output, except during the first pulse, as there is a finite time to saturate and de-saturate. The *transient* performance should consider the DC component of the fault current, as it has far more effect in producing severe saturation of the current transformer than the AC component, Ref. [25].

As the CT saturation increases, so does the secondary harmonics, before the CT goes into a completely saturated mode. Harmonics of the order of 50% third, 30% fifth, 18% seventh and 15% ninth and higher order may be produced. These can cause improper operation of the protective devices. This situation can be avoided by proper selection and application of current transformers.

Figure3.27 shows partial saturation of a CT on asymmetrical current. The CT comes out of saturation with time delay as the fault current becomes more symmetrical. Such a situation can affect the timing of the relay operation, see Chapter 8.

3.8   FERRORESONANCE

In the presence of system capacitance, certain transformer and reactor combinations can give rise to ferroresonance phenomena, due to nonlinearity and saturation of the reactance. This is characterized by peaky current surges of short duration that generate overvoltages, and have harmonics or sub harmonics and reach amplitude

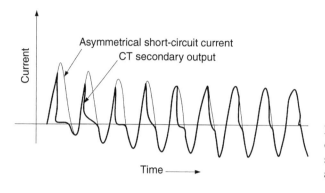

Figure 3.27 Waveforms of partial saturation of CT secondary current on asymmetrical fault.

of 2 per unit. The phenomena may be initiated by some system changes or disturbances and the response may be stable or unstable, which may persist indefinitely or cease after a few seconds. The ferroresonance is documented in some cases as follows:

- Transformers feeders on double circuits becoming energized through the mutual capacitance between lines and going into ferroresonance when one transformer feeder is switched out.

- Transformers losing a phase, say due to operation of a current limiting fuse on ungrounded systems.

- Grading capacitors of high voltage breakers (provided across multi-breaks per phase in HV breakers for voltage distribution), remaining in service when the breaker opens.

- Possibility of ferroresonance with a CVT (Capacitor Voltage Transformer) or electromagnetic PT (Potential Transformer)under certain operating conditions.

Figure 3.28 (a) depicts the basic circuit, a nonlinear inductor, that may be PT or transformer windings is energized through a resistor and capacitor. The phasor diagram is shown in Fig. 3.28(b), *as long as the reactor operates below its saturation level.* In Fig. 3.28(c), point A and C are stable operating points, while point B is not. Point C is characterized by large magnetizing current and voltage. This overvoltage appears between line and ground. The changeover from point A to point C is initiated by some transient in the system, and can be a nuisance due to its random nature.

Figure 3.28(d) shows two lines with mutual capacitance couplings. The transformer forms a series ferroresonance circuit with the line mutual capacitance. The ferroresonance can occur at lower frequency than the fundamental and overheat the transformers. Switching transients will be superimposed upon the ferroresonance voltages.

This phenomenon of switching a nonlinear reactor can occur in other applications too, for example, shunt reactors in transmission systems.

High overvoltages of the order of 4.5 per unit, distorted wave shapes and harmonics can be generated.

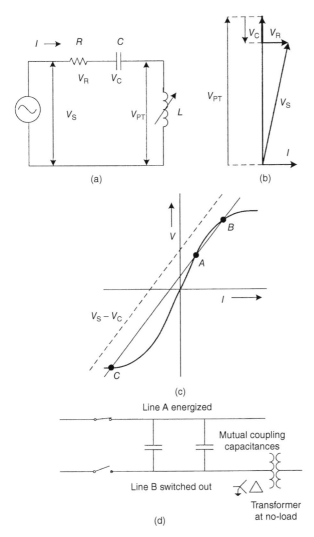

Figure 3.28   (a) Equivalent circuit for energization of a nonlinear reactor, (b) phasor diagram, (c) possible point of ferroresonance depending on saturation and voltage characteristics, (d) ferroresonance due to capacitive coupling between two transmission lines when the coupled line B energizes a transformer at no- load.

- Resonance can occur over a wide range of X_c/X_m. Ref. [26,27] specify the range as:

$$0.1 < X_c/X_m < 40 \tag{3.34}$$

- Resonance occurs only when the transformer is unloaded, or very lightly loaded. Transformers loaded to more than 10% of their rating are not susceptible to ferroresonance.

The capacitance of cables varies between 40 and 100 nF per 1000 ft, depending upon conductor size. However the magnetizing reactance of a 35 kV transformer is several times higher than that of a 15 kV transformer; the ferroresonance can be more damaging at higher voltages. For delta connected transformers the ferroresonance can occur for less than 100 ft of cable. Therefore, the grounded wye–wye transformer connection has become the most popular in underground distribution system in North America. It is more resistant, though not totally immune to ferroresonance.

3.8.1   Series Ferroresonance

Figure 3.29(a) shows a basic circuit of ferroresonance. Consider that current limiting fuses in one or two lines operate. X_c is the capacitance of the cables and transformer bushings to ground. Also the switch may not close simultaneously in all the three phases or while opening the phases may not open simultaneously. We can draw the equivalent circuits with one phase closed and also with two phases closed as shown in Figs. 3.29(b) and (c), respectively.

Similar equivalent circuit can be drawn for wye–wye ungrounded transformer. The minimum capacitance to produce ferroresonance can be calculated from $X_c/X_m = 40$, say.

$$C_{m\,res} = \frac{2.21 \times 10^{-7}\text{MVA}_{transf}I_m}{V_n^2} \tag{3.35}$$

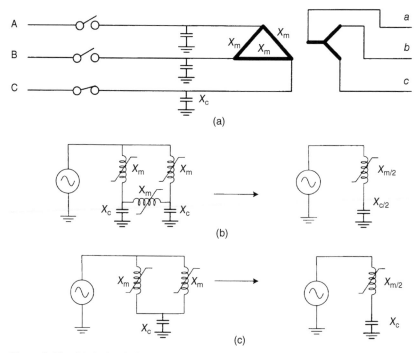

Figure 3.29   (a) A circuit for possible ferroresonance and (b) one phase closed; (c) two phases closed.

TABLE 3.6 Capacitance Limits for Ferroresonance, C_{mf} in pF for Ferroresonance

Transformer kVA		System Voltage kV		
Single Phase	Three-Phase	8.32/4.8	12.5/7.2, 13.8/7.99	25/14.4, 27.8/16
		$C_{m\,res}$	$C_{m\,res}$	$C_{m\,res}$
5	15	72	26	16
10	25	119	43	11
	50	339	86	21
25	75	358	129	32
	100	477	172	43
50	150	719	258	64
75	225	1070	387	97
100	300	1432	516	129
167	500	2390	859	215

where

I_m = magnetizing current of the transformer as a percentage of the full load current. Typically magnetizing current of a distribution transformer is 1% to 3% of the transformer full load current.

MVA_{trasf} = rating of transformer in MVA

$C_{m\,res}$ = minimum capacitance for resonance in pF

V_n = line to neutral voltage in kV.

Table 3.6 gives the approximate values of $C_{m\,res}$.

3.8.2 Parallel Ferroresonance

A less common condition of ferroresonance is illustrated in Figs. 3.30(a) and (b). This involves the mutual X_m formed between the windings at the same voltage level in 4 or 5 leg core type transformers. These may resonate even when connected in grounded wye configuration. However, the overvoltages are limited to 1.5 pu because equivalent circuit is a parallel LC combination, and Z_m limits the voltage by saturation.

Example 3.6: An EMTP simulation of ferroresonance in a nonlinear inductor is examined. The reactor is energized through a 60 Hz voltage of 240 volts and a parallel capacitor of 1.7 μF in three cases:

(a) with no prior magnetic flux,

(b) and (c) with some prior trapped flux. The current-flux characteristics are represented by a two-piece curve.

The simulation results of the voltage across the reactor and the current through it are shown in Figs. 3.31 and 3.32, respectively. Without a prior flux, there are no current or voltage transients. With initial flux the voltage rises to 4 pu of the applied

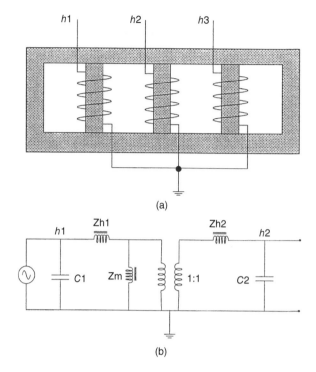

Figure 3.30 (a) Mutual coupling ferroresonance; (b) equivalent circuit for mutual coupling ferroresonance.

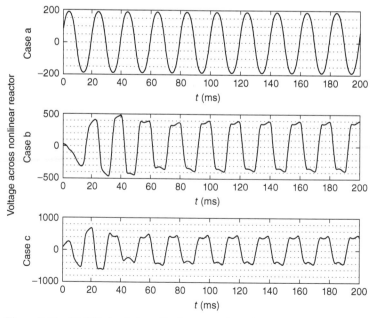

Figure 3.31 EMTP simulation of voltage transients in three cases of a nonlinear reactor (Example 3.6).

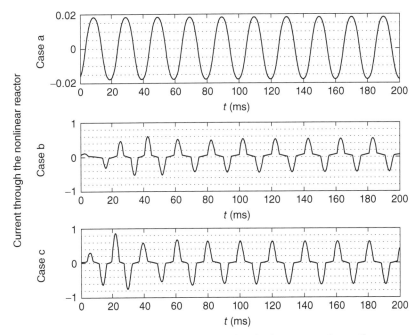

Figure 3.32   EMTP simulation of current transient in three cases of a nonlinear reactor (Example 3.6).

voltage for case(c). Also the current increases many times the base current without prior flux.

The ferroresonance in power systems is avoided by: (1) proper protective devices, for example a negative sequence relay can detect the failure of a fuse in a three-phase circuit and avoid damage to the rotating machines, (2) provision of surge arresters, (3) secondary loading resistors and capacitors in the protective relaying circuits, (4) loading resistors across open delta potential transformer windings. Ref. [28] documents a case of plant shutdown due to ferroresonance in potential transformer (PT) circuits.

The protective relays for detecting ground fault in ungrounded systems, which operate on the principle of neutral shift when a ground fault occurs are provided with damping resistors/ surge arresters and are tuned to fundamental frequency with a parallel capacitor circuit.

3.9   POWER CAPACITORS

Power capacitors are used in a number of applications (Chapter 11), and application in harmonic filters is of special interest in this book. Though these will not produce harmonics by themselves in sinusoidal circuits, but can greatly magnify existing harmonics by bringing about a resonant condition (Chapter 9). Capacitors are applied in

HV systems for series compensation of HV transmission lines (Chapter 5). These can give rise to subsynchronous resonance which can give rise to torsional vibrations and damage to rotating machines can occur. Power capacitor in appropriate system configurations, when applied as filters, will limit the harmonics in the power systems, by acting as a sink to the desired harmonics. The inrush currents of the power capacitors are at much higher frequencies, mainly dependent on the reactance in the switching circuits; switching transients are discussed in Chapter 11. These aspects are subjects of analyses and discussions in this book in appropriate chapters.

3.10 TRANSMISSION LINES

If the impedance versus frequency of long transmission line (frequency scan) using distributive line parameters is examined, it will show a number of natural resonant frequencies, even when no harmonic or nonlinear loads are being served (Chapter 12). This does not generate harmonics in itself, but nonlinear loads are invariably present. Harmonics due to DC links, interconnecting two AC power systems operating at different frequencies, say 50 Hz and 60 Hz are discussed in Chapter 5.

REFERENCES

1. M. Heathoter. J&P Transformer Book, 13th Edition, Newnes, New York, 2007.
2. J. C. Das, Power System Analysis- Short-Circuit Load Flow and Harmonics, 2nd Edition, CRC Press, Boca Raton, 2012
3. ANSI/IEEE Std. C57.12, General requirements for liquid immersed distribution, power and regulating transformers, 2006.
4. C. E. Lin, J. B. Wei, C. L. Huang, C. J. Huang, "A new method for representation of hysteresis loops," IEEE Transactions on Power Delivery. vol. 4, pp. 413–419, 1989.
5. J. C. Das, Transients in Electrical Systems, McGraw-Hill, New York 2011.
6. J. Grainger and W. Stevenson, Jr. Power System Analysis, McGraw-Hill, New York, 1994.
7. GE, "B30 Bus Differential Relay," GE Instruction Manual GEK-1133711, 2011.
8. J. L. Blackburn. Symmetrical Components for Power System Engineering, Marcel & Dekker, New York 1993.
9. C. O. Calabrase. Symmetrical Components Applied to Electrical Power Networks. Ronald Press Group, New York 1959.
10. J. D. Green, C. A. Gross, "Non-linear modeling of transformers," IEEE Transactions on Industry Applications, vol. 24, pp. 434–438, 1988.
11. S. Lu, Y. Liu, J. D. R. Ree, "Harmonics generated from a DC biased transformer," IEEE Transactions on Power Delivery, vol. 8, pp. 725–731, 1993.
12. D. L. Dickmander, S. Y. Lee, G. L. Désilets and M. Granger. "AC/DC harmonic interaction in the presence of GIC for the Quebec-New England phase II HVDC transmission," IEEE Transactions on Power Delivery, vol. 9, no. 1, pp. 68–75, 1994.
13. A. E. Fitzgerald, Jr. S. D. Umans, and C. Kingsley, Electrical Machinery, McGraw Hill Higher Education, New York, 2002.
14. C. Concordia, Synchronous Machines, John Wiley, New York, 1951.
15. R. H. Park, Two reaction theory of synchronous machines, part-1," AIEEE Transactions vol. 48, pp. 716–730, 1929.
16. NEMA. Large machines—synchronous generators. MG-1, Part 22.
17. R. H. Park. Two reaction theory of synchronous machines, part I. AIEE Transactions vol. 48, pp. 716–730, 1929.

18. R. H. Park. Two reaction theory of synchronous machines, part II. AIEE Transactions vol. 52, pp. 352–355, 1933.
19. C. V. Jones. The Unified Theory of Electrical Machines. Pergamon Press, NY, 1964.
20. A. T. Morgan, General Theory of Electrical Machines, Heyden & Sons Ltd., London, 1979.
21. IEEE Standard115, Test proceedure for synchronous machines Part I-acceptance and performance testing, Part II-test proceedures and parameter determination for dynamic analysis, 2009.
22. IEEE C37.102. IEEE guide for AC generator protection.
23. J. C. Das. "13.8 kV selective high-resistance grounding system for a geothermal generating plant--a case study," IEEE Transactions on Industry Applications, vol. 49, no. 3, pp. 1234–1343, 2013.
24. ANSI/IEEE Standard C57.13. Requirements for instrument transformers, 1993.
25. J. R. Linders, "Relay performance considerations with low-ratio CTs and high fault currents," IEEE Transactions on Industry Applications, vol. 31, no. 2, pp. 392–405, 1995.
26. R. H. Hopkinson, "Ferroresonance during single-phase switching of three-phase distribution transformer banks," IEEE Transactions on PAS, vol.84, pp. 289–293, 1965.
27. D. R. Smith, S. R. Swanson, and J. D. Borst, "Overvoltages with remotely switched cable fed grounded wye–wye transformers," IEEE Transactions on PAS, vol. PAS-94, pp. 1843–1853, 1975.
28. D. R. Crane, G. W. Walsh, "Large mill outage caused by potential transformer ferroresonance," IEEE Transactions on Industry Applications, vol. 24, no. 4, pp. 635–640, 1988.

CHAPTER 4

HARMONIC GENERATION – II

4.1 STATIC POWER CONVERTERS

The primary sources of harmonics in the power system are power converters, rectifiers, inverters, diacs, triacs, GTO's and adjustable speed drives. The *characteristic* harmonics are those produced by the power electronic converters during normal operation and these harmonics are integer multiples of the fundamental frequency of the power system. The static converters do produce some *noncharacteristic* or uncharacteristic harmonics, as ideal conditions of commutation and control (discussed further in this chapter) are not achieved in practice. The ignition delay angles may not be uniform, and there may be unbalance in the supply voltages and the bridge circuits. The harmonic voltages will be unbalanced in each phase and three-phase computer models are required. Harmonic filters are mostly provided for characteristic harmonics, and uncharacteristic harmonics can cause considerable problems.

4.2 SINGLE-PHASE BRIDGE CIRCUIT

The single-phase rectifier full-bridge circuit of Fig. 4.1(a) is first considered to establish relations on the AC and DC sides and origin of harmonics. It is assumed that there is no voltage drop or leakage current, the switching is instantaneous, the voltage source is sinusoidal, and the load is resistive.

Define:

V_{dc}, V_{rms} = average and rms DC output voltage

I_{dc}, I_{rms} = average and rms DC output current = load current

V_m, V = peak and rms value of input voltage

For full wave conduction, the waveforms of input and output currents are then as shown in Fig. 4.1(b) and (c). The *average* DC current is

$$I_{dc} = \frac{1}{2\pi} \int_0^{2\pi} \frac{V_m}{R} \sin \omega t \, d\omega t = \frac{2}{2\pi} \int_0^{\pi} \frac{V_m}{R} \sin \omega t \, d\omega t = \frac{2V_m}{\pi R} \qquad (4.1)$$

Power System Harmonics and Passive Filter Designs, First Edition. J.C. Das.

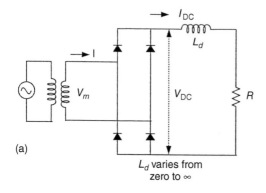

(a)

L_d varies from
zero to ∞

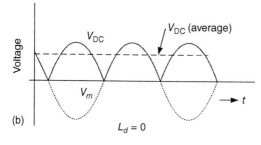

(b)

$L_d = 0$

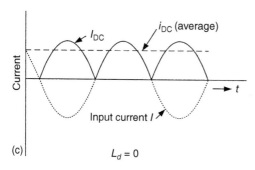

(c)

$L_d = 0$

Figure 4.1 (a) A single phase full wave bridge rectifier circuit with resistive load; (b) and (c) waveforms with no dc reactor.

and the rms value or the effective value of the output current, including all harmonics, is

$$I_{rms} = \sqrt{\frac{2}{2\pi} \int_0^\pi \left(\frac{V_m}{R}\right)^2 \sin^2 \omega t \; d\omega t} = \frac{V_m}{\sqrt{2}R} \tag{4.2}$$

The *input current has no harmonics* (Fig. 4.1(c)). The average DC voltage is given by

$$V_{dc} = \frac{1}{\pi} \int_0^\pi \sqrt{2}V \sin \omega t \, d(\omega t) = \frac{2\sqrt{2}}{\pi} V = 0.9 \; V = 0.637 \; V_m$$

$$V_{\text{rms}} = \sqrt{\frac{1}{2\pi} \int_0^\pi (\sqrt{2}V \sin \omega t)^2 d\omega t} = 0.707 \ V_m$$

$$I_{\text{rms}} = \frac{0.707 \ V_m}{R}$$

The output AC power is defined as:

$$P_{\text{AC}} = V_{\text{rms}}I_{\text{rms}} = \frac{(0.707V_{\text{rms}})^2}{R} \qquad (4.3)$$

where V_{rms} considers the effect of harmonics on the output. The DC output power is

$$P_{\text{DC}} = V_{\text{dc}}I_{\text{dc}} = \frac{(0.4057V_m^2)}{R} \qquad (4.4)$$

The efficiency of rectification is given by $P_{\text{DC}}/P_{\text{AC}}$ (1%). The *form factor* is a measure of the shape of the output voltage or current and it is defined as

$$\text{FF} = \frac{I_{\text{rms}}}{I_{\text{dc}}} = \frac{V_{\text{rms}}}{V_{\text{dc}}} = 1.11 \qquad (4.5)$$

The *ripple factor*, which is a measure of the ripple content of the output current or voltage, is defined as the rms value of output voltage or current, including all harmonics, divided by the average value:

$$\text{RF} = \sqrt{\left(\frac{I_{\text{rms}}}{I_{\text{dc}}}\right)^2 - 1} = \sqrt{\text{FF}^2 - 1} \qquad (4.6)$$

For the single-phase bridge circuit with resistive load, the ripple factor is

$$\text{RF} = \sqrt{\left(\frac{I_{\text{rms}}}{I_{\text{dc}}}\right)^2 - 1} = 0.48 \qquad (4.7)$$

This shows that the ripple content of the DC output voltage or current is high (Fig. 4.1(b)). This is not acceptable even for the simplest of applications. Let a series reactor be added in the DC circuit. The load current is no longer a rectified sine wave but the average current is still equal to $2V_m/\pi R$. The AC line current is no longer sinusoidal, but approximates a poorly defined square wave with superimposed ripples (Fig. 4.2(a) and (b)). The inductance has reduced the harmonic content of the load current by increasing the harmonic content of the AC line current. When the inductance is large, the ripple across the load is insignificant, and load current can be assumed constant. The AC current wave is now a square wave (Fig. 4.2(c) and (d)).

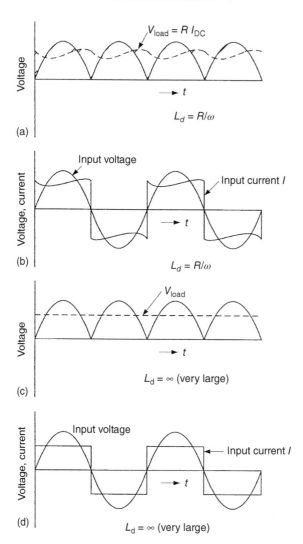

Figure 4.2   (a) and (b). Waveforms of a single phase full wave bridge rectifier with small dc reactor; (c) and (d) with large DC output reactor.

4.2.1   Phase Control

A silicon-controlled rectifier (SCR) can be turned on by applying a short pulse to its gate and turned off due to natural or line commutation. Figure 4.3(a) shows a fully controlled single phase bridge circuit with a DC reactor and resistive load. The angle by which the conduction is delayed after the input voltage starts to go positive until the thyristor is fired is called the delay angle, angle α in Fig. 4.3(b), which also shows waveforms with a large DC reactor. Thyristors 1 and 2 and 3 and 4 are fired in pairs as shown in Fig. 4.3(b). Even when the polarity of the voltage is reversed, the current keeps flowing in thyristors 1 and 2 until thyristors 3 and 4 are fired, Fig. 4.3(b). Firing of thyristors 3 and 4 reverse biases thyristors 1 and 2 and turns them off. (*This is referred to as class F type forced commutation or line commutation.*) The average

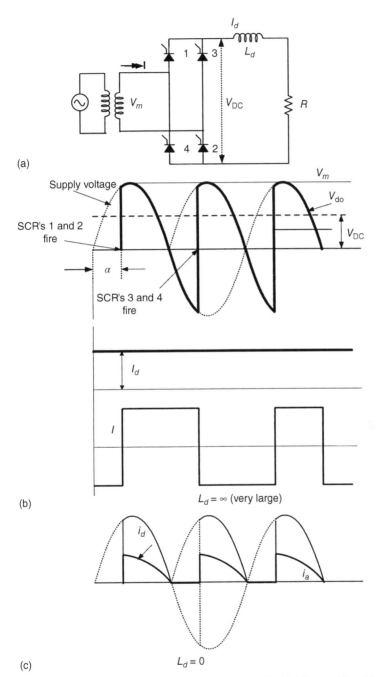

Figure 4.3   (a) Circuit of a single-phase fully controlled bridge rectifier; (b) and (c) waveforms with large DC reactor and with no DC reactor.

DC voltage is

$$V_{dc} = \frac{2}{2\pi} \int_{\alpha}^{\pi+\alpha} V_m \sin \omega t \ d\omega t = \frac{2V_m}{\pi} \cos \alpha \tag{4.8}$$

and the Fourier analysis of the rectangular current wave in Fig. 4.3(b) (see Chapter 2) gives

$$a_h = -\frac{4I_a}{h\pi} \sin \ h\alpha, \qquad h = 1,3,5, \ ... $$
$$= 0 \ \ h = 2,4,6, \ ... \tag{4.9}$$

$$b_h = \frac{4I_a}{h\pi} \cos \ h\alpha \qquad h = 1,3,5, \ ... $$
$$= 0 \ \ h = 2,4,6, \ ... \tag{4.10}$$

(Note that the order of harmonic denoted as "n" in Chapter 2 is denoted by letter "h" in this chapter)

Since:

$$I = \sum_{h=1,2, \ ...}^{\infty} [a_h \cos(h\omega t) + b_h \sin(h\omega t)] \tag{4.11}$$

The input current is given by:

$$I = \frac{4}{\pi} I_d \left[\sin(\omega t - \alpha) + \frac{1}{3} \sin 3(\omega t - \alpha) + \frac{1}{5} \sin 5(\omega t - \alpha) + \ ... \right] \tag{4.12}$$

When the output reactor is small, the input current wave is no longer rectangular, and the line harmonics increase. Figure 4.3(c) depicts the current wave shapes when L_d is zero.

Triplen harmonics are present. Figure 4.4 shows harmonics as a function of the delay angle for a resistive load, Ref. [1]. The overlap angle (defined further) decreases the magnitude of harmonics.

We can write the instantaneous input current as:

$$i_{input} = \sum_{h=1,2, \ ...}^{\infty} \sqrt{2}I_h \sin(\omega t + \phi_h) \tag{4.13}$$

where

$$\phi_h = \tan^{-1} \left(\frac{a_h}{b_h} \right) = -h\alpha \tag{4.14}$$

$\varphi_h = -h\alpha$ is the displacement angle of the nth harmonic current. The rms value of the h-th harmonic input current is:

$$I_h = \frac{1}{\sqrt{2}}(a_h^2 + b_h^2)^{1/2} = \frac{2\sqrt{2}}{h\pi} I_d \tag{4.15}$$

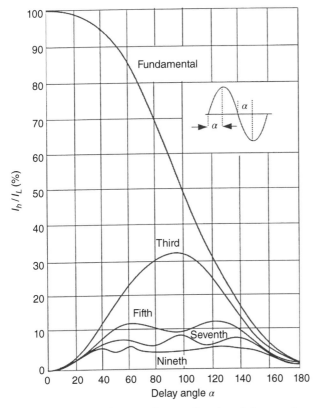

Figure 4.4 Harmonic generation as a function of phase angle control, with resistive load, source Ref. [1].

The rms value of the fundamental is:

$$I_1 = \frac{2\sqrt{2}}{\pi} I_d \tag{4.16}$$

Thus, the rms value of the input current is:

$$I = \left(\sum_{h=1,2,..}^{\infty} I_h^2 \right)^{1/2} \tag{4.17}$$

The harmonic factor is:

$$\text{HF} = \left[\left(\frac{I_{\text{rms}}}{I_1} \right)^2 - 1 \right]^{1/2} = 0.4834 \tag{4.18}$$

The displacement factor is:

$$\text{DF} = \cos \phi_1 = \cos(-\alpha) \tag{4.19}$$

The power factor is:

$$\text{PF} = \frac{V_{\text{rms}}I_1}{V_{\text{rms}}I_{\text{rms}}}\cos\phi_1 = \frac{2\sqrt{2}}{\pi}\cos\alpha \qquad (4.20)$$

4.3   REACTIVE POWER REQUIREMENTS OF CONVERTERS

We discussed the concepts of power factor, total power factor and distortion power factor in Chapter 1. The fundamental input power factor angle is equal to the firing angle α. For the single-phase bridge circuit, the input active power and reactive power are

$$P = \frac{4}{2\pi}I_d V_m \cos\alpha \qquad (4.21)$$

$$Q = \frac{4}{2\pi}I_d V_m \sin\alpha \qquad (4.22)$$

The power factor becomes depressed for large firing angles. This is the case whenever large phase control is used in the converter circuits.

The maximum reactive power input for a half-controlled bridge (formed with the replacement of bottom half SCRs with diodes, that is, in Fig. 4.3(a), SCR's 2 and 4 are replaced with diodes) will be one-half of that of a fully controlled bridge. In a half controlled bridge, the output voltage and input current will be zero for period from π to $(\pi + \alpha)$. This single-phase circuit and waveforms are shown in Fig. 4.5(a) and (b). A Fourier analysis similar to that of a fully controlled bridge can be performed. Figure 4.5(c) shows the relation of active power input to the fundamental reactive power for fully controlled and half-controlled bridge circuit. For both the circuits the output current is assumed to be steady. The maximum fundamental reactive power input for 1/2 controlled bridge is 50% of that for the fully controlled bridge.

However, half-controlled bridge circuits are not used in practice; these can give rise to even harmonics, see Section 4.7.1. The reactive power requirements of converters becomes important in many installations and this can be reduced by limiting the amount of phase control, reducing reactance of converter transformers, which limits μ (overlap angle, defined further) and *sequential control* of converters, which has been popular in HVDC transmission systems. In the sequential control two or more converter sections can operate in series, with one section fully phased-on and the other sections adding or subtracting from the voltage of the first section (Fig. 4.6).

The reactive power consumed by the converters can be supplied by shunt capacitors and filters. In these cases, line commutated converters are economical and for HVDC transmission, these have been used extensively. The thyristor based converters become economical in large power handling capabilities as per device basis, the thyristors can handle two to three times more power than GTOs, gate turn off thyristor) IGCTs (integrated gate bipolar transistor) and MTOs (MOS turn-off thyristor). There are variations in the current source converters such as resonant converters,

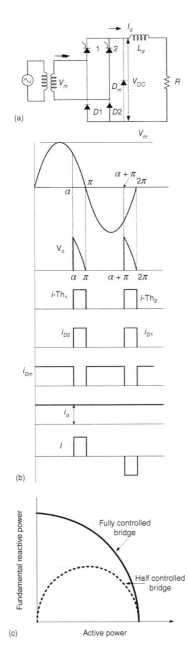

(a)

(b)

(c)

Figure 4.5 (a) Circuit of a single phase half-controlled bridge, (b) waveforms of a single phase half controlled bridge; (c) fundamental reactive power consumption of half controlled bridge and fully controlled bridge (also applicable to three-phase bridges).

hybrid converters, and artificial commutation converters, some details of which are included in Chapter 6. New converter topologies are being developed. With GTOs, the forced commutation can improve the power factor and reduce the input harmonic levels. The forced commutation techniques are:

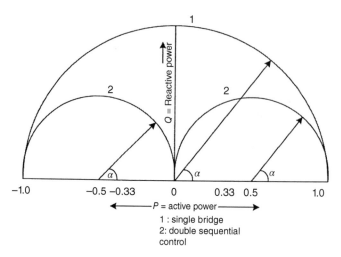

Figure 4.6   Reduction in reactive power requirements with sequential control of converters.

- Extinction angle control
- Pulse-width modulation (PVM), and sinusoidal PVM, see Section 4.13.
- Symmetrical angle control

Reference [2] describes a PWM converter with sinusoidal AC currents and minimum filter requirements. A full-range four-quadrant operation is described, with control of input power factor. Reference [3] describes experimental test results on a three-phase bipolar transistor controlled current PWM modulator with leading power factor. Near to unity power factor converter topologies are available, Ref [4]. The current trend lies in reducing the line harmonics and simultaneously improving the power factor.

4.4   THREE-PHASE BRIDGE CIRCUIT

A three-phase bridge has two forms: (1) half-controlled and (2) fully controlled. The three-phase fully controlled bridge is first described, as it is most commonly used.

Figure 4.7(a) shows a three-phase fully controlled bridge circuit and Fig. 4.7(b) shows its current and voltage waveforms. The firing sequence of thyristors is shown in Table 4.1. At any time two thyristors are conducting. The firing frequency is six times the fundamental frequency and the firing angle can be measured from point O as shown in Fig. 4.7(b). With a large output reactor the output DC current is continuous and the input current is a rectangular pulse of $2\pi/3$ duration and amplitude i_d. The average DC voltage is:

$$V_{dc} = 2\left[\frac{3}{2\pi}\int_{-\pi/3+\alpha}^{\pi/3+\alpha} V_m \cos \omega t \; d\omega t\right] = \frac{3\sqrt{3}}{\pi}V_m \cos \alpha \qquad (4.23)$$

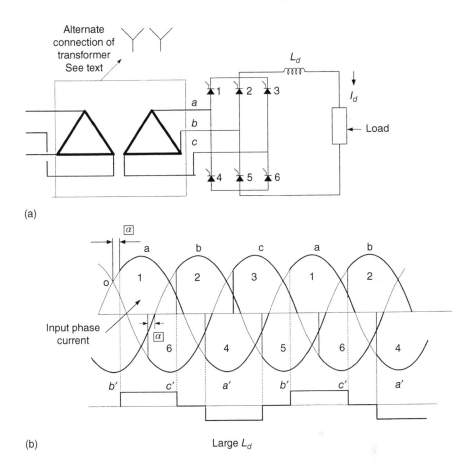

Figure 4.7   (a) Circuit of a three-phase fully controlled bridge; (b) voltage and current waveforms for a delay angle of α.

TABLE 4.1   Firing Sequence of Thyristors in Six-Pulse Full Converter

Conducting thyristors	5,3	1,5	6,1	2,6	4,2	3,4
Thyristor to be fired	1	6	2	4	3	5
Thyristor turning off	3	5	1	6	2	4

where V_m is the peak value of line to neutral voltage. For firing angles $> \pi/2$, the circuit can work as an inverter, that is, DC power is fed back into the AC system. This requires that a DC source with opposite polarity is connected at the output. The power factor is lagging for rectifier operation and leading for inverter operation.

Figure 4.7 shows a connection diagram and waveforms for a three-phase fully controlled bridge, with delta–delta or wye-wye connection of the rectifier transformer. This means that the phase shift in the three-phase transformer windings is zero degrees. We can also have any three phase transformer winding connection which has zero degree phase shift between primary and secondary windings (Chapter 3).

   If we consider firing angle $\alpha = 0$ then the input current is rectangular and it's Fourier analysis for current in phase a gives:

$$i_a = \frac{2\sqrt{3}}{\pi}I_d\left[\cos \omega t - \frac{1}{5}\cos 5\omega t + \frac{1}{7}\cos 7\omega t - \frac{1}{11}\cos 11\omega t + \frac{1}{13}\cos 13\omega t - .....\right]$$

(4.24)

Thus, the maximum fundamental frequency current is

$$\frac{2\sqrt{3}}{\pi}I_d \quad \text{peak} \quad = \frac{\sqrt{6}}{\pi}I_d \quad \text{rms}$$

(4.25)

If we have a delta–wye rectifier transformer connection, the input current is stepped (not shown) and the resulting Fourier series for the current waveform is:

$$i_a = \frac{2\sqrt{3}}{\pi}I_d\left(\cos \omega t + \frac{1}{5}\cos 5\omega t - \frac{1}{7}\cos 7\omega t - \frac{1}{11}\cos 11\omega t + \frac{1}{13}\cos 13\omega t-, \ ...\right)$$

(4.26)

   From these equations, the following observations can be made:

1. The line harmonics are of the order:

$$h = pm \pm 1, \quad m = 1, 2, \ ...$$

(4.27)

   where p is the pulse number. The pulse number is defined as the total number of successive nonsimultaneous commutations occurring within the converter circuit during each cycle when operating without phase control. This relationship also holds for a single-phase bridge converter, as the pulse number for a single phase bridge circuit is 2. The harmonics given by Eq. (4.27) are an integer of the fundamental frequency and are called characteristic harmonics, while all other harmonics are called noncharacteristic.

2. The triplen harmonics are absent. This is because an ideal rectangular wave shape and instantaneous transfer of current at the firing angle are assumed. In practice, some noncharacteristic harmonics are also produced.

3. The rms magnitude of the hth harmonic is I_f / h, that is, the fifth harmonic is a maximum of 20% of the fundamental:

$$I_h = \frac{I_f}{h}$$

(4.28)

4. Fourier series of the input current is given by:

$$a_h = -\frac{4I_d}{h\pi} \sin\frac{h\pi}{3} \sin(h\alpha) \quad h = 1, 3, 5 \dots$$

$$b_h = -\frac{4I_d}{h\pi} \sin\frac{h\pi}{3} \cos(h\alpha) \quad h = 1, 3, 5 \tag{4.29}$$

Therefore:

$$I = I_h \sin(h\omega t + \phi_h) \text{ where } \phi_h = \tan^{-1}\frac{a_h}{b_h} = -h\alpha \tag{4.30}$$

The rms value of the nth harmonic input current is

$$I_h = (a_h^2 + b_h^2)^{1/2} = \frac{2\sqrt{2}}{h\pi} \sin\frac{h\pi}{3} \tag{4.31}$$

The rms value of the fundamental current is

$$I_1 = \frac{\sqrt{6}}{\pi}I_d = 0.779I_d \tag{4.32}$$

The rms input current (including harmonics) is

$$\left[\frac{2}{\pi} \int_{(-\pi/3)+\alpha}^{(\pi/3)+\alpha} I_d^2 d(\omega t)\right]^{1/2} = I_d\sqrt{\frac{2}{3}} = 0.8165 \ I_d \tag{4.33}$$

5. The lowest harmonic in the output of the converter is the sixth. As the pulse number of the converter increases, the ripple in the DC output voltage and the harmonic content in the input current are reduced. Also, for a given voltage and firing angle, the average DC voltage increases with the pulse number.

6. The ripple factor of a six pulse fully controlled rectifier is found as follows:

$$V_{rms}^2 = \frac{3}{\pi} \int_{-\pi/6+\alpha}^{\pi/6+\alpha} (\sqrt{3}V_m \cos\theta)^2 d\alpha$$

$$= \frac{9V_m^2}{2\pi}\left[\frac{\pi}{3} + \frac{\sqrt{3}}{2}\cos 2\alpha\right]$$

$$RF = \frac{\sqrt{V_{rms}^2 - V_{dc}^2}}{V_{dc}} = \frac{\sqrt{\frac{\pi}{2}\left[\frac{\pi}{3} + \frac{\sqrt{3}}{2}\cos 2\alpha\right] - 3\cos^2\alpha}}{\sqrt{3}\cos\alpha} \tag{4.34}$$

At $\alpha = 0$, RF = 0.0418, minimum.

7. From Eq. (4.33) rms current including harmonics is $0.8165I_d$ and from Eq. (4.32) fundamental current is $0.7797\ I_d$. Then the total harmonic current is:

$$I_h = [(0.8165I_d)^2 - (0.7797I_d)^2]^{1/2} = 0.24I_d \qquad (4.35)$$

In the earlier analysis, we assumed that the commutation is instantaneous. It may take some time before the current is commutated, through the inductive circuit of the AC system. This is discussed in Section 4.4.2.

Example 4.1: Consider that a three-phase fully controlled bridge serves a resistive load, then the following relations are applicable:

$$V_{dc} = 1.6542\ V_m$$

$$I_{dc} = \frac{1.6542\ V_m}{R}$$

$$V_{rms} = 1.6554\ V_m$$

$$I_{rms} = \frac{1.6554\ V_m}{R}$$

$$FF = \frac{1.6554}{1.6542} = 1.0007$$

$$RF = \sqrt{1.0007^2 - 1} = 0.0374 = 3.74\%$$

Also from Eq. (4.34), the RF is approximately 3.74%.

Example 4.2: Prove equation (4.26)

The phase shift in the winding connections of transformers manufactured according to ANSI/IEEE standards [5] is discussed in Chapter 3. Recall that the low-side voltage whether in wye or delta connection, has a phase shift of 30° lagging with respect to high side voltage phase-to-neutral voltage vectors. *Also for negative sequence this phase shift angle becomes negative.*

Therefore, Eq. (4.24) becomes:

$$i_a = \frac{2\sqrt{3}}{\pi}I_d \left[\sin\left(\omega t - 30°\right) - \frac{1}{5}\sin(5\omega t - 150°) + \frac{1}{7}\sin(7\omega t - 210°)...... \right]$$

The positive sequence voltages or currents undergo a shift of plus 30° while the negative sequence undergoes a phase shift of −30°.

Then:

$$i_a = \frac{2\sqrt{3}}{\pi}I_d \left[\sin\left(\omega t - 30° + 30°\right) - \frac{1}{5}\sin(5\omega t - 150° - 30°) \right.$$

$$\left. + \frac{1}{7}\sin\left(7\omega t - 210° + 30°\right)...... \right]$$

$$= \frac{2\sqrt{3}}{\pi}I_d \left[\sin\left(\omega t\right) + \frac{1}{5}\sin(5\omega t) - \frac{1}{7}\sin(7\omega t)...... \right]$$

4.4.1   Cancellation of Harmonics Due to Phase Multiplication

Equations (4.24) and (4.26) show that the harmonics 5th, 7th, and 17th are of opposite sign. We know that there is a 30° phase shift between the primary and secondary voltage vectors of a delta–wye transformer, while for a delta–delta or wye–wye connected transformer, this phase shift is 0°. If the load is equally divided into two transformers, one with delta–delta connections and the other with wye–delta or delta–wye connections, harmonics of the order of 5th, 7th, 17th, ... are eliminated, and the system behaves like a 12-pulse circuit. This is called *phase multiplication*. The circuit is shown in Fig. 4.8(a) and the waveform in the time domain is shown in Fig. 4.8(b). Extending this concept, 24-pulse operation can be achieved with four transformers with 15° mutual phase shifts. As the magnitude of the harmonic is inversely proportional to the pulse number, the troublesome lowerorder harmonics of larger magnitude are eliminated. This cancellation of harmonics, though, is not 100% as the ideal conditions of operation are rarely met in practice. The transformers should have exactly the same ratios and same impedances, the loads should be equally divided and converters should have exactly the same delay angle. Approximately 75% cancellation may be achieved in practice, and in harmonic analysis studies 25% residual harmonics are modeled.

See Chapter 6 for some practical examples of phase multiplications. An improved input waveform is obtained due to phase displacement in the secondary windings of the transformers.

4.4.2   Effect of Source Impedance

The commutation of current from one SCR to and other will take place instantaneously if the source inductance is zero. The commutation is delayed by an angle μ due to source inductance, and during this period a short-circuit occurs through the conducting devices, the AC circulating current being limited by the source impedance; μ is called the overlap angle. When α is zero, the short-circuit conditions are those corresponding to maximum asymmetry and μ is large, that is, slow initial rise, see Fig. 4.9 for overlap angle μ. At $\alpha = 90°$, the conditions are of zero asymmetry with its fast rate of rise of current. Commutation produces two primary notches per cycle and four secondary notches of lesser amplitude, which are due to notch reflection from the other legs of the bridge, (Fig. 4.10). For a purely

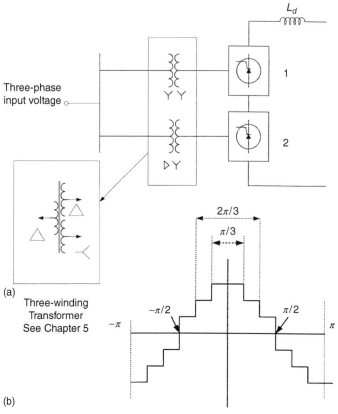

(a)
Three-winding
Transformer
See Chapter 5

(b)

Figure 4.8   Harmonic elimination with phase multiplication, circuit diagram; (b) stepped input current waveform.

inductive source impedance, the average DC voltage is reduced and is given by

$$V_d = V_{do} - \frac{3\omega L_s}{\pi} I_d \tag{4.36}$$

where L_s is the source inductance, and for a six-pulse fully controlled bridge V_{do} is given by Eq. (4.23), same as V_{dc}, and it is called the internal voltage of the rectifier.

Figure 4.9 shows that the overlap helps to reduce the harmonic content in the input current wave, which is rounded off and is more close to a sinusoid. Alternating-current harmonics at the overlap are given by the following equations from Ref [6]:

$$I_h = I_d \left[\sqrt{\frac{6}{\pi}} \frac{\sqrt{A^2 + B^2 - 2AB \cos(2\alpha + \mu)}}{h[\cos\alpha - \cos(\alpha + \mu)]} \right] \tag{4.37}$$

where

$$A = \frac{\sin\left[(h-1)\frac{\mu}{2}\right]}{h-1} \tag{4.38}$$

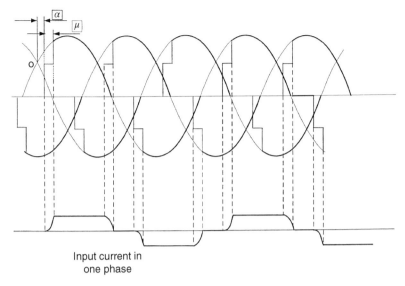

Figure 4.9   Effect of the overlap angle on the current input waveform.

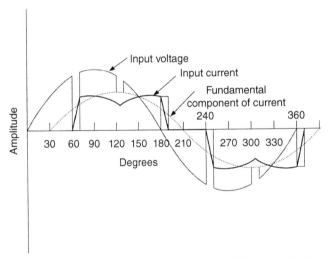

Figure 4.10   Voltage notching in a three-phase fully controlled bridge, with output DC reactor.

$$B = \frac{\sin\left[(h+1)\frac{\mu}{2}\right]}{h+1} \tag{4.39}$$

The depth of the voltage notch is calculated by the *IZ* drop and is a function of the impedance. The width of the notch is the commutation angle:

$$\mu = \cos^{-1}[\cos\alpha - (X_s + X_t)I_d] - \alpha \tag{4.40}$$

$$\cos \mu = 1 - 2E_x/V_{do} \tag{4.41}$$

where X_s is the system reactance in per unit on converter base, X_t is the transformer reactance in per unit on converter base, I_d is the DC current in per unit on converter base, and E_x is the DC voltage drop caused by commutating reactance. Notches cause EMI (electromagnetic interference) problems and misoperation of electronic devices that sense the true zero crossing of the voltage wave (Chapter 8).

Note that the assumption of a flat-topped wave is not correct. The waveform will be trapezoidal with superimposed sinusoidal ripple content [6]. References [6–8] provide estimation of harmonics with ripple content (see Chapter 7 for further calculations). The interharmonics produced by the converters are discussed in Chapter 5.

Example 4.3:   Calculate the total power factor of a six-pulse converter, the pertinent data are maximum rating 5 MVA, load voltage = 2.4 kV, load current = 500A, angle $\alpha = 30°$, 13.8 kV rectifier transformer rating 5 MVA, percentage impedance 5.5%, X/R ratio = 8. What is the distortion power factor?

First calculate the overlap angle. Based on the data provided:

pu (per unit) impedance of transformer on converter base $X_t = 0.05457$ pu and source reactance = X_s = neglect

Then from Eq. (4.40), $\mu = 5.7°$.

Neglecting overlap angle, for a six-pulse converter, the rms input current (including effect of harmonics) is $0.8165\, I_d$.

$$\mathrm{HF} = \left(\left[\frac{I_{\mathrm{input,rms}}}{I_{\mathrm{fund,rms}}} \right]^2 - 1 \right)^{1/2} = 31.08\%$$

$$\mathrm{DF} = \cos(-\alpha)$$

$$\mathrm{PF} = \frac{3}{\pi} \cos \alpha = 0.9549 \ \mathrm{DF}$$

Substituting α, DF = 0.866, and PF = 0.827.

Equation (4.37) considers overlap angle and ignores ripple content. Using this equation, the values are listed in Table 4.2.

Then from Eq. (1.38), the distortion PF is:

$$\mathrm{PF}_{\mathrm{disortion}} = \frac{1}{\sqrt{1 + \left(\frac{\mathrm{THD}_I}{100} \right)^2}}$$

$$\%\mathrm{THD}_I = \sqrt{(0.2543)^2 + (0.2154)^2 + (0.1195)^2 + (0.0970)^2} = 0.364$$

TABLE 4.2 Calculations of Harmonics, Example 4.2

Harmonic Order	A	B	Harmonic %
5	0.049414	0.041025	25.43
7	0.058808	0.048439	21.54
11	0.047712	0.046840	11.95
13	0.046840	0.04582	9.70

Therefore distortion power factor is 0.9396. Note that only harmonics up to 13th have been included. If we include higher order harmonics, THD_I will increase and distortion power factor will reduce. For $THD_I = 0.62$, the calculated results approach $DF = 0.866$, calculated earlier.

The DC output current is of 500 A is given, then the fundamental component of the AC input current is 389.85A, Eq. (4.32).

From the current distortion factor calculated earlier the rms input current = 414.8A.

The DC output voltage is 2400V, the AC voltage with $\alpha = 30°$ is 1185 rms volts, Eq. (4.23).

Then input volt-ampere = 1474.6 kVA

Therefore total power factor = 0.814.

4.5 HARMONICS ON OUTPUT (DC) SIDE

For a *single phase circuit*, the output waveform (on the DC side) contains even harmonics of the input frequency and the Fourier expansion is

$$V_{d0} = V_{dc} + e_2 \sin 2\omega t + e_2' \cos 2\omega t + e_4 \sin 4\omega t + e_4' \cos 4\omega t + \dots \qquad (4.42)$$

where e_m and $e_m' (m = 2, 4, 6, \dots)$ are given by

$$e_m = \frac{2V_m}{\pi} \left[\frac{\sin (m + 1)\alpha}{m + 1} - \frac{\sin(m - 1)\alpha}{m - 1} \right] \qquad (4.43)$$

$$e_m' = \frac{2V_m}{\pi} \left[\frac{\cos (m + 1)\alpha}{m + 1} - \frac{\cos(m - 1)\alpha}{m - 1} \right] \qquad (4.44)$$

The converter can be considered as a harmonic current source; the even harmonics go into the load and the odd harmonics into the supply source (Fig. 4.11). The harmonics fed into the supply system propagate into the power system. These can either be magnified or attenuated.

The harmonics in the load circuit have an adverse effect on the loads, but are, generally, localized to the loads to which these connect. There is harmonic power

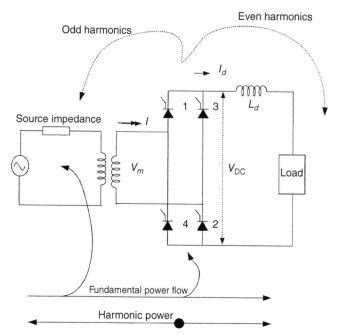

Figure 4.11 Single phase two- pulse controlled rectifier as a source of harmonic generation and harmonic power flow.

associated with harmonic currents, which is a function of relative system impedances and load side impedance (Chapter 1).

For a *three-phase fully* controlled 6-pulse bridge circuit, only triplen harmonics of the input frequency are present in the output DC voltage. The order is 6th, 12th, 18th ... Thus, for the harmonics on the DC side we can write:

$$h = pm \quad m = 1, 2, \ldots \tag{4.45}$$

Figure 4.12 (a) shows the ripple content in the output DC voltage of a six-pulse converter. With the effect of firing and overlap angles the waveform becomes as shown in Fig. 4.12(b).

Figure 4.13(a) and (b) show the DC output harmonics with the effect of firing angle α and overlap angle μ. The rms value of the harmonic voltages is given by [9]:

$$V_h = \frac{V_{c0}}{\sqrt{2}(h^2 - 1)} \left\{ \begin{array}{l} (h-1)^2\cos^2\left[(h+1)\frac{\mu}{2}\right] + (h+1)^2\cos^2\left[(h-1)\frac{\mu}{2}\right] - \\ 2(h-1)(h+1)\cos\left[(h+1)\frac{\mu}{2}\right]\cos\left[(h-1)\frac{\mu}{2}\right]\cos(2\alpha+\mu) \end{array} \right\}^{1/2} \tag{4.46}$$

where V_{c0} is the commutating phase-to-phase rms voltage. Similar plots for $h = 18$ and 24 are provided in Ref. [9].

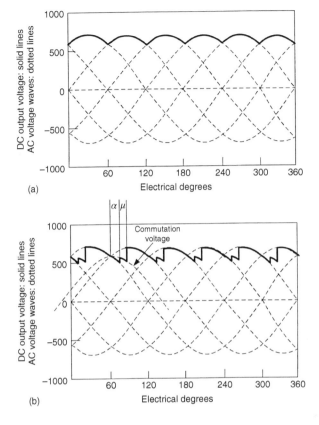

(a)

(b)

Figure 4.12   (a) Ripple content of DC output voltage of a three-phase six-pulse bridge rectifier; (b) output wave shape modified by firing angle and overlap angle.

For a three-pulse, half-controlled bridge, the lowest harmonic will be the third. As an application of Figures 4.13(a) and (b), consider an $a = \pi/4$ and $\mu = 10°$. Then the output harmonics in six-pulse converter can be read from these figures:

$$6\text{th} = 0.19 \text{ pu}$$
$$12\text{th} = 0.055 \text{ pu.}$$

The ripple factor is:

$$RF = \frac{\sqrt{0.19^2 + 0.055^2}}{1} = 19.7\% \tag{4.47}$$

4.6   INVERTER OPERATION

From Eq. (4.23) the DC voltage is zero for a 90° delay angle. As the delay angle is further increased, the average DC voltage becomes negative. This can be examined by drawing output voltage waveform with advancing delay angle α in Fig. 4.14. If instantaneous commutation is assumed, then at 180° the negative voltage is as large

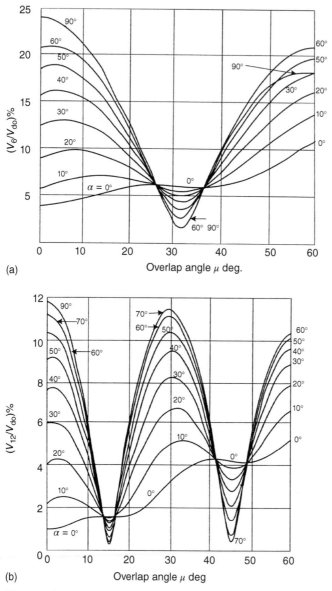

Figure 4.13   (a) DC output 6th harmonic with firing angle and overlap angle; (b) DC output 12th harmonic with firing angle and overlap angle. Source Ref. [9].

as that for a rectifier at zero delay angle. However, operation at 180° is not possible, as some delay, denoted by γ (called the margin angle) is needed for commutation of current and some additional time, angle δ, is needed for the outgoing thyristors to turn off before voltage across it reverses. Otherwise the outgoing thyristor will commutate the current back and lead to commutation failure. The firing angle is limited to $(\pi - \gamma)$. In HVDC transmission, two modes of operation are recognized-mode

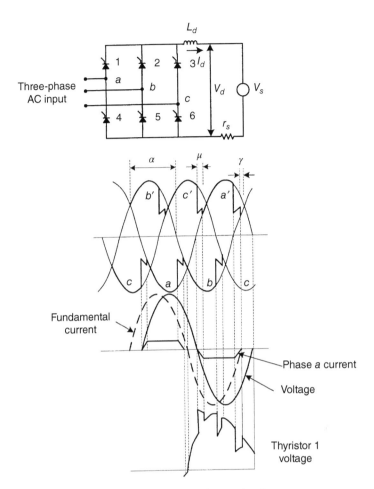

Figure 4.14   Circuit diagram and waveforms of an inverter.

1 and mode 2. In mode 1 the firing angle is varied and current remains constant. The duration for which the thyristor is reverse biased will reach a minimum value at some level of internal voltage and any further increase will result in commutation failure. Thus, in the second mode the inverter operates at a constant margin angle or extinction angle.

For a six-pulse bridge, internal no-load voltage is given by Eq. (4.23).

Then, considering DC voltage drops in the source inductance, (commutating inductance), the voltage as given by Eq. (4.23), can be modified and written as:

$$V_d = \frac{3\sqrt{3}V_m}{\pi}\cos\alpha - \frac{3\omega L_s I_d}{\pi}$$

$$= \frac{3\sqrt{3}V_m}{\pi}\cos(\alpha + \mu) + \frac{3\omega L_s I_d}{\pi} \qquad (4.48)$$

We can write these equations as:

$$V_d = V_{do} - \frac{3}{\pi} X_c I_d = V_{do} - R_c I_d \tag{4.49}$$

where R_c is the equivalent commutating resistance, which does not represent a real resistance and does not consume any real power.

For rectifier *or inverter* operation, a thyristor experiences a reverse voltage for $(\pi - \alpha - \mu)$ after the commutation has taken place. However, for inverter operation, angle α is $> 90°$. This means that for inverter operation the thyristors are reverse biased for a shorter duration as compared to rectifier operation. Therefore firing angle should be limited so that angle $\gamma = (\pi - \alpha - \mu)$ is adequate for proper commutation. Angle $(\pi - \gamma)$ is called the *extinction angle*.

Figure 4.14 shows the inverter waveforms for $\alpha = 5\pi/6$. The voltage across one of the thyristors is also shown. Unless proper control is exercised to maintain margin angle commutation failure can occur. Equation (4.48) can also be written as:

$$\frac{\pi}{3\sqrt{3}} \left(V_d + \frac{3\omega L_s}{\pi} I_d \right) = V_m \cos \alpha \tag{4.50}$$

To determine the instant of firing, replace α with ωt and replace $(\alpha + \mu)$ with $(\pi - \gamma_{min})$:

$$\frac{2\omega L_s}{\sqrt{3}} I_d - V_m \cos \gamma_{min} = V_m \cos \omega t \tag{4.51}$$

where γ_{min} is constant. The signal $V_m \cos \omega t$ can be easily generated in a three-phase system. Firing angle α will get adjusted for different values of V_d and till γ_{min} is reached. Therefore, left-hand side of (4.51) depends on I_d. The right-hand side is the negative value of AC phase c voltage for firing thyristor 2. Although the right-hand side is the same for all thyristors, the control voltage will be different for each thyristor and separate firing circuits are required.

Figure 4.15 shows definitions of the angles related to rectifier and inverter operation.

- α is the ignition delay angle
- μ is the overlap angle
- δ is the extinction delay angle $= \alpha + \mu$
- $\beta = \pi - \delta$ is the ignition advance angle
- $\gamma = \pi - \alpha$ is the extinction advance angle
- $\mu = \delta - \alpha = \beta - \gamma$ is the overlap angle.

Also see Ref. [10] for further details.

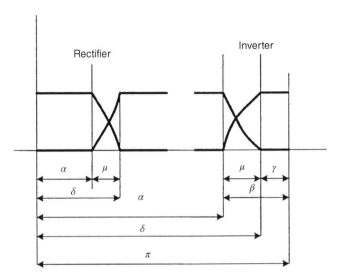

Figure 4.15   Angles associated with rectifier and inverter operation, see text.

4.7   DIODE BRIDGE CONVERTERS

The current source converters as described earlier can have the following parameters: (1) front end diodes, without any DC control, (2) converters with an output DC reactor and front-end thyristors follow the DC link voltage: a full converter controls the amount of DC power from zero to full DC output, (3) the current source converters may also have turn-off devices, (IGBTs). The harmonic injection into the supply system may be represented by a Norton equivalent. This type of converter is used at the front end of current source inverters.

The full-wave diode bridge with capacitor load, as shown in Fig. 4.16(a), is the second type of converter. It converts from AC to DC and does not control the amount of DC power. This type of converter does not cause line notching, but the current drawn is more similar to a pulse current rather than the approximate square-wave current of the full converter. The voltage and current waveform are shown in Fig. 4.16(b). This circuit is better represented by a Thévenin equivalent and the source impedance has a greater impact.

Typical current harmonics for comparison are shown in Table 4.3. In the diode converter with DC link capacitor, the fifth harmonic is higher by a factor of 3–4 times and the seventh harmonic by a factor of 3 when compared with compared to current source converters. This type of converter with DC link capacitor is used in voltage source inverters (VSIs). Sometimes, a controlled bridge may replace the diode bridge preceding the DC link capacitor.

4.7.1   Half Controlled Bridge-Three-Phase Semi-Converters

Referring to Fig. 4.7(a) only three (top) control devices are used and the bottom controlled devices in the three-phase fully controlled bridge are replaced with

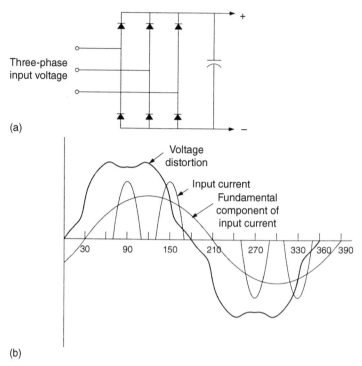

(a)

(b)

Figure 4.16   (a) Circuit of a three-phase diode bridge with DC link capacitor; (b) input
current waveform and voltage distortion.

TABLE 4.3   Typical Current Harmonics for a Six-Pulse Converter and Diode Bridge
Converter, as a Percentage of Fundamental Frequency Current

Current Harmonic	Six-Pulse Converter	Diode Bridge Converter
5	17.94	64.5
7	11.5	34.6
11	4.48	5.25
13	2.95	5.89

Source Ref. [6]

diodes; which are less expensive. Half controlled three-phase bridge circuits,
semi-converters, are used for industrial applications requiring relatively small output
power levels, say up to 200 kW or more. Examples of applications are printing press
drives, plastic extruders, paper mills and other applications requiring a limited range
of DC voltage control. Figure 4.17(a) shows a circuit diagram with the free wheeling
diode, and Fig. 4.17(b) is a signature of the current on the delta primary side. Note
the asymmetry with respect to positive and negative half cycles. Even harmonics are
generated when there is a firing delay. Both odd and even harmonics vary with the
firing angle. Figure 4.18 shows the variation of even harmonics (2nd and 4th) with

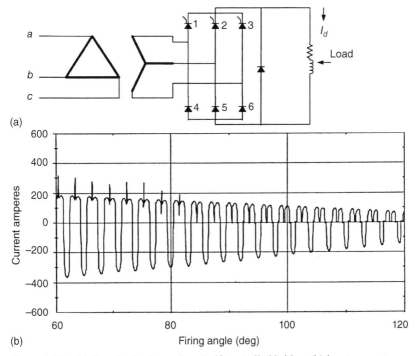

(a)

(b)

Figure 4.17   (a) Circuit of a three-phase half controlled bridge; (b) input current unsymmetrical waveform.

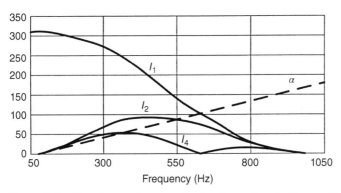

Figure 4.18   Even harmonics generated in a three-phase half-controlled bridge circuit versus frequency.

frequency and firing delay angle. For $\alpha \geq \pi/3$, and discontinuous output voltage:

$$V_{dc} = \frac{3\sqrt{3}}{2\pi} V_m \cos(1 + \alpha) \qquad (4.52)$$

For a three-phase semi-converter, connected to an input voltage of three-phase 480V, maximum $V_{dc} = 324$ V.

The rms output voltage is given by:

$$V_{rms} = \sqrt{3}V_m \left[\frac{3}{4\pi} \left(\pi - \alpha + \frac{1}{2}\sin 2\alpha \right) \right]^{1/2} \tag{4.53}$$

The rms voltage of a three-phase semi-converter, connected to an input voltage of three-phase 480V, and operating at $\alpha = \pi/2$, $V_{rms} = 415.7$ V.

The Fourier series expansion is:

$$I = \frac{-2I_d}{h\pi} \left[\left(\cos\frac{h\pi}{6} - \cos(h\alpha)\cos\frac{h\pi}{6} \right) \sin(h\omega t) + \sin(h\alpha)\cos\frac{h\pi}{6}\cos(h\omega t) \right]$$

$$h = \text{even}$$

$$I = \frac{-2I_d}{h\pi} \left[\left(\sin\frac{h\pi}{6} - \cos(h\alpha)\sin\frac{h\pi}{6} \right) \cos(h\omega t) + \sin(h\alpha)\sin\frac{h\pi}{6}\sin(h\omega t) \right]$$

$$h = \text{odd} \tag{4.54}$$

Half controlled bridge cannot operate as an inverter, since the polarity of DC voltage cannot reverse.

4.8   SWITCH-MODE POWER (SMP) SUPPLIES

Single-phase rectifiers are used for power supplies in copiers, computers, TV sets, and household appliances. In these applications the rectifiers use a DC filter capacitor and draw impulsive current from the AC supply. The harmonic current is worse than that given by Eq. (4.12). Figures 4.19(a) and (b) show conventional and switch mode power supplies (SMPSs), respectively. In the conventional power supply system, the main ripple frequency is 120 Hz, and the current drawn is relatively linear. Capacitors C_1 and C_2 and inductor act as a passive filter. In the SMPSs, the incoming voltage is rectified at line voltage and the high DC voltage is stored in capacitor C_1. The transistorized switcher and controls switch the DC voltage from C_1 at a high rate (10–100 kHz). These high-frequency pulses are stepped down in a transformer and rectified. The switcher eliminates the series regulator and its losses in conventional power supplies. There are four common configurations used with the switched mode operation of the DC to AC conversion stage, and these are fly back, push–pull, half-bridge, and full bridge. The input current wave for such an SMPS is highly nonlinear, flowing in pulses for part of the sinusoidal AC voltage cycle (Fig. 4.19(c)). The spectrum of an SMPS is given in Table 4.4 and it shows high magnitude of the third and fifth harmonics. Almost all electronic equipment used in office environment fall in this category, such as, PCs, printers, copiers, workstations, and peripheral equipment.

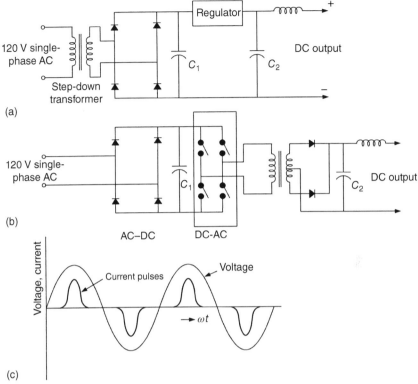

Figure 4.19   (a) Circuit of a conventional power supply; (b) circuit of a switch mode power supply; (c) input current pulsed waveform.

4.9   HOME APPLIANCES

A number of households are served from the same utility transformer. The harmonics can be measured at the metering point of each residential house. The nonlinear loads in households are increasing because of replacement of incandescent lighting with fluorescent lighting, and use of variable speed air conditioning systems for energy efficiency. Reference [11,12] document the results of harmonic measurements of fluorescent lighting (CLFs), television sets, PCs and air conditioners on medium voltage feeders. There are three main considerations with respect to harmonic emissions from home appliances:

- The harmonic emissions from various load types will not arithmetically sum up. Generally there will a reduction of harmonics when measured at a common bus. This situation is not unique to residential loads and can occur, in general, in all power systems, see Chapter 6.

- In the harmonic analysis it is assumed that the supply source is sinusoidal ignoring the effect of harmonic generation. The waveforms and emissions may considerably change with distorted supply inputs.

TABLE 4.4   Spectrum of Typical Switch Mode Power Supply

Harmonic	Magnitude	Angle	Harmonic	Magnitude	Angle
1	100	−37	14	0.1	65
2	0.2	65	15	1.9	−51
3	67	−97	17	1.8	−151
4	0.4	−72	19	1.1	84
5	39	−166	21	0.6	−41
6	0.4	−154	23	0.8	−148
7	13	113	25	0.4	64
8	0.3	0.3	27	0.2	−25
9	4.4	−46	29	0.2	−122
11	5.3	−158	31	0.2	102
12	0.1	142	33	0.2	56
13	2.5	92			

Source: Ref. [13].

- Referring to Eq. (4.28) the characteristic harmonics are given by the inverse of the harmonic order. This is because the odd harmonic waveform is approximated as a square wave. For home appliances this type of characterization has not been established. The following equation is proposed:

$$I_h = \frac{I_1}{h^\alpha} \tag{4.55}$$

where α is a parameter that determines the decline rate of the current spectrum and is estimated by performing curve fitting on the normalized spectra of home appliances.

Table 4.5 shows the residential loads and harmonic characteristics (Ref. [13]). The wide variations are noteworthy.

Ref. [14] is mainly devoted to harmonics in household loads.

4.10   ARC FURNACES

Arc furnaces may range from small units of a few ton capacities, power rating 2–3 MVA, to larger units having 400-ton capacity and power requirement of 100 MVA. The harmonics produced by electric arc furnaces are not definitely predicted due to variation of the arc feed material. The arc current is highly nonlinear, and reveals a continuous spectrum of harmonic frequencies of both integer and noninteger order (interharmonics).

The arc furnace load gives the worst distortion, and due to the physical phenomenon of the melting with a moving electrode and molten material, the arc current wave may not be same from cycle to cycle. The low-level integer harmonics predominate over the noninteger ones. There is a vast difference in the harmonics produced

TABLE 4.5   Residential Loads and Whole House Harmonic Current Characteristics

Type of Load	RMS Load Current	THDi (%)	h_3	h_5	h_7	h_9
Clothes dryer	25.3	4.6	3.9	2.3	0.3	0.3
Stovetop	24.3	3.6	3.0	1.8	0.9	0.2
Refrigerator #1	2.7	13.4	9.2	8.9	1.2	0.6
Refrigerator #2	3.2	10.4	9.6	3.7	0.8	0.2
Desktop computer/printer	1.1	140.0	91.0	75.2	58.2	39.0
Conventional heat pump#1	23.8	10.6	8.0	6.8	0.5	0.6
Conventional heat pump#2	25.7	13.2	12.7	3.2	0.7	0.2
ASD heat pump#1	14.4	123.0	84.6	68.3	47.8	27.7
ASD heat pump#2	27.7	16.1	15.0	4.2	2.3	1.9
ASD heat pump#3	13.0	53.6	48.8	6.3	17.0	10.1
Color television	0.7	120.8	85.0	60.6	34.6	14.6
Microwave #1	11.7	18.2	15.7	5.1	3.2	2.1
Microwave #2	11.7	26.4	23.3	9.6	2.2	1.6
Vehicle battery charger	0.5	51.7	43.2	26.9	2.6	4.2
Light dimmer	1.6	49.7	41.2	16.0	12.1	10.0
Electric dryer	25.3	4.6	3.9	2.3	0.3	0.3
Fluorescent ceiling light	2.5	39.5	36.9	13.8	2.3	1.4
Fluorescent desk lamp	0.6	17.6	17.1	3.4	2.0	0.6
Vacuum	6	25.9	25.7	2.7	1.8	0.4
House #1	72.0	4.8	3.6	3.2	0.2	
House #2	51.3	7.7	7.2	2.6	0.8	0.2
House #3	41.6	10.9	8.6	6.5	1.5	0.2
House #4	19.9	6.4	5.5	2.7	1.1	1.2
House #5	6.6	16.2	11.7	10.2	4.1	1.2
House #6	60.8	8.5	6.9	4.9	0.6	0.2
House #7	30.4	11.8	10.7	5.0	0.3	0.3
House #8	62.6	31.6	29.5	6.9	6.8	5.0

between the melting and refining stages. As the pool of molten metal grows, the arc becomes more stable and the current becomes steady with much less distortion. Figure 1.3 shows erratic arc current in a supply phase during the scrap melting cycle, and Table 4.6 shows typical harmonic content of two stages of the melting cycle in a typical arc furnace, from IEEE standard 519 [6]. The values shown in this table cannot be generalized. Both odd and even harmonics are produced. Arc furnace loads are harsh loads on the supply system, with attendant problems of phase unbalance, flicker, harmonics, impact loading, and possible resonance. Though the table terminates at 7th harmonic, higher order harmonics, of the order of 8th, 9th, and 10th of 0.5% can be considered in the models for harmonic analysis study. Furthermore, rather than modeling current harmonics, it is preferable to model the voltage harmonics typically generated in arc furnaces (Chapter 16).

TABLE 4.6 Harmonic Content of Arc Furnace Current as Percentage of Fundamental

h	Initial Melting	Refining
2	7.7	0.0
3	5.8	2.0
4	2.5	0.0
5	4.2	2.1
7	3.1	

Source Ref. [6]

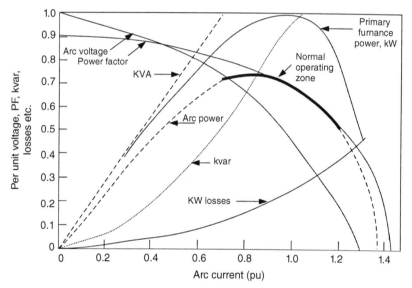

Figure 4.20   Typical performance curves of an arc furnace showing the normal operating zone in thick circular arc.

Figure 4.20 illustrates that the arc furnace presents a load of low lagging power factor. Large erratic reactive current swings cause voltage drops across the reactive impedance of the AC system, resulting in irregular variation of the terminal voltage. These voltage variations cause variation in the light output of the incandescent lamps and are referred to as flicker, based on the sensitivity of the human eye to the perception of variation in the light output of the incandescent lamps, the discussion is continued in Chapter 5.

4.10.1   Induction Heating

Induction heating is another major source of harmonics. Induction heaters typically employ a six-pulse SCR rectifier circuit at the front end which results in generation of significant harmonic currents. The harmonic spectrum shown in Table 4.7 and

TABLE 4.7   Harmonic Spectrum Primary of Delta-Wye 5000 KVA , 13.8-4.16 kV Induction Heater Load

Harmonic Order	Magnitude	Angle (deg.)	Harmonic Order	Magnitude	Angle (deg)
FUND	100	−38	2	0.1	−59
3	1.5	−7	4	0.1	20
5	20.2	174	6	0.2	−161
7	13.6	101	10	0.2	75
9	0.9	122	12	0.2	8
11	8.2	−44	18	0.1	7
13	7.1	−118	20	0.3	151
15	0.8	−101	24	0.2	128
17	4.7	99	26	0.1	−77
19	4.7	25	30	0.3	−112
21	0.8	42	32	0.2	46
23	3.1	−120			
25	3.4	164			
27	0.6	170			
29	2.0	16			
31	2.4	−57			
33	0.5	−37			
35	1.4	165			
37	1.6	85			

the waveform in Fig. 4.21 are generated from the primary side of the delta–wye transformer feeding a 3000 kW induction heater (Ref [13]). Note that even harmonics are produced. Also interharmonics are produced in induction furnaces, (Chapter 5).

4.11   CYCLOCONVERTERS

Cycloconverters are used in a wide spectrum of applications from ball-mill, rock crushers, rolling mill drives in steel industry and linear motor drives to static var generators. The range of application for synchronous or induction motors varies from 1000 to 50,000 hp.

Figure 4.22(a) shows the circuit of a three-phase /single phase cycloconverter, which synthesizes a 12-Hz output, and Fig. 4.22(c) shows the output voltage waveform with resistive load. The positive converter operates for half the period of the output frequency and the negative converter operates for the other half. The output voltage is made of segments of input voltages (Fig. 4.22 (b)) and the average value of a segment depends on the delay angle for that segment; α_p is the delay angle of the positive converter and $\pi - \alpha_p$ is the delay angle of the negative converter. The output voltage contains harmonics and the input power factor is poor. The control of AC motors requires a three-phase voltage at a variable frequency. The circuit of

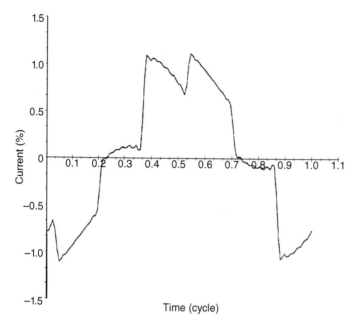

Figure 4.21   Waveform of input current to an induction heating load, see text.

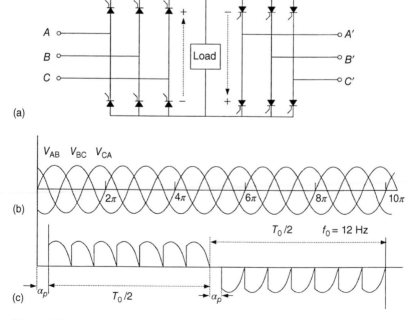

Figure 4.22   Principle of a cycloconverter operation, see text.

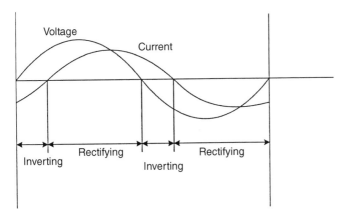

Figure 4.23   Operation of a cycloconverter with lagging load current.

Fig. 4.22(a) can be extended to provide three-phase output, that is, a total of 36 thyristors are required for fully controlled operation. A 12-pulse cycloconverter requires 72 SCRs. It is obtained by connecting 2 six-pulse cycloconverters in series with appropriate transformer connections to introduce phase shift (Ref. [15,16]). The higher is the pulse number, the closer will be the generated waveform to sinusoidal wave-shape. Figure 4.23 shows operation with lagging load current, the circulating currents in converters can be controlled by blocking the converter which is not delivering the load.

The line current spectrum of a three-phase, six-pulse cycloconverter operating from an input frequency of 60 Hz and the output frequency of 5 Hz is illustrated in Fig. 4.24 (Ref. [17]).

If the delay angles of the segments are varied so that the average value of the segment corresponds as closely as possible to the variations in the desired sinusoidal output voltage, the harmonics in the output are minimized. Such delay angles can be generated by comparing a cosine signal at source frequency with an output sinusoidal voltage.

Independent of pulse number, for a cycloconverter with three-phase balanced output, the characteristic harmonics are:

$$f_{ch} = [f \pm 6mf_0] \quad m \geq 1 \tag{4.56}$$

where f_0 is the output frequency of the cycloconverter, f_{ch} are the characteristic harmonics, $m = 1$, 2, 3, ... (see Chapter 5 for details of output and input harmonics). Cycloconverters generate a large spectrum of harmonics including subharmonics; For example, a three-phase three-pulse cycloconverter operating from a 60 Hz source frequency and output frequency of 35 Hz will have a 40-Hz harmonic.

The harmonics are a function of the following parameters:

• Pulse number,
• Ratio of output frequency to input frequency
• The relative level of output voltage

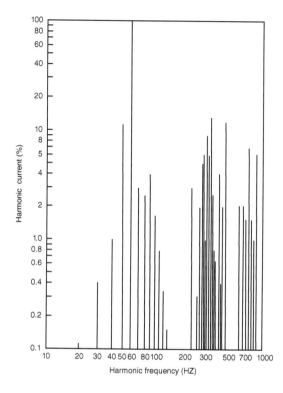

Figure 4.24   Typical harmonic spectrum of a cycloconverter, Ref. [17].

- Load displacement angle
- The method of control of firing instants.

As the output frequency varies, so does the spectrum of harmonics. The harmonic distortion components have sums of and differences between multiples of output and input frequencies. Therefore, control of harmonics with single tuned filters becomes ineffective. Ref. [18] is entirely devoted to cycloconverters. The discussions are continued in Chapter 5.

4.12   THYRISTOR-CONTROLLED REACTOR

Figure 4.25 shows the circuit diagram of an FC-TCR (fixed-capacitor-thyristor-controlled reactor). The arrangement provides discrete leading vars from the capacitors and continuously lagging vars from thyristor-controlled reactors. The capacitors are used as tuned filters, as considerable harmonics are generated by thyristor control.

The steady-state characteristics of an FC-TCR are shown in Fig. 4.26. The control range is AB with a positive slope, determined by the firing angle control:

$$Q_\alpha = |b_c - b_{1(\alpha)}|V^2 \tag{4.57}$$

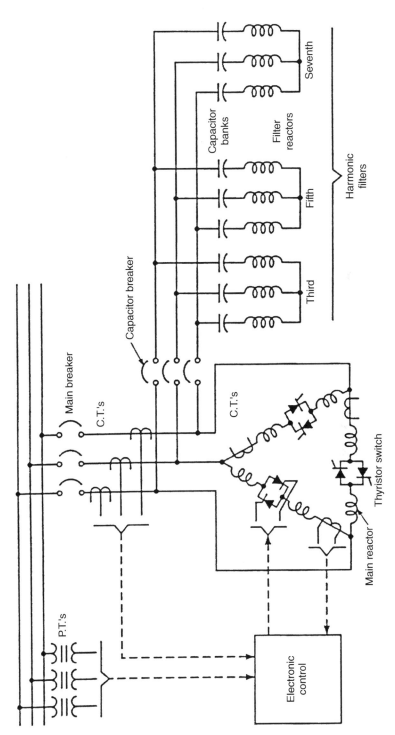

Figure 4.25 Circuit of a thyristor controlled reactor with passive harmonic filters.

151

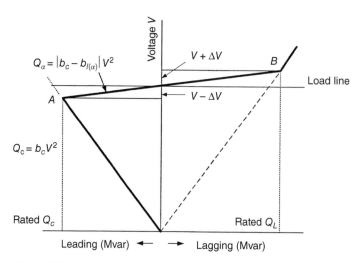

Figure 4.26 Steady state V-Q characteristic of an FC-TCR.

where b_c is the susceptance of the capacitor, and $b_{1(\alpha)}$ is the susceptance of the inductor at firing angle a. As the inductance is varied, the susceptance varies over a large range. The voltage varies within limits $V \pm \Delta V$. Outside the control interval AB, the FC-TCR acts as an inductor in the high-voltage range and as a capacitor in the low-voltage range. The response time is of the order of one or two cycles. The compensator is designed to provide emergency reactive and capacitive loading beyond its continuous steady-state rating.

Consider a TCR, controlled by two thyristors in an antiparallel circuit as shown in Fig.4.27. If both thyristors are gated at maximum voltage, there are no harmonics and the reactor is connected directly across the voltage, producing a 90° lagging current, ignoring the losses. If the gating is delayed, the waveforms as shown in Fig. 4.27(b) result. The instantaneous fundamental current through the reactor is

$$I_{LF}(\alpha) = \frac{V}{X}\left(1 - \frac{2}{\pi}\alpha - \frac{1}{\pi}\sin 2\alpha\right) \tag{4.58}$$

The admittance varies with a and is given by

$$\frac{V}{X}\left(1 - \frac{2}{\pi}\alpha - \frac{1}{\pi}\sin 2\alpha\right) \tag{4.59}$$

where V is the line-to-line fundamental rms voltage, α is the gating angle, and β is the conduction angle and $X = \omega L$. The current in the reactor can be expressed as

$$i_L(\alpha) = \frac{V}{X}(\sin \omega t - \sin \alpha) \tag{4.60}$$

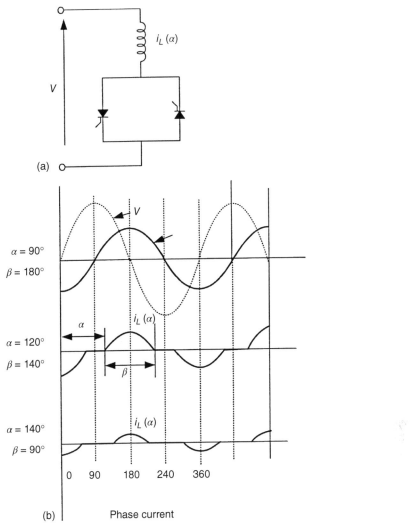

Figure 4.27   (a) Equivalent circuit of a TCR; (b) current waveforms with varying conduction and firing angles.

Appreciable amount of harmonics are generated. Assuming balanced gating angles only odd harmonics are produced. The rms value is given by

$$I_h = \frac{4V}{\pi X} \left[\frac{\sin \alpha \cos (h\alpha) - h \cos \alpha \sin(h\alpha)}{h(h^2 - 1)} \right] \tag{4.61}$$

where $h = 3, 5, 7, \ldots$ unequal conduction angles will produce even harmonics including a DC component. The triplen currents circulate in the delta connected transformer windings.

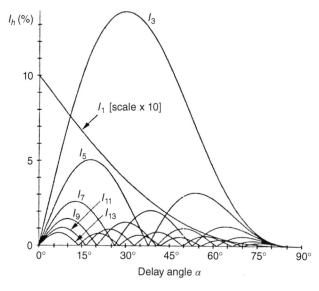

Figure 4.28 Harmonic generation is a TCR, see text.

Figure 4.28 shows the amplitude of harmonic components in TCR versus firing angle α. Triplen harmonics circulate in delta and do not appear in the lines. Another method of reducing harmonics is to use the number of parallel connected TCRs with sequential control. *Parallel connected (m numbers) TCRs are employed, each with* $1/m$ of the total rating required. The reactors are sequentially controlled; which implies that only one of the m reactors is delay angle controlled and the remaining m-1 reactors are fully "on". In this way amplitude of each harmonic is reduced by a factor m with respect to the rated fundamental current. See Ref [19].

4.13   PULSE WIDTH MODULATION

Pulse width modulated AC/DC inverters are used in many applications such as induction and synchronous motor ASDs (from fractional hp to several hundreds of hp), static var compensators, UPS (uninterruptible power supply systems) and FACTs (flexible AC transmission systems). Typically there are three supply methods:

- VSI (voltage source inverter)
- CSI (current source inverter) and
- ZSI, (impedance source inverters), discussed in Chapter 6.

In VSI and CSI for ASDs the front end for DC power supply is a usually an AC/DC rectifier. In three phase circuits full wave VSI and CSI are commonly applied. Multistage PWM inverters can be constructed either by multistage or multilevel technology, Chapter 6.

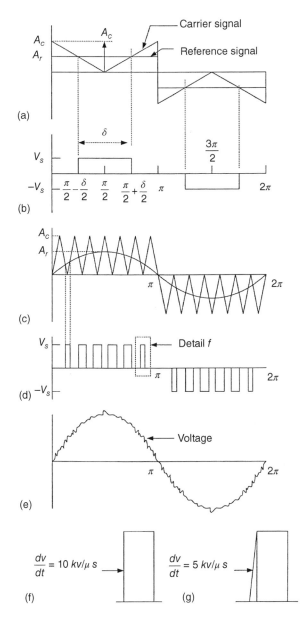

Figure 4.29   Pulse width modulation, (a) and (b) single pulse width modulation, (c) and (d) sinusoidal pulse width modulation, (e) reflection of switching transients on input current wave, (f) high dv/dt due to high frequency switching, (g) reduced dv/dt due to soft switching.

Figure 4.29 shows the basic principles of PWM. The voltage source invertors using IGBTs operate from a DC link bus. The inverter synthesizes a variable voltage, of variable-frequency waveform (V/f = constant), by switching the DC bus voltage at high frequencies (5−20 kHz). The inverter output line-to-line voltage is a series of voltage pulses with constant amplitude and varying widths. IGBTs have become popular as these can be turned on and off from simple low-cost driver circuits. Motor low-speed torque can be increased and improved low-speed stability is obtained. The

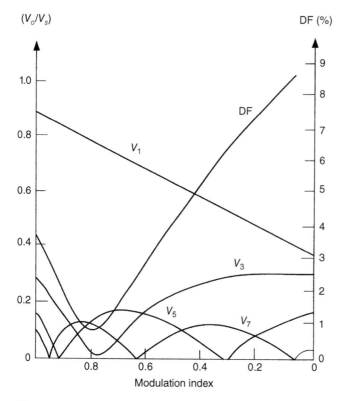

Figure 4.30   Harmonic generation, single-pulse width modulation.

high-frequency switching results in high dv/dt and the effects on motor insulation, connecting cables, and EMI (electromagnetic interference) are discussed in Chapter 8. Recent trends in soft switching technology reduce the rise time (Fig. 4.29(f) and (g)).

Techniques of PWM are as follows:

- Single pulse width modulation
- Multiple pulse width modulation
- Sinusoidal pulse width modulation
- Modified sinusoidal pulse width modulation

4.13.1   Single Pulse Width Modulation

In a single-pulse width modulation technique, there is one pulse per half-cycle and the width of the pulse is varied to control the inverter output voltage (Fig. 4.29(a) and (b)). The gating signal is generated by comparing a reference rectangular signal of amplitude A_r, with a triangular carrier wave of amplitude A_c. By varying A_r from 0 to A_c, the pulse width δ can be varied from 0° to 180°. The modulation index is defined as A_r/A_c. The harmonic content is high (Fig. 4.30).

4.13.2   Multiple Pulse Width Modulation

In a multiple pulse width modulation, the harmonics can be reduced by using several pulses in each half cycle of output voltage. This type of modulation is also known as *uniform pulse-width modulation*. The number of pulses per half-cycle is $N = f_c/2f_0$, where f_c is the carrier frequency and f_0 is the output frequency.

If δ is the width of each pulse, the output voltage is

$$V_0 = \left[\frac{2p}{2\pi} \int_{(\pi/p-\delta)/2}^{(\pi/p+\delta)/2} V_s^2 d\omega t \right]^{1/2} \tag{4.62}$$

$$V_0 = V_s \sqrt{\frac{p\delta}{\pi}} \tag{4.63}$$

where V_0 is the output voltage and V_s is source voltage.

And the Fourier series of the instantaneous voltage output is:

$$v_0 = \sum_{h=1,3,5..}^{\infty} (A_h \cos(h\omega t) + B_h \sin(h\omega t)) \tag{4.64}$$

where constants A_h and B_h can be calculated assuming that the positive pulse starts at $\omega t = \alpha_m$ and ends at $\omega t = \alpha_m + \pi$.

4.13.3   Sinusoidal Pulse Width Modulation

In sinusoidal PWM, the pulse width is varied in proportion to the amplitude of the sine wave at the center of the pulse (Fig. 4.29(c) and (d)). The distortion factor and the lower-order harmonic magnitudes are reduced considerably. The gating signals are generated by comparing a reference sinusoidal signal with a triangular carrier wave of frequency f_c. The frequency of the reference signal f_r determines the inverter output frequency f_0, and its peak amplitude A_r controls the modulation index and output voltage V_0. This type of modulation eliminates harmonics and generates a nearly sinusoidal voltage wave. The current or voltage waveform has a sinusoidal shape due to pulse width shaping; however, harmonics at switching frequency are superimposed (Fig. 4.29(e)). Figure 4.31 shows the harmonic profile for $p = 5$. This eliminates all harmonics less than or equal to $2p-1$. The lowest harmonic is nine.

Figure 4.29(d) shows that between 60° and 120°, the width of the pulse does not change much. In modified sinusoidal pulse width modulation (MSPWM) the pulses are applied during 0°–60° and 120°–180°.

Figure 4.32(a) shows harmonic current spectrum of an ASD calculated based upon specified system impedance at the point of interconnection. The drive system is 18-pulse PWM. Note the difference in the harmonic emission in phases *a*, *b*, and *c*. Generally, for harmonic load flow three-phase models are not used; though it may be necessary to do so where harmonic phase unbalance exists. The emission is small and meets the requirements of IEEE 519 [6] without additional filters. Figure 4.32(b) is a long term measurement patterns of harmonic voltage distortions in three phases - again the distortion levels differ in the three phases at each interval of time.

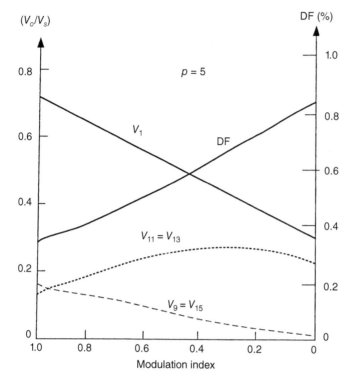

Figure 4.31 Harmonic generation, sinusoidal pulse-width modulation, $p = 5$.

4.14 VOLTAGE SOURCE CONVERTERS

The FACTS (flexible AC transmission systems), Ref. [19], use voltage source bridges. The current source converters have been extensively used for HVDC transmission. In the 1990's HVDC transmission using voltage source converters with PWM was introduced, commercially called HVDC-*light,* Ref [20]. There are two basic categories of self commutating converters:

- Current source converters in which DC current has always one polarity, and power reversal takes place through reversal of DC voltage polarity.
- Voltage source converters in which DC voltage always has one polarity and the power reversal takes place through reversal of DC current polarity.

Conventional thyristors have turn-on control only. These can be only current source converters. Switching devices such as GTO, IGBT, MTO, and IGCT have both turn on and turn off control. Turn-off device based converters can be of either type. The requirements of voltage source converters are:

- The converter should be able to act as an inverter or rectifier with leading or lagging reactive power, that is, four-quadrant operation is required, as compared to

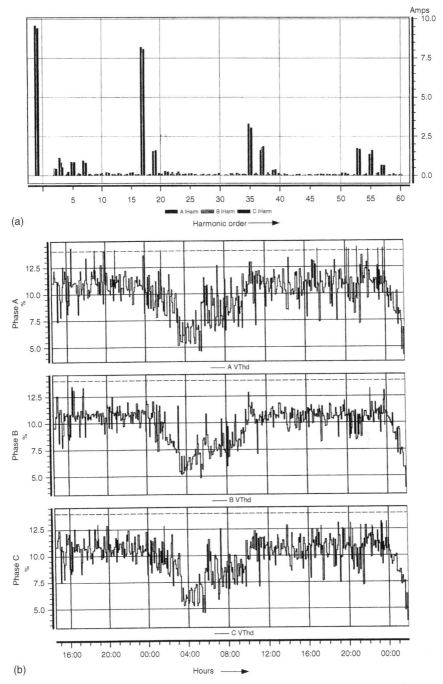

Figure 4.32   (a) Spectrum of harmonic currents in phases a, b, and c, 18-pulse medium voltage drive system, (b) long term measurement of harmonic voltage distortion patterns in three phases.

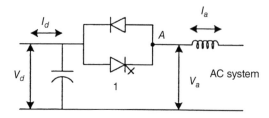

Figure 4.33   To illustrate the principle of a voltage source converter.

current source line commutated converter which has two quadrant operations. Turn-off devices are required.

- The active power and reactive power should be independently controllable with control of phase angle.

The principle is illustrated with respect to single valve operation (Fig. 4.33). Consider that DC voltage remains constant and the turn-off device is turned on by gate control. Then the positive of DC voltage is applied to terminal A, and the current flows from $+V_d$ to A, that is, inverter action. If the current flows from A to $+V_d$, *even when device 1 is turned on*, it will flow through the parallel diode, rectifier action. Thus, the power can flow in either direction. A valve with a combination of turn-off device and diode can handle power in either direction. The magnitude, phase angle and frequency of the output voltage can be controlled.

4.14.1   Three-Level Converter

Figure 4.34 shows the circuit of a three level converter, and associated waveforms. In Fig. 4.34(a) each half of the phase leg is split into two series connected circuits, mid point connected through diodes, which ensure better voltage sharing between the two sections.

Waveforms in Figs 4.34(b) through (e) are obtained corresponding to 1 three-phase leg. Waveform (b) is obtained with 180° conduction of the devices. Waveform (c) is obtained, if device 1 is turned off and 2A is turned on at an angle α *earlier* than for 180° conduction. The AC voltage V_a is clamped to zero with respect to mid point N of the two capacitors. This voltage occurs because devices 1A and 2A conduct and in combination with diodes clamp the voltage to zero. This continues for a period of 2α, till 1A is turned off and 2 is turned on and voltage is now $-V_d/2$, with both the 2 and 2A turned off and 1 and 1A turned on. The angle α is variable and output voltage V_a is square waves $\sigma = 180° - 2\alpha°$. The converter is called a three-level converter as DC voltage has three levels, $V_d/2$, 0, and $-V_d/2$.

The magnitude of the AC is varied without changing the magnitude of the DC voltage by varying angle α. Figure 4.34(d) shows the voltage V_b and Fig. 4.34(e) shows the phase-to-phase voltage V_{ab}.

The harmonic and fundamental rms voltages are given by:

$$V_h = \frac{2\sqrt{2}}{\pi}\left(\frac{V_d}{2}\right)\frac{1}{2}\sin\frac{h\alpha}{2}$$

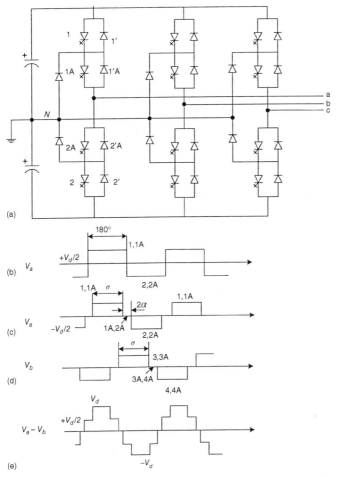

Figure 4.34   A three-phase, three-level voltage source converter; (b) through (e) operational waveforms, see text.

$$V_f = \frac{2\sqrt{2}}{\pi} \left(\frac{V_d}{2} \right) \sin \frac{\alpha}{2} \qquad (4.65)$$

$V_h = 0$ at $\alpha = 0°$ and maximum at $\alpha = 180°$.

A STATCOM (static synchronous compensator, see Ref. [19]) may use many six pulse converters outputs phase shifted and combined magnetically to give a pulse number of 24 or 48 for the transmission systems. The output waveform is nearly sinusoidal and the harmonics present in the output current and voltage are small. This ensures waveform quality without passive filters (Chapter 6).

The PWM signals are produced using two carrier triangular signals; two different offsets of opposite sign are added to the triangular waveform to give the required carriers. These are compared with the reference voltage waveform and the output and its inverse is used for the gating signals - four gating signals one for each IGBT in a leg. The five-level converters are discussed in Ref. [21,22].

4.15 WIND POWER GENERATION

The renewable energy sources are gaining momentum in the present times. These invariably generate harmonics, and Chapter 16 has some typical case studies of harmonic generation from wind and solar plants. The harmonic generation in wind farms depends on a number of factors, like the type of wind generator, power electronics, presence of filters and the grid connections. Figure 4.35 shows some basic system configurations and grid connections.

4.15.1 Direct Coupled Induction Generator

The direct coupled induction machine is generally of four-pole type, a gear box transforms the rotor speed to a higher speed for generator operation above synchronous speed. It requires reactive power from grid or ancillary sources, and starting after a blackout may be a problem. Wind dependent power surges produce voltage drops and flicker. The connection to the grid is made through thyristor switches that are by-passed after start. A wound rotor machine has the capability of adjusting the slip and torque characteristics by inserting resistors in the rotor circuit and the slip can be increased at an expense of more losses and heavier weight (Fig. 4.35(a)). The system will not meet the current regulations of connection to grid and may be acceptable for isolated systems. As the thyristor soft start is bypassed during normal operation, the harmonic emission is not of concern.

4.15.2 Induction Generator Connected to Grid through Full Sized Converter

The induction generator is connected to the grid through two back-to-back voltage source converters. Because of full power rating of the inverter, the cost of electronics is high. The wind dependent power spikes are damped by the DC link. The grid side inverter need not be switched in and out so frequently and harmonic pollution occurs.

4.15.3 Doubly Fed Induction Generator

The stator of the induction machine is directly connected to the grid, while the rotor is connected through voltage source converter (Fig. 4.35(b)). The energy flow over the converter in the rotor circuit is bidirectional. In subsynchronous mode, the energy flows to the rotor, in super synchronous mode it flows from rotor to the grid. The ratings of the converter are much reduced, generally one-third of the full power and depend upon the speed range of turbine. The power rating is:

$$P = P_s \pm P_r \qquad (4.66)$$

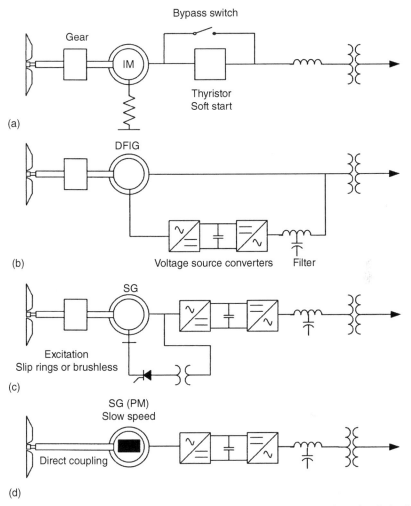

Figure 4.35 Grid connections of wind generators. (a) Direct connection of an induction generator stall regulated; (b) connection of DFIG, pitch-regulated; (c) synchronous generator brush type or brushless, voltage source converters, pitch-regulated; (d) gearless connection of a low-speed permanent magnet synchronous generator.

where P_s and P_r are the stator and rotor powers. But the rotor has only the slip frequency induced in its windings, therefore, we can write:

$$P_r = P_a \times s \tag{4.67}$$

where s is the slip and P_a is the air gap power. For a speed range of $\pm 30\%$, the slip is ± 0.3, and a third of converter power is required. Also we can write:

$$n_s = \frac{f_r \pm f}{p} 120 \tag{4.68}$$

where p are the number of pair of poles.

This method of connection is most popular, but gives rise to high harmonic pollution.

Synchronous generators can be of brush type or brushless type of permanent magnet excitation systems. These are also connected to the grid much alike asynchronous machines. The excitation power has to be drawn from the source, unless the generator is of permanent magnet type. Figures 4.35(c) and (d) show typical connections of synchronous generators.

4.15.4 Harmonics in Wind Farms

The harmonic emission will depend on converter topology, applied harmonic filters and short-circuit current at PCC. Even harmonics in wind generation can arise due to unsymmetrical half waves and may appear at fast load changes. Subharmonics can be produced due to periodical switching with variable frequency. Interharmonics can be generated when the frequency is not synchronized to the fundamental frequency, which may happen at low- and high-frequency switching.

The interharmonics due to back-to-back configuration of two converters can be calculated according to IEC [23].

$$f_{n,m} = [(p_1 k_1) \pm 1]f \pm (p_2 k_2)F \qquad (4.69)$$

where $f_{n,m}$ is the interharmonics frequency,

f is the input frequency,

F is the output frequency,

p_1, and p_2 are the pulse numbers of the two converters, and k_1 and $k_2 = 1, 2.3....$

Interharmonics are also generated due to speed dependent frequency conversion between rotor and stator of DFIG (doubly fed induction generator), and as side bands of characteristic harmonics of PWM converters.

Noncharacteristic harmonics can be generated due to grid unbalance. Even harmonics can be generated due to asymmetries caused by control faults.

Reference [24] describes a current source converter providing reactive power control and reduced harmonics for multi-megawatt wind turbines. It utilizes two-series connected three-phase inverters that employ fully controllable switches and an interconnection transformer on the utility side. Two converters feed into two, three-phase wye connected windings, with a delta connected winding on the utility source side. A three phase active harmonic current compensator is used on the output of the converter.

The greater the number of turbines the lower is the magnitude of the harmonics and subharmonics, especially of the lower order. For the turbines coupled to the grid with six pulse thyristor power converters, the amplitude of harmonics is higher over the entire range. Figure 4.36 shows the voltage spectra of the wind turbines with variable speed generators and grid connection through line commutated converters.

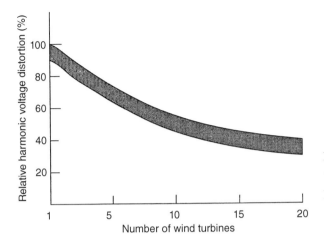

Figure 4.36   Voltage spectra of wind turbine generators connected through line-commutated converters; the band shows upper and lower limits.

A survey of the harmonic emissions of a commercially operated wind farm is in Ref. [25]. This shows 40 wind turbine units each connected through a step up transformer of 1.75 MVA to a voltage of 34.5 kV. The substation grid transformer is rated 82 MVA, 34.5-138 kV (Fig. 4.37). The wind generators are of DIFG type. Short term measurements are taken at bus 3 and measure the harmonic emission of the wind farm for 12 second. The long term measurements are taken at bus 4. The measurements are taken using FFT with rectangular window width of 12 cycles, which provides a 5 Hz resolution (Chapter 2). To avoid spectral leakage IEC 6100-4-7 suggests grouping method. The group magnitude G_n of the n-th order integer harmonic is obtained by the summation of FFT results at frequency nf_1 and its side bands at $nf_1 \pm 5$ Hz. The formulation for the interharmonics is:

$$C_{\text{isg},n}^2 = \sum_{i=2}^{10} C_{k+1}^2 \tag{4.70}$$

where C_{k+1} is the rms value of corresponding spectral components obtained from DFFT that exceed harmonic order n and $C_{\text{isg},n}$ is the rms value of the interharmonic group centered subgroup of order n.

Figure 4.38 shows the integer harmonic spectrum for phase A. The spectrum varies slightly between phases. Figure 4.39 is the interharmonic spectrum for phase A. Again, the spectrum varies somewhat between phases. A fifth harmonic current-level spectral graph and the calculated probability distribution function from long time measurements are illustrated in Fig. 4.40.

4.16   FLUORESCENT LIGHTING

The harmonic current generated by fluorescent lighting is strongly influenced by the type of lamp ballast used. Magnetic ballasts have third harmonic current of the

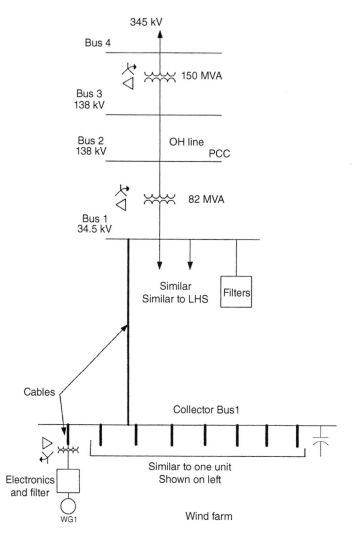

Figure 4.37   System configuration for measurement of harmonics in a wind farm.

order of 20% of the fundamental. The harmonic generating capability of the electronic ballasts varies over a range from 8% to 32%. ANSI Standard C82.11-1993 limits the electronic ballasts current harmonics to less than 32% and triplens less than 30% [26]. ANSI standard does not apply to compact fluorescents and higher harmonic levels can occur. Harmonic current generation in electronic ballasts is due to operation of a single phase diode bridge rectifier. Passive and active power factor correction circuitry is used to reduce distortion levels in the input current. A typical spectrum of the fluorescent lighting is shown in Table 4.8 and its waveform is shown in Fig. 4.41.

Lighting ballasts may produce large harmonic distortions and third harmonic currents in the neutral. The newer rapid start ballast has a much lower harmonic

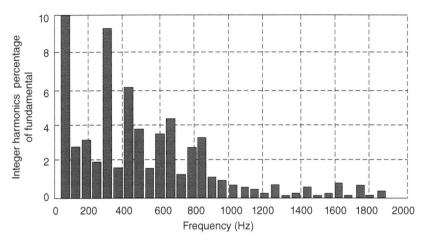

Figure 4.38   Harmonic spectrum of the wind farm in phase *a* at Bus 3 in Fig. 4.37, Ref. [25].

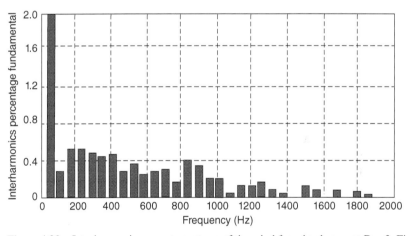

Figure 4.39   Interharmonic current spectrum of the wind farm in phase *a* at Bus 3, Fig. 4.37, Ref. [25].

distortion. The current harmonic limits for lighting ballasts are given in Tables 4.9 and 4.10. Table 4.9 shows that the limits for the newer ballasts are much lower when compared to earlier ballasts (Table 4.10). This also compares distortion produced by the lighting ballasts with other office equipment.

4.17   ADJUSTABLE SPEED DRIVES

Adjustable speed drives account for the largest percentage of nonlinear loads in the industry. We can divide the ASD's into the following major categories:

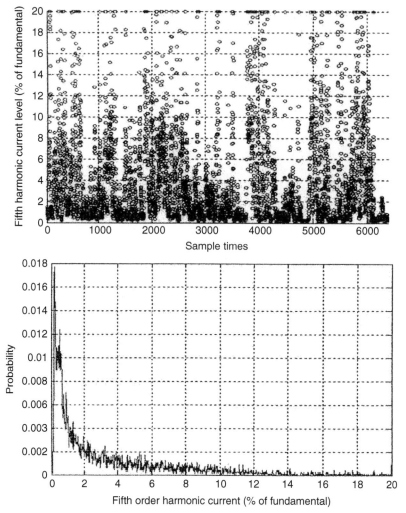

Figure 4.40   Fifth harmonic current level and calculated probability distribution function of the wind farm, long term measurements, Ref. [25].

- Induction motors-squirrel cage type
- Induction motors, wound rotor type
- Synchronous motors
- Permanent magnet motors
- DC motors

The horsepower ranges from fractional horsepower motors to motors of several- thousands of hp. The 135,000-hp synchronous motor ASD installed at NASA National Transonic Facility in Virginia is the largest drive in the world. Another example of a large 18,800-hp synchronous motor drive is at Boeing Wind

TABLE 4.8 Typical Harmonic Current Spectrum of Fluorescent Lighting with Electronic Ballasts

Harmonic Order	Magnitude	Angle (deg)
FUND	100	−124
2	0.2	136
3	19.9	144
5	7.4	62
7	3.2	−39
9	2.4	−171
11	1.8	111
13	0.8	17
15	0.4	−93
17	0.1	−164
19	0.2	−99
21	0.1	160
23	0.1	86
27	0.1	161
32	0.1	156

Tunnel fan, which produces wind speeds up to 435 km/hour. The operation, control and power electronics for the drive systems is a vast subject; we are interested in the harmonic generation.

An induction motor may have:

- variable voltage constant frequency control,
- variable voltage and variable frequency control,
- variable current and variable frequency control,
- variable voltage control or regulation of slip power.

The selection of power electronics will depend upon a number of factors; harmonic line pollution is one such considerations. Torque pulsations, voltage reflections, torque and speed control, noise are some other considerations. The common mode voltages are high frequency voltages that can damage motor insulation, chapter 8. The output filters can overcome this problem.

The following are the major topologies that are described in detail.

4.17.1 Voltage Fed Inverters

The system consists of a constant DC voltage source, 6/12/24-pulse rectifiers can be used to control source harmonics, the capacitor in the DC link smoothes the DC voltage and supplies reactive power to the motor, the self-commutated inverter unit uses gate turn-off thyristors, high-voltage IGBTs, or IGCTs. The inverters can be of multilevel depending on the output requirements and neutral clamping type (NPC) (Chapter 6). In PWM techniques, frequency and amplitude of motor voltages are

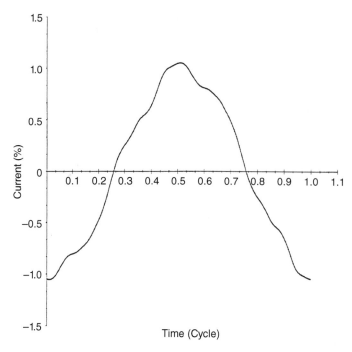

Figure 4.41   Current waveform of fluorescent lighting, electronic ballasts.

TABLE 4.9   Current Harmonic Limits for Lighting Ballasts

Harmonic	Maximum value (%)
Fundamental	100
Second harmonic	5
Third harmonic	30
Individual harmonics > 11th	7
Odd triplens	30
Harmonic factor	32

controlled while in DTC (direct torque control) motor flux and torque are controlled directly. With output filter and NPC, nearly sinusoidal output voltages are obtained. The power range with three-level inverters is up to approximately 10,000-hp squirrel cage induction motors.

4.17.2   Current Source Inverter

The system consists of current source inverter with DC link reactor, which can become fairly bulky. The amplitude of the motor current can be adjusted by

TABLE 4.10   THD Ranges for Different Types of Lighting Ballasts

Device Type	THD(%)
Older rapid start magnetic ballast	10–29
Electronic IC-based ballast	4–10
Electronic discrete based ballast	18–30
Newer rapid start electronic ballast	< 10
Newer instant start electronic ballast	15–27
High intensity discharge ballast	15–27
Office equipment	50–150

controlled rectifier at the input and the frequency is controlled by the inverter. The power range is up to 10,000-hp squirrel cage induction motors.

4.17.3   Load Commutated Inverter

Load-commutated inverters (LCI) are only applied to synchronous motors. A DC link is formed from line-commutated controlled rectifier and DC reactor. The LCI operates with variable machine voltage and frequency to the synchronous motor. The synchronous motor behaves like a DC motor, the inverter operating as a static commutator. The controls prevent motor from falling out of step.

4.17.4   Cycloconverters

The cycloconverter may be 6-pulse or 12-pulse type. The input frequency is converted to output load frequency without an intermediate DC voltage, but the load frequency is generally limited to 40% of the line frequency.

Slip frequency recovery schemes, and Scheribus drives are discussed in Section 4.20.

Table 4.11 provides a summary of the various drive technologies, which also shows relative harmonic pollution and power factor. When slow speeds are required, a PWM inverter is preferred to six step voltage source inverter for smoother waveform and improved power factor. Selection of higher pulse numbers even for smaller drives can reduce harmonic emission. Direct torque control (DTC) is a new technology which gives accurate speed and torque control, even at low speeds. With DTC motor flux and motor torque are used as primary control variables and direct control can be achieved without speed feedback device. The actual values are compared with reference values with fast 25μs control loop and static speed error is only 0.1 to 0.5% of nominal speed. Drive system controls are not discussed in this book.

We referred to forced commutation or line commutation is Section 4.2.1. It is also called class F commutation and is applied for phase-controlled bridge circuits for both rectification and inversion. All types of AC to variable voltage DC converters used for motor control and regulated power supplies; with DC smoothing reactor,

TABLE 4.11 Adjustable Speed Drives

Drive Type	DC Drive	Current Source Inverter	Voltage Source Inverter	Wound Rotor Slip Recovery	Load-Commutated Inverter	Cycloconverters
Motor type	DC	Squirrel cage induction	Squirrel cage induction	Wound rotor slip ring	Synchronous, brushless, or slip ring excitation	Synchronous or squirrel cage induction
hp range (typical)	1–10,000 hp	100–10,000 hp	100–10,000 hp	500–30,000 hp	1000–100,000 hp	1000–40,000 hp
Speed range (typical)	50–1	0 ± 75 Hz	0 ± 200 Hz	3–1 (subsynchronous)	Maximum speed 7500 rpm	Maximum speed 600–720 rpm for 50/60 Hz supply
Converter type	Phase controlled, line commutated	Current link, force commutated	Voltage link, self-commutated Multilevel converters Sometimes, forced-commutated inverters (early designs)	Current link, line commutated Kramer drive	Current link, load commutated	Cycloconverter three-phase current source
Features	Low converter cost, wide speed range	Simple and reliable control	Simple and reliable control	Lower cost for small speed ranges	Simple control and wide speed range	Wide speed range and fast response

Harmonics	See section for line harmonics	Line harmonics dependent on pulse number	Line harmonics dependent on pulse number, section	High harmonic pollution	Line harmonics dependent on pulse number	Voltage and current waves nearly sinusoidal
Interharmonics	Not generated	Can be generated	Can be generated	Greater possibility of interharmonics	Can be generated	Greater possibility of interharmonics
Power factor	Low due to phase control	Low	Can be high, close to unity	Lower than that for drives in previous columns	Low. Machine PF slightly leading	Line power factor is poor and not constant over entire speed range Motor power factor of unity is possible
Applications	Extruders, conveyors, machine tools, welders, general-purpose industrial drives	Pumps, fans, compressors, industry drives in medium power	Conveyors, machine tools, general-purpose industrial drive	Large pumps and fans with limited speed range in subsynchronous region Crushers and mills, wood chippers	Large pumps and blowers High-speed compressors, wind tunnel fans, rolling mills, extruders; coupling of variable speed generators to constant frequency utility networks	High-power low-speed drives, ball mills, cement mills, mine hoists, wind tunnel fans, propulsion drives for ships, rolling mills

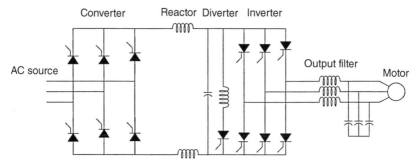

Figure 4.42   Circuit diagram of a SCI (self commutated inverter) for drive systems with diverter and output filter, see text.

make use of line commutation without external commutating components. The other types of commutation circuits are as follows:

- Class A: resonant commutation,
- Class B: Self commutation,
- Class C: Auxiliary commutation,
- Class D: Complementary commutation; not discussed.

The circuit of a self commutated inverter (SCI) for induction motor ASDs is shown in Fig. 4.42. It differs from LCI, used for synchronous motors, with respect to the diverter circuit. This forces commutation of the inverter at low frequencies and the output filter smoothes the waveforms. As with the conventional LCI, for normal operation the front end converter is commutated by the line voltage and the inverter by the load. The diverter circuit commutates at low frequency operation. With appropriate filter designs, the current distortion is less than 2% at 60 Hz (see Ref [16,27]).

4.18   PULSE BURST MODULATION

Typical applications of pulse burst modulation (PBM) are ovens, furnaces, die heaters, and spot welders. Three-phase PBM circuits can inject DC currents into the system, even when the load is purely resistive. A solid-state switch is kept turned on for an integer number γn of half-cycles out of a total of n cycles (Fig. 4.43(a) and (b)). The control ratio $0 < \gamma < 1$ is adjusted by feedback control. The integral cycle control minimizes EMI, yet the circuit may inject significant DC currents into the power system. Neutral wire carries pulses of current at switch off and switch on, which have high harmonic content, depending on the control ratio γ. Harmonics in the 100–400 Hz band can reach 20% of the line current. Loading of neutrals with triplen and fifth harmonics is a concern. The spectrum is deficient in high-order harmonics. Figure 4.43(c) shows measured harmonic currents in neutral as a percentage of V/R, for $\gamma = 0.5$ (Ref. [28]).

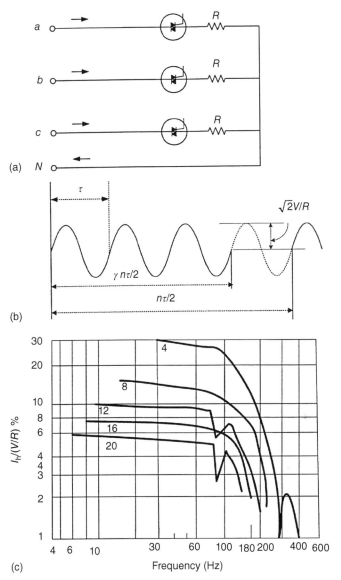

Figure 4.43   (a) Circuit for pulse burst modulation, (b) pulse burst control, (c) harmonic generation

4.19   CHOPPER CIRCUITS AND ELECTRIC TRACTION

The DC traction power supply is obtained in the rectifier substations by unsmoothed rectification of utility AC power supply, and 12-pulse bridge rectifiers are common. Switching transients from commutation occur and harmonics are injected into the supply system. Auxiliary converters in the traction vehicles also generate harmonics,

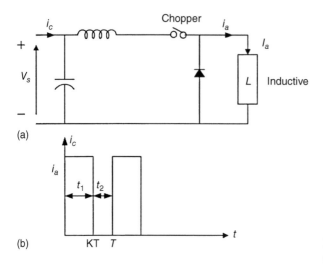

(a)

(b)   KT T

Figure 4.44   (a) Chopper circuit with input filter; (b) current waveform.

while EMI radiation is produced from fast current and voltage changes in the switching equipment.

A chopper with high inductive load is shown in Fig. 4.44; the input current is pulsed and assumed as rectangular. The Fourier series is

$$i_c(t) = \frac{I_a}{h\pi} \sum_{h=1}^{h=\infty} \sin 2h\pi k \cos 2h\pi f_c t + \frac{I_a}{h\pi} \sum_{h=1}^{h=\aleph} (1 - \cos 2h\pi k) \sin 2h\pi f_c t \qquad (4.71)$$

where f_c is the chopping frequency, and k is the mark-period ratio (duty cycle of the chopper $= t_1/T$). The fundamental component is given for $h = 1$. In railway DC fed traction drives, thyristor choppers operate up to about 400 Hz. The chopper circuit is operated at fixed frequency and the chopping frequency is superimposed on the line harmonics. An input low-pass filter, Chapter 15, is normally connected to filter out the chopper-generated harmonics and to control the large ripple current. The filter has physically large dimensions, as it has a low resonant frequency. The worst case harmonics occur at a mark-period ratio of 0.5. Imperfections and unbalances in the chopper phases modify the harmonic distribution and produce additional harmonics at all chopper frequencies. Transient inrush current to filter occurs when the train starts and may cause interference if it contains critical frequencies.

In a VSI fed induction motor traction drive from a *DC traction system*, the inverter fundamental frequency increases from zero to about 120 Hz as the train accelerates. Up to the motor base speed the inverter switches many times per cycle. To calculate source current harmonics this variable-frequency operation must be considered in addition to three-phase operation of the inverter. Harmonics in the DC link current depend on the spectrum of the switching function. Optimized PWM with quarter-wave symmetry is used in traction converters and though each DC link current component contains both odd and even harmonics, the positive and negative sequence components cancel in the DC link waveform, leaving only zero sequence and triplen harmonics in the spectrum.

Multilevel VSI drives or step-down chopper drives are used with GTOs. VSIs may use a different control strategy, other than PWM, such as torque band control with asynchronous switching, and the disadvantage is that relatively high $6h$ harmonics are produced, together with third harmonic and some subharmonics. In the chopper inverter drive the harmonics due to chopper and inverter combine. The input harmonic current spectrum consists of multiples of chopper frequency with side bands at six times the inverter frequency:

$$f_h = kf_c \pm 6hf_i \tag{4.72}$$

where k and h are positive integers, f_c is the chopper frequency, and f_i is the inverter frequency.

In AC *fed* traction drives, the harmonics can be calculated depending on the drive system topology. Drives with dual semi-controlled converters and DC motors are rich sources of harmonics. In drives fed with a line pulse converter and voltage or current source inverter, the line converter is operated with PWM to regulate the demand to the VSI, while maintaining nearly unity power factor. The source current harmonics are mainly derived from the line operation of the pulse converter.

4.20   SLIP FREQUENCY RECOVERY SCHEMES

The slip frequency power of large induction motors can be recovered and fed back into the supply system. Figure 4.45 shows an example of a subsynchronous cascade. The rotor slip frequency voltage is rectified, and the power taken by the rotor is fed into the supply system through a line commutated inverter. These schemes are called slip-power recovery drives (SPRDs) and are applied to blowers, fans and pumps. The speed of the induction motor can be adjusted as desired throughout the subsynchronous range, without losses, though the reactive power consumption of the motor cannot be corrected in the arrangement shown in Fig.4.45. The slip frequency current is rectified, and converted into supply system frequency. A step up transformer may be interposed as shown. The firing angle of inverter is controlled so that the motor electromagnetic torque is zero at synchronous speed.

Reference [29] reports investigation of harmonic generation on a small 415-V, 10-hp 60-Hz motor of certain parameters. The rotor current is not rectangular and has high ripple content, the peak value can be 25–33% of the average DC current (Fig. 4.46(a)). The stator current is nonperiodic, Fig. 4.46(b) . It contains harmonics whose frequency varies with the rotor speed and the current wave shape changes from one cycle to the next cycle. The harmonics of the rotor current produce harmonics in the stator current as follows:

$$[1 + (h-1)s]f \quad h = 7, 13, \ldots \quad \text{(positive sequence harmonics)}$$

$$[1 - (h+1)s]f \quad h = 5, 11, .. \quad \text{(negative sequence harmonics)} \tag{4.73}$$

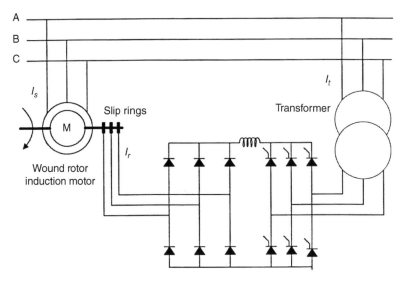

Figure 4.45   Kramer drive-slip frequency recovery scheme.

For a slip of 0.057, the stator fifth and seventh harmonics are 2.42f and 4.42f, respectively. Table 4.12 shows the harmonic distortion in percent for rotor voltage and current (I_r) and also for transformer current (I_t). The level of distortion is fairly high. The stator and net supply current harmonic components cannot be easily predicted due to their nonperiodic waveforms. Even harmonics result due to noninteger multiple-frequency components. Thus, the system can cause subharmonics in the AC system.

Torsional oscillations can be excited if the first or second natural torsional frequency of the mechanical system is excited, resulting in shaft stresses [10,30]. The AC harmonics for this type of load cannot be reduced by phase multiplication as DC current ripple is independent of the rectifier ripple.

In Scherbius drive scheme the diode rectifier is replaced with thyristor bridge, permitting slip power to flow in either direction.. The speed can be controlled in both subsynchronous and supersynchronous region. If slip power is fed into the line, it is acts like a Kramer drive, but if slip power is fed into the motor by reversal of rectifier and inverter action, the motor will operate in the supersynchronous region. The dual converter system can also be replaced with phase-controlled line commutated cycloconverter.

4.21   POWER SEMICONDUCTOR DEVICES

We referred to a number of semiconductor devices in this chapter. It is pertinent to provide some explanations, though it is not the intention to go in to their characteristics. The symbols and common nomenclature in use is shown in Fig. 4.47. The nominal rating of large power devices is in the range of 1-5kA, and 5−10 kV per

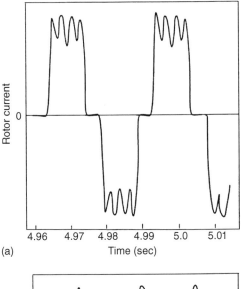

(a)

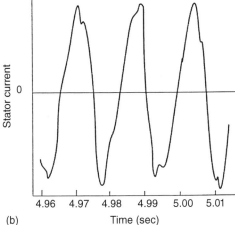

(b)

Figure 4.46   (a) Rotor current waveform; (b) stator current waveform.

device, and the usable rating is only 25–50% of the rating. A number of devices are connected in series/parallel combination for higher ratings. Some description of these devices is:

- Diodes are key components in most power semiconductor applications.
- Transistors are a family of three layer devices. IGBT (Insulated gate bipolar transistor), a type of transistor has became a choice in a wide range of medium power applications up to 10 MW. Switching frequency is up to 30 kHz or more. MOSFET (metal-oxide semiconductor field effect transistor) is good for low voltage and high switching frequency, > 100 kHz. Power bipolar junction transistors (BJT) have been replaced with MOSFET and IGBTs.

TABLE 4.12   Motor Slip, Firing Angle and THD

α (deg)	Slip	THD (%)		
		Rotor voltage	Rotor current I_R	Transformer current I_t
100	0.20	24.3	25.1	30.1
120	0.57	13.6	24.3	31.4
140	0.86	11.2	23.6	29.2

Source Ref. [28]

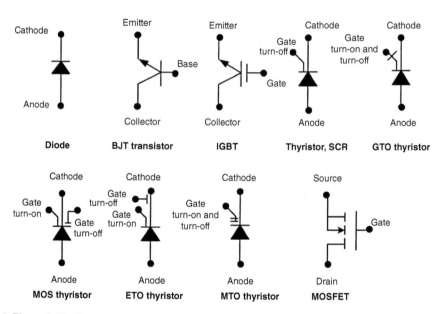

Figure 4.47   Power semiconductor devices symbols in common use.

- Thyristors and SCR (Silicon controlled rectifier, commercially pioneered by General Electric Company) are a family of four layer devices. A thyristor latches into full conduction in its forward direction when a turn-on pulse is applied to the gate. It returns to its nonconducting state when the current is brought to zero by external means or when the current falls to zero. These are applied at high voltage and current levels.

- Gate turn off thyristor invented by GE is now called GTO. It has the turn on capability like a thyristor and will return to nonconducting stage when the current falls to zero, but it can also be turned off by application of a pulse to the gate in reverse direction.

- MTO is a MOS turn-off thyristor invented at Silicon Power Corporation by Harshad Mehta and uses transistors to assist in turn-off. The turn off losses are lowered and fast turn-off capability is achieved. The MOS devices have a vertically diffused structure.

- Integrated gate commutated thyristor (GCT or IGCT) was developed by Mitsubishi and ABB. Similar to MTO, it achieves fast turn off and low switching losses.

- MOS controlled thyristor (MCT), developed at GE, is perhaps the ultimate in thyristor family, and includes MOS structure for both turn-off and turn-on.

Generation of harmonics from various topologies and their control is of major concern. It is not practical to provide description of harmonic emission from all power electronic device topologies in practical use. This chapter is indicative of the wide variations in the harmonic emissions, depending upon topologies. *Thousands of such topologies exist, while every year some new ones are added.* For the purpose of harmonic load flow and harmonic filter designs the harmonic emission from loads should be first estimated. This is not an easy task due to varied topologies. Mostly, appropriate spectrums and the harmonic angles should be ascertained from the manufacturers, better still, by measurements where applicable. Furthermore the emission is also a function of load as well as system source impedance, which aspects are further discussed in this book.

REFERENCES

1. Westinghouse Training Center. Power system harmonics, course C/E 57, 1987.
2. L. Malesani, P. Tenti. "Three-phase AC/DC PWM converter with sinusoidal AC currents and minimum filter requirements," IEEE Transactions on Industry Applications, vol. IA-23, no. 1, pp. 71–78, 1987.
3. B. T. Ooi, J. C. Salmon, J. W. Dixon, A. Kulkarni. "A three-phase controlled-current PWM converter with leading power factor," IEEE Transactions on Industry Applications, vol. IA-23, no. 1, pp. 78–84, 1987.
4. J. Cardosa, T. Lipo. Current stiff converter topologies with resonant snubbers, IEEE Industry Application Society Annual Meeting, New Orleans, LA, pp. 1322–1329, 1997.
5. IEEE Standard C57.1200, General requirements of liquid immersed distribution power and regulating transformers, 2006.
6. IEEE Standard 519, IEEE recommended practices and requirements for harmonic control in power systems, 1992.
7. J. C. Read. "The calculations of rectifier and inverter performance characteristics," Journal of the Institution of Electrical Engineers (UK), pp. 495–509, 1945.
8. M. Grotzbech and R. Redmann. "Line current harmonics of VSI-fed ASDs," IEEE Transactions on Industry Applications, pp. 683–690, 2000.
9. E. W. Kimbark. Direct Current Transmission, Vol. 1, John Wiley & Sons, New York, 1971.
10. J. C. Das. Transients in Electrical Systems, McGraw Hill, New York, 2012.
11. A. E. Emanuel, J. A. Orr, D. Cyganski and E. M. Gulachenski, "A survey of harmonic voltages and currents at the consumer's bus," IEEE Transactions on Power Delivery, vol. 8, no. 1, pp. 411–421, 1993.
12. A. E. Emanuel, et al., "Voltage distortions in distribution feeders with non-linear loads," IEEE Transactions on Power Delivery, vol. 9, no. 1, pp. 79–87, 1994.
13. IEEE P519.1. Draft guide for applying harmonic limits on power systems, 2004.
14. A. Nasif, Harmonics in Power Systems. VDM Verlag, 2010.
15. M. H. Rashid. Power Electronics. Prentice Hall, New Jersey, 1988.
16. F. L. Luo, H. Ye and M. H. Rashid. Digital Power Electronics and Applications. Academic Press, Elsevier, San Diego, California, 2005.
17. IEEE Working Group on Power System Harmonics. "Power system harmonics: An Overview," IEEE Trans on Power Apparatus and Systems PAS, vol. 102, pp. 2455–2459, 1983.

18. B. R. Pelly. Thyristor Phase-Controlled Converters and Cycloconverters, Operation Control and Performance. John Wiley, New York, 1971.

19. N. G. Hingorani, L. Gyugyi. Understanding FACTS. IEEE Press, NJ, 2001.

20. B. Jackson, Y. J. Hafner, P. Rey, and G. Asplund. HVDC with voltage source converters and extruded cables for up ±to 300 kV and 1000 MW, in Proceedings on CIGRE, pp. 84–105, 2006.

21. N. Hatti, Y. Kondo and H. Akagi. "Five level diode-clamped PWM converters connected back-to-back for motor drives," IEEE Trans. Industry Applications, vol. 44, no. 4, 1268–1276, 2008.

22. H. Akagi, H. Fujita, S. Yonetani, and Y. Kondo. "A 6.6 kV transformer-less STATCOM based on a five level diode-clamped PWM converter: System design and experimentation of 200-V, 10 kVA laboratory model," IEEE Trans. Industry Applications, vol. 44, no. 2, pp. 672–680, 2008.

23. IEC 61000-2-4, Electromagnetic compatibility, Part 2. Environmental Section 4: compatibility levels in industrial plants for low-frequency conducted disturbances.

24. P. Tenca, A. A. Rockhill and T.A. Lipo. "Wind turbine current-source converter providing reactive power control and reduced harmonics," IEEE Trans. Industry Applications. vol. 43, no. 4, pp. 1050–1060, 2007.

25. S. Liang, Q. Hu, W. J. Lee. "A survey of harmonic emissions of a commercially operated wind farm," IEEE Trans. Industry Applications, vol. 48, no. 3, pp. 1115–1123, 2012.

26. ANSI Standard C82.11 High-frequency Fluorescent Lamp Ballasts, 2011.

27. B. K. Bose (ed.) Adjustable Speed AC Drive Systems, IEEE Press, New York, 1981.

28. A. E. Emanuel, B. J. Pileggi. "Disturbances generated by three-phase pulse-burst-modulated loads and the remedial methods," IEEE Transactions on PAS, vol. 100, no. 11, pp. 4533–4539, Nov. 1981.

29. Y. Baghzzouz, M. Azam. "Harmonic analysis of slip-power frequency drives," IEEE Transactions on Industry Applications, vol. 28, no. 1, pp. 50–56, Jan./Feb. 1992.

30. H. Flick. "Excitation of subsynchronous torsional oscillations in turbine generator sets by a current-source converter," Siemens Power Engineering vol. 4, no. 2, pp. 83–86, 1982.

CHAPTER 5

INTERHARMONICS AND FLICKER

5.1   INTERHARMONICS

IEC [1] defines interharmonics as "Between the harmonics of the power frequency voltage and current, further frequencies can be observed, which are not an integer of the fundamental." They appear as discrete frequencies or as a wide-band spectrum. A more recent definition is any frequency that is not an integer multiple of the fundamental frequency (Chapter 1).

5.1.1   Subsynchronous Interharmonics (Subharmonics)

A group of harmonics which are characterized by $h < 1$, that is, these groups have periods larger than the fundamental frequency and have been commonly called subsynchronous frequency components or subsynchronous interharmonics. In earlier documents, these were called subharmonics. The term subharmonic is popular in the engineering community, but it has no official definition. Also see IEEE task force on harmonic modeling and simulation [2]. IEEE Standard 519 does not address interharmonics but there are many publications on this subject.

5.2   SOURCES OF INTERHARMONICS

Cycloconverters are a major source of interharmonics. Large mill motor drives using cycloconverters ranging up to 8 MVA appeared in the 1970s. In 1995, rolling mill drives of 56 MVA were installed. These are also used in 25-Hz railroad traction power applications. See Chapter 4 for waveforms and an expression for harmonic generation. The cycloconverters can be thought of as frequency converters. The DC voltage is modulated by the output frequency of the converter and interharmonic currents appear in the input, as discussed further.

   Electrical arc furnaces (EAF) are another major source. Arcing devices give rise to interharmonics. These include arc welders. These loads are associated with low-frequency voltage fluctuations giving rise to flicker. These also exhibit higher frequency interharmonic components. The interharmonic limits must be accounted for in filter designs. Other sources are as follows:

Power System Harmonics and Passive Filter Designs, First Edition. J.C. Das.
© 2015 The Institute of Electrical and Electronics Engineers, Inc. Published 2015 by John Wiley & Sons, Inc.

- Induction furnaces
- Integral cycle control
- Low-frequency power line carrier; ripple control
- HVDC
- Traction drives
- ASDs
- Slip frequency recovery schemes

5.2.1   Imperfect System Conditions

Practically, the ideal conditions of operation are not obtained, which give rise to non-characteristic harmonics and interharmonics. Consider the following:

- The AC system three-phase voltages are not perfectly balanced. The utilities and industrial power systems have some single phase loads also, which give rise to unbalance. A 1% lower voltage in one phase will give approximately 7% third and 4.3% fifth harmonics in a standard 12-pulse converter.
- The impedances in the three phases are not exactly equal, especially unequal commutation reactances or unequal phase impedances of the converter transformer.
- With a commutating reactance of 0.20 pu and variation of 7.5% in each phase, firing angle 15° will give a fifth harmonic of 33% of fundamental.
- DC modulation and cross modulation. Addition of a harmonic h on DC side will transfer to AC side; the harmonic will be of different order but of same phase sequence. The back-to-back frequency conversion represents the worst case of interharmonic generation. Consider that two AC systems are interconnected through a DC link and operate at different frequencies. An equivalent circuit can be shown as in Fig. 5.1. The source of ripple in this figure is voltage from the remote end of the DC link and the distortion caused by the converter itself, the ripple current depending on the DC-link reactor. The system can represent HVDC link, the AC power systems operating at different frequencies, see Section 5.2.3.

A conventional six-pulse three-phase bridge circuit can be conceived as a combination of a switching function and a modulating function. The switching function can be termed as

$$s(t) = k \left[\cos \omega_1 t - \frac{1}{5} \cos 5\omega_1 t + \frac{1}{7} \cos 7\omega_1 t - .... \right] \tag{5.1}$$

and the modulating function as the sum of DC current and superimposed ripple content:

$$i(t) = I_d \sum_{z=1}^{\infty} a_z \, \sin(\omega_z t + \varphi) \tag{5.2}$$

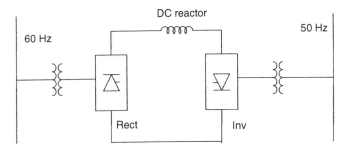

Figure 5.1   Two AC systems connected through a DC-link equivalent circuit.

where a_z is peak magnitude of sinusoidal components and ω_z can be of any value and not an integer multiple of ω_1. Then the harmonics on AC side are as follows:

$$i_{AC}(t) = i(t)s(t) \tag{5.3}$$

For a 12-pulse operation, following equations can be written for the harmonics on the AC side:

$$i_{AC} = ki_d(12\text{th}, \quad 13\text{th}, \quad 23\text{rd}, \ \dots)$$

$$+ \frac{kb}{2}[\sin(\omega_1 t + 12\omega_2 t + \phi_{12}) - \sin(\omega_1 t - 12\omega_2 t - \phi_{12})]$$

$$- \frac{kb}{22}[\sin(11\omega_1 t + 12\omega_2 t + \phi_{12}) - \sin(11\omega_1 t - 12\omega_2 t - \phi_{12})]$$

$$+ \frac{kb}{26}[\sin(13\omega_1 t + 12\omega_2 t + \phi_{12}) - \sin(13\omega_1 t - 12\omega_2 t - \phi_{12})]$$

$$- \frac{kc}{46}[\sin(23\omega_1 t + 24\omega_2 t + \phi_{24}) - \sin(23\omega_1 t - 24\omega_2 t - \phi_{24})]$$

$$+ \frac{kc}{50}[\sin(25\omega_1 t + 24\omega_2 t + \phi_{24}) - \sin(25\omega_1 t - 24\omega_2 t - \phi_{24})]$$

$$- \frac{kd}{70}[\sin(35\omega_1 t + 36\omega_2 t + \phi_{36}) - \sin(35\omega_1 t - 36\omega_2 t - \phi_{36})]$$

etc ... $\qquad\qquad\qquad (5.4)$

where $k = 2\sqrt{3}\pi$, b, c, and d are the magnitudes of the 12th, 24th, and 36th harmonic currents on the DC side.

In a rectifier–DC link–converter system linking two isolated AC systems if the frequency on two sides differ by Δf_0, then for a 12-pulse system, the DC-side voltage at frequency of $12n(f_0 + \Delta f_0)$ will be modulated by another converter:

$$(12m \pm 1)f_0 + 12n(f_0 + \Delta f_0) \tag{5.5}$$

On the AC side, among other frequencies, includes the frequency

$$f_0 + 12n\Delta f_0 \qquad (5.6)$$

This will beat with the fundamental component at a frequency of $12n\Delta f_0$, which will allow flicker-producing currents to flow.

The control systems and gate control of the electronic switching devices are not perfectly symmetrical; these concepts are continued in Chapter 14. These system conditions will give rise to noncharacteristics and interharmonics, which would have been absent if the systems were perfectly symmetrical; a discussion continues in Chapter 14.

5.2.2 Interharmonics from ASDs

The interharmonics can originate from the converters by interaction of a harmonic from the DC link into the power source. A harmonic of the order 150 Hz reacting with fundamental frequency of 60 Hz produces a current wave shape as shown in Fig. 5.2, the subtraction and addition of two components occur periodically (EMTP simulation).

Consider an ASD with the motor running at 44 Hz; this frequency will be present at the DC link as a ripple of 44 Hz times the pulse number of the inverter. The current on the DC link contains both the 60 Hz and 44 Hz ripples. The 44 Hz ripples will pass on to the supply side and present themselves as interharmonics because 44 times the pulse number is not an integer of 60 Hz.

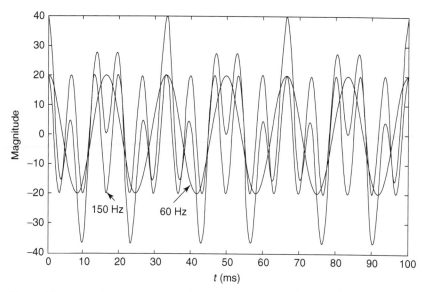

Figure 5.2   Interaction of two harmonic frequencies of differing magnitude, EMTP simulation.

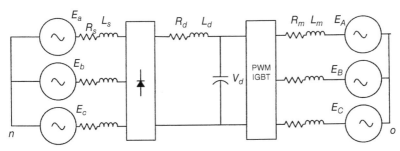

Figure 5.3   An ASD circuit with front-end diode-bridge circuit, DC-link reactor, and PWM inverter.

TABLE 5.1   Interharmonic Current Level and Load Current Unbalance Measured in a Small Typical Drive

Load Current Unbalance (%)	Ratio of Source Interharmonic Current to Fundamental current
0.019	0.000
0.166	0.037
0.328	0.046
0.511	0.065
0.551	0.075

Source: Ref. [3].

If m is the mth motor harmonic and nth is the PWM inverter harmonic and ω is the inverter operating frequency, then significant components of inverter input current will exist at frequencies $(n \pm m)\omega$ for n with a significant switching frequency component and m with a significant motor harmonic current component.

Consider an ASD system with front-end diode-bridge rectifier, DC-link reactor, and a PWM inverter (Fig. 5.3). Harmonic currents of the inverter create interharmonics in the power system when these propagate through the DC link.

For balanced cases with linear modulation of the inverter, the DC-link harmonics are of the high order, which are blocked by the DC inductor.

With unbalanced loads or overmodulation, significant amount of interharmonics are generated. A relationship between interharmonic current level and load unbalance is shown in Table 5.1. The load current unbalance can be defined as the difference between the maximum and minimum phase current magnitudes divided by the average of the phase current magnitudes. The frequency modulation index m_f was chosen so that the switching frequency is in the range 1.8–2 kHz and amplitude modulation ratio $m_a = V_{AN}/(0.5V_d)$. With balanced loads and linear modulation, the motor harmonics begin at switching frequency, and as these were well above DC-link resonant frequency of 92 Hz, no significant amount of inverter harmonics are present in the power system.

The unbalance causes low-order harmonics particularly the second and 12th, Ref [3]. The second and 12th inverter current harmonics in the DC link cause interharmonics when reflected to the AC side of the rectifier:

$$f_h = |\mu f_1 \pm k f_s| \tag{5.7}$$

f_h is the frequency of the interharmonic, μ is the order of current harmonic, typically 2 or 12.

f_I is the inverter operating frequency, $k = 1, \ 5, \ 7, \ \ldots$

f_s is the source frequency $= 60 \ \text{Hz}$.

The most significant values of interharmonics will occur with $\mu = 2$ and $k = 1$ (Fig. 5.4). For inverter frequencies of 25, 37.5, and 48 Hz and source frequency of 60 Hz, the side band pairs are 10 and 110, 15 and 135, and 36 and 156 Hz, respectively. The frequency modulation rate m_f was chosen so that switching frequency is in the range 1.8-2 kHz.

With overmodulation, $m_a > 1$, lower order harmonics appear on the DC link in balanced case, most dominant being sixth harmonic. When it is reflected to the AC side, interharmonics at 228 and 348 Hz occur with inverter operating frequency

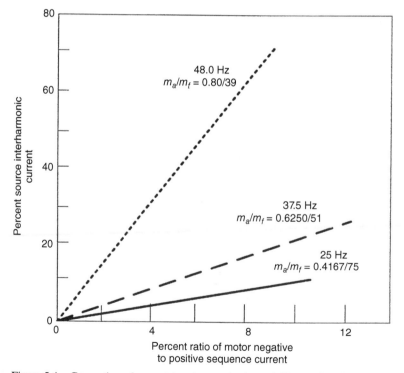

Figure 5.4 Generation of current interharmonics in an ASD as a function of unbalance ratio. Source: Ref. [3].

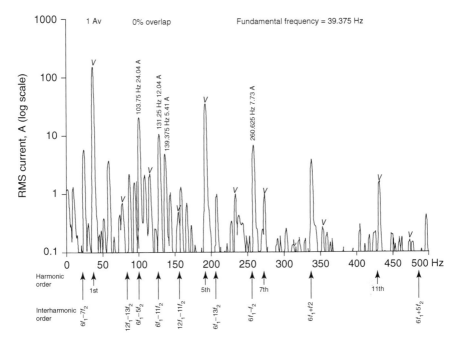

Figure 5.5   Harmonic spectrum from an actual ASD. Source: Ref. [4].

of 48 Hz. This assumes that the DC reactance is substantial and source inductance is negligible. As L_d is reduced, the rectifier current harmonics rise till the rectifier output current becomes discontinuous (Chapter 4). The effect of the source inductance will be to change apparent DC-link inductance and therefore the tuning of the DC-link components [3].

An example of harmonic spectrum from an actual operating system is shown in Fig. 5.5 [4]. The motor is fed at 39.4 Hz (50-Hz source frequency). If the harmonics or interharmonics coincide with the natural frequency of the motor/shaft/load mechanical system, then shaft damage is possible.

5.2.3   HVDC Systems

HVDC systems are another possible source of interharmonics. Interharmonics of the order of 0.1% of the rated current can be expected in HVDC systems, when two ends are working at even slightly different frequencies [5,6]. Referring to Fig. 5.1, following harmonics are generated considering six-pulse converters at either end.

The modulation theory has been used in harmonic interactions in HVDC systems (see also Chapters 12 and 14). When AC networks operate at different frequencies, interharmonics will be produced.

DC Side   The voltage harmonics will contain frequency groups $6n\omega_1$ and $(6n\omega_1 + \omega_m)$:

where ω_m is the mth harmonic frequency, which may be an integer harmonic of either of the two AC systems – call it a disturbing frequency.

The characteristic harmonics in DC voltage ($= 6n\omega_1$) appear in DC current. Harmonic frequencies will be $\omega_m = 6m\omega_2$. This gives a new group of harmonics:

$$6n\omega_1 \pm 6m\omega_1 = 6(n \pm m)\omega_1 = 6k\omega_1 \qquad (5.8)$$

where n, m, and k are integers.

All characteristic harmonics from the inverter will appear in the DC current and the frequency will be $6m\omega_2$. Thus, second group of harmonics on DC side are as follows:

$$6n\omega_1 \pm 6m\omega_2 = 6(n\omega_1 \pm m\omega_2) \qquad (5.9)$$

The third set of frequencies from Eq. (5.9) will also appear in the DC current, so that $\omega_m = 6(n\omega_1 \pm p\omega_2)$; and therefore, the frequencies are

$$6n\omega_1 \pm 6(m\omega_1 \pm p\omega_2) = 6[(n \pm m)\omega_1 \pm p\omega_2)] = 6(k\omega_1 \pm p\omega_2) \qquad (5.10)$$

where n, m, p, k are integers.

AC Side The harmonics on the AC side will be the following:

Those caused by DC characteristic harmonics, $\omega_m = 6m\omega_1$. The harmonics transferred to AC side are as follows:

$$6m\omega_1 \pm (6n \pm 1)\omega_1 = (6k \pm 1)\omega_1 \qquad (5.11)$$

Those caused by characteristic DC voltage harmonics generated at far end $\omega_m = 6m\omega_2$

$$6m\omega_2 \pm (6n \pm 1)\omega_1 \qquad (5.12)$$

Those caused by

$$\omega_m = 6(m\omega_1 \pm p\omega_2) \qquad (5.13)$$

The harmonics transferred to AC side will be

$$6(m\omega_1 \pm p\omega_2) \pm (6n \pm 1)\omega_1 \qquad (5.14)$$

This assumes low DC-side impedance (also see Ref. [7,8]). Practically, DC-link reactor and AC and DC filters are used to mitigate the harmonics (Chapter 15).

The interharmonics due to Kramer drives, wind power generation, and electric traction are discussed in Chapter 4.

5.2.4 Cycloconverters

Cycloconverters are discussed in Chapter 4. Certain relationship exists between the converter pulse numbers, the harmonic frequencies present in the output voltage, and the input current. The harmonic frequencies in *the output voltage* are the integer multiple of pulse number and input frequency, $(np)f$, to which are added and subtracted integer multiple of output frequency, that is,

$$h_{\text{output voltage}} = (np)f \pm mf_0 \qquad (5.15)$$

Here, n is any integer and not the order of the harmonic, and m is also an integer as described later.

For cycloconverter with *single-phase output*, the harmonic frequencies present in the input current are related to those in the output voltage. There are two families of input harmonics:

$$h_{\text{input current}} = |[(np) - 1] f \pm (m - 1)f_0|$$

$$h_{\text{input current}} = |[(np) + 1] f \pm (m - 1)f_0| \qquad (5.16)$$

where m is odd for (np) even and m is even for (np) odd.

In addition, the characteristic family of harmonics independent of pulse number is given by

$$|f \pm 2mf_0| \quad m \geq 1 \qquad (5.17)$$

For a cycloconverter with a balanced three-phase output, for each family of output voltage harmonics, $(np)f \pm mf_0$, there are two families of input current harmonics:

$$h_{\text{input current}} = |[(np) - 1] f \pm 3(m - 1)f_0|$$

$$h_{\text{input current}} = |[(hp) + 1]f \pm 3(m - 1)f_0| \qquad (5.18)$$

where m is odd for (np) even and m is even for (np) odd.

In addition, the characteristic family of harmonics independent of pulse number is given by

$$|f \pm 6mf_0| \quad m \geq 1 \qquad (5.19)$$

Figure 5.6 shows a chart of relationships between the predominant harmonic frequencies present in a three-phase *input current* waveform of the cycloconverter with a balanced three-phase output, and output to input frequency ratio. For input current waveforms with higher pulse numbers, certain harmonic families are eliminated as shown [9].

The magnitude of the harmonic is a function of the output voltage and the load displacement angle, but is independent of the frequency of the component. Thus, for a given output voltage ratio and load displacement angle, those harmonic components that are present always have the same relative magnitude independent of pulse number or the number of output phases.

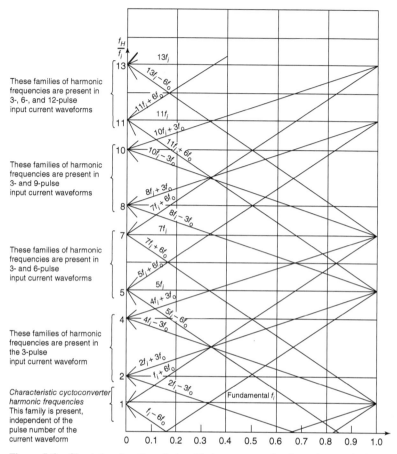

These families of harmonic frequencies are present in 3-, 6-, and 12-pulse input current waveforms

These families of harmonic frequencies are present in 3- and 9-pulse input current waveforms

These families of harmonic frequencies are present in 3- and 6-pulse input current waveforms

These families of harmonic frequencies are present in the 3-pulse input current waveform

Characteristic cyctoconverter harmonic frequencies This family is present, independent of the pulse number of the current waveform

Figure 5.6   Chart showing the relationship between predominant harmonic frequencies present in a three-phase input current waveform of the cycloconverter with a balanced three-phase output, and the output-to-input frequency ratio. For input current waveforms with higher pulse numbers, certain harmonic families are eliminated as indicated. Source: Ref. [9].

5.3   ARC FURNACES

A schematic diagram of an EAF installation is shown in Fig. 5.7. The furnace is generally operated with static var compensation systems and passive shunt harmonic filters (Chapter 15). The installations can compensate rapidly changing reactive power demand, arrest voltage fluctuations, reduce flicker and harmonics, and simultaneously improve the power factor to unity. Typical harmonic emissions from IEEE Standard 519 are shown in Table 4.6. In practice, a large variation in harmonics is noted. For example, maximum to minimum limits of *voltage distortions* at second, third, and fourth harmonics may vary from 17% to 5%, 29% to 20%, and 7.5% to 3%, respectively. The tap-to-tap time (the time for one cycle operation, melting, refining, tipping, and recharging) may vary between 20 and 60 min depending on the processes, and

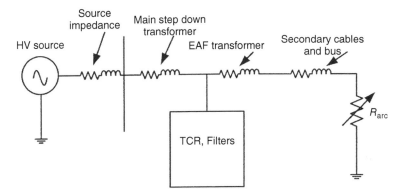

Figure 5.7   Schematic diagram of an EAF installation.

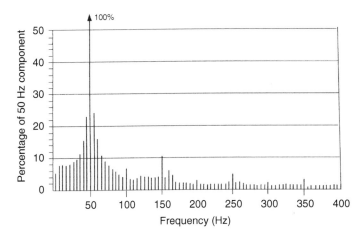

Figure 5.8   Typical spectrum of harmonic and interharmonic emissions from an EAF.

the furnace transformer is de-energized and then re-energized during this operation. This gives rise to additional harmonics during switching; saturation of transformer due to DC and second harmonic components, dynamic stresses, can bring about resonant conditions with improperly designed passive filters. New technologies such as STATCOM, see Section 5.8.1, and active filters can be applied.

Figure 5.8 shows a typical spectrum of harmonic and interharmonic emission from an arc furnace (50 Hz power supply frequency) and Fig. 5.9 is a plot of interharmonic emission from EAF [10].

A typical filter configuration to avoid magnifying interharmonics is illustrated in Fig. 5.10. The resistors provide damping to prevent magnification of interharmonics components. Type C filters are commonly employed (see Chapter 15 for the filter types).

Figure 5.11 is based on Ref. [11]. The second harmonic filter is essentially a type C filter (see Chapter 15). The resistor R_D remains permanently connected.

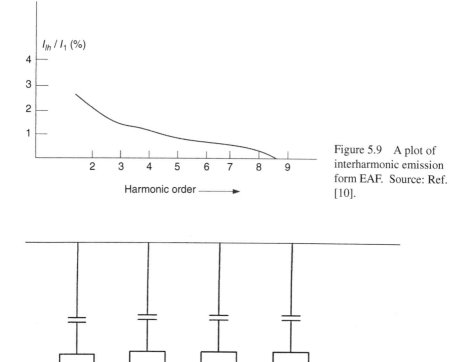

Figure 5.9   A plot of interharmonic emission form EAF. Source: Ref. [10].

Third order filter
$h_T = 3.1$

High pass filter
$h_T = 5.1$

High pass filter
$h_T = 7.0$

High pass filter
$h_T = 10.5$

Figure 5.10   A typical harmonic filter configuration to avoid magnification of interharmonics in EAF.

High damping is needed during transformer energization in order to reduce stresses on the elements of harmonic filters. This is achieved by connecting a low resistance R_{TS} in parallel with R_D during energization for a short time. The damping of transients becomes an important consideration, also see Chapter 16 for a case study.

The following harmonic restrictions from EAFs are from Ref. [12]. It assumes a PCC < 161 kV and SCR < 50.

- Individual integer harmonic components (even and odd) should be less than 2% of the specified demand current for the facility, 95% point on the cumulative probability distribution.

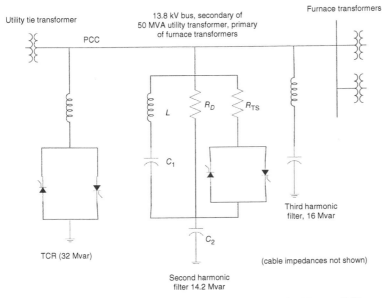

Figure 5.11   A configuration for harmonic emission control with type C filter and damping of transformer inrush current harmonics. Source: Ref. [11].

- Individual noninteger distortion components (interharmonics) should not exceed 0.5% of the specified demand current of the facility, 95% point on the cumulative probability distribution.

- The total demand distortion at the point of common coupling (PCC) should be limited to 2.5%, 95% point on the cumulative probability distribution. Additional restrictions should be applied to limit shorter duration harmonic levels, if it is determined that they could excite resonance in the power supply system or cause problem at local generators. This can be done by specifying separate limits that are only exceeded 1% of the time (usually not necessary).

5.3.1 Induction Furnaces

A system configuration of the induction furnace is shown in Fig. 5.12. It depicts a 12-pulse rectification with H-bridge inverter (see Chapter 6) and induction furnace load. The measurement data in this configuration are for 25-t, 12-MVA IMF (induction melting furnace) from Ref. [13]. A typical time-varying supply-side current during one melting cycle is shown in Fig. 5.13. The harmonics and interharmonics may interact with the industrial loads or passive filters and may be amplified by resonance with the passive filters. A model is generated, and the variable frequency operation is represented by a time-varying R–L circuit in parallel with a current source. The results of the measurements are shown in Fig. 5.14. Here, type A measurements show interharmonics due to cross modulation of the fundamental power supply frequency f and the inverter output frequency referred to the DC link, $2f_o$. Type B measurements are cross modulation of the fundamental frequency at the DC link at $2kf_o$, where f_o is

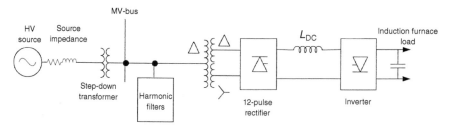

Figure 5.12   A system configuration for an induction furnace.

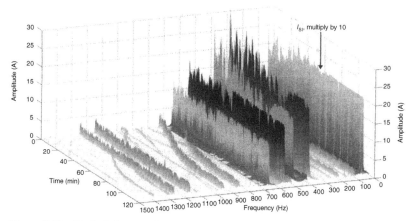

Figure 5.13   Typical time-varying supply-side current during melting cycle of a 12-MVA, 25-t induction furnace. Source: Ref. [13].

the output frequency of the inverter. Type C is the cross modulation of the harmonic current frequencies at $(12h \pm 1)f$ and the DC-link harmonic frequency $2f_o$.

5.4   EFFECTS OF INTERHARMONICS

The interharmonic frequency components greater than the power frequency produce heating effects similar to those produced by harmonics. The low-frequency voltage interharmonics cause significant additional loss in induction motors stator windings.

The impact on light flicker is important. Modulation of power system voltage with interharmonic voltage introduces variations in system voltage rms value.

The IEC flickermeter is used to measure the light flicker indirectly by simulating the response of an incandescent lamp and human-eye-brain response to visual stimuli, IEC Standard [14]. The other impacts of concern are as follows:

- Excitation of subsynchronous conditions in turbogenerator shafts.
- Interharmonic voltage distortions similar to other harmonics.
- Interference with low-frequency power line carrier control signals.

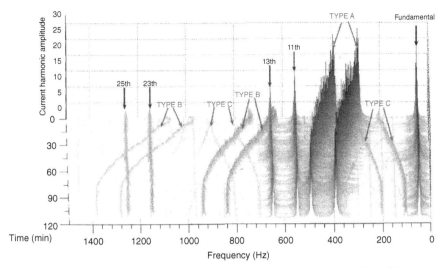

Figure 5.14   Interharmonics due to cross modulation, see text for types A, B, and C harmonic emissions shown in this figure. Source: Ref. [13].

- Overloading of conventional-tuned filters. See Chapter 15 for the limitations of the passive filters, the tuning frequencies, the displaced frequencies, and possibility of a resonance with the series-tuned frequency. As the interharmonics vary with the operating frequency of a cycloconverter, a resonance can be brought where none existed before making the design of single-tuned filters impractical.

The distortion indices for the interharmonics can be described similar to indices for harmonics. The total interharmonic distortion (THID) factor (voltage) is

$$\text{THID} = \frac{\sqrt{\sum_{i=1}^{n} V_i^2}}{V_1} \tag{5.20}$$

where i is the total number of interharmonics being considered including subharmonics, and n is the total number of frequency bins. A factor exclusively for subharmonics can be defined as total subharmonic distortion factor:

$$\text{TSHD} = \frac{\sqrt{\sum_{s=1}^{S} V_s^2}}{V_1} \tag{5.21}$$

An important consideration is that torsional interaction may develop at the nearby generating facilities (see Section 5.10). In this case, it is necessary to impose severe restrictions on interharmonic components. In other cases, interharmonics need not be treated any different from integer harmonics.

5.5   REDUCTION OF INTERHARMONICS

The interharmonics can be controlled by

- Higher pulse numbers
- DC filters, active or passive, to reduce the ripple content
- Size of the DC-link reactor
- Pulse width modulated drives

5.6   FLICKER

Voltage flicker occurs due to operation of rapidly varying loads, such as arc furnaces that affect the system voltage. This can cause annoyance by causing visible light flicker on tungsten filament lamps. The human eye is most sensitive to light variations in the frequency range 5–10 Hz and voltage variations of less than 0.5%, and this frequency can cause annoying flicker from tungsten lamp lighting.

5.6.1   Perceptible Limits

The percentage pulsation of voltage related to frequency, at which it is most perceptible, from various references, is included in Fig. 5.15 of Ref [15]. In this figure, the solid lines are composite curves of voltage flicker by General Electric Company (General Electric Review, 1925); Kansas Power and Light Company, Electrical World, May 19, 1934; T7D Committee EEI, October 14, 1934, Chicago; Detroit Edison Company; West Pennsylvania Power Company; and Public Service Company

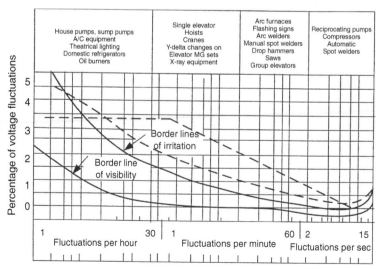

Figure 5.15   Maximum permissible voltage fluctuations see explanations in text. Source: Ref. [15].

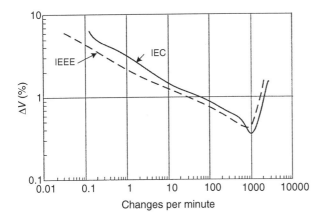

Figure 5.16   Comparison of IEC and IEEE standards with respect to flicker tolerance. Source: Ref. [16].

of North Illinois. Dotted lines show voltage flicker allowed by two utilities, reference Electrical World November 3, 1958, and June 1961. The flicker depends on the whole chain of "voltage fluctuations-luminance-eyes-brain."

Though this figure has been in use for a long time, it was superseded in IEEE Standard 1453, Ref [16] with Fig. 5.16. The solid-state compensators and loads may produce modulation of the voltage magnitude that is more complex than what was envisaged in the original flicker curves. This standard adopts IEC Standard 61000-3-3 [17] in total. Define

$$P_{lt} = \sqrt[3]{\frac{1}{12} * \sum_{j=1}^{12} P_{stj}^{3}} \qquad (5.22)$$

where P_{lt} is a measure of long-term perception of flicker obtained for a 2-h period. This is made up of 12 consecutive P_{st} values, where P_{st} is a measure of short-term perception of flicker for 10-min interval. This value is the standard output of IEC flickermeter. Further qualification is that IEC flickermeter is suited to events that occur once per hour or more often. The curves in Fig. 5.15 are still useful for infrequent events similar to a motor start, once per day, or even as frequent as some residential air conditioning equipment. Figure 5.16 depicts comparison of IEEE and IEC for flicker irritation.

The short-term flicker severity is suitable for accessing the disturbances caused by individual sources with a short duty cycle. When the combined effect of several disturbing loads operating randomly is required, it is necessary to provide a criterion for long-term flicker severity, P_{lt}. For this purpose, the P_{lt} is derived from short-term severity values over an appropriate period related to the duty cycle of the load, over which an observer may react to flicker.

For acceptance of flicker causing loads to utility systems, IEC standards [17–19] are recommended. The application of shape factors allows the effect of loads with voltage fluctuations other than the rectangular to be evaluated in terms of P_{st} values. Further research is needed to investigate effects of interharmonics on flicker and flicker transfer coefficients from HV to LV electrical power systems [20,21].

5.6.2   Planning and Compatibility Levels

Two levels: Planning level and compatibility levels are defined. Compatibility level is the specified disturbance level in a specified environment for coordination in setting the emission and immunity limits. Planning level, in a particular environment, is adopted as a reference value for limits to be set for the emissions from large loads and installations, in order to coordinate these limits with all the limits adopted for equipment intended to be connected to the power supply system.

As an example, planning levels for P_{st} and P_{lt} in MV (voltages > 1 kV and < 35 kV), HV (voltages > 35 kV and < 230 kV), and EHV (voltages > 230 kV) are shown in Table 5.2, and compatibility levels for LV and MV power systems are shown in Table 5.3.

5.6.3   Flicker Caused by Arcing Loads

Arc furnaces cause flicker because the current drawn during melting and refining periods is erratic and fluctuates widely and the power factor is low (Chapter 4). An EAF current profile during melting and refining is depicted in Fig. 5.17; see also Fig. 1.3 for the erratic nature of the current. Figure 5.18(a) shows flicker perception level P_{fs}, with respect to voltage variation, while Fig. 5.18(b) shows P_{st} for a certain source impedance, assumed constant, and Fig. 5.19 depicts P_{fs}, P_{st}, P_{lt}, and voltage variations.

There are certain other loads that can also generate flicker, for example, large spot welding machines often operate close to the flicker perception limits. Industrial processes may comprise a number of motors having rapidly varying loads or starting at regular intervals, and even domestic appliances such as cookers and washing machines can cause flicker on weak systems. However, the harshest load for flicker

TABLE 5.2   Planning Levels for Pst and Plt in MV, HV, and EHV Power Systems

	Planning Levels	
	MV	HV-EHV
Pst	0.9	0.8
Plt	0.7	0.6

Source: Ref. [16].

TABLE 5.3   Compatibility Levels for P_{st} and P_{lt} in LV and MV Systems

	Compatibility Level
Pst	1.0
Plt	0.8

Source: Ref. [16].

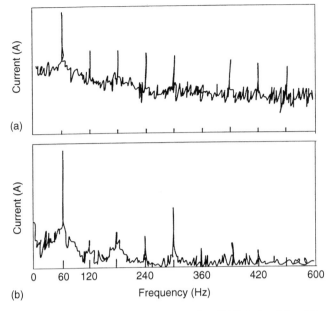

Figure 5.17   Erratic current spectrum of an EAF during (a) melting and (b) refining.

is an arc furnace. During the melting cycle of a furnace, the reactive power demand is high. Figure 5.17 shows that an arc furnace current is random and no periodicity can be assigned, yet some harmonic spectra have been established, Table 4.6, from IEEE 519. Note that even harmonics are produced during melting stage. The high reactive power demand and poor power factor causes cyclic voltage drops in the supply system. Reactive power flow in an inductive element requires voltage differential between sending end and receiving ends, and there is reactive power loss in the element itself. When the reactive power demand is erratic, it causes corresponding swings in the voltage dips, much depending on the stiffness of the system behind the application of the erratic load. This voltage drop is proportional to the short-circuit MVA of the supply system and the arc furnace load.

For a furnace installation, the short-circuit voltage depression (SCVD) is defined as

$$SCVD = \frac{2MW_{furnace}}{MVA_{SC}} \tag{5.23}$$

where the installed load of the furnace in MW is $MW_{furnace}$ and MVA_{SC} is the short-circuit level of the utility's supply system. This gives an idea whether potential problems with flicker can be expected. An SCVD of 0.02–0.025 may be in the acceptable zone, between 0.03 and 0.035 in the borderline zone, and above 0.035 objectionable [22]. When there are multiple furnaces, these can be grouped into one equivalent MW. A case study in Chapter 16 describes the use of tuned filters to compensate for the reactive power requirements of an arc furnace installation. The worst flicker occurs during the first 5–10 min of each heating cycle and decreases as the ratio of the solid to liquid metal decreases.

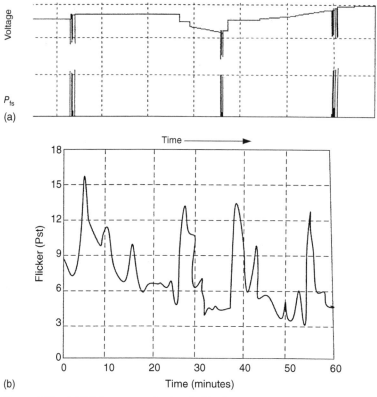

Figure 5.18 (a) Flicker perception level; (b) Measured short-term flicker of an EAF, source impedance considered time invariant.

The significance of $\Delta V/V$ and number of voltage changes are illustrated with reference to Fig. 5.20 from IEC [14]. This shows a 50-Hz waveform, having a 1.0 average voltage with a relative voltage change $\Delta v/\bar{v} = 40\%$ and with 8.8-Hz rectangular modulation. It can be written as

$$v(t) = 1 \times \sin(2\pi \times 50t) \times \left\{ 1 + \frac{40}{100} \times \frac{1}{2} \times \mathrm{signum}\,[2\pi \times 8.8 \times t] \right\} \qquad (5.24)$$

Each full period produces two distinct changes: one with increasing magnitude and one with decreasing magnitude. Two changes per period with a frequency of 8.8 Hz give rise to 17.6 changes per second.

5.7 FLICKER TESTING

The European test of flicker is designed for 230-V, 50-Hz power and the limits specified in IEC are based on the subjective severity of flicker from 230-V/60-W

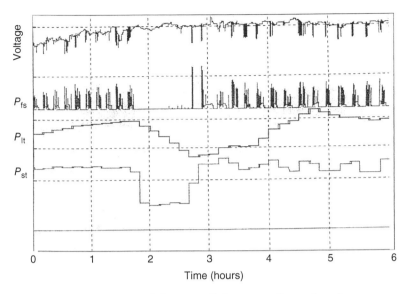

Figure 5.19   Measurement of short-term flicker at a medium voltage bus.

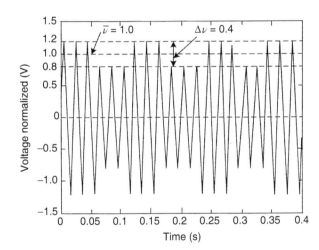

Figure 5.20   Modulation with rectangular voltage change $\Delta V/V = 40\%$, 8.8 Hz, 17.6 changes per second. Source: Ref. [14].

coiled–coil filament lamps and fluctuations of the supply voltage. In the United States, the lighting circuits are connected at 115–120 V. For a three-phase system, a reference impedance of $0.4 + j0.25$ ohms, line to neutral, is recommended, and IEC Standard 61000-3-3 is for equipments with currents ≤ 16 A. The corresponding values in the United States could be 32 A; a statistical data of the impedance up to the PCC (Point of Common Coupling) with other consumers is required. As for a 115-V system, the current doubles, the authors in Ref [23] recommend a reference impedance of $0.2 + j0.15$ ohms.

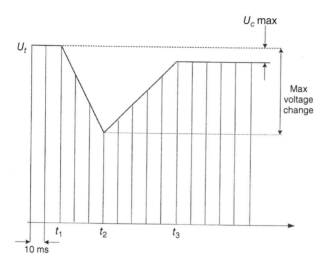

Figure 5.21   To explain the IEC terminology for the calculation of Pst.

An explanation of some terms used in IEC terminology is

$\Delta U(t)$ — Voltage change characteristics. The time function of change in rms voltage between periods when the voltage is in a steady-state condition for at least 1 s.

$U(t)$ — rms, voltage shape. The time function of the rms voltage evaluated stepwise over successive half-periods of the fundamental voltage.

ΔU_{max} — Maximum voltage change. The difference between maximum and minimum rms voltage change characteristics.

ΔU_c — Steady-state voltage change. The difference between maximum and minimum rms values of voltage change characteristics.

$d(t), d_{max}, d_c$ — Ratios of the magnitudes of $\Delta U(t), \Delta U_{max}, \Delta U_c$ to the phase voltage.

See Fig. 5.21.

The expression shown below is used for P_{st} based on an observation period $T_{st} = 10$ min.

$$P_{st} = \sqrt{0.0314P_{0.1} + 0.0525P_{1s} + 0.0657P_{3s} + 0.28P_{10s} + 0.08P_{50s}} \qquad (5.25)$$

where percentiles $P_{0.1}, P_{1s}, P_{3s}, P_{10s}, P_{50s}$ are the flicker levels exceeded for 0.1%, 1%, 3%, 10%, and 50% of the time during the observation period. The suffix "s" indicates that the smoothed value should be used, which is obtained as follows:

$$P_{50s} = (P_{30} + P_{50} + P_{80})/3$$

$$P_{10s} = (P_6 + P_8 + P_{10} + P_{13} + P_{17})/5$$

$$P_{3s} = (P_{2.2} + P_3 + P_4)/3$$

$$P_{1s} = (P_{0.7} + P_1 + P_{1.5})/3 \qquad (5.26)$$

5.8 CONTROL OF FLICKER

The response of the passive compensating devices is slow. When it is essential to compensate load fluctuations within a few milliseconds, SVCs are required. Referring to Fig. 4.25, large TCR flicker compensators of 200 MW have been installed for arc furnace installations. Closed-loop control is necessary due to the randomness of load variations, and complex circuitry is required to achieve response times of less than one cycle. Significant harmonic distortion may be generated, and harmonic filters will be required. TSCs (thyristor-switched capacitors [23]) have also been installed and these have inherently one cycle delay as the capacitors can only be switched when their terminal voltage matches the system voltage. Thus, the response time is slower. SVCs employing TSCs do not generate harmonics, but the resonance with the system and transformer needs to be checked.

If the voltage to the load is corrected fast and maintained constant, the flicker can be eliminated. Figure 5.22 shows an equivalent circuit diagram and the phasor diagram. The objective is to keep V_L constant by injecting variable voltage V_c to compensate for the voltage drop in the source impedance due to flow of highly erratic current I_s.

5.8.1 STATCOM for Control of Flicker

It has been long recognized that reactive power can be generated without use of bulk capacitors and reactors and STATCOM (static compensator) also called STATCON (static condenser) makes it possible. It is capable of operating with leading or lagging power factors. Its operation can be described with reference to Fig. 5.23, which depicts a synchronous voltage source. A solid-state synchronous voltage source abbreviated as SS is analogous to a synchronous machine and can be implemented with a voltage source inverter using GTOs (gate turn-off thyristors). The reactive power exchange between the inverter and AC system can be controlled by varying the amplitude of the three-phase voltage produced by the SS. Similarly, the real power exchange between the inverter and AC system can be controlled by phase shifting the output voltage of the inverter with respect to the AC system. Figure 5.23(a) shows a rotating synchronous condenser. No reversal of power is possible, it can absorb only limited capacitive current from the system, filed current cannot be reversed. Figure 5.23(b) shows a STATCOM, the coupling transformer, the inverter, and an energy source that can be a DC capacitor, battery, or superconducting magnet. The reactive and real power generated by the SS can be controlled independently and any combination of real power generation/absorption with var generation and absorption is possible, as shown in Fig. 5.23(c). The real power supplied/absorbed must be supplied by the storage device, while the reactive power exchanged is internally generated in the SS. This bidirectional power exchange capability of the

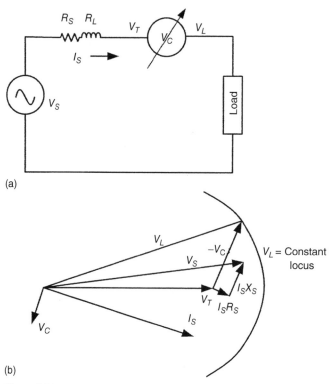

(a)

(b)

Figure 5.22   (a) Equivalent circuit diagram for compensation of load voltage; (b) phasor diagram.

SS makes complete temporary support of the AC system possible. STATCOM can be considered as an SS with a storage device as DC capacitor. A GTO-based power converter produces an AC voltage in phase with the transmission line voltage. When the voltage source is greater than the line voltage ($V_L < V_0$), leading vars are drawn from the line and the equipment appears as a capacitor; when voltage source is less than the line voltage ($V_L > V_0$), a lagging reactive current is drawn. Using the principle of harmonic neutralization, the output of n basic six-pulse inverters, with relative phase displacements, can be combined to produce an overall multiphase system. The output waveform is nearly sinusoidal and the harmonics present in the output voltage and input current are small, though not zero, see Table 6.1 for typical harmonic emission.

With design of high bandwidth control capability, STATCOM can be used to force three-phase currents of arbitrary wave shape through the line reactance. This means that it can be made to supply nonsinusoidal, unbalanced, randomly fluctuating currents demanded by the arc furnace. With a suitable choice of DC capacitor, it can also supply the fluctuating real power requirements, which cannot be achieved with SVCs.

The instantaneous reactive power on the source side is the reactive power circulating between the electrical system and the device, while reactive power on the

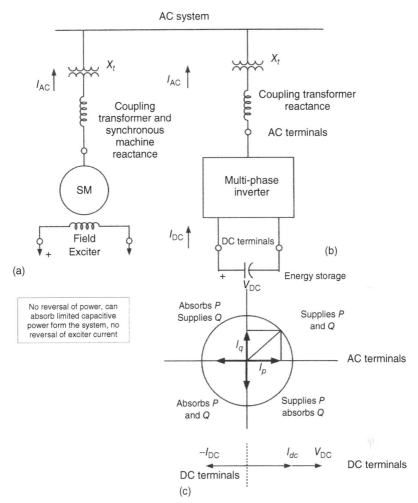

Figure 5.23 (a) A rotating condenser; (b) a shunt-connected synchronous voltage source; (c) possible modes of operation for real and reactive power generation.

output side is the instantaneous reactive power between the device and its load. There is no relation between the instantaneous reactive powers on the load and source side, and the instantaneous imaginary power on the input is not equal to the instantaneous reactive power on the output (Chapter 1). The STATCOM for furnace compensation may use vector control based on the concepts of instantaneous active and reactive powers, i_α and i_β (Chapter 1).

Figure 5.24 shows flicker reduction factor as a function of flicker frequency, STATCOM versus SVC [24]. Flicker mitigation with a fixed reactive power compensator and an active compensator - a hybrid solution for welding processes is described in [25]. Flicker compensation with series active filters (SAF) and parallel active filters is also applicable, Ref [26, 27] and Chapter 6. A combination of SAF and shunt

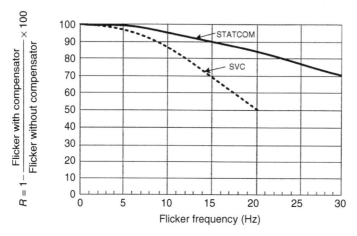

Figure 5.24   Flicker factor R for a STATCON and SVC.

passive filters is possible, in which SAF behaves like an isolator between the source and the load. A series capacitor can compensate for the voltage drop due to system impedance and fluctuating load demand, thus stabilizing the system voltage and suppress flicker and noise.

5.9   TRACING METHODS OF FLICKER AND INTERHARMONICS

Determining each source of flicker and ascertaining how much it contribute to the flicker at PCC has been studied by many authors [28–32].

5.9.1   Active Power Index Method

The interharmonic active power can be obtained by measurements of voltage and current at PCC:

$$P_{\text{IH}} = |V_{\text{IH}}||I_{\text{IH}}|\cos(\phi_{\text{IH}}) \qquad (5.27)$$

Consider that a number of consumers are connected to a source served by the utility. Representing each possible polluting source as a Norton equivalent circuit and carrying out the measurements on the feeder serving the consumer as well supply source, measurement at point A in Fig. 5.25, the polluting source, can be identified. Any of the sources can be the polluting source. For point A, if $P_{\text{IH}} > 0$, the interharmonic component is from the supply system, and if the measured $P_{\text{IH}} < 0$ the interharmonic component comes from the respective consumer.

Practically, the active power of the interharmonics is small, the angle ϕ_{IH} may be close to plus minus 90° and the measurements may oscillate. The interharmonic emissions may not be stable during the measurements.

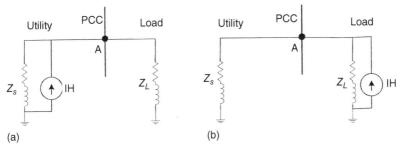

Figure 5.25 Determination of interharmonic source by Norton's equivalent: (a) interharmonic source at PCC is utility; (b) interharmonic source at PCC is the consumer load.

5.9.2 Impedance-Based Method

The harmonic impedance at the metering point can be obtained by

$$Z_{IH} = \left|\frac{V_{IH}}{I_{IH}}\right| \cos(\varphi_{IH}) + j\left|\frac{V_{IH}}{I_{IH}}\right| \sin(\varphi_{IH}) = R_{IH} + jX_{IH} \qquad (5.28)$$

The concept is that the interharmonic impedance is either upstream or downstream of the measuring point. The system impedance is generally much smaller than the load-side impedance, almost 1/5th of the source impedance. The magnitude of Z_{IH} can be checked if it is source-side or load-side impedance. The source impedance is given by the short-circuit calculations. It can be corrected for the interharmonic by multiplying with a factor of f_{IH}/f (neglecting resistance) and the load impedance by V_1/I_1.

If the source impedance after correction for the interharmonic frequency and Z_{IH} are not of the comparable values, the measured impedance is likely to be downstream impedance and the interharmonic source is the supply system.

5.9.3 Reactive Load Current Component Method

The active power and impedance-based methods focus on a major source of flicker, yet, how much each source contributes to flicker level at PCC is of much interest. The reactive load current component method is based on the concept that variation of the fundamental component of the voltage waveform in time causes the amplitude modulation affect which causes flicker. The background flicker contributed by the supply source adds to the flicker caused by loads at the PCC.

The system impedance and its angle are not constant quantities, and the values can be based on measurements.

Also source voltage cannot be assumed constant. Then, a time-varying relation can be written as

$$e_s = iR_s + L_s\frac{di}{dt} + v_{pcc} \qquad (5.29)$$

where v_{pcc} is the voltage at PCC.

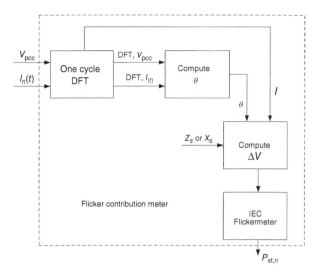

Figure 5.26   A block circuit diagram of individual flicker contribution meter. Source: Ref [30].

This can be approximated as

$$E_s \approx V_{PCC} + X_s I \sin \theta \tag{5.30}$$

where θ is the impedance angle.

A block circuit diagram of this method is shown in Fig. 5.26 and the IEC flickermeter block circuit diagram is shown in Fig. 5.27. Reference [30] illustrates flicker contribution measurements of some sample plants, such as steel plants and EAFs by the proposed flicker contribution meter.

Reverse power flow procedure to identify the source of harmonics can be used. Line and bus data at several points in the network are used with a least square estimator to calculate the injection spectrum at buses expected to be harmonic sources. When energy at harmonic frequencies is found to be injected into the network, then that bus is identified as a harmonic source.

5.10   TORSIONAL ANALYSIS

Torsional vibrations are responsible for failure of drive system components and can stress or shear the turbine blades in generating units. Figure 5.28(a) shows a simple torsional model in steady-state torqued condition at rest or at constant speed. The electrical torque and the load torque are constant and in balance. There is no relative motion between the masses, but there is a twist in the shaft, with a spring constant of K. Note the relative positions of the angles of twist.

If the steady-state torques were removed, the two inertias will vibrate about the zero torque axes (Fig. 5.28(b)). In the absence of any damping, these vibrations will continue with peak torque and twist equal to initial steady-state values. The resonant frequency will be given by

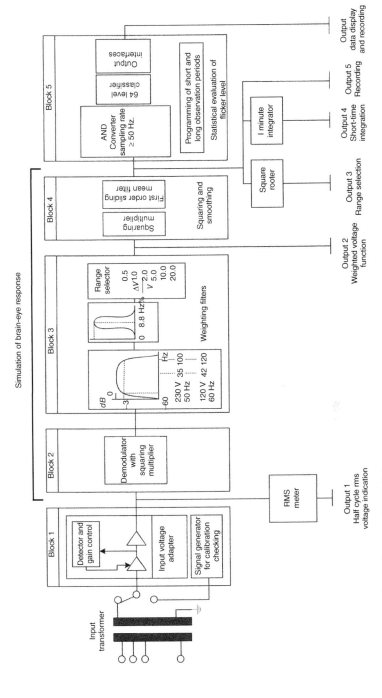

Figure 5.27   Block circuit diagram of IEC flickermeter.

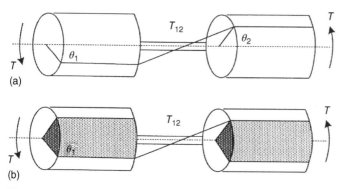

Figure 5.28   (a) Torsional model with two shaft-coupled rotating masses under steady state, constant torque; (b) forced oscillations with torque removed.

$$f_0 = \sqrt{\frac{K(J_1 + J_2)}{J_1 J_2}} \qquad (5.31)$$

The stored energy in the system is converted to kinetic energy two times per cycle, and the two inertias oscillate in opposition to each other. If one of the torques is removed, an oscillation with smaller amplitude will occur.

5.10.1   Steady-State Excitation

Consider that a steady-state excitation of frequency f_0 is applied to the system shown in Fig. 5.28(a). A torsional vibration will be excited and it may continue to grow in magnitude, till the energy loss per cycle is equal to the energy that the small disturbance adds to the system during a cycle. If the excitation frequency varies at a certain rate, the torsional vibrations will be amplified as the system passes through the resonant point [33–35].

In an ASD during normal operation, there are multiples of converter pulse outputs, that is, for a 12-pulse converter 12×, 24×, 36×, and 48× electrical output frequency. Figure 5.29 from Ref. [36] identifies where the excitation frequencies intersect with the torsional natural frequencies. This analysis is for a 15,000-hp, 6000-rpm synchronous motor drive in a petrochemical industry.

The driven load may have a positive slope, that is, the load increases with the speed. This occurs for fans and blowers. The load may have a negative slope, that is, conveyers and crushers. If the motor torque is removed, the negative load slope tends to give increasing torque pulsations.

An induction motor produces transient torques during starting. A synchronous motor, in addition to the initial fixed-frequency excitation such as an induction motor, produces a slip frequency excitation that varies from 120 Hz at starting to 0 Hz at synchronism. When a synchronous machine pulls out, it will produce a sinusoidal excitation at the pull out frequency, till it is disconnected or resynchronized. The critical speed with twice the slip frequency during starting cannot be avoided.

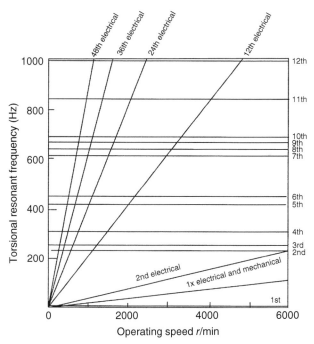

Figure 5.29   Torsional interference diagram of a 15000-hp synchronous motor. Source: Ref. [36].

The critical speed with the slip frequency excitation following pull out can be avoided, if the machine is de-energized in time.

5.10.2   Excitation from Mechanical System

There can be excitations from the mechanical system too, which are proportional to speed and may occur at a frequency of multiple of shaft revolution or integral multiple of gear tooth passing frequency. These are due to imperfections in the mechanical system and are generally of smaller magnitude. Excitations can also occur from the load system. A mechanical jam may produce severe dynamic torques.

It may be difficult to totally avoid some amplification of the torques during stating; however, it should show damping trend as the drive train quickly passes through the vibration mode. The torsional analysis requires host of motor and load data, starting characteristics, inertias, and spring constants. These are summarized in Table 5.4. A torsional analysis may discover many natural frequencies of the system.

Torsional analysis should also be carried out during starting and short circuit. Reference [37] illustrates that during starting of a 6000-hp induction motor drive for a high-speed compressor at 16 000 rpm, it passes through two resonant frequencies. The motor can even stall during acceleration, and its acceleration time will be higher as it passes through two low-level torque points during acceleration.

TABLE 5.4   Data Required for Torsional Vibration Analysis

Parameter	Description
M_m	Maximum transient shaft torque during starting
M_s	Breakaway torque refiner
F_1	Transferred thrust load to motor at zero end gap in thrust bearings, both directions
P_1	Power loss in the refiner during idling
Critical damping	Critical damping in the shaft system in %
Fatigue analysis	Data include shaft diameter, speed ratio, material, shear stress, and stress concentration factor due to step change in shaft diameter
J_1, J_2, J_3, J_4	Rotating inertias in kg-m^2 of lb ft^2
K_1, K_2, K_3	Spring constants, Nm/rad or lb-in/rad
Motor	Starting torque–speed characteristics, average and oscillating torques, effect of variation of system voltage and starting conditions
Load	Starting torque–speed characteristics

To illustrate the impact of harmonics on starting a 3000-hp, 4.16-kV, single-cage induction motor, connected to a step-down transformer is started under two conditions:

- Normal balanced power supply conditions, devoid of any harmonics.
- The power supply system polluted with seventh harmonic.

Figure 5.30(a) shows normal torque–speed starting curve, while Fig. 5.30(b) shows starting with supply system polluted with seventh harmonics. This shows that though the motor is able to start, serious torque oscillations continue to occur due to harmonic torque. These can be much damaging to the motor shaft. The motor is started with low inertial load to reduce starting time.

5.10.3   Analysis

An n-spring connected rotating masses can be described by the equations:

$$J\frac{d\overline{\omega}_m}{dt} + \overline{D}\overline{\omega}_m + \overline{H}\,\overline{\theta}_m = \overline{T}_{\text{turbine}} - \overline{T}_{\text{generator}} \qquad (5.32)$$

$\overline{H}$ is diagonal matrix of stiffness coefficients, $\overline{\theta}_m$ is the vector of angular positions, $\overline{\omega}_m$ is the vector of mechanical speeds, $\overline{T}_{\text{turbine}}$ is the vector of torques applied to turbine stages, and $\overline{D}$ is the diagonal matrix of damping coefficients. The moment of inertia and damping coefficients are available from design data. The spring action creates a torque proportional to the angle twist.

Figure 5.31(a) shows a torsional system model for the steam turbine generator. The masses will rotate at one or more of the turbine mechanical natural frequencies called torsional mode frequencies. When the mechanical system oscillates under such

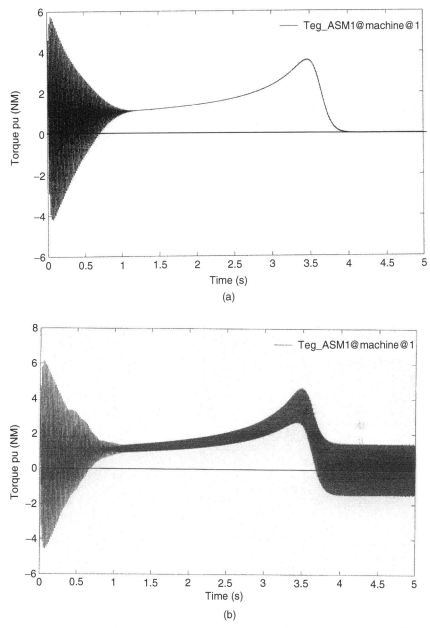

Figure 5.30   (a) Simulated normal starting torque–speed characteristics of a 4.16-kV, 3000-hp, single-cage induction motor from balanced supply system; (b) starting characteristics with supply system polluted with 7th harmonic, EMTP simulations.

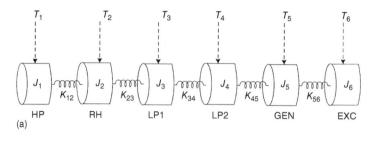

(a)

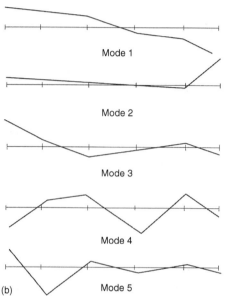

(b)

Figure 5.31   (a) Rotating mass model of steam turbine generator; (b) oscillation modes. Source: Ref. [38].

steady state at one or more natural frequencies, the relative amplitude and phase of individual turbine-rotor elements are fixed and are *called mode shapes of torsional motion*, Fig. 5.31(b) [38].

Torsional mode damping quantifies the decay of torsional oscillations. The ratio of natural log of the successive peaks of oscillation is called *logarithmic decrement*. The decrement factor is defined as the time in seconds to decay from the original point to $1/e$ of its value.

The modal spring–mass model is a mathematical representation of Fig. 5.31 for oscillation in mode n given by

$$
\begin{vmatrix} J_1 & & & \\ & J_2 & & \\ & & \cdot & \\ & & & J_n \end{vmatrix} \begin{vmatrix} \ddot{\theta}_1 \\ \ddot{\theta}_2 \\ \cdot \\ \ddot{\theta}_n \end{vmatrix} + \begin{vmatrix} K_{12} & -K_{12} & & \\ -K_{12} & K_{12} & K_{23} & -K_{23} \\ & -K_{23} & \cdot & \cdot & -K_{n1,n} \\ & & \cdot & \\ & & -K_{n1,n} & K_{n1,n} \end{vmatrix} \begin{vmatrix} \theta_1 \\ \theta_2 \\ \cdot \\ \cdot \\ \theta_n \end{vmatrix} = \begin{vmatrix} T_1 \\ T_2 \\ \cdot \\ \cdot \\ T_n \end{vmatrix}
$$

(5.33)

The derivation follows from eigenvectors and frequencies of the spring–mass model. It is seen that there are *n* second-order differential equations of motion for an *n* mass model and coupled to one another by spring elements.

Diagonalization of the stiffness term would yield *n* decoupled equations called the *modal spring–mass models.* This diagonalization can be accomplished by coordinate transformation from a reference frame in the rotors to a reference frame of the eigenvectors.

5.11 SUBSYNCHRONOUS RESONANCE

We have defined the SSR in the opening paragraph of this chapter. The exchange of the energy with a turbine generator takes place at one or more of the natural frequencies of the combined system, and these frequencies are below the synchronous frequency of the system.

The turbine generator shaft has natural modes of oscillations, which can be at subsynchronous frequencies. If the induced subsynchronous torque coincides with one of the shaft natural modes of oscillation, the shaft will oscillate at this natural frequency, sometimes with high amplitude. This may cause shaft fatigue and possible failure. The interactions can be caused by

1. *Induction generator effect:* The resistance of rotor to subsynchronous currents is negative and network presents a resistance that is positive. If the negative resistance of the generator is greater than the positive system resistance, there will be sustained subsynchronous currents.

2. Torsional interaction has been described earlier.

3. Transient torques that result from a system disturbance cause changes in the network, resulting in sudden changes in the current that will oscillate at the natural frequency of the network.

The series compensation of the transmission lines is the most common cause of subsynchronous resonance.

5.11.1 Series Compensation of Transmission Lines

Series compensation of HV transmission lines is used for (1) voltage stability, as it reduces the series reactive impedance to minimize the receiving end voltage variations and the possibility of voltage collapse, (2) improvement of transient stability by increasing the power transmission by maintaining the midpoint voltage during swings of the machines, and (3) power oscillation damping by varying the applied compensation so as to counteract the accelerating and DC-accelerating swings of the machines. A fixed type of series compensation can, however, give rise to subsynchronous oscillations as we will discuss.

An implementation schematic of the series capacitor installation is not shown here, see Ref. [38].

A series capacitor has a natural resonant frequency given by

$$f_n = \frac{1}{2\pi\sqrt{LC}} \qquad (5.34)$$

f_n is usually less than the power system frequency. At this frequency, the electrical system may reinforce one of the frequencies of the mechanical resonance, causing *subsynchronous resonance* (SSR). If f_r is the subsynchronous resonance frequency of the compensated line, then at resonance

$$2\pi f_r L = \frac{1}{2\pi f_r C}$$

$$f_r = f\sqrt{K_{sc}} \qquad (5.35)$$

This shows that the subsynchronous resonance occurs at frequency f_r, which is equal to normal frequency multiplied by the square root of the degree of compensation, it is typically between 15 and 30 Hz. As the compensation is in 25-75% range, f_r is lower than f. The transient currents at subharmonic frequency are superimposed upon power frequency component and may be damped out within a few cycles by the resistance of the line. Under certain conditions, subharmonic currents can have a destabilizing effect on rotating machines. If the electrical circuit oscillates, then the subharmonic component of the current results in a corresponding subharmonic field in the generator. This field rotates backward with respect to the main field and produces an alternating torque on the rotor at the difference frequency $f - f_r$. If the mechanical resonance frequency of the shaft of the generator coincides with this frequency, damage to the generator shaft can occur. A dramatic voltage rise can occur if the generator quadrature axis reactance and the system capacitive reactance are in resonance. There is no field winding or voltage regulator to control quadrature axis flux in a generator. Magnetic circuits of transformers can be driven to saturation and surge arresters can fail. The inherent dominant subsynchronous frequency characteristics of the series capacitor can be modified by a parallel-connected TCR.

If the series capacitor is thyristor or GTO controlled (TCSC), then the whole operation changes. It can be modulated to damp out any subsynchronous as well as low-frequency oscillations. Thyristor-controlled series capacitors have been employed for many HVDC projects.

5.11.2　Subsynchronous Resonance HVDC Systems

Subsynchronous resonance can occur in HVDC systems due to interaction between oscillations in transmission systems and mechanical torsional vibrations in generator turbine set. This is mainly brought out by *negative damping* in HVDC control loop. By designing HVDC controls with positive damping, the situation can be avoided. This torsional interaction is significant near the converter substations and is negligible

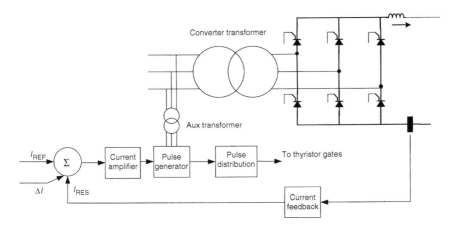

Figure 5.32   Control circuit diagram for HVDC IPC control.

for generators away from the converter stations. The negative damping increases with increased HVDC power flow and increased delay angle control of the thyristors. The short-circuit levels in the AC system have an impact - the higher short-circuit levels have higher damping effects.

The firing angle control system can include a subsynchronous damping controller to secure positive damping. It detects torsional mode of oscillations in rotational velocity of generator by frequency modulation of converter AC voltage. The torsional-mode oscillations are counteracted by the modulation of converter firing angles.

The AC- and DC-side harmonics are controlled by having AC and DC filters (Chapter 15). The harmonic voltages in AC systems are of positive and negative sequences and have three-phase unbalance. At a harmonic resonance, the harmonic voltages can be magnified. There are two methods of firing angle controls:

- Individual phase control (IPC)
- Equidistant phase control (EPC)

Individual phase control is not much in use now. The control pulses are derived from commutation voltage. As discussed in Chapter 4, the start of conduction of individual thyristors is delayed with respect to phase angle of zero crossing. The control circuit diagram is shown in Fig. 5.32. The control function (say V_{CF}) is derived from the reference current I_{REF}, current margin ΔI, and feedback current I_{RES} (I response). It is seen that the instant of control pulse and the firing delay angle α depends on the phase voltage derived from the auxiliary transformer and the control function, V_{CF}. In this method, the distortion in the AC supply waveform can cause variation of firing angle α and lead to instability.

In EPC, the pulses are derived from a pulse generator at a frequency of $6f$ (six-pulse converter) or $12f$ (12-pulse converter) where f is the fundamental frequency. These pulses are separated in a pulse distribution unit and applied to

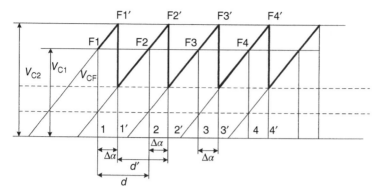

Figure 5.33 Control circuit operation of Fig. 5.32.

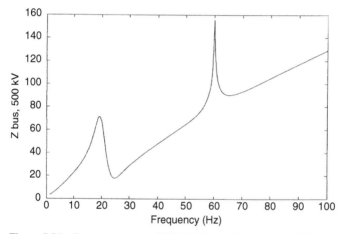

Figure 5.34 Frequency scan of 500-kV bus at the secondary (500 kV) of the transformer, EMTP simulation.

individual thyristors. If the power source frequency is considered stable and constant, the control pulses are equidistant with constant frequency. The pulses are delivered to converter via a ring counter, which has required number of stages (6 or 12) with only one stage active at any time. The stages are sequentially switched giving a short output pulse, one per cycle. For a 12-pulse converter, the pulses are obtained at an interval of $2\pi/12$.

Figure 5.33 illustrates that the control function V_{CF} are pulses at a constant slope, generated at the intersection of the controller voltage V_C. These points of intersections are marked F1, F2, ... for V_{C1} and F1′, F2′, ... for the voltage V_{C2}. The distance d or d' between consecutive control pulses in same and determined by the slope of the control function V_{CF}. This control function ramp is selected so that pulse interval is exactly $2\pi f/p$, where p is the pulse number of converter. If the control function is increased from V_{C1} to V_{C2}, the points of intersections are shifted and the firing angle is increased by $\Delta\alpha$.

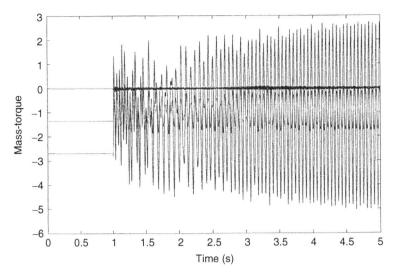

Figure 5.35   Shaft torque transients, mass 1, EMTP simulation (Example 5.1).

Example 5.1: Consider a 600-MVA, 22-kV generator connected to a step-up transformer of 600 MVA, delta–wye connected, 22–500 kV, wye windings solidly grounded, which feeds into a 400-mile-long 500-kV line. A CP model of the transmission line is modeled in EMTP. A series capacitor compensation of 50% at the terminal point of the transmission line is provided. For subsynchronous oscillations, the shaft mass system of steam turbine generators is modeled with four masses of certain inertia constants connected together through spring constants (HP and LP sections of turbine, rotor, and exciter). The line serves receiving end loads. External torques can be applied to each of the masses, for example, turbine, generator, and exciter masses. An EMTP simulation of the frequency scan at the 500-kV side of the step-up transformer is shown in Fig. 5.34. This shows one resonance at 19 Hz and the other close to the fundamental frequency. A three-phase fault occurs at the secondary of the transformer at 1 s and cleared at 1.1 s, fault duration = 6 cycles. The resulting torque transients in the 500-MVA synchronous machine mass 1 are shown in Fig. 5.35, with a total simulation time of 5 s. It is seen that these transients do not decay even after 5 s and diverge, imposing stresses on the generator shaft and mechanical systems. The angular frequency of mass 1 (zero external torque that will give maximum swings) is plotted in Fig. 5.36. This shows violent speed variations. The frequency relays or vibration probes may isolate the generator from the system. The generator parameters for the EMTP model are as follows:

  Field current at rated voltage = 1200A, $R_a = 0.0045, X_0 = 0.12, X_d = 1.65, X'_d = 0.25, X''_d = 0.20, X'_q = 0.46, X''_q = 0.20$ all in pu.
  $T'_{qo} = 0.55, T''_{qo} = 0.09, T_{do} = 4.5, T''_{do} = 0.04$ all in seconds.
  The generator is modeled with AVR and PSS (power system stabilizer). The transformer is rated 22–500 kV, 600 MVA, %Z = 10% (also see Ref [38]).

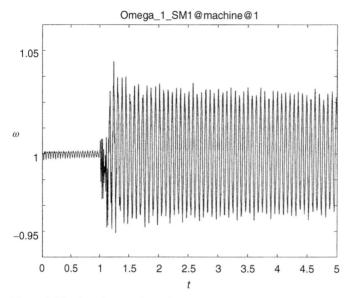

Figure 5.36   Angular speed transients, mass 1, with no external torque, EMTP simulation (Example 5.1).

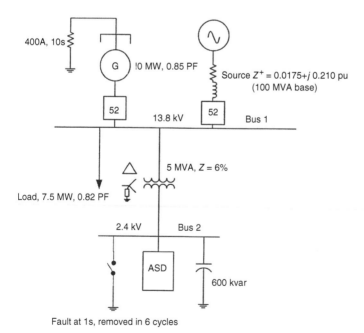

Figure 5.37   A circuit configuration for study of subsynchronous oscillation due to an ASD cascade (Example 5.2).

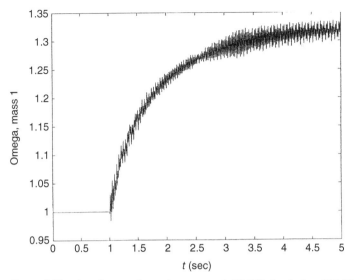

Figure 5.38   Angular speed transients, mass 1, EMTP simulation, 10 MW generator operating alone (Example 5.2).

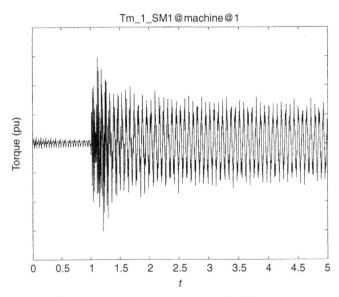

Figure 5.39   Shaft torque transients, mass 1, 10-MW generator operating alone, EMTP simulation (Example 5.2).

The NGH-SSR (after Narain Hingorani Subsynchronous Resonance Suppressor), Ref. [39,40], scheme can minimize subsynchronous electrical torque and hence mechanical torque and shaft twisting, limit build up of oscillations due to subsynchronous resonance, and protect series capacitors from overvoltages, not discussed here.

5.11.3 Subsynchronous Resonance Drive Systems

In Section 5.2.2, the interharmonics due to drive systems are discussed. In this section, we stated that for inverter frequencies of 25, 37.5, and 48 Hz and source frequency of 60 Hz, the side band pairs are 10 and 110, 15 and 135 Hz, and 36 and 156 Hz, respectively. These can create subsynchronous resonance, though a number of conditions and parameters must coincide for such an event.

Example 5.2: To illustrate subsynchronous resonance in an ASD, an EMTP simulation of the simple drive system shown in Fig. 5.37 is carried out. A 10-MVA generator supplies loads connected to its bus and may operate in synchronism with utility source. It supplies 12-pulse ASD load connected through a step-down transformer of 5 MVA. To compensate the load voltage dip at 2.4-kV bus, a 600-kvar capacitor bank is provided. Characteristic harmonics of the order of 11th and 13th are modeled. Also pair of harmonics 36 and 156 are modeled. The turbine generator

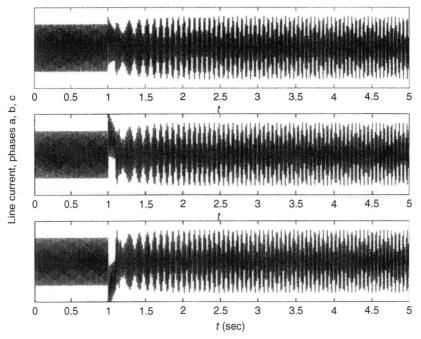

Figure 5.40   Transient line currents of 10-MW generator in three phases, EMTP simulation (Example 5.2).

train is modeled with four rotating masses and spring constants. The disturbance is modeled as a three-phase fault on 2.4-kV bus at 1 s, cleared in six cycles.

The resonant frequency of the system as calculated on 13.8-kV bus 1 occurs at 45.8 Hz when the load is entirely supplied by the generator. When the generator is operating in synchronism with the utility, there is only a small shift in the resonant frequency.

Figures 5.38 and 5.39 illustrate the speed and torque oscillations of mass 1, respectively. The transients have only decayed slightly over a period of 5 s. When the generator is operated in synchronism with the utility, the torque oscillations of mass 1 are slightly lower. Figure 5.40 shows the line current oscillations of the generator. Ref. [41] provides further reading.

REFERENCES

1. IEC Standard 61000-2-1, Part 2. Part 2, Environment, Section 1. description of the environment. electromagnetic environment for low frequency conducted disturbances and signaling in public power supply systems. 1990.
2. IEEE Task Force on Harmonic Modeling and Simulation, "Interharmonics: Theory and modeling," IEEE Transactions on Power Delivery, vol. 22, no. 4, pp. 2335–2348, 2007.
3. M. B. Rifai, T. H. Ortmeyer and W. J. McQuillan, "Evaluation of current interharmonics from AC drives," IEEE Transactions on Power Delivery, vol. 15, no. 3, pp. 1094–1098, 2000.
4. R. Yacamini, "Power system harmonics-Part 4, Interharmonics," IEE Power Engineering Journal, pp. 185–193, 1996.
5. J. D. Anisworth. "Non-characteristic frequencies in AC/DC converters," International conference on Harmonics in Power Systems, UMIST, pp. 76–84, Manchester 1981.
6. L. Hu and R. Yacamini, "Harmonic transfer through converters and HVDC links," IEE Transactions, vol. PE-7, no. 3, pp. 514–525, 1992.
7. L. Hu and L. Ran, "Direct method for calculation of AC side harmonics and interharmonics in an HVDC system" IEE Proceedings on Generation Transmission and Distribution, vol 147, no. 6, pp. 329–335, 2000.
8. L. Hu, R. Yacamani, "Calculation of harmonics and interharmonics in HVDC systems with low DC side impedance," IEE Proceedings-C, vol. 140, no.6, pp. 469–476, 1993.
9. B. R. Pelly. Thyristor Phase-Controlled Converters and Cycloconverters, Operation Control and Performance. John Wiley, New York, 1971.
10. E.W. Gunther, "Interharmonics recommended updates to IEEE 519," IEEE Power Engineering Society Summer Meeting, pp. 950–954, July 2002.
11. C. O. Gercek, M. Ermis, A. Ertas, K. N. Kose, and O. Unsar, "Design implementation and operation of a new C-type 2nd order harmonic filter for electric arc and ladle furnaces," IEEE Transactions on Industry Applications, vol. 47, no. 4, pp. 1545–1557, 2011.
12. IEEE P519.1. Draft guide for applying harmonic limits on power systems, 2004.
13. I. Yilmaz, Ö. Salor, M. Ermiş, I. Çadirci. "Field-data-based modeling of medium –frequency induction melting furnaces for power quality studies," IEEE Trans. Industry Applications, vol. 48, no. 4, pp. 1215–1224, 2012.
14. IEC Standard 61000-4-15. Part 4–15. Testing and measurement techniques; Flicker meter-Functional and design specifications, 2010.
15. IEEE Standard 519. IEEE recommended practices and requirements for harmonic control in power systems, 1992.

16. IEEE Standard 1453. Recommended practice for measurement and limits of voltage fluctuations and associated light flicker on AC power systems, 2004.

17. IEC Standard 6100-3-3. Electromagnetic compatibility (EMC) part 3–3: Limits-Section 3. Limitations of voltage changes voltage fluctuations and flicker in public low-voltage supply systems, for equipment with rated current ≤ 16A per phase and not subjected to conditional connection, 2008.

18. IEC Standard 61000-3-8. Electromagnetic compatibility (EMC) part 3–8: Limits-Section 8. signaling on low-voltage electrical installations—emission levels, frequency bands and electromagnetic disturbance levels, 1997.

19. IEC Standard 61000-3-11. Electromagnetic compatibility (EMC) part 3-11: Limits-limitations of voltage fluctuations and flicker in low-voltage power supply systems for equipment with rated current ≤ 75A, 2000.

20. S. M. Halpin and V. Singhvi, "Limits of interharmonics in the 1–100 Hz range based upon lamp flicker considerations," IEEE Transactions on Power Delivery, vol.22, no.1, pp. 270–276, 2007.

21. M. Göl et al. "A new field data –based EAF model for power quality studies," IEEE Transactions on Industry Applications, vol. 46, no. 3, pp. 1230–1241, 2010.

22. S. R. Mendis and D. A. González. "Harmonic and transient overvoltages analyses in arc furnace power systems," IEEE Transactions on Industry Applications, vol. 28, no. 2, pp. 336–342, 1992.

23. R.W. Fei, J.D. Lloyd, A.D. Crapo, and S. Dixon. "Light flicker test in the United States," IEEE Trans. Industry Applications, vol. 36, no. 2, pp. 438-443, March/April 2000.

24. C. D. Schauder and L. Gyugyi, "STATCOM for electric arc furnace compensation," EPRI Workshop, Palo Alto, 1995.

25. M. Routimo, A. Makinen, M. Salo, R. Seesvuori, J. Kiviranta, and H. Tuusa, "Flicker mitigation with hybrid compensator," IEEE Transactions on Industry Applications, vol. 44, no. 4, pp. 1227–1238, 2008.

26. A. Nabae and M. Yamaguchi. "Suppression of flicker in an arc furnace supply system by an active capacitance-A novel voltage stabilizer in power systems," IEEE Transactions on Industry Applications, vol. 31, no. 1, pp. 107–111, 1995.

27. F. Z. Peng et al., "A new approach to harmonic compensation in power systems—a combined system shunt and series active filters," IEEE Transactions on Industry Applications, vol. 26, no. 6, pp. 983–990, 1990.

28. D. Zhang, W. Xu, and A. Nassif, Flicker source identification by interharmonic power direction, in Proceedings of Canadian Conference on Electrical and Computer Engineering, pp. 549–552, May1–4, 2005.

29. P. G. V. Axelberg, M. H. J. Bollen, and I. Y. H Gu, "Trace of flicker sources by using the quantity of flicker power," IEEE Transactions on Power Delivery, vol. 23, no.1, pp. 465–471, 2008.

30. S. Perera, D. Robinson, S. Elphick, D. Geddy, N. Browne, V. Smith, and V. Gosbell. "Synchronized flicker measurements for flicker transfer evaluation in power Systems," IEEE Transactions on Power Delivery, vol. 21, no. 3, pp. 1477–1482, 2006.

31. E. Altintas, O. Sailor, I. Cadirci, and M. Ermis, "A new flicker tracing method based on individual reactive current components of multiple EAFs at PCC," IEEE Transactions on Industry Applications, vol. 46, no. 5, pp. 1746–1754, 2011.

32. T. Heydt. "Identification of harmonic sources by a state estimation technique," IEEE Transactions on Power Delivery, vol. 4, no. 1, pp. 569–576, 1989.

33. E. L. Owen, "Torsional coordination of high speed synchronous motors, Part1," IEEE Transactions on Industry Applications, vol. 17, pp. 567–571, 1981.

34. E. L. Owen, H. D. Snively and T. A. Lipo, "Torsional coordination of high speed synchronous motors, Part1," IEEE Trans. Industry Applications, vol. 17, pp. 572–580, 1981.

35. C. B. Meyers, "Torsional vibration problems and analysis of cement industry drives," IEEE Transactions on Industry Applications, vol. 17, no. 1, pp. 81–89, 1981.

36. B. M. Wood, W. T. Oberle, J. H. Dulas, and F. Steuri, "Application of a 15,000-hp, 6000 r/min adjustable speed drive in a petrochemical facility," IEEE Transactions on Industry Applications, vol. 31, no. 6, pp. 1427–1436, 1995.

37. W. E. Lockley, T. S. Driscoll, W. H. Wharran, and R. H. Paes. "Harmonic torque considerations of applying 6000-hp induction motor and drive to a high speed compressor," IEEE Transactions on Industry Applications, vol. 31, no. 6. pp. 1412–1418, 1995.

38. J. C. Das. Transients in Electrical Systems Analysis Recognition and Mitigation. McGraw-Hill, New York, 2010.

39. N. G. Hingorani. A new scheme for subsynchronous resonance damping of torsional oscillations and transient torques—Part I IEEE PES summer meeting, Paper no. 80 SM687-4, Minneapolis, 1980.

40. N. G. Hingorani, K. P. Stump. A new scheme for subsynchronous resonance damping of torsional oscillations and transient torques—Part II IEEE PES summer meeting, Paper no. 80 SM688-2, Minneapolis, 1980.

41. J. C. Das, "Subsynchronous resonance-series compensated HV lines and converter cascades," International Journal of Engineering Applications, vol. 2, no.1, pp. 1–10, 2014.

CHAPTER 6

HARMONIC REDUCTION AT THE SOURCE

Many technologies are available and constantly being advanced to control the harmonics at source. Where the harmonic levels are low, the equipment can be designed to withstand the effect of harmonics, for example, transformers, cables, and motors can be derated. Motors for PWM inverters can be provided with special insulation to withstand high du/dt, and the relays can be rms sensing. This may not be cost effective in most cases, and the deleterious effects of harmonics cannot be entirely mitigated. Also the harmonic injection at the PCC (Point of Common Coupling, Chapter 10) may exceed IEEE limits. Some form of harmonic mitigation to control harmonic injection at PCC is often required, unless the harmonic loads are small (see Chapter 10 for a calculation). Other than that, there are three major methodologies for limitation of harmonics:

1. Passive filters at suitable locations, preferably close to the source of harmonic generation can be provided so that the harmonic currents are trapped at the source and the harmonics propagated in the system are reduced. For large harmonic producing loads, that is, HVDC systems, FACTS controllers, SVCs, TCRs, and passive filters are commonly applied.

2. Active filtering techniques, generally, incorporated with the harmonic producing equipment itself can reduce the harmonic generation at source. Hybrid combinations of active and passive filters are also a possibility.

3. Alternative technologies can be adopted to limit the harmonics at source, for example, phase multiplication, operation with higher pulse numbers, converters with interphase reactors, active wave-shaping techniques, multilevel converters, and harmonic compensation built into the harmonic producing equipment itself to reduce harmonic generation.

The most useful strategy in a given situation largely depends on the currents and voltages involved, the nature of loads, and the specific system parameters, for example, short-circuit level at the PCC.

Whatever technology or option is adopted, the harmonic emissions at PCC must meet the requirements of standards.

Power System Harmonics and Passive Filter Designs, First Edition. J.C. Das.
© 2015 The Institute of Electrical and Electronics Engineers, Inc. Published 2015 by John Wiley & Sons, Inc.

6.1 PHASE MULTIPLICATION

Section 4.4.1 showed that by proper choice of winding connections of the input three-phase transformers, 12-pulse operation can be obtained, and the harmonics will be of the order of $12n \pm 1$, though the cancellation of the lower order harmonic pairs 5th and 7th, 11th and 13th, 17th and 19th, ... will not be perfect.

Figure 4.8 shows that two separate three-phase transformers with 30° phase shift are connected to the same bus. This is not desirable because harmonics of the order fifth, seventh, ... can circulate through the transformer primary windings through the bus connection, and depending on the impedance of the transformers to these harmonics, these harmonic circulating currents can be excessive and should be avoided (though these harmonics fifth, seventh, ... will not appear in the service lines feeding the bus). From this consideration, the transformer windings can be connected in series, which require special nonstandard transformers.

Alternatively, a three-winding transformer of appropriate winding connections can be used.

The concept of 12-pulse operations can be extended to provide 18-, 24-, or 48-pulse operation by appropriate phase shifts in the transformer windings. For example, if we have 7.5° phase shift on the two transformers of one 12-pulse converter and −7.5° phase shift on the two transformers of the second 12-pulse converter, a 24-pulse operation will be obtained.

Two examples of phase multiplication are described:

1. An example of phase multiplication is discussed in Section 5.8.1 in connection with the application of STATCOM for flicker control. Its power circuit is essentially shown in Fig. 6.1; and Fig. 6.2 shows the output voltage and current waveform of a 48-pulse STATCOM generating reactive power, which are almost sinusoids, Ref. [1]. A STATCOM of ± 100 Mvar is installed at TVA system at Sullivan [2]. It has eight (8) basic six-pulse, three-phase converters, giving 48 pole operations. The harmonic emission from STATCOM is small due to phase multiplication. Table 6.1 shows the harmonic emission from a STATCOM based on a vendor's data.

2. A multicell PWM VSI is shown in Fig. 6.3. Multistage PWM inventors may consist of many cells. Details of a cell with three-phase input and single-phase output are shown in Fig. 6.4. In Fig. 6.3, there are three such cells per phase. The drive input transformer has nine isolated secondaries. A greatly improved voltage waveform is obtained due to phase displacements in the transformer secondary windings, and the harmonic distortion levels with the drive operating in isolation may meet the IEEE 519 [3] requirements of harmonic emission without additional filters.

6.2 VARYING TOPOLOGIES

Chapter 5 shows that for a given application, it is possible to select a topology that will give reduced harmonic emission. For a 1000-hp drive system, it is possible to select

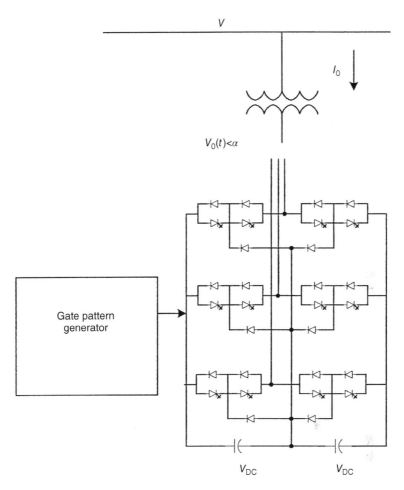

Figure 6.1   Three-phase, three-level, 12-pulse bridge, control logic not fully shown.

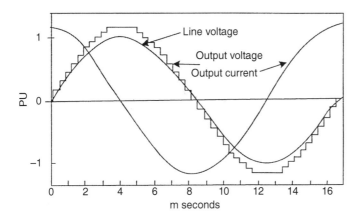

Figure 6.2   Fortyeight-pulse operation for harmonic emission control in a STATCOM.

TABLE 6.1 Typical Harmonic Emission from STATCOM (Based on a
Manufacturer's Data)

Harmonic Order	Percentage of Var Output	Harmonic Order	Percentage of Var Output
3	1.84	19	0.20
5	3.89	41	0.20
6	0.20	42	0.20
7	2.05	46	0.20
9	0.2	48	0.20
11	0.41	49	0.20
13	0.20		
17	0.20		

6-pulse, 12-pulse, or even 18-pulse operation with CSI or VSI converters, which will give varied amount of harmonics in the input and output. See Table 6.2 for THD values in a typical industrial distribution system – all impedance and load parameters remaining identical, the THD values vary over a wide range depending on the magnitude of nonlinear load and the selected converter types.

6.3 HARMONIC CANCELLATION: COMMERCIAL LOADS

On the basis of the harmonic spectrum of pulse-mode power supplies, fluorescent lighting, and other electronic loads, that is, HVAC loads, which are being converted into ASDs to improve efficiency, it may seem that at the supply point high harmonic distortion levels will occur. However, this is rarely the case [4]. Cancellation between the various types of loads occurs, especially for commercial consumers. The harmonic current limits specified in IEEE 519 are rarely exceeded. The commercial nonlinear loads consist of the following:

- Fluorescent lighting is the major load from 30% to 60%
- ASDs (about 5–10%)
- Electronic power supplies and UPS (uninterruptible power supplies) systems (20–40%).

  The cancellation occurs due to

- Delta–wye connection of transformers
- Transformer phase shift for single-phase loads connected at two different voltage levels
- Harmonic phase angle difference between similar electronic load types due to differences in network and load parameters. A diversity factor of 0.90, 0.59, and 0.31 has been calculated for 5th, 7th, and 11th harmonics

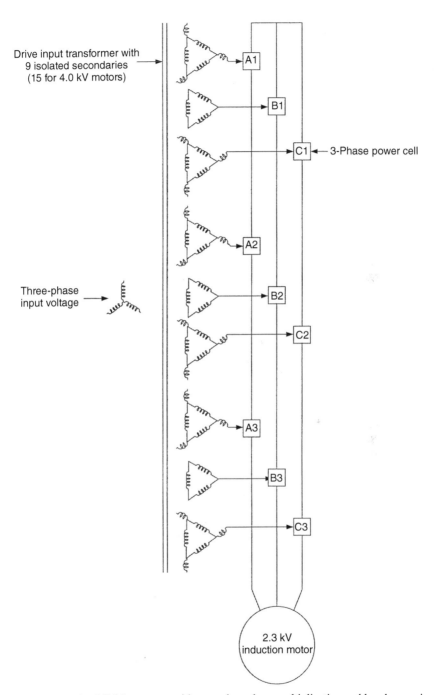

Figure 6.3   An AC drive system with secondary phase multiplication and low harmonic distortion.

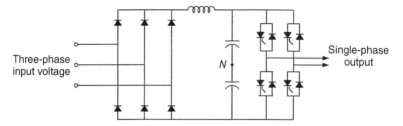

Figure 6.4   Three-phase input and single-phase output PWM cell.

TABLE 6.2   Current Harmonic Distortion, Various Converter Technologies

Type of VFD Converter	Current THD
6-pulse, 480-V, 250-hp CSI, IEEE	22
6-pulse, 2400-V, 2500-hp synchronous LCI	25
12-pulse, 1100-V, 1500-hp LCI	13
12-pulse, IEEE, typical	7
18-pulse, IEEE, typical	3.7
24-pulse, IEEE, typical	3.0
6-pulse, 200-kVA rectifier	45
6-pulse, 1200-hp thyristor DC drive	34
12-pulse, 1200-hp thyristor DC drive	8
6-pulse, 150-hp PWM with line reactors	15

- Cancellation due to phase angle diversity of different loads
- The harmonic spectra of nonlinear loads have different characteristics.

As the voltage becomes more distorted due to increasing number of nonlinear loads, the harmonic current drawn by each nonlinear load *decreases*. This attenuation can be 50% or greater.

Ref. [5] shows distortion measurements at the individual loads and the supply source for a commercial installation and cancellation effects due to different waveforms of the nonlinear loads.

Example 6.1:   To demonstrate cancellation between different load types, consider a 13.8–0.48 kV, 2.0-MVA transformer, primary 13.8-kV windings in delta connection and secondary 480-V windings in wye connection. A 40% electronic load, 30% fluorescent lighting load, and 10% ASD loads for HVAC are applied, and rest 20% load is linear. Table 6.3 shows the load models and the corresponding harmonic spectrum tables. The spectra of electronic loads and fluorescent loads are shown in Tables 4.4 and 4.8, while HVAC load harmonic spectrum is shown in Table 6.4. On the basis of a harmonic load flow study, using an iterative Newton–Raphson method, the resulting waveforms of the loads and the waveform on the transformer primary, that is, on

TABLE 6.3   Load Composition for Example 6.1

Harmonic Load Type	Percentage of Total Load (%)	Harmonic Spectrum, Table
Electronic load	40	4.4
ASD loads for HVAC	10	6.4
Fluorescent lighting	30	4.8
Linear load	20	–

TABLE 6.4   Harmonic Spectrum ASD Input Current Harmonic Spectrum HVAC Commercial Applications

Harmonic	Percentage of Fundamental	Phase Angle Degrees	Harmonic	Percentage of Fundamental	Phase Angle Degrees
Fundamental	100.0	−14	2nd	3.8	−85
3rd	8.5	−114	4th	3.5	−105
5th	79.5	145	6th	0.3	25
7th	66.0	124	8th	2.5	55
9th	2.7	11	10th	1.7	68
11th	36.0	−9.2	12th	1.2	132
13th	21.8	−118	14th	1.2	156
15th	2.4	22	16th	0.3	−136
17th	10.4	−23	18th	0.8	−92
19th	8.0	−79	20th	0.9	−117
21st	1.4	131	2nd	0.5	−105
23rd	6.7	39	24th	0	
25th	4.5	−2	26th	0.3	−12
27th	0.9	143	28th	0.2	76
29th	3.7	83	30th	0.3	42
31st	3.1	29	32nd	0.4	10
33rd	0.4	−110	34th	0.1	31

13.8 kV are shown in Fig. 6.5. It is seen that the combination of loads results in much cancellation of the harmonics.

6.4   INPUT REACTORS TO THE PWM ASDs

By adding a choke (reactor) rated at 3% of the drive system kVA, the ASD current distortion is considerably reduced. Figure 6.6 is adapted from IEEE guide [5] and shows the reduction in the current distortion, a transformer impedance of 5% is assumed.

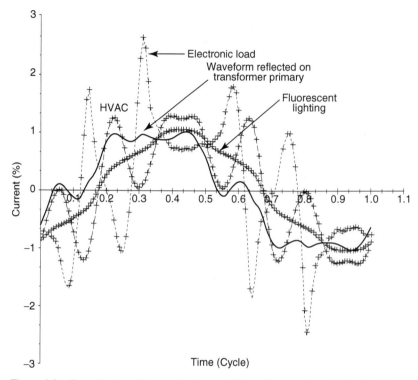

Figure 6.5   Cancellation of harmonics due to different harmonic producing loads (Example 6.1).

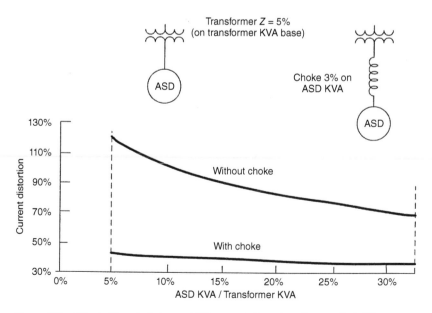

Figure 6.6   Effect of input choke on ASD current distortion. Source: Ref. [5].

6.5   ACTIVE FILTERS

The passive filters are discussed in Chapter 15. This section provides an overview of the active filters. By injecting harmonic distortion into the system, which is equal to the distortion caused by the nonlinear load, but of opposite polarity, the waveform can be corrected to a sinusoid. The voltage distortion is caused by the harmonic currents flowing in the system impedance. If a nonlinear current with opposite polarity is fed into the system, the voltage will revert to a sinusoid.

Active filters can be classified according to the way these are connected in the circuit [6–8]:

- in series connection,
- in parallel shunt connection, and
- hybrid connections of active and passive filters.

6.5.1   Shunt Connection

As we have seen, the voltage distortion in a weak system is very much dependent on harmonic current, while a stiff system of zero impedance will have no voltage distortion. Thus, provided that the system is not too stiff, a nonsinusoidal voltage can be corrected by injecting proper harmonic current. A harmonic current source is represented as a Norton equivalent circuit, and it may be implemented with a PWM inverter to inject a harmonic current of the same magnitude as that of the nonlinear load into the system, but of harmonics of opposite polarity. A shunt connection is shown in Fig. 6.7(a). The load current will be sinusoidal, so long as the load impedance is higher than the source impedance.

In Chapter 4, we studied two basic types of converters: CSI and VSI. A converter with DC output reactor and constant DC current is a current harmonic source. A converter with a diode front end and DC capacitor has a highly distorted current depending on the AC source impedance, but the voltage at rectifier input is less dependent on AC impedance. This is a voltage harmonic source. It presents low impedance, and shunt connection will not be effective. A shunt connection is more suitable for current source controllers where the output reactor resists the change of current. If a shunt connection is used to compensate a diode rectifier or when the power system contains passive filters or capacitor banks, the current injected by the parallel filter will flow into the diode rectifier; as a result, the source side harmonics cannot be cancelled.

6.5.2   Series Connection

Figure 6.7(b) shows a series connection. A voltage V_f is injected in series with the line and it compensates the voltage distortion produced by a nonlinear load. A series active filter is more suitable for harmonic compensation of diode rectifiers where the DC voltage for the inverter is derived from a capacitor, which opposes the change of the voltage.

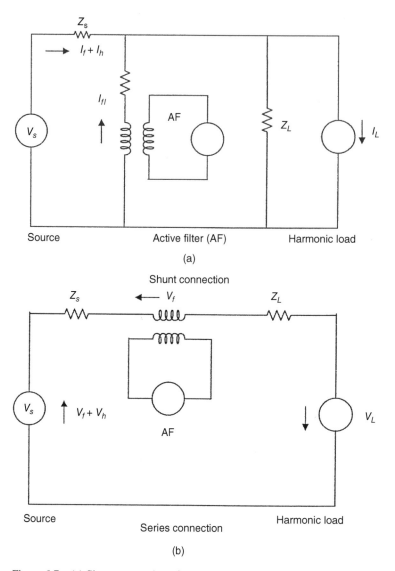

Figure 6.7   (a) Shunt connection of an active filter; (b) series connection of an active filter.

Thus, the compensation characteristics of the active filters are influenced by the system impedance and load. This is very much akin to passive filters; however, active filters have better harmonic compensation characteristics against the impedance variation and frequency variation of harmonic currents.

Figure 6.8(a) and (b) show the equivalent circuit of parallel active filter for harmonic current and voltage source nonlinear loads, and Figs. 6.8(c) and (d) show similar circuits for series filters. G is the equivalent transfer function of the filter including the detection circuit for harmonics. We can write $I_C = GI_L$. Then in

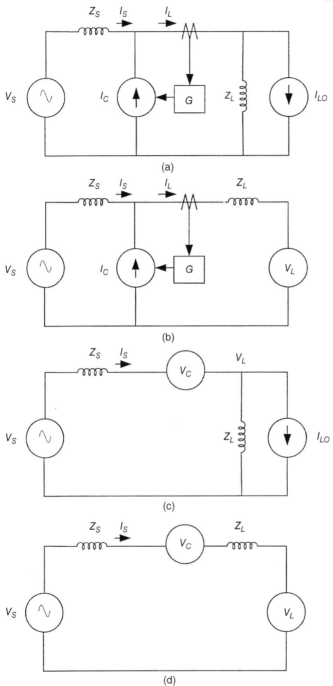

Figure 6.8 (a) and (b) Parallel active filter for harmonic current and voltage source, respectively; (c) and (d) series active filter for harmonic current and voltage sources, respectively.

Fig. 6.8(a), equation for I_s can be written as

$$I_s = \frac{Z_L I_{L0}}{Z_s + \frac{Z_L}{1-G}} + \frac{V_s}{Z_s + \frac{Z_L}{1-G}}$$

If

$$\left| \frac{Z_L}{1-G} \right|_h \gg |Z_s|_h \tag{6.1}$$

is satisfied, that is, $|1 - G|_h \approx 0$, the source current becomes sinusoidal and I_{Sh} (source harmonic current) is nearly zero. Only G can be predesigned for the active filter and mainly dominated by the detection circuit of harmonics, while Z_s and Z_L are determined by the power system.

When a parallel passive filter or shunt capacitor is placed in parallel with shunt active filter, the load impedance is much reduced. In this case, the current flowing into passive filter can be very large:

$$I_{Lh} - I_{L0h} = \left(\frac{V_{Sh}}{Z_L} \right) \tag{6.2}$$

The subscript h signifies harmonic component.

In Fig. 6.8(b), a parallel active filter is shown compensating a harmonic voltage source load. The load impedance is represented by a Thévenin equivalent. If the following equation is satisfied:

$$\left| Z_s + \frac{Z_L}{1-G} \right|_h \gg 1 \ \text{pu} \tag{6.3}$$

the source current will become sinusoidal, but it is difficult for a parallel active filter to satisfy Eq. (6.3) because a harmonic voltage source represents low internal impedance. Considering a diode rectifier with large smoothing DC capacitor, Z_L is nearly zero. The source impedance is usually 0.1 pu and Eq. (6.3) cannot be satisfied with source impedance alone.

Figure 6.8(c) shows a series active filter compensating a harmonic current source, and Fig. 6.8(d) shows a series active filter compensating a harmonic voltage source. V_c represents the output voltage of the *series filter*.

In Fig. 6.8(c), if the series active filter is controlled so that

$$V_C = KGI_S \tag{6.4}$$

then the source current is

$$I_S = \frac{Z_L I_L + V_s}{Z_s + Z_L + KG} \tag{6.5}$$

In order that the source current becomes sinusoidal:

$$K \gg |Z_L|_h, K \gg |Z_s + Z_L|_h \tag{6.6}$$

Then

$$V_C = Z_L I_{Lh} + V_{Sh}$$

$$I_s = 0 \tag{6.7}$$

However, these conditions cannot be satisfied. K should be large, and impedance on the load side should be small for harmonics in order to suppress the source harmonic current. This cannot be satisfied for a conventional phase-controlled thyristor rectifier, and Z_L is almost infinite. The required output voltage V_c also becomes infinite.

In Fig. 6.8(d), for series filter compensating a harmonic voltage source, the current is

$$I_S = \frac{V_s - V_L}{Z_S + Z_L + KG} \tag{6.8}$$

When K is much greater than 1 pu, I_S is zero. To realize a large gain, a hysteresis or ramp-comparison control method can be used (see references [8,9]).

Table 6.5 shows comparison of shunt and series connections.

The control systems of the active filters have a profound effect on the performance, and a converter can have even a negative reactance. The active filters by themselves have the limitations that initial costs are high and do not constitute a

TABLE 6.5   Comparison of Shunt and Series Connections of Active Filters

Parameter	Shunt Connection	Series Connection
Operation	Current source, Norton's equivalent	Voltage source, Thévenin equivalent
Loads	Inductive or current source loads, or harmonic current sources, that is, phase-controlled thyristor rectifiers of DC drives	Capacitive or voltage source loads or harmonic voltage sources, that is, diode rectifiers with direct smoothing capacitors for AC drives
Operating conditions	Z_L should be high and $\lvert 1 - G \rvert_h \ll 1$	Z_L should be low and $\lvert 1 - G \rvert_h \ll 1$
Compensation characteristics	Independent of source impedance, for current source loads, but will depend on Zs when the load impedance is low	Independent of Zs and Z_L for voltage source loads, but depends on Z_L when the loads are current source type
Application considerations	Injected current flows into the load side and may cause overcurrent when applied to a capacitor or voltage source load	A low-impedance parallel branch (power factor improvement capacitor or passive filter) is required when applied to inductive or current source load

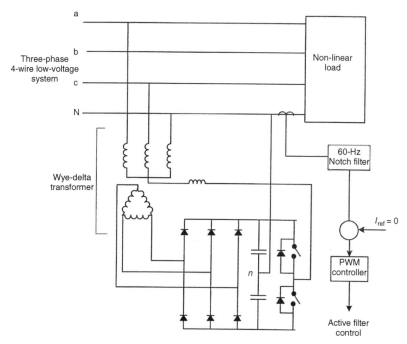

Figure 6.9   An active filter and its control for elimination of harmonics in the neutral circuit of a three-phase, four-wire low-voltage system.

cost-effective solution for nonlinear loads above approximately 500 kW, though further developments will lower the costs and extend applicability.

An active filter for the elimination of harmonics in three-phase, four-wire systems serving single-phase, nonlinear loads is shown in Fig. 6.9. The delta-connected windings of the wye–delta connected transformer serve a three-phase rectifier to maintain DC voltage across the capacitors. The neutral current is sensed through a CT, and 60-Hz notch filter removes the fundamental component of the current. The filtered signal is compared with I_{ref} set to zero, and the resulting error signal fed to a PWM controller to inject an equal and opposite current. This neutralizes the harmonic current flowing in the neutral. If the active filter can cancel 100% of the neutral current, then the flow of all harmonic currents in the neutral is prevented [10].

6.5.3   Combination of Active Filters

A combination of series and shunt active filters is shown in Fig. 6.10. This looks similar to the unified power controller for power transmission lines [1]. For applications in power transmission lines, a unified power controller consists of two voltage source switching converters, a series and a shunt converter, and a DC-link capacitor. The arrangement functions as an ideal AC-to-AC power converter in which real power can flow in either direction between AC terminals of the two converters, and each inverter can independently generate or absorb reactive power at its own terminals. Here, its

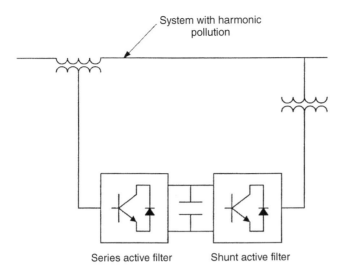

Figure 6.10   Connections of a unified power quality conditioner. Source: Ref. [8].

operation is different [8]. A series filter blocks harmonic currents flowing in and out of the distribution feeders. It detects the supply current and is controlled to present zero impedance to the fundamental frequency and high impedance to the harmonics. The shunt filter absorbs the harmonics from the supply feeders and detects the bus voltage at the point of connection. It is controlled to present infinite impedance to the fundamental frequency and low impedance to the harmonics. The harmonic currents and voltages are extracted from the supply system in the time domain.

6.5.4 Active Filter Configurations

The electronics and power devices used in both types of converters for filters are quite similar (Fig. 6.11(a) and (b)), which show three-phase voltage source and current source PWM converters. The current source active filter has a DC reactor with a constant DC current, whereas the voltage source active filter has a capacitor on the DC side with constant DC voltage. An output filter is provided to attenuate the inverter switching effects. In a current source type, LC filters are necessary (Fig. 6.11(b)). Transient oscillations can appear because of resonance between filter capacitors and inductors. The controls are implemented so that the inverter outputs a harmonic current equivalent but opposite to that of the load. The source side current is therefore sinusoidal, but the voltage will be sinusoidal only if the source does not generate any harmonics. Bipolar junction transistors are used with switching frequencies up to 50 kHz for modest ratings. SCRs and GTOs are used for higher power outputs.

6.5.5 Active Filter Controls

The basic control circuit block diagram in shown in Fig. 6.12(a). The converter switching is controlled to maintain converter voltage or current to the desired

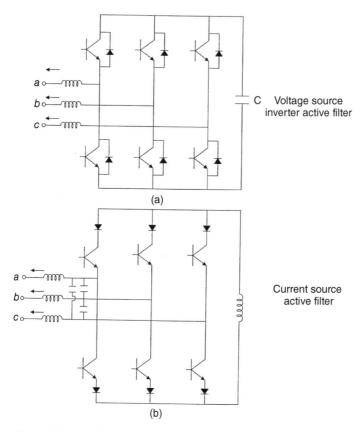

(a)

C   Voltage source
    inverter active filter

Current source
active filter

(b)

Figure 6.11   (a) Voltage source inverter active filter; (b) current source inverter active filter.

reference signal. Several methods are applied to calculate reference signal for proper harmonic elimination and power factor correction. Corrections in the time domain are based on holding instantaneous voltage or current within reasonable tolerance of a sine wave. The error signal can be the difference between actual and reference waveforms. The error function can be instantaneous reactive power (IRP) or EXT (extraction of fundamental frequency component).

- Signal method based on input current or voltage.
- Instantaneous reactive power compensation based on active and reactive components of input signal, $p-q$ theory, Chapter 1.

Consider the equivalent circuit of a shunt active filter (Fig. 6.12(b)), which is considered an ideal current source drawing current I_F. The harmonic producing load is also considered an ideal current source injecting current I_L. To eliminate harmonics from the source current, the load current I_L is sampled. These samples are passed through passive filters, which take out the fundamental component of the current I_{Lf}.

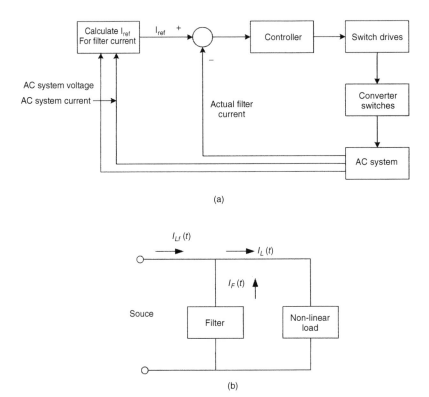

Figure 6.12   (a) A basic control circuit for the active filters; (b) equivalent circuit.

Then, the desired current to be supplied by the shunt active filter is

$$I_F(t) = I_L(t) - I_{Lf}(t) \tag{6.9}$$

Depending on the point of application in the power system, instability can occur. The phase margin is in the range $10°-90°$. A shunt active filter based on voltage detection is very stable as the phase margin is over $90°$, irrespective of point of installation [11−13]. If it is assumed that extraction of harmonic current or voltage is represented by first-order lag system, then these detection methods give

$$I_F(s) = \frac{sK_S}{1 + sT} I_{Sh}(s)$$

$$I_F(s) = \frac{K_V}{1 + sT} V_{Sh}(s) \tag{6.10}$$

where K_S and K_V are feedback gains.

6.5.6   Instantaneous Reactive Power Compensation

The desired current is calculated so that instantaneous active and reactive powers in a three-phase system are kept constant. This means that the active filter compensates for the variation in instantaneous power. In the active filter control, first the values of p and q are computed (see Chapter 1). Then, the reference current signals are calculated. The instantaneous real power p_L and IRP q_L on the load side can be defined as (Eq. 1.66):

$$\begin{vmatrix} p_L \\ q_L \end{vmatrix} = \begin{vmatrix} e_\alpha & e_\beta \\ -e_\beta & e_\alpha \end{vmatrix} \begin{vmatrix} i_{L\alpha} \\ i_{L\beta} \end{vmatrix} \tag{6.11}$$

The dimensions of q_L are not watt, volt-ampere, or var because $e_\alpha i_\beta$ and $e_\beta i_\alpha$ are products of current and voltage in different phases:

$$\begin{vmatrix} i_{L\alpha} \\ i_{L\beta} \end{vmatrix} = \begin{vmatrix} e_\alpha & e_\beta \\ -e_\beta & e_\alpha \end{vmatrix}^{-1} \begin{vmatrix} p_L \\ q_L \end{vmatrix} \tag{6.12}$$

Write

$$p_L = p_{DC} + p_{AC}$$

$$q_L = q_{DC} + q_{AC} \tag{6.13}$$

where

  p_{DC} is the DC component of the instantaneous real power, fundamental frequency

  p_{AC} is the AC component of the instantaneous real power, harmonic frequencies

  q_{DC} is the DC component of the instantaneous imaginary power, fundamental frequency

  q_{AC} is the AC component of the instantaneous imaginary power, harmonic frequencies.

In the circuit of compensation reference currents, following expression results

$$\begin{vmatrix} i_u^* \\ i_v^* \\ i_w^* \end{vmatrix} = \sqrt{\frac{2}{3}} \begin{vmatrix} -1 & 0 \\ -\frac{1}{2} & \frac{\sqrt{3}}{2} \\ -\frac{1}{2} & -\frac{\sqrt{3}}{2} \end{vmatrix} \begin{vmatrix} e_\alpha & e_\beta \\ -e_\beta & e_\alpha \end{vmatrix}^{-1} \begin{vmatrix} p* + p_{av} \\ q* \end{vmatrix} \tag{6.14}$$

where p_{av} is the instantaneous real power for loss in the active filter, and p^*, q^* are given by following equations for proper harmonic filtering.

$$p^* = -p_{AC}$$

$$q^* = -q_{AC} \tag{6.15}$$

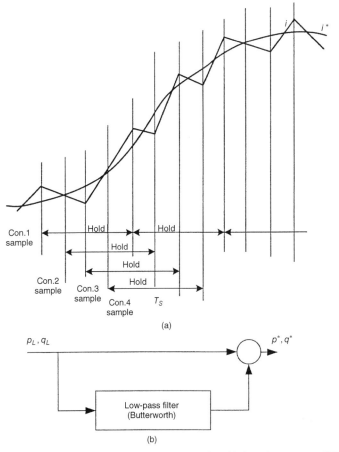

Figure 6.13   (a) Current control scheme of multiple voltage source PWM converters; (b) low-pass filter. Source: Ref. [14].

p_{AC} and q_{AC} are generated by higher order harmonics other than the fundamental components of voltage or current. By canceling these power components, the corresponding harmonics will be eliminated. A high-pass filter configuration using a Butterworth low-pass filter can be used to cancel the AC components of p and q. The design of the low-pass filter is most important in control circuit (see Ref [14] for further details). A converter control scheme is described (Fig. 6.13(a)). The reference current is directly compared to the actual current. Then the output signal of the comparator is sampled and held at a regular interval T, synchronized with the clock frequency equal to $1/T_s$. Note that 12 external clocks are applied to each converter, and phases in one converter do not overlap. Harmonic currents are reduced considerably as if switching frequency was increased. Figure 6.13(b) shows low-pass filter for calculation of p^* and q^* (see Chapter 15 for filter designs).

Other time-domain techniques can be classified into three main categories [7]:

- Triangular wave

- Hysteresis
- Deadbeat

The triangular wave method is easiest to implement and can be used to generate two-state or three-state switching functions. A two-state function can be connected positively or negatively, while a three-state function can be positive, negative, or zero (Fig. 6.14).

In the two-state system, the inverter is always *on* (Fig. 6.14(a)). The extracted error signal is compared to a high-frequency triangular carrier wave, and the inverter switches each time the waves cross. The result is an injected signal that produces equal and opposite distortion.

In a three-state system (hysteresis method), preset upper and lower limits are compared to an error signal (Fig. 6.14(b)). So long as the error is within a tolerable band, there is no switching and the inverter is off.

The advantages of time-domain methods are fast response, though these are limited to one-node application, to which these are connected and from which the measurements are taken.

6.5.7 Corrections in the Frequency Domain

Fourier transformation is used to determine the harmonics to be injected. The error signal is extracted using a 60-Hz filter, and the Fourier transform of the error signal is taken. The cancellation of M harmonics method allows for compensation up to the Mth harmonic, where M represents the highest harmonic to be compensated. A switching function is constructed by solving a set of nonlinear equations to determine the precise switching times and magnitudes. Quarter-wave symmetry is assumed to reduce the computations. Because an error function is used, the system can easily accommodate system changes, but requires intense calculations and the time delays associated with it. The computations increase with M and the increased computational requirements are the main disadvantage, though these can be applied in dispersed networks.

The predetermined frequency method injects specific frequencies into the system, which are decided in the design stage of the system, much similar to passive harmonic filtering. This eliminates the need for real-time commutation of switching signals, but the harmonic levels present must be carefully evaluated beforehand and each filter designed for the specific requirements (see Refs. [15,16]).

6.6 ACTIVE CURRENT SHAPING

By using proper control systems, the input current of converters can be forced to follow a sinusoid in phase with voltage, addressing the need for reactive power compensation as well as harmonic elimination [7]. The desired current is calculated by multiplying sinusoidal voltage by a factor K, which is determined by load power [8].

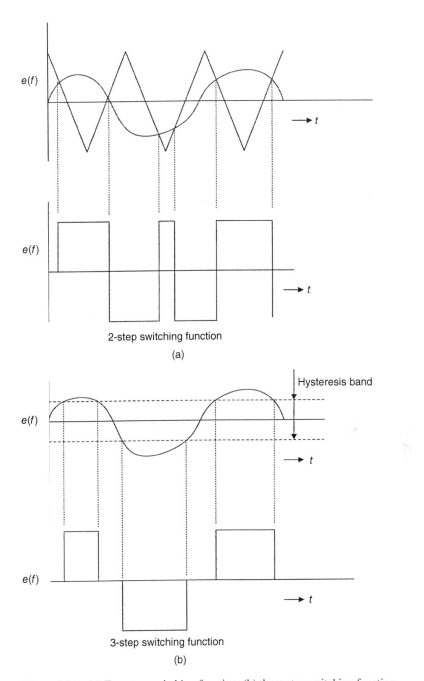

$e(f)$

t

$e(f)$

t

2-step switching function

(a)

Hysteresis band

$e(f)$

t

$e(f)$

t

3-step switching function

(b)

Figure 6.14   (a) Two-step switching function; (b) three-step switching function.

The load current can be written as

$$I_L(t) = Kv(t) + i_q(t) \tag{6.16}$$

where $Kv(t)$ is the active component of the load current in which K is a coefficient that can be calculated in the control circuit, and $i_q(t)$ is the nonactive component of the current. The nonactive current must be compensated to have maximum power factor and harmonic rejection. The desired reference current in the active filter is

$$i_q(t) = I_L(t) - Kv(t) \tag{6.17}$$

Figure 6.15(a) shows that a diode bridge, and a DC/DC converter (chopper) is placed between the bridge rectifier and DC load. The DC output voltage can be regulated as AC input voltage changes, and the poor power factor and high harmonic contents of the input current wave are compensated. The low-frequency harmonics generated by the nonlinear loads are shifted to higher frequencies by switching frequency modulation. (The switching frequency is limited by the switching losses.)

Figure 6.15(b) is a control tolerance band current-mode control block diagram. The current is controlled so that the peak-to-peak ripple in I_L remains constant. This means that using a pre-elected value of ripple, I_L is forced to be lie within the hysteresis band determined by

$$I_L^* + \frac{I_{ripple}}{2}$$

$$I_L^* - \frac{I_{ripple}}{2} \tag{6.18}$$

This is achieved by controlling the switching status. In continuous-mode conduction, the "on" and "off" intervals are given by

$$t_{on} = \frac{L_d I_{ripple}}{|v_s|} \tag{6.19}$$

$$t_{off} = \frac{L_d I_{ripple}}{V_d - |v_s|} \tag{6.20}$$

where V_d is the constant output voltage of the converter and $|v_s|$ is the value of input AC voltage. The switching frequency f_s is given by

$$f_s = \frac{1}{t_{on} + t_{off}} \tag{6.21}$$

Figure 6.15(c) shows the control waveform. The current literature is rich in the control and application of active filters [17–22].

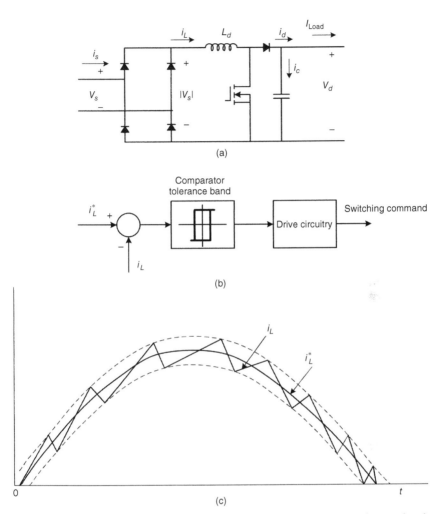

Figure 6.15   (a) Sinusoidal current rectifier using boost preregulator; (b) tolerance-band current-mode block diagram; (c) i_L^*, i_L waveforms.

6.7   HYBRID CONNECTIONS OF ACTIVE AND PASSIVE FILTERS

Hybrid connections of active and passive filters are shown in Fig. 6.16. Figure 6.16(a) is a combination of shunt active and shunt passive filters. Figure 6.16(b) shows a combination of a series active filter and a shunt passive filter while Fig. 6.16(c) shows an active filter in series with a shunt passive filter. The combination of shunt active and passive filters has already been applied to harmonic compensation of large steel mill drives. Addition of a large shunt capacitor will reduce the load resistance, and shunt passive filter will draw a large source current from a stiff system and may act

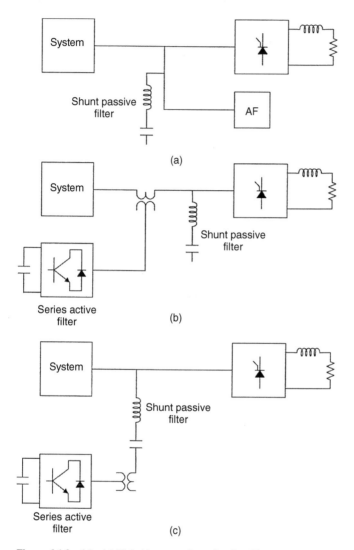

Figure 6.16   (a)–(c) Hybrid connection of active filters.

as a sink to the upstream harmonics. It is required that in a hybrid combination the filters share compensation properly in the frequency domain.

In a series connection, the active filter is connected in series with the passive filter, both being in parallel with the load, as shown in Fig. 6.16(c). With suitable control of the active filter, it is possible to avoid resonance and improve filter performance. The active filter can be either voltage or current controlled. In current-mode control, the inverter is a voltage source to compensate for current harmonics. In voltage-mode control, the converter is a voltage source inverter controlled to compensate for the voltage harmonics. The advantage is that the converter itself is far smaller, only about 5% of the load power. The active filter in such schemes regulates the effective source

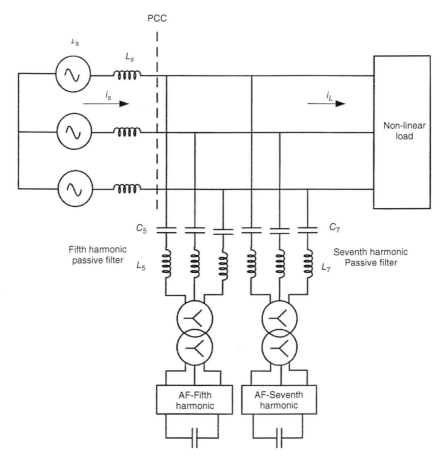

Figure 6.17   A system configuration for DHAF.

impedance as experienced by the passive filter, and the currents are forced to flow in the passive filter rather than in the system. This makes the passive filter characteristics independent of the actual source impedance, and a consistent performance can be obtained.

Hybrid series and shunt filters implemented with pulse width modulation are practical for large nonlinear loads below 10 MW. For higher loads, these are not cost effective because of higher bandwidth and high ratings. A dominant harmonic active filter (DHAF) using square-wave active filter inverters is described in Ref. [23], and the authors claim its applicability to high-power nonlinear loads (10–100 MW). Figure 6.17 shows the system configuration. The active compensation required is only a fraction (2–3%) of the required kvar rating.

A synchronous reference frame (SRF)-based controller is used. It achieves fifth harmonic isolation by using closed-loop control on fifth harmonic component of the supply source current. The three-phase supply source currents are measured and transformed into SRF ($(d^{e5} - q^{e5})$ axis rotating at fifth harmonic. The fifth harmonic current of the supply source is transformed into DC quantities in d^{e5}, q^{e5} axes and extracted by the subsequent low-pass filters. The supply currents i_{sd}^{-e5} and i_{sq}^{-e5} are compared with reference currents. (These should be zero for complete cancellation of fifth harmonic.) The errors are fed into a PI controller to generate the required voltage command for active filter inverter. A reverse transformation from d^{e5}, q^{e5} is applied to phase quantities $a-b-c$ to convert inverter voltage command to three-phase quantities [24].

An experimental set-up of active and passive filters connected in series is shown in Fig. 6.18, Ref. [25] (see Eqs. (6.1)–(6.8) for the active filters). The calculated harmonic current in each phase is amplified by gain K and input to a PWM controller as a voltage reference:

$$v_C^* = KI_{Sh} \tag{6.22}$$

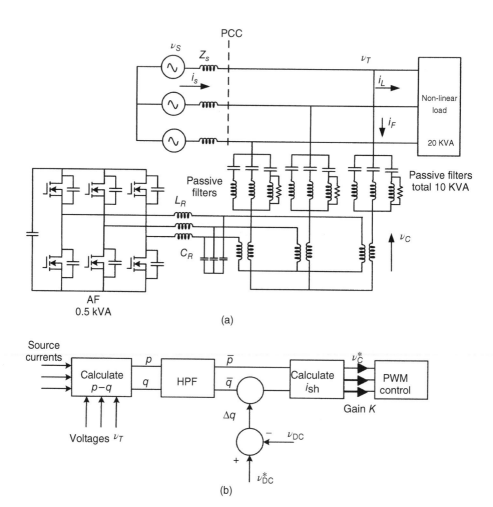

Figure 6.18   (a) Series connection of active and passive filters; (b) control circuit diagram.

Also

$$I_{\text{Sh}} = \frac{Z_F}{Z_S + Z_F} I_{\text{Lh}} \qquad (6.23)$$

When no active filter is connected and the source impedance is small or unless the passive filter is tuned to harmonic frequencies generated by the load, desirable filtering will not be obtained. When the active filter is connected and controlled as a voltage source, it forces all harmonics contained in the load current to flow in the passive filter so that no harmonic current flows in the source. No fundamental voltage is applied to the active filter, and this results in great reduction of the voltage rating of the active filter [14]. Refs [25–27] provide further reading.

6.8 IMPEDANCE SOURCE INVERTERS

The new technologies for harmonic mitigation are on the rise, and these also overcome some of the limitations of the conventional converters.

Impedance source inverters (ZSI) are a new approach in DC/AC conversion [28–30]. It consists of a network formed by two capacitors and two inductors and provides buck–boost characteristics. It has been used in many industrial systems for ASDs and also for distributed generation. The traditional VSI and CSI technologies have some limitations, which are as follows:

- These are either boost or buck converters.
- VSI and CSI circuits are not interchangeable.
- These are vulnerable to EMI noise in terms of reliability.

For example, in CSI, one of the upper and one of the lower devices have to be gated-on and maintained, otherwise an open circuit of the DC inductor will occur and destroy the devices. The open-circuit problem due to misgating is a major concern of the converter reliability.

In the ZSI circuit in Figs. 6.19 and 6.20(a), if the two inductors have zero inductance the ZSI becomes a VSI, and if the two capacitors have zero capacitance, the

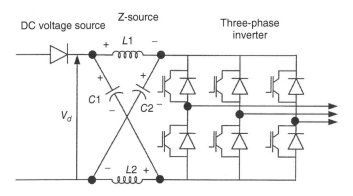

Figure 6.19 Configuration of a ZSI.

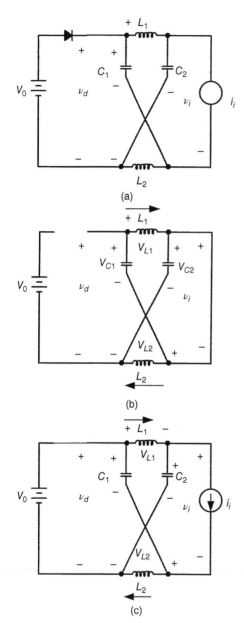

(a)

(b)

(c)

Figure 6.20 (a) Equivalent circuit of a Z-source inverter viewed from DC link; (b) equivalent circuit of a Z-source inverter viewed from DC link, with the inverter in shoot-through zero state; (c) equivalent circuit viewed from DC link when inverter is in one of eight nonshoot-through states.

ZSI is a CSI. The Z circuit can restrict overvoltages and overcurrent. The legs in the main bridge circuit can operate in short circuit or open circuit for a short duration. It suppresses EMI noise and has relatively lower current and voltage surges. The misgating will not damage the devices. All the traditional PWM schemes can be used to control the ZSI. It has nine permissible states (vectors): six active vectors as a VSI has and three zero vectors when the load terminals are short circuited through both

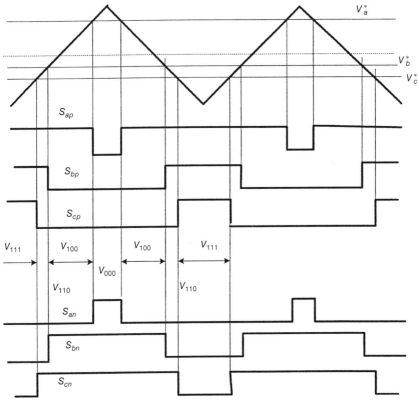

Figure 6.21   Traditional carrier-based PWM control without shoot-through zero state. Traditional zero states (vectors) are generated in every switching cycle and determined by the references. Source: Ref. [29].

upper and lower devices in any one phase. This zero state vectors are impossible for VSI, because these will cause shoot through. In Fig. 6.20(b), the inverter bridge is equivalent to short circuit at the terminals of Z-source when in the shoot-through zero state; and it becomes an equivalent current source when in one of the eight nonshoot-through switching states (Fig. 6.20(c)). Figure 6.21 shows traditional PWM switching sequence based on triangular carrier wave. In every switching cycle, two nonshoot-through states are used along with two adjacent active states to synthesize the desired voltage. Modified shoot-through zero states as shown in Fig. 6.22 are used to boost the DC-link voltage. The equivalent switching frequency viewed from the Z-source network is six times the switching frequency of the main inverter.

Assuming symmetrical Z-source network

$$V_{C_1} = V_{C_2} = V_c, \quad v_{L_1} = v_{L_2} = v_L \tag{6.24}$$

In the shoot-through zero state, for an interval T_0, during a switching cycle $T = T_1 + T_0$:

$$v_L = V_C, \quad V_d = 2V_C, \quad v_i = 0 \tag{6.25}$$

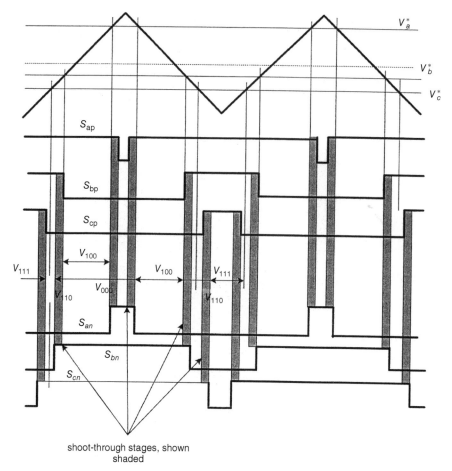

Figure 6.22   Modified carrier-based PWM control with shoot-through zero states that are evenly distributed among the three phase legs, while the active vectors are unchanged. Source: Ref [29].

When the inverter is in one of the eight nonshoot-through states for an interval T_1 during switching cycle T,

$$v_L = V_0 - V_C, \quad V_d = V_0, \quad v_i = V_c - v_L = 2V_C - V_0 \qquad (6.26)$$

The average voltage of inductors over one switching period is zero in steady state:

$$V_L = \bar{v}_L = \frac{T_0 V_C + T_1(V_0 - V_C)}{T} = 0$$

or

$$\frac{V_C}{V_0} = \frac{T_1}{T_1 - T_0} \qquad (6.27)$$

The peak DC-link voltage across the inverter bridge is

$$\widehat{v}_i = V_c - v_L = \frac{T}{T_1 - T_0} V_0 = BV_0 \qquad (6.28)$$

where

$$B = \frac{T}{T_1 - T_0} = \frac{1}{1 - 2(T_0/T)} \geq 1 \qquad (6.29)$$

B is the boost factor resulting from the shoot-through state.

Figure 6.23 shows a three-level diode-clamped converter, which produces a much less distorted three-level output waveform. On its output end, it has a three-level DC/AC inverter (NPC–Neutral Point Clamped). The neutral potential needed for the three-level inverter is tapped from the wye-connected input filter. It has the capability to ride through deep voltage sags [30].

6.9   MATRIX CONVERTERS

Matrix converter (MC) is based on bidirectional switches, incorporating PWM voltage control, developed by Venturine in 1980 [31]. It consists of an array of nine power semiconductor switches, which directly connect a three-phase AC source to a three-phase AC load. There are no DC-link passive components such as reactors or capacitors, and input power factor can be controlled independent of load current. All switches require a bidirectional switch capable of blocking voltage and conducting current in either direction. The basic circuit diagram is illustrated in Fig. 6.24(a). Any input phase can be connected to output phases at any instant by switching logic, while the current in any phase of load is drawn from any phase or phases of the input power supply. For the bidirectional switches, a combination of reverse blocking self-controlled devices, such as power MOSFET or IGBTs or transistor – embedded diode bridge, has been used (Fig. 6.24(b)). It has, therefore, inherent bidirectional power flow *and sinusoidal input/output waveforms*. The disadvantage is voltage transfer ratio of 0.866 and nonavailability of bidirectional high-frequency switch integrated in a silicon chip.

The switching logic is shown in Fig.6.24(c). The switches are controlled so that at any point of time, only one of the three switches connected to an output phase is closed to prevent short circuit of the supply lines. Thus, with a possibility of 2^9 (= 512) switch combinations, only 27 are permissible.

The output voltages can, therefore, be written in terms of input voltages through a switch matrix representing nine switches:

$$\begin{vmatrix} v_{an} \\ v_{bn} \\ v_{cn} \end{vmatrix} = \begin{vmatrix} S_{Aa} & S_{Ba} & S_{Ca} \\ S_{Ab} & S_{Bb} & S_{Cb} \\ S_{Ac} & S_{Bc} & S_{Cc} \end{vmatrix} \begin{vmatrix} V_{AO} \\ V_{BO} \\ V_{CO} \end{vmatrix} \qquad (6.30)$$

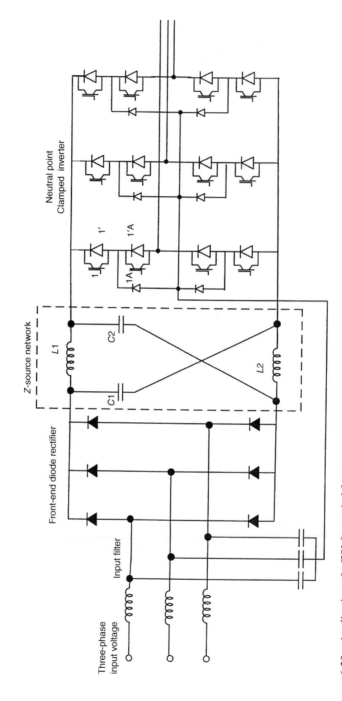

Figure 6.23 Application of a ZSI for a wind farm.

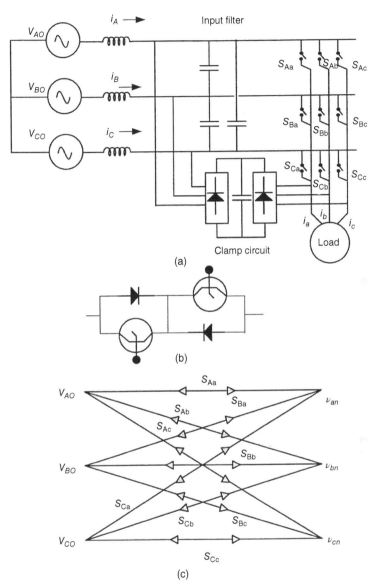

Figure 6.24   (a) Structure of a matrix converter; (b) implementation of a bidirectional switch; (c) switching matrix symbol for the converter.

A similar matrix equation applies for the input and output currents:

$$\begin{vmatrix} i_A \\ i_B \\ i_C \end{vmatrix} = \begin{vmatrix} S_{Aa} & S_{Ab} & S_{Ac} \\ S_{Ba} & S_{Bb} & S_{Bc} \\ S_{Ca} & S_{Cb} & S_{Cc} \end{vmatrix} \begin{vmatrix} i_a \\ i_b \\ i_c \end{vmatrix}$$ (6.31)

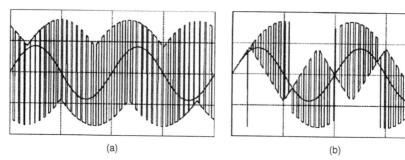

(a)   (b)

Figure 6.25 (a) Modulation method for the conventional matrix converter – maximum envelope modulation method; (b) SEM method.

Note that in Eq. (6.31), the matrix is transpose of matrix in Eq. (6.30). The matrix converter in theory can operate at any frequency at the output or input, including zero; and thus, it can be termed as a universal converter and can be employed as three-phase AC/DC converter, DC/three-phase AC converter, or even a buck–boost chopper.

The control methods are complex. In Venturini method, switching function is calculated involving the duty cycle of each of the nine bidirectional switches and generates a three-phase output voltage by sequential piecewise sampling of the input waveforms. These follow a set of reference or target voltage waveform. A transfer function approach is employed by relating input and output voltages and currents through a modulation matrix.

There are commutation and protection issues in the matrix converters. The converter does not have free-wheeling diodes. To maintain continuity of output current, the next switch in sequence must be turned on immediately. A momentary short circuit may develop. One solution is to use a semi-soft current commutation using multistepped switching procedure, which adds to complexity. A clamp capacitor connected through two full-wave rectifiers at the input and output lines of the MC serves as a voltage clamp for voltage spikes and a three-phase, single-stage input filter is used to attenuate the high-order harmonics and render sinusoidal current.

Subenvelope modulation method (SEM) to reduce THD of MC converters is of interest from harmonic mitigation considerations. As any of the three input phases can be connected to any output phase, it is possible to modulate the output phase between two *adjacent* input phases, Fig. 6.25(a) shows maximum envelope modulation method and Fig. 6.25(b) shows SEM method. The pulse magnitude of the output voltage can be low, and the high-frequency components of the output voltage can be reduced. The THD and dv/dt are reduced. The input line current pulses are smaller and wider, and therefore THD of input line current is reduced [32].

6.10 MUTILEVEL INVERTERS

In medium- and high-voltage applications, PWM-based two-level inverters are limited due to current and voltage ratings of the switching devices. The concept of multilevel converters was published in Ref. [33]. Multilevel converters are a new breed, Ref. [34], which overcomes some of the disadvantages of PWM inverters, namely:

- The carrier frequency is high, usually between 2 and 20 kHz.
- In a normal PWM waveform, the pulse height is the DC-link voltage.
- The pulse width will be narrow when output voltage is low.
- dv/dt is high and causes a strong EMI. This introduces a number of harmonics and possible ringing. PWM needs rigorous switching conditions and resultant switching losses. The inverter control circuitry is complex.

The multilevel inverters emerged as a solution to high-power applications. The switching frequencies are low, equal to, or only a few times the output frequency. The pulse heights are low. For an m-level inverter, the pulse height is V_m/m where V_m is the output voltage amplitude. This results in much smaller dv/dt and EMI as compared to PWM inverters. The harmonics and THD are further reduced. Smooth switching conditions are obtained with much lower switching power losses.

Multilevel converters have been applied to HVDC, large motor drives, railway traction applications, UPFC, STATCOM, and SVCs. The quality of output voltage increases as the number of voltage level increases. The applications have been extended to active power filters, voltage sag compensators, and photovoltaic systems.

The various types of multilevel converters are as follows:

- Diode-clamped multilevel inverters (DCMI) were proposed by Nabae in 1980 [33]. It is also called the neutral point-clamped (NPC) inverter, because the NPC inverter effectively doubles the device voltage without precise voltage matching.
- Capacitor-clamped (flying capacitors) multilevel inverters appeared in the 1990s.
- Cascaded multilevel inverters (CMIs) with separate DC source applications prevailed in the mid-1990s for motor drives and utility applications. It has drawn great interest for medium-voltage high-power inverters. It is also used for regenerative-type motor applications [35].
- Recently some new topologies of multilevel inverters have been emerged, such as generalized multilevel inverters (GMIs), hybrid multilevel inverters, and soft-switched multilevel inverters. These are applied at medium-voltage levels for mills, conveyors, pumps, fans, compressors, and so on. The applications are also extended to low-power applications [36,37].

Figure 6.26(a) shows a single-phase, five-level, diode-clamped circuit with four DC bus capacitors, C_1, C_2, C_3, and C_4. The staircase voltage wave is synthesized from several levels of DC capacitor voltages. An m-level diode-clamped converter consists of $m - 1$ capacitors on the DC bus and produces m levels of the phase voltage by appropriate switching. The voltage across each capacitor is $V_{DC}/4$. The staircase voltage shown in Fig. 6.26(b) is generated by the five switch combinations shown in

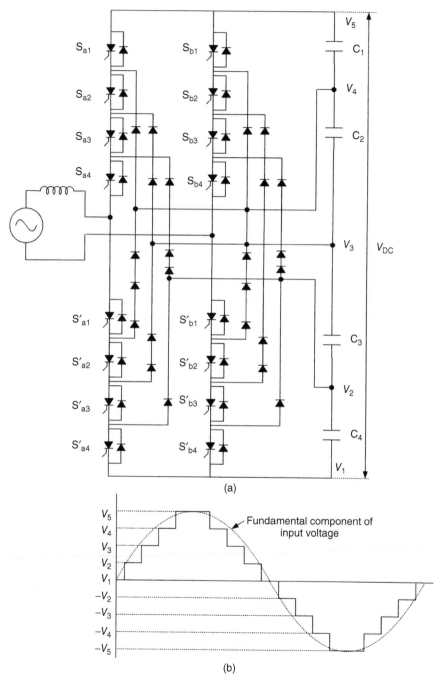

Figure 6.26   (a) Single-phase full bridge, five-level, diode-clamped circuit; (b) Stepped voltage generation; (c) two diode-clamped multilevel converters for back-to-back intertie system.

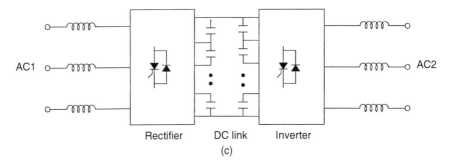

Rectifier   DC link   Inverter

(c)

Figure 6.26   (*Continued*)

Fig. 6.26(a) and the switching matrix shown below:

$$
\begin{array}{cccccccc}
 & S_{a1} & S_{a2} & S_{a3} & S_{a4} & S'_{a1} & S'_{a2} & S'_{a3} & S'_{a4} \\
V_4 & 1 & 1 & 1 & 1 & 0 & 0 & 0 & 0 \\
V_3 & 0 & 1 & 1 & 1 & 1 & 0 & 0 & 0 \\
V_2 & 0 & 0 & 1 & 1 & 1 & 1 & 0 & 0 \\
V_1 & 0 & 0 & 0 & 1 & 1 & 1 & 1 & 0 \\
V_0 & 0 & 0 & 0 & 0 & 1 & 1 & 1 & 1 \\
\end{array}
\tag{6.32}
$$

With high switching levels, the harmonic content is low enough and filters are not needed. The disadvantages are large clamping diodes, unequal switching ratings, and real power control. An application in a back-to-back intertie connection is depicted in Fig. 6.26(c). The resulting harmonic distortion is within IEEE limits without filters. The THD versus m is in Fig. 6.27. The THD is also a function of the switching angle. Figure 6.27 illustrates the THD obtained with best switching angles.

6.10.1   Flying Capacitor (Capacitor-Clamped) Inverters

The clamping diodes can be replaced with capacitors called flying *capacitor-based control* or multilevel converters using cascade converters with DC sources. Figure 6.28 depicts the phase leg of a five-level flying capacitor-clamped inverter. The output voltage levels can be controlled by appropriate switching combinations. And the inverter has greater flexibility than a diode-clamped inverter. Table 6.6 shows switching combinations for output voltages. Note that the output voltage can be obtained with a number of combinations.

6.10.2   Multilevel Inverters Using H-Bridge Converters

Figure 6.29 illustrates the basic structure of multilevel inverters using H-bridge converters. This shows one phase leg with three HBs. Each HB is supplied from a separate

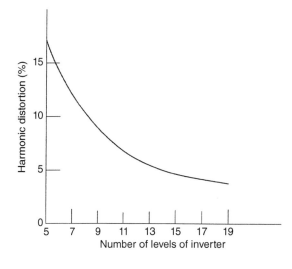

Figure 6.27   Harmonic distortion versus number of level of multilevel inverters, with optimized switching angles.

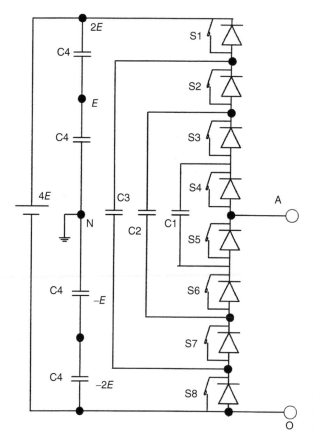

Figure 6.28   Five-level, capacitor-clamped multilevel inverter circuit.

TABLE 6.6   Switching Operation – Flying Capacitor Inverter, Fig. 6.28

Output Voltage	S1 Upper	S2 Upper	S3 Upper	S4 Upper	S5 Lower	S6 Lower	S7 Lower	S8 Lower
2E	X	X	X	X				
E	X	X	X		X			
		X	X	X				X
	X		X	X			X	
0	X	X			X			X
		X	X				X	X
	X		X		X		X	
	X			X	X			X
		X		X		X		X
		X	X			X	X	
-E	X				X	X	X	
			X		X	X	X	X
		X		X	X	X	X
−2E					X	X	X	X

DC source. The resulting voltage is synthesized by the addition of voltage generated by three HBs per leg.

When the DC-link voltages are the same, the multilevel inverter is called a CMI (cascaded multilevel inverter), and when the voltages are different it is called a binary hybrid multilevel inverter.

In binary hybrid multilevel inverter, the DC-link voltage of HBi is

$$V_{\mathrm{DC}i} = 2^{i-1}E \tag{6.33}$$

That is, for a three-level inverter of Fig. 6.29, $V_{\mathrm{DC1}} = E$, $V_{\mathrm{DC2}} = 2E$, and $V_{\mathrm{DC3}} = 4E$. The operation is listed in Table 6.7, and Fig. 6.30 shows that the output waveform has 15 levels. Note that the HB with higher DC-link voltage has a lower number of commutations.

For a cascade equal voltage multilevel inverter (CEMI), we have

$$V_{\mathrm{DC1}} = V_{\mathrm{DC2}} = V_{\mathrm{DC3}} = E \tag{6.34}$$

The output waveform is shown in Fig. 6.31.

6.10.3   THMI Inverters

A trinary hybrid inverter (THMI) has a circuit similar to Fig. 6.29, except that the ratio of DC-link voltages is $1{:}3{:}\ldots 3^{h-1}$, where h is the number of HBs. Thus, a maximum of 3^h voltage levels can be synthesized. The DC line voltages are as follows:

$$V_{\mathrm{DC}i} = 3^{i-1}E \tag{6.35}$$

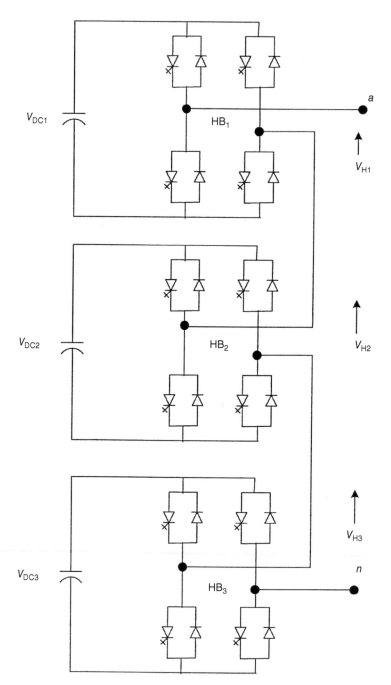

Figure 6.29 Multilevel inverter based on series connection of HBs.

TABLE 6.7   Operation of Multilevel Inverter, Fig. 6.29

E	v_{H1}	v_{H2}	v_{H3}
0	0	0	0
+1E	E	0	0
+2E	0	2E	0
+3E	E	2E	0
+4E	0	0	4E
+5E	E	0	4E
+6E	0	2E	4E
+7E	E	2E	4E
−E	−E	0	0
−2E	0	−2E	0
−3E	−E	−2E	0
−4E	0	0	−4E
−5E	−E	0	−4E
−6E	0	−2E	−4E
−7E	−E	−2E	−4E

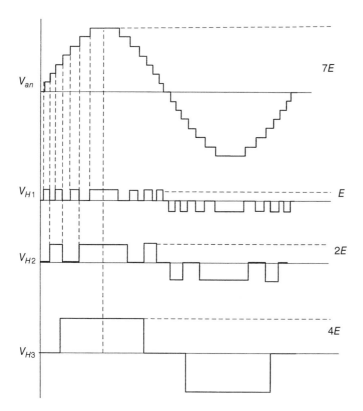

Figure 6.30   Waveform of BHMI, 15-level.

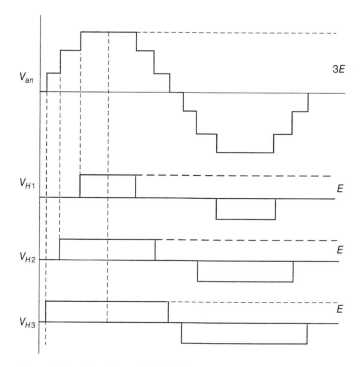

Figure 6.31 Waveform of a CMI.

That is, for three-HB one phase leg, $V_{DC1} = E$, $V_{DC2} = 3E$, and $V_{DC3} = 9E$. The output voltage, as before, is

$$v_{an} = \sum_{i=1}^{h} v_{Hi} = \sum_{i=1}^{h} F_i V_{DCi} \qquad (6.36)$$

where F_i is a switching function. In a three-HB per phase leg, v_{an} has 27 levels and by proper switching angles, the odd and even harmonics can practically be eliminated. It can be shown that THMI has the greatest number of output voltage levels. A number of modulation schemes can be applied. The harmonic generation is dependent on the modulation strategy and modulation index, which considers switching angles.

The GMI can balance each DC voltage level automatically at any number of levels.

6.11 SWITCHING ALGORITHMS FOR HARMONIC CONTROL

There are a number of modulation strategies that impact the output waveform and the harmonic distortion. We examined that a staircase voltage is generated with DC sources and proper control of switching angles. The switches in the inverters need to be switched on and off during one fundamental cycle; however, low-frequency harmonics occur. The optimization of the switching angles for selected harmonic elimination is one method of harmonic limitation, the other being increasing the switching frequency and space vector, that is, PWM for two-level inverters or multicarrier-based PWM for multilevel inverters [38,39].

The Fourier series expansion of the stepped output voltage waveform of a stepped multilevel inverter with nonequal DC sources is

$$V = \sum_{h=1,3,5..}^{\infty} \frac{4V_{DC}}{h\pi} \{(V_1 \cos(h\theta_1) + ...V_S \cos(h\theta_s))\} \sin(h\omega t) \qquad (6.37)$$

where h is the harmonic order and s is the number of DC sources. The product $V_1 V_{DC}$ is the value of ith DC source. If all DC sources have the same value V_{DC}, then $V_1 = V_2 = V_s = 1$. The voltage waveform of a 11-level cascade inverter with five full bridges and five DC sources (which are all of equal magnitude) is illustrated in Fig. 6.32. The output phase voltage is given by $V_{an} = v_1 + v_2 + v_3 + v_4 + v_5$. By controlling the switching angles, the harmonics of certain order are eliminated.

To calculate the angles for harmonic elimination, the following group of polynomials can be written as

$$\frac{4V_{DC}}{\pi}[\cos \theta_1 + \cos \theta_2 + \cos \theta_3 + \cos \theta_4 + \cos \theta_5] = V_F$$

$$\cos 5\theta_1 + \cos 5\theta_2 + \cos 5\theta_3 + \cos 5\theta_4 + 5 \cos \theta_5 = 0$$

$$\cos 7\theta_1 + \cos 7\theta_2 + \cos 7\theta_3 + \cos 7\theta_4 + \cos 7\theta_5 = 0$$

$$\cos 11\theta_1 + \cos 11\theta_2 + \cos 11\theta_3 + \cos 11\theta_4 + \cos 11\theta_5 = 0$$

$$\cos 13\theta_1 + \cos 13\theta_2 + \cos 13\theta_3 + \cos 13\theta_4 + \cos 13\theta_5 = 0 \qquad (6.38)$$

The first equation provides the desired fundamental component V_F. The switching angles $0 \le \theta_1 < \theta_2 < ... < \theta_s \le \pi/2$ are chosen to make the first harmonic equal to the fundamental voltage V_F. The other equations are utilized to eliminate 5th, 7th, 11th, and 13th harmonics.

There are different approaches to solve these equations. One method of solving these equations is Newton–Raphson iterative method.

6.12   THEORY OF RESULTANTS OF POLYNOMIALS

The basic concepts of mathematical theory of resultants of polynomials are as follows:

Given two polynomials $a(x_1, x_2)$ and $b(x_1, x_2)$, the common zeros satisfy the relation:

$$a(x_{10}, x_{20}) = b(x_{10}, x_{20}) = 0 \qquad (6.39)$$

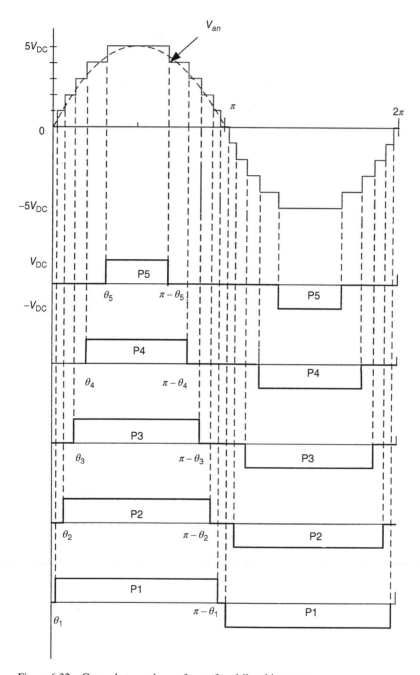

Figure 6.32 General stepped waveform of multilevel inverters.

If $a(x_1, x_2)$ and $b(x_1, x_2)$ are considered polynomials in x_2, whose coefficients are polynomials in x_1, then there is always a polynomial $r(x_1)$, called the resultant polynomial such that

$$\alpha(x_1, x_2)a(x_1, x_2) + \beta(x_1, x_2)b(x_1, x_2) = r(x_1) \qquad (6.40)$$

Then, if Eq. (6.39) is satisfied, then $r(x_1) = 0$. This means that if (x_{10}, x_{20}) is a common zero of the pair $a(x_1, x_2)$, $b(x_1, x_2)$, then the first coordinate x_{10} is a zero of $r(x_1) = 0$. The $r(x_1)$ can be found by the following manipulation:

$$a(x_1, x_2) = a_3(x_1)x_2^3 + a_2(x_1)x_2^2 + a_1(x_1)x_2 + a_0(x_1)$$

$$b(x_1, x_2) = b_2(x_1)x_2^2 + b_1(x_1)x_2 + b_0(x_1) \qquad (6.41)$$

Check if the polynomials of the form:

$$\alpha(x_1, x_2) = \alpha_1(x_1)x_2 + \alpha_0(x_1)$$

$$\beta(x_1, x_2) = \beta_2(x_1)x_2^2 + \beta_1(x_1)x_2 + \beta_0(x_1) \qquad (6.42)$$

can be found so that Eq. (6.40) is satisfied.

Equating powers of x_2, this equation can be written as

$$
\begin{vmatrix}
a_0(x_1) & 0 & b_0(x_1) & 0 & 0 \\
a_1(x_1) & a_0(x_1) & b_1(x_1) & b_0(x_1) & 0 \\
a_2(x_1) & a_1(x_1) & b_2(x_1) & b_1(x_1) & b_0(x_1) \\
a_3(x_1) & a_2(x_1) & 0 & b_2(x_1) & b_1(x_1) \\
0 & a_3(x_1) & 0 & 0 & b_2(x_1)
\end{vmatrix}
\begin{vmatrix}
\alpha_0(x_1) \\
\alpha_1(x_1) \\
\beta_0(x_1) \\
\beta_1(x_1) \\
\beta_2(x_1)
\end{vmatrix}
=
\begin{vmatrix}
r(x_1) \\
0 \\
0 \\
0 \\
0
\end{vmatrix}
\qquad (6.43)
$$

The matrix on the left-hand side is called the Sylvester matrix denoted by $S_{a,b}(x_1)$. The inverse of this matrix can be found by matrix manipulation techniques. Solving for $\alpha_i(x_1), \beta_i(x_1)$ gives

$$
\begin{vmatrix}
\alpha_0(x_1) \\
\alpha_1(x_1) \\
\beta_0(x_1) \\
\beta_1(x_1) \\
\beta_2(x_1)
\end{vmatrix}
= \frac{\text{adj } S_{a,b}(x_1)}{\det S_{a,b}(x_1)}
\begin{vmatrix}
r(x_1) \\
0 \\
0 \\
0 \\
0
\end{vmatrix}
\qquad (6.44)
$$

The resultant polynomial defined by $r(x_1)$ is det. of Sylvester matrix. See Ref. [40–43] for further discussions. This theory can be applied for optimization of the switching angles for minimum harmonic emission.

6.12.1 A Specific Application

Continuing with Eq. (6.38), let us consider three DC levels as an example of the calculation procedure. The set of equations is first converted into a polynomial system by setting:

$$x_1 = \cos \theta_1$$
$$x_2 = \cos \theta_2$$
$$x_3 = \cos \theta_3 \tag{6.45}$$

Also the following trigonometric identities exist:

$$\cos 5\theta = 5 \cos \theta - 20 \cos^3 \theta + 16 \cos^5 \theta$$
$$\cos 7\theta = -7 \cos \theta + 56 \cos^3 \theta - 112 \cos^5 \theta + 64 \cos^7 \theta \tag{6.46}$$

Then the equivalent equations are as follows:

$$p_1(x) \cong V_1 x_1 + V_2 x_2 + V_3 x_3 - m = 0$$

$$p_5(x) \cong \sum_{i=1}^{3} V_i(5x_i - 20x_i^3 + 16x_i^5) = 0$$

$$p_7(x) = \sum_{i=1}^{3} V_i(-7x_i + 56x_i^3 - 112x_i^5 + 64x_i^7) = 0 \tag{6.47}$$

where $x = (x_1, x_2, x_3)$ and

$$m \cong \frac{V_1}{(4V_{DC}/\pi)} \tag{6.48}$$

This gives a set of three polynomial equations in three unknowns. The modulation index $m_a = m/s = V_f/(s4V_{DC}/\pi)$. Because each converter has a DC voltage of V_{DC}, the maximum fundamental output possible is $4sV_{DC}/\pi$. The constraint

$$0 \le x_3 < x_2 < x_1 \le 1 \tag{6.49}$$

applies. The goal is to find the simultaneous solutions of $p_5(x_1, x_2)$ and $p_7(x_1, x_2)$ as discussed above by elimination using resultants. There are more than one set of solutions as m is varied, and a solution that gives the minimum THD can be chosen. The variations of source voltages can be accounted for.

When the number of DC levels increase, the number of equations, their order, and the number of variables increase. Finding solution to these equations requires advanced mathematical algorithms and is complex.

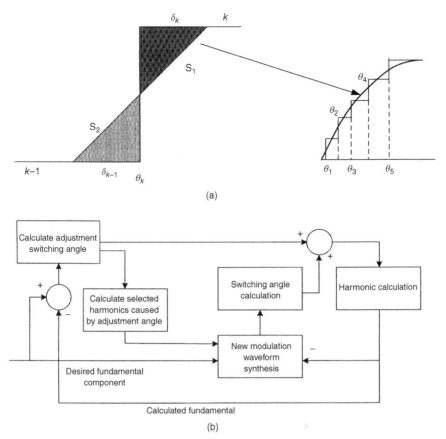

(a)

(b)

Figure 6.33   (a) Equal area criteria; (b) control circuit diagram with adjustment of switching angle for the highest voltage level.

To simplify these calculations, Ref. [44] proposes a four-equation-based harmonic elimination method, mathematically simpler to solve. It is based on equal area criteria for the switching angle.

$$S_1 = S_2 \tag{6.50}$$

In Fig. 6.33(a), an enlarged view of a staircase step is shown. The equal area criterion alone will not eliminate harmonics, and staircase waveform thus generated will have harmonics. By taking $(h_1 - h_5 - h_7 \ldots)$ as the modulation waveform, the harmonic content in the resulting waveform will be

$$(h_5 + h_7 + h_{11} \ldots) - (h_5' + h_7' + h_{11}' \ldots) \tag{6.51}$$

where $(h_5 + h_7 + h_{11} \ldots)$ is generated by h_1 and $-(h_5' + h_7' + h_{11}' \ldots)$ is generated by $-(h_5 + h_7 + h_{11} \ldots)$. Because of equal area criteria, $-(h_5' + h_7' + h_{11}' \ldots)$ will nearly follow $(h_5 + h_7 + h_{11} \ldots)$. The harmonic elimination is thus realized and by iteration the desired harmonics can be eliminated.

To use equal area criterion, δ_k, which is the junction point of modulation waveform at voltage level k (Fig. 6.33(a)), is found by Newton–Raphson method:

$$\delta_k = \tan^{-1}\left(\frac{kV_{\mathrm{DC}} + h_5 \sin\left(5\delta_k\right) \ldots h_m \sin(m\delta_k)}{V_F \cos(\delta_k)}\right) \qquad (6.52)$$

TABLE 6.8   Switching Angles, MI = 0.84

Angle	Radians	Angle	Radians
θ_1	0.5995	θ_5	0.50503
θ_2	0.18863	θ_6	0.63771
θ_3	0.28101	θ_7	0.87771
θ_4	0.36322	θ_8	1.0889

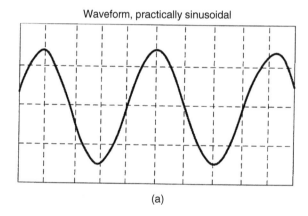

Waveform, practically sinusoidal

(a)

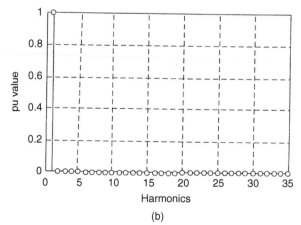

Harmonics

(b)

Figure 6.34   (a) Output line voltage, $M1 = 0.84$: FFT analysis results.
Source: Based on Ref. [34].

After $\delta_k's$ are found, the switching angles can be calculated

$$\theta_k = k\delta_k - (k-1)\delta_{k-1} + V_F(\cos(\delta_k) - \cos(\delta_{k-1})) - \frac{h_5}{5}(\cos(5\delta_k) - \cos(5\delta_{k-1}))$$

$$- ..... - \frac{h_m}{m}(\cos(m\delta_k) - \cos(m\delta_{k-1})) \tag{6.53}$$

With new sets of $\delta_k's$, new values of harmonic currents are found and the process is repeated. A control circuit diagram is shown in Fig. 6.33(b).

Test results of verification for a 17-level cascade inverter designed to deliver 1 MVA at 6000 V are provided in Ref. [44]. Each phase of the inverter contains a three-level diode-clamped H-bridge; eight switching angles are needed to create 17 voltage levels. These angles are shown in Table 6.8. Figure 6.34(a) illustrates output line-to-line voltage at MI of 0.84 and Fig. 6.34(b) illustrates its FFT. The aimed harmonic distortion is only 1 Pico pu.

The technical literature on harmonic elimination is vast and this subject is of much research. Much advancement in power electronic topologies and controls has taken place in the recent years. This chapter is not exhaustive and, practically, all such topologies cannot be covered. The techniques described in this chapter form an introduction; references are provided for the interested reader to probe further.

REFERENCES

1. N. G. Hingorani and L. Gyugyi. Understanding FACTS, IEEE Press, NJ, 2001.
2. C. Schauder, M. Gernhard, E. Stacey, T. Lemak, L. Gyugi, T. W. Cease, and A. Edris. "Development of a 100 Mvar static condenser for voltage control of transmission systems," IEEE Transactions on PWRD, vol. 10, pp. 1486–1496, 1995.
3. IEEE Standard 519. IEEE recommended practice and requirements for harmonic control in electrical systems, 1992.
4. A. Mansoor, W. M. Grady, et al, "An investigation of harmonic attenuation and diversity among distributed single phase power electronic loads," Proceedings of the 1994 IEEE T&D Conference, Chicago, IL, April 10–15, 1994, pp. 110–116.
5. IEEE P519.1. Draft guide for applying harmonic limits on power systems, 2004.
6. H Akagi. "Trends in active power line conditioners," IEEE Transactions on Power Electronics vol. 9, pp. 263–268, 1994.
7. W. M. Grady, M. J. Samotyi, A. H. Noyola. "Survey of active line conditioning methodologies," IEEE Transactions on Power Delivery vol. 5, pp. 1536–1541, 1990.
8. H Akagi. "New trends in active filters for power conditioning," IEEE Transactions on Industry Applications, vol. 32: 1312–1322, 1996.
9. F. Z. Peng, "Application issues of active power filters" IEEE Industry Application Magazine, pp. 21–30, September/October 1998.
10. P. N. Enjeti, W. Shireen, P. Packebush and I. J. Pitel. "Analysis and design of new active power filter to cancel neutral current harmonics in three-phase four-wire electric distribution systems," IEEE Transactions on Industry Applications, vol. 30, no. 6, pp. 1565–1571, 1994.
11. F. Z. Peng, H. Akagi and A. Nabe, "A new approach to harmonic compensation in power systems-A combined system of passive and active filters," IEEE Transactions on Industry Applications, vol. 26, no. 6, 1990.
12. Y. Sato, T. Kawase, M. Akiyama, and T. Kataoka. "A control strategy for general-purpose active filters based on voltage detection," IEEE Transactions on Industry Applications, vol. 36, no. 5, pp. 1405–1412, 2000.

13. H. Akagi, "Control strategy and site selection of a shunt active filter for damping of harmonic propagation is distribution systems," in conf record, IEEE/PES Winter Meeting, 96.
14. H. Akagi, A. Nabae and S. Atoh, "Control strategy of active power filters using multiple voltage-source PWM converters," IEEE Transactions on Industry Applications, vol. 22, no. 3, pp. 460–465, 1986.
15. A. Cavallini, G. C. Montanarion. "Compensation strategies for shunt active filter control," IEEE Transactions on Power Electronics vol. 9, pp. 587–593, 1994.
16. C. V. Nunez-Noriega, G. G. Karady. "Five Step-low frequency switching active filter for network harmonic compensation in substations," IEEE Transactions on Power Delivery, vol. 14, pp. 1298–1303, 1999.
17. P. Jintakosonwit, H. Fujita, H. Akagi and S. Ogasawara. "Implementation of performance of cooperative control of shunt active filters for harmonic damping throughout a power distribution system," IEEE Transactions on Industry Applications, vol. 39, no. 2, pp. 556–563, 2003.
18. S. Buso, L. Malesani, P. Mattavelli, and R. Veronese, "Design of fully digital control of parallel active filters for thyristor rectifiers to comply with IEC-1000-3-2 standards," IEEE Transactions on Industry Applications, vol. 34, no. 3, pp. 508–517, 1998.
19. S. Fuuda and T. Yoda. "A novel current-tracking method for active filters based on sinusoidal internal model," IEEE Transactions on Industry Applications, vol. 37, no. 3, pp. 888–895, 2001.
20. S. Buso, S. Fasolo, L. Malesani, and P. Mattavelli. "A dead-beat adaptive hysteresis current control," IEEE Transactions on Industry Applications, vol. 36, no. 4, pp. 1174–1180, 2000.
21. Y. Hayashi, N. Sato and K. Takahashi. "A novel control of a current source active filter for ac power system harmonic compensation," IEEE Transactions on Industry Applications, vol. 27, no. 2, pp. 380–385, 1991.
22. H. Fujita and H. Akagi. "An approach to harmonic current-free AC/DC power conversion for large industrial loads: The integration of series active filter with a double-series diode rectifier." IEEE Transaction on Industry Applications, vol. 33, no. 5, pp. 1233–1240, 1997.
23. P. T. Cheng, S. Bhattacharya, and D. M. Divan. "Operations of dominant harmonic active filter (DHAF) under realistic utility conditions," IEEE Transactions on Industry Applications, vol. 37, no. 4, pp. 1037–1044, 2001.
24. P. Mattvelli. "Synchronous frame harmonic control for high-performance AC power supplies," IEEE Transactions on Industry Applications, vol. 37, no. 30, pp. 864–872, 2001.
25. H. Fujita and H. Akagi. "A practical approach to harmonic compensation in power systems-series connection of passive and active filters," IEEE Transactions on Industry Applications, vol. 27, no. 6, pp. 1020–1025, 1991.
26. M. Rastogi, R. Naik and N. Mohan. "A comparative evaluation of harmonic reduction techniques in three-phase utility interface of power electronic loads," IEEE Transactions on Industry Applications, vol. 30, no. 5, pp. 1149–1155. 1994.
27. S. Bhattacharya, P. -T. Cheng and D. M. Divan. "Hybrid solutions of improving passive filter performance in high power applications," IEEE Transactions on Industry Applications, vol. 33, no. 3, pp. 732–747, 1997.
28. F.Z. Peng, "Z-source inverter," IEEE Transactions on Industry Applications, vol. 39, pp. 504–510, 2003
29. J. Anderson, F. Z. Peng, "Four quasi Z-source inverters," Proceedings of IEEE PESC, pp. 2743–2749, 2008.
30. P. C. Loh, F. Gao, P. C. Tan and F. Blabjerg, "Three-level AC-DC-AC Z-source converter using reduced passive component count," Proceedings of the IEEE PESC, pp. 2691–2697, 2007.
31. M. Venturine. "A new sine-wave in sine-wave out converter technique eliminated reactor elements," Proceedings of Powercon, pp. E3-1–E3-15, 1980.
32. F. L. Luo and Z. Y. Pan. "Sub-envelope modulation method to reduce total harmonic distortion of AC/DC matrix converters," Proceedings of IEEE Conference PESC, pp. 2260–2265, 2006.
33. A. Nabae, I. Takahashi and H. Akagi. "A neutral point clamped PWM inverter," IEEE Trans. Industry Applications, vol. 17, pp. 518–523, 1981.
34. J. S. Lai and F. Z. Peng. "Multilevel converters-A new breed of power converters," IEEE Transactions on Industry Applications, vol. 32, pp. 509–517, 1996.
35. P. W. Hammond. "New approach to enhance power quality for medium voltage AC drives," IEEE Transactions on Industry Applications. vol. 33, pp. 202–208, 1997.

36. M. Trzymadlowski. Introduction to Modern Power Electronics. John Wiley, New York 1998.

37. F. L. Luo and H. Ye. Power Electronics-Advanced Conversion Technologies. CRC Press, Boca Raton, FL, 2010.

38. L. Li, D. Czarkowski, Y. G. Liu and P. Pillay, "Multilevel selective harmonic elimination PWM technique based on phase shift harmonic suppression," IEEE Transactions on Industry Application, vol. 36, no. 1, pp. 160–170, 2000.

39. H. S. Patel and R. G. Hoft, "Generalized techniques of harmonic elimination and voltage control in thyristor inverters: Part 1-Harmonic elimination," IEEE Transactions on Industry Applications, vol. IA-9, no. 3, pp. 310–317, 1973.

40. C. Chen, Linear System Theory and Design, Third Edition, Oxford Press, 1999.

41. D. Cox, J. Little, and D. O'Shea, An Introduction to Computational Algebraic Geometry and Cumulative Algebra, Second Edition, Springer-Verlag, 1996.

42. L. M. Tolbert, J. N. Chiasson, Z. Du and K. J. McKenzie. "Elimination of harmonics in a multilevel converter with non-equal DC sources," IEEE Transactions on Industry Applications, vol. 41, no. 1, 2005.

43. J. N. Chiasson, L. M. Tolbert, J. Du, and K. J. McKenzie. "The use of power sums to solve the harmonic elimination equations for multi-level converters," European Power Electric Drives, vol. 15, no. 1, pp. 9–27, 2005.

44. J. Wang, D. Ahmadi, "A precise and practical harmonic elimination method for multilevel converters," IEEE Transactions on Industry Applications, vol. 46, no. 2, pp. 857–865, 2010, Chapter 6.

CHAPTER 7

ESTIMATION AND MEASUREMENTS OF HARMONICS

We have seen that wide variations in the harmonic emissions occur depending on topologies. For the purpose of harmonic load flow and harmonic filter designs, the harmonic emission from the nonlinear loads should be accurately estimated. *This is the first step*, and it is not an easy task due to varied topologies and harmonic limitation at source. There are three possibilities:

- For small systems and 6-pulse converters, enough publications and data exist to estimate the harmonics analytically. The harmonic emissions from 6-pulse or 12-pulse current source or voltage source converters have been widely discussed in the literature.

- For the electrical systems where the harmonic loads are in operation, the harmonics can be estimated by online measurements. These should be done under various operating conditions as the emissions will vary depending on the variations in the processes. The harmonic resonance, if any, the time stamp of harmonics, and the current and voltage distortions at the various buses and point of common coupling (PCC) can be captured.

- The harmonic spectra and angle of each harmonic including noncharacteristic harmonics can be obtained from the vendors. As the harmonic emission is also a function of the operating load and system impedance, it is necessary to obtain this data under various operating conditions and system impedance variations at the point of application of the nonlinear loads.

Online measurement of harmonics will provide reliable results when this analysis is conducted properly over a period of time; however, the measurement techniques cannot be applied for the power systems in the design stage. More often than not, the harmonic generation needs to be estimated at the design stage of the power systems, so that appropriate measures could be taken to meet the requirements of harmonic limitations as per standards. When shunt power capacitors are applied along with harmonic producing loads, further complications of harmonic resonance can occur at

Power System Harmonics and Passive Filter Designs, First Edition. J.C. Das.
© 2015 The Institute of Electrical and Electronics Engineers, Inc. Published 2015 by John Wiley & Sons, Inc.

one of the load-generated frequencies. These important issues must be studied at the design stage.

Enough data have been gathered for the harmonic emissions from various sources in typical installations, yet the system impact must be considered, the system source impedance at the point of connection of harmonic producing load impacts the harmonic emission.

Six-pulse current source converters are analyzed. Figure 7.1 shows the line current waveforms. The theoretical or textbook waveform is rectangular and it considered instantaneous commutation (Fig. 7.1(a)). The effect of commutation delay and firing angle still retains the flat-top assumption (Fig. 7.1(b)), but the waveform has lost the even symmetry with respect to the center of the idealized rectangular pulse. The DC current is not flat-topped and the actual waveform has a ripple (Fig. 7.1(c)). For lower values of the DC reactor and large phase-control angles, the current is discontinuous (Fig. 7.1(d)).

7.1   WAVEFORM WITHOUT RIPPLE CONTENT

The harmonic emissions from six-pulse current source converters have been investigated in Ref [1–4]. The first approximation to the wave with overlap angle is to regard it as a trapezoidal wave shape in which the leading and trailing edges are assumed to be linear. The waveform becomes symmetrical and easy to analyze; Fig. 7.2(a) shows the actual wave shape while Fig. 7.2(b) depicts its approximation.

The classical work was done by Read [2] and Figs. 7.3–7.10 illustrate the harmonic emissions based on the firing angle α for rectifiers, and $\beta - \mu$ for the inverters for harmonics of the order of 5th, 7th, 11th, 13th, 17th, 19th, 23rd, and 25th, respectively. The X-axis scales the parameter $X_T + X_S$, where X_T is the percentage reactance of the rectifier transformer and X_S is the percentage impedance of the supply system. These are referred to rms current in the transformer primary lines equal to that corresponding to DC current I_d, assuming rectangular shaped currents and no magnetizing current. Also,

$$X_T + X_S = \frac{2\varepsilon}{V_{\text{do}}} \times 100 \tag{7.1}$$

where ε is the change of DC voltage from no load to load current I_d due to reactance drop in volts.

Example 7.1:   Consider a 13.8 kV three-phase 60 Hz power source with a 2 MVA step-down transformer of 13.8–0.48 kV, serving a six-pulse current source converter. The transformer impedance is 5.75% on 2 MVA base. If converter base load is considered equal to the transformer KVA rating, then $X_T = 5.75\%$. The supply source impedance $X_S = 0.004$ per unit (= 0.4%). The transformer impedance predominates.

If the converter is being operated at a firing angle $\alpha = 45°$, then for abscissa = 6.15, the various harmonics can be directly read from Figs. 7.3 to 7.10. For example, fifth harmonic from Fig. 7.3 can be read as approximately = 19% of the fundamental.

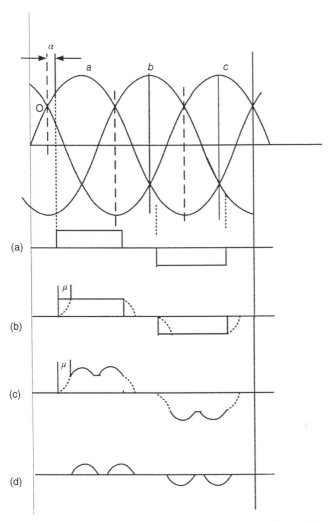

Figure 7.1   (a) Rectangular current waveform; (b) waveform with commutation angle; (c) waveform with ripple content; (d) discontinuous waveform due to large delay angle control.

7.1.1   Geometric Construction for Estimation of Harmonics

The following geometric construction is also provided in Ref [2]. Referring to Fig. 7.11,

$$OR = \frac{\sin mp.\mu/2}{mp}$$

$$OS = \frac{\sin(mp + 2).\mu/2}{mp + 2} \qquad (7.2)$$

$$OT = \frac{\sin(mp - 2).\mu/2}{mp - 2}$$

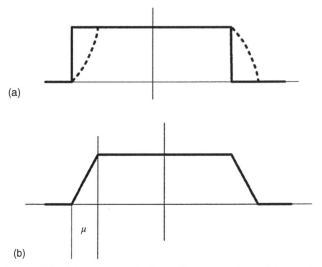

(a)

(b)

Figure 7.2 (a) Unsymmetrical waveform due to large delay angle; (b) approximation to symmetrical trapezoidal waveform.

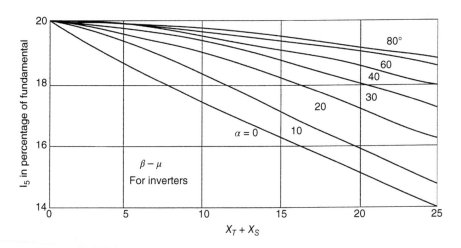

Figure 7.3 Fifth harmonic of primary line current $p = \leq 6$.

where m is any integer $= 1, 2, 3 \ldots$ and p is the pulse number. Then from Fig. 7.11,

$$I_{h(mp+1)} = \frac{I_1}{2(mp+1)\sin\theta.\sin\mu/2} \times \text{RS or RS}'$$

$$I_{h(mp-1)} = \frac{I_1}{2(mp-1)\sin\theta.\sin\mu/2} \times \text{RT or RT}' \qquad (7.3)$$

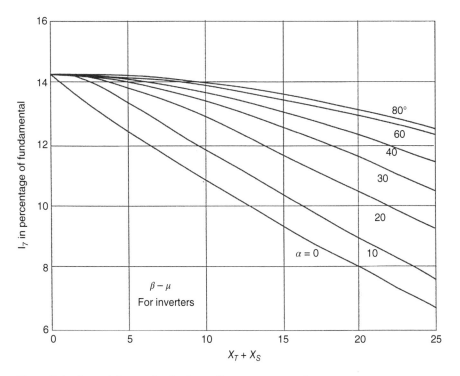

Figure 7.4   Seventh harmonic of primary line current $p =\leq 6$.

where

$$\theta = \alpha + \mu/2 \quad \text{Rectifiers}$$

$$= \beta - \mu/2 \quad \text{Inverters}$$

$$I_{h(mp+1)} = \text{harmonic current, harmonic of order } mp + 1$$

$$I_{h(mp-1)} = \text{harmonic current, harmonic of order } mp - 1 \tag{7.4}$$

Example 7.2: Calculate the fifth harmonic in Example 7.1 with the analytical/geometric expressions in Eq. (7.3).

First, calculate the overlap angle:

$$\mu = \cos^{-1}[\cos\alpha - (X_S + X_t)I_d] - \alpha$$

Substituting the values,

$$\mu = 4.8°. \text{ Note that } I_d \text{ is considered in per unit} = 1.$$

Then, from the geometric construction, OR = 0.04145, OS = 0.04111, OT = 0.041692, RT = 0.057, $\theta = 47.4°$, and fifth harmonic from Eq. (7.3) = 18.6%.

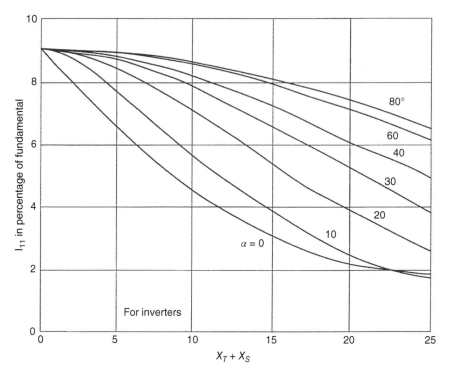

Figure 7.5 Eleventh harmonic of primary line current $p =\leq 12$.

7.1.2 Harmonic Estimation Using IEEE 519 Equations

An estimation of the harmonics ignoring waveform ripple is provided by Eqs. (4.37)–(4.39) from IEEE 519. These equations are reproduced below for ease of reference:

$$I_h = I_d \sqrt{\frac{6}{\pi}} \frac{\sqrt{A^2 + B^2 - 2AB \cos(2\alpha + \mu)}}{h[\cos \alpha - \cos(\alpha + \mu)]} \qquad (7.5)$$

where

$$A = \frac{\sin\left[(h-1)\frac{\mu}{2}\right]}{h-1} \qquad (7.6)$$

and

$$B = \frac{\sin\left[(h+1)\frac{\mu}{2}\right]}{h+1} \qquad (7.7)$$

Equation (7.5) can also be written as

$$\frac{I_h}{I_1} = \frac{\sqrt{A^2 + B^2 - 2AB \cos(2\alpha + \mu)}}{h[\cos \alpha - \cos(\alpha + \mu)]} \qquad (7.8)$$

where I_1 is the fundamental frequency current.

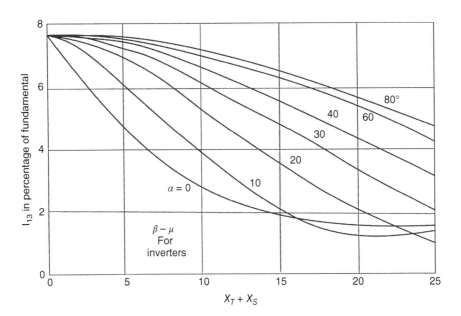

Figure 7.6   Thirteenth harmonic of primary line current $p = \leq 12$.

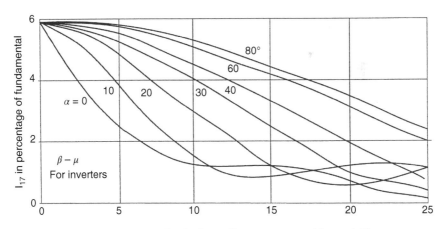

Figure 7.7   Seventeenth harmonic of primary line current $p = \leq 18$ or $p \neq 12$.

Example 7.3:   We will calculate the magnitude of the fifth harmonic in Example 7.2. From Eqs. (7.6)–(7.8), $A = 0.0417$, $B = 0.0414$. Substituting these values into Eq. (7.8), we have a fifth harmonic current equal to 18.95% of the fundamental current.

Thus, we see that calculations in Examples 7.1–7.3 give consistent results.

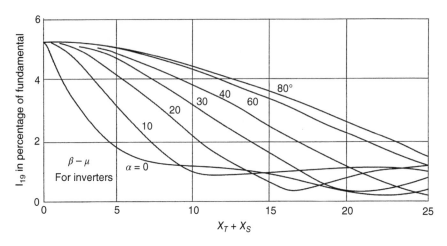

Figure 7.8 Nineteenth harmonic of primary line current $p = \leq 18$ or $p \neq 12$.

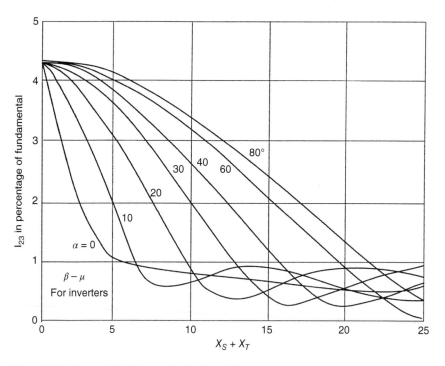

Figure 7.9 Twenty-third harmonic of primary line current $p = \leq 21$ or $p \neq 18$.

7.2 WAVEFORM WITH RIPPLE CONTENT

The above-mentioned equations ignored ripple content. Figure 7.12, from IEEE 519, considers a ripple content that is sinusoidal, and a sine half-wave is superimposed on the trapezoidal waveform. There is linear rate of rise and fall at the leading and trailing

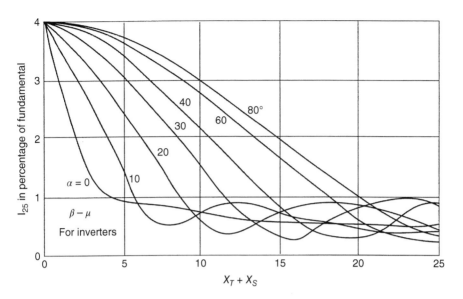

Figure 7.10   Twenty-fifth harmonic of primary line current $p = \leq 24$ or $p \neq 18$.

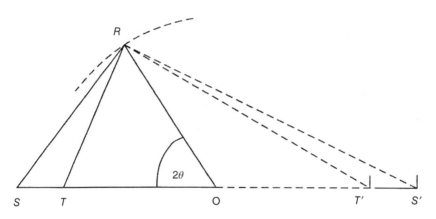

Figure 7.11   Construction for values of two harmonics in line currents.

edges. The two current lobes are equally displaced, which form the center. These are represented by tops of sine waves appropriately displaced from system zero. For the interval between the lobes, the current is assumed constant and equal to commutation current. This current is defined as I_c.

The change of the current between the end of commutation and peak of each sine lobe is represented by Δi, the peak ripple current.

This ripple is expressed as a fraction of the commutation current I_c and is defined as ripple coefficient r_c:

$$r_c = \frac{\Delta i}{I_c} \tag{7.9}$$

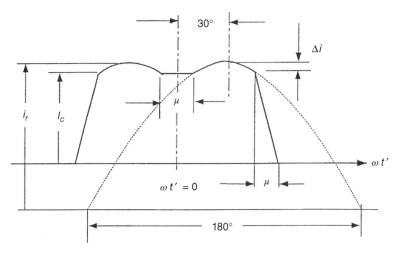

Figure 7.12   Trapezoidal current waveform with superimposed sinusoidal AC ripple.

Also the ripple ratio r is defined

$$r = \frac{\Delta i}{I_d} \tag{7.10}$$

where I_d is the average DC current.

7.2.1   Graphical Procedure for Estimating Harmonics with Ripple Content

Figures 7.13(a)–(h) provide the relative values of the harmonics in rms as I_h/I_1. To apply these figures, following steps of calculation are required:

- Short-circuit current based on supply side and transformer inductance is calculated.
- Decide the level of commutation current I_c.
- The DC voltage at the converter terminals is decided and expressed as a function of the maximum average DC voltage.
- The firing angle and the overlap angles can be read from Fig. 7.14.
- Figure 7.15 is used to read off voltage ripple area, A_r.
- The ripple coefficient r_c is calculated:

$$r_c = \frac{A_r(V_{do}/X_r)}{I_c} \tag{7.11}$$

where

$$X_r = \omega L_d + 2X_c \tag{7.12}$$

- The harmonics can be read off from Fig. 7.13(a)–(h).

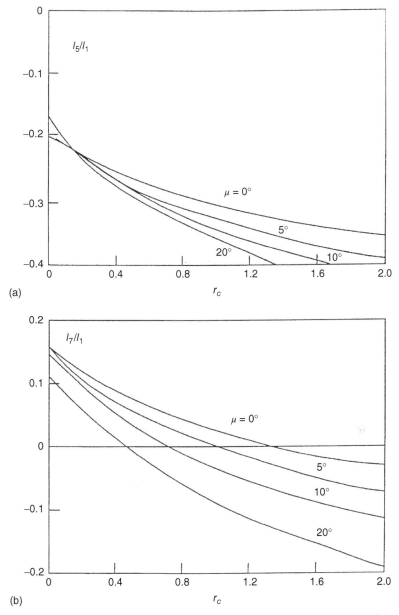

(a)

(b)

Figure 7.13   (a)–(h) Harmonic currents in per unit of fundamental frequency line current as a function of r_c.

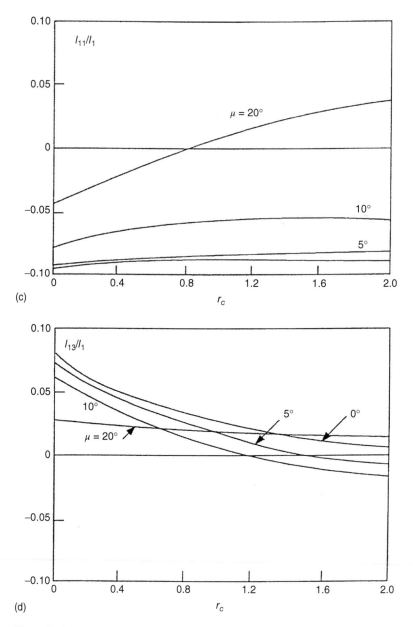

Figure 7.13   (*Continued*)

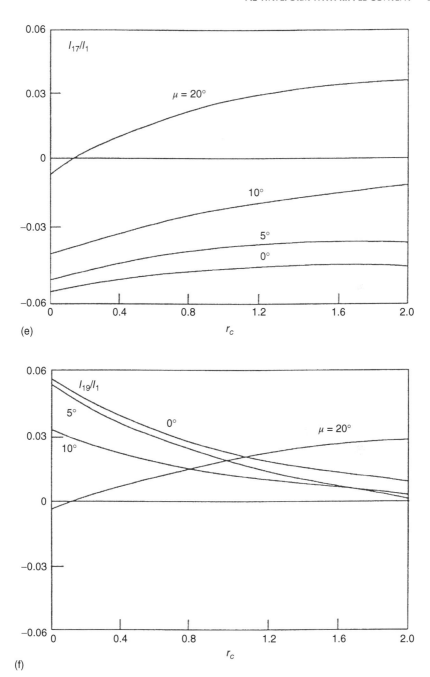

Figure 7.13   (*Continued*)

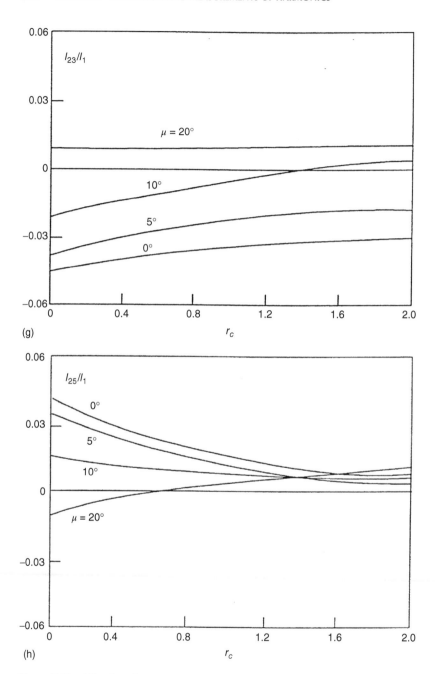

Figure 7.13   (*Continued*)

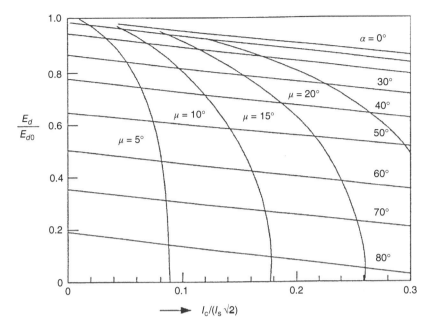

Figure 7.14 Converter load curves for a six-pulse current source converter.

- The harmonics can be converted into amperes by multiplying with a factor I_1/I_c from Fig. 7.16 and then by I_c.

Example 7.4: We will continue with the earlier examples to illustrate the method of calculation. Consider that $V_{do} = 2.34 \ (480/\sqrt{3}) = 648$ V. The short-circuit current on the 480 V bus is designated as $I_s = 39$ kA. Let $I_c = 2$ kA, and $V_d/V_{do} = 0.8$, where V_d is the DC operating voltage and V_{do} is the no-load voltage; then, $I_c/I_s\sqrt{2} = 0.036$. Using these values and from Fig. 7.14 for $a = 45°$, the overlap angle μ can be read as approximately 4.5°. Let us say it is 4.8°, as calculated before.

Entering the values of μ and $V_d/V_{do} = 0.8$ in Fig. 7.15, we read $A_r = 0.1$, where A_r is the voltage-ripple integral or the ripple area.

Calculate the ripple current $\Delta i = A_r V_{do}/X_r$:

$$X_r = \omega L_d + 2X_C$$

where X_r is ripple reactance, L_d is the inductance in the DC circuit in H, and X_C is the commutating reactance.

First consider that there is no DC inductance, then $X_r = 2X_C = 2 \times 0.0071$ ohms (the reactance of the transformer and the source referred to the 480 V side) and calculated $\Delta i = 4563$, then the ripple coefficient $r_c = \Delta i/I_c = 2.28$.

Now, the harmonics as a percentage of the fundamental current can be calculated graphically from Figs. 7.13(a)–(h) from Ref. [1]. Negative values indicate a phase shift of π, and I_1 is the fundamental frequency current.

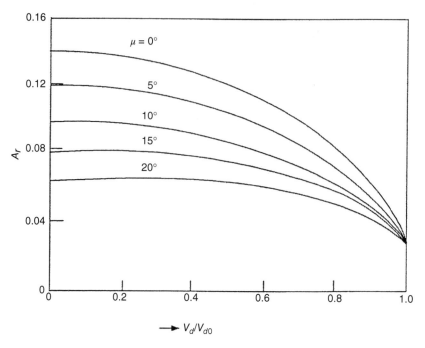

Figure 7.15 V_d/V_{do} versus A_r for various overlap angles.

Continuing with the example, and entering Fig. 7.13(a) for $\mu = 4.8$ and $r_c = 2.28$, we find that it is outside the range on the X-axis. An approximate value of 40% fifth harmonic can be read.

Note that the harmonics on the line side are impacted by the ripple content.

7.2.2 Analytical Calculations

The following equations are applicable from IEEE Standard 519 [5] (also see Ref [1]).

$$I_h = I_c \frac{2\sqrt{2}}{\pi} \left[\frac{\sin\left(\frac{h\pi}{3}\right) \sin\frac{h\mu}{2}}{h^2 \frac{\mu}{2}} + \frac{r_c g_h \cos\left(h\frac{\pi}{6}\right)}{1 - \sin\left(\frac{\pi}{3} + \frac{\mu}{2}\right)} \right] \tag{7.13}$$

where

$$g_h = \frac{\sin\left[(h+1)\left(\frac{\pi}{6} - \frac{\mu}{2}\right)\right]}{h+1} + \frac{\sin\left[(h-1)\left(\frac{\pi}{6} - \frac{\mu}{2}\right)\right]}{h-1}$$

$$- \frac{2\sin\left[h\left(\frac{\pi}{6} - \frac{\mu}{2}\right)\sin\left(\frac{\pi}{3} + \frac{\mu}{2}\right)\right]}{h} \tag{7.14}$$

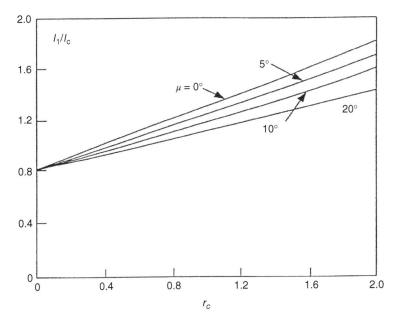

Figure 7.16 Fundamental line current and commutation current as a function of r_c.

where I_c is the value of the DC current at the end of the commutation and r_c is the ripple coefficient ($= \Delta i / I_c$).

In Fig. 7.12, the time zero reference is at $\omega t' = 0$, at the center of the current block. This is even symmetry and, therefore, only cosine terms are present. The instantaneous current is then

$$i_h = I_h \sqrt{2} \ \cos n\omega t' \tag{7.15}$$

More accurately, Eqs. (7.13) and (7.14) can be used.

Example 7.5: From Eq. (7.14), and substituting the numerical values:

$$g_h = \frac{\sin 165.6°}{6} + \frac{\sin 110.4°}{4} - \frac{2(\sin 138° \ \sin 62.4°)}{5} = 0.03875$$

and from Eq. (7.13):

$$I_h = \frac{2\sqrt{2}}{\pi} \left[\frac{\sin 300° \ \sin 12°}{1.0472} + \frac{(2.28)(0.03857)\cos 150°}{0.113} \right] = -0.726 I_c$$

The ratio of currents I_1 / I_c is given in Fig. 7.16. For $r_c = 2.28$, it is 1.8; therefore, the fifth harmonic in terms of the fundamental current is 42.3%. This is close to 40% estimated from the graphs.

7.2.3   Effect of DC Reactor

The ripple content is a function of the source reactance (commutating reactance) and also the L_d in the DC-link circuit. For zero DC output voltage, the DC current is

$$I_{dm} = 0.218 \frac{V_{\text{ln}}}{X_d + 2X_N} \tag{7.16}$$

where X_d is the reactance in the DC circuit and X_N is the total source reactance including that of the rectifier transformer.

Figure 7.17 shows the variations of the positive and negative sequence harmonics with various ripple contents in terms of I_{dm}/I_{d1}. Note that the harmonics at ratio $I_{dm}/I_{d1} = 1$, that is, no ripple are considered as the base. From this figure, fifth harmonic of approximately 2.4 times the base value occurs at maximum ripple content, which is equal to approximately 48% of the fundamental. Also the higher order harmonics such as 13th and 17th with ripple content are less than that for a rectangular waveform (also see Ref. [6]).

Example 7.6:   Consider that a DC inductor of 1 mH is added in Example 7.5. Recalculate the magnitude of the fifth harmonic.

Following the same procedure as in example, $X_r = 0.3912$ ohms, $\Delta i = 165.6$, the ripple coefficient $r_c = \Delta i / I_c = 0.08$, and the fifth harmonic reduces to approximately 20% of the fundamental current.

7.3   PHASE ANGLE OF HARMONICS

For simplicity, all the harmonics may be considered cophasial. This does not always give the most conservative results, unless the system has one predominant harmonic, in which case only harmonic magnitudes can be represented. The phase angles of the current sources are functions of the supply voltage phase angle and are expressed as

$$\theta_h = \theta_{h,\text{spectrum}} + h(\theta_1 - \theta_{1,\text{spectrum}}) \tag{7.17}$$

where θ_1 is the phase angle obtained from fundamental frequency load flow solution and $\theta_{h,\text{spectrum}}$ is the typical phase angle of harmonic current source spectrum. The phase angles of a three-phase harmonic source are rarely 120° apart, as even a slight unbalance in the fundamental frequency can be reflected in a considerable unbalance in the harmonic phase angle.

When a predominant harmonic source acts in isolation, it may not be necessary to model the phase angles of the harmonics. For multiple harmonic sources, phase angles should be modeled. Figure 7.18(a) shows the time–current waveform of

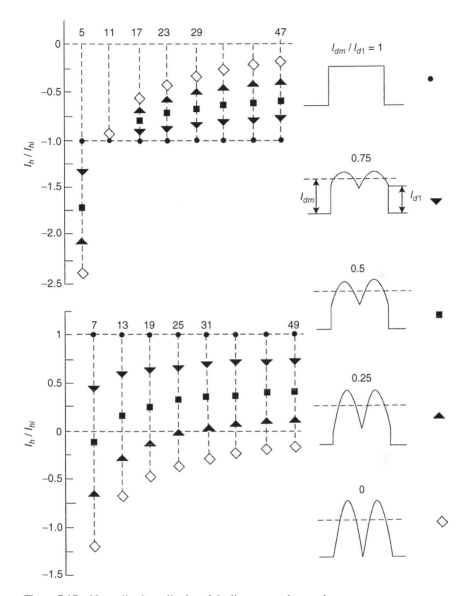

Figure 7.17   Normalized amplitudes of the line current harmonics.

a six-pulse current source converter when the phase angles are represented, and it is recognizable as the line current of a six-pulse converter, with overlap and no ripple content; however, the waveform of Fig. 7.18(b) has exactly the same spectra, but all the harmonics are cophasial. It can be shown that the harmonic current flow and the calculated distortions for a single-source harmonic current will be almost identical for these two waveforms. Table 7.1 shows the spectra with phase angles.

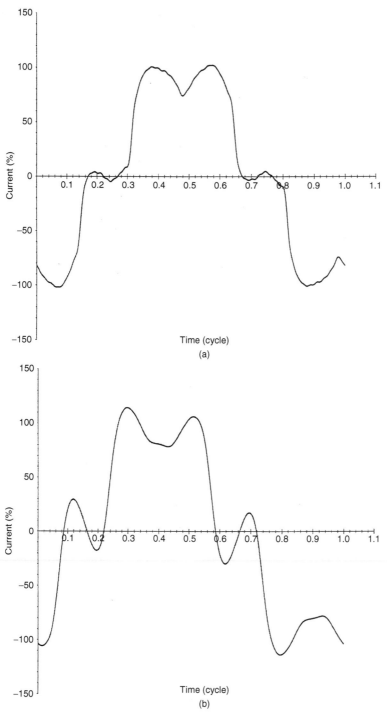

Figure 7.18 (a) Six-pulse current source converter current waveform, harmonics modeled at proper angles; (b) waveform with all harmonics cophasial.

TABLE 7.1   DC Drive Current Harmonic Spectrum

Harmonic Order	Magnitude	Phase Angle
1	100.0	−43
2	0.3	68
3	0.4	−126
4	0.2	30
5	25.3	−30
7	5.5	−122
9	0.6	37
11	8.2	−102
13	3.9	170
17	5.0	−179
19	2.9	193
23	3.4	109
25	2.2	14
29	2.7	39
31	1.9	−59
35	2.3	−35
37	1.7	−135
41	1.8	−110
43	1.5	150
47	1.7	177
49	1.4	70

Figure 7.19 shows the time–current waveform of a pulse width modulated or voltage source converter, with harmonic phase angles. The harmonic spectrum and phase angles are shown in Table 7.2. The phase angle of the fundamental is shown to be zero. Equation (7.17) applies and for a certain phase angle of the fundamental, the phase angle of the harmonics is calculated by shifting the angle column by $h\theta_1$ (harmonic order multiplied by fundamental frequency phase angle).

For multiple source assessment, the worst-case combination of the phase angles can be obtained by performing harmonic studies with one harmonic producing element modeled at a time. The worst-case harmonic level, voltage, or current is the arithmetical summation of the harmonic magnitudes in each harmonic model. This will be a rather lengthy study. Alternatively, all the harmonic sources can be simultaneously modeled with proper phase angles. The fundamental frequency angles are known by prior load flow.

Example 7.7:   Figure 7.20 shows a simple distribution system to show the impact on the modeling of proper phase angles of the harmonics on the harmonic distortion. Three different types of nonlinear loads are applied to distribution transformers T1, T2, and T3. Transformers T1 and T2 have 1.2 MVA fluorescent lighting and

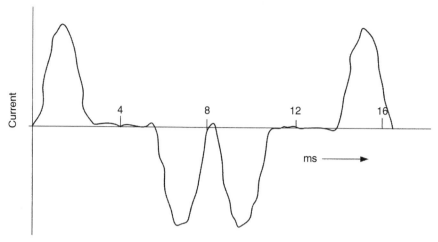

Figure 7.19   PWM voltage source converter current waveform, harmonics modeled with proper phase angle.

TABLE 7.2   Harmonic Spectrum of a Voltage Source PWM ASD

Harmonic Order	Magnitude	Phase Angle
1	100	0
3	0.35	−159
5	61.0	−175
7	33.8	−172
9	0.50	158
11	3.84	166
13	7.78	−177
15	0.41	135
17	1.28	32
19	1.60	179
21	0.35	110
23	1.10	38
25	0.18	49

switch-mode power supply loads, respectively, while 2.4 kV transformer T3 carries 2 MVA PWM ASD load.

First perform a fundamental frequency load flow analysis and calculate the magnitude and angles of bus voltages. These are shown in Fig. 7.20. It is seen that the fundamental frequency angle at the buses carrying nonlinear loads is −4.1° or −4°.

Now calculate the angle of the harmonics with fundamental frequency load flow angles as the base. These calculations are shown in Tables 7.3–7.5.

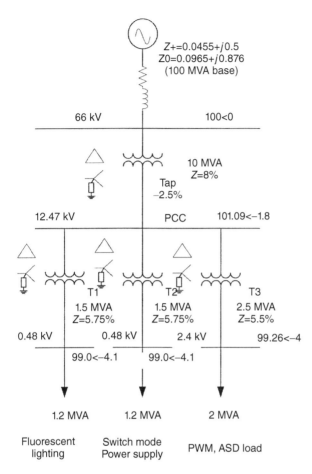

Figure 7.20   A distribution system for calculations of current and voltage harmonics, with and without proper phase angles, multiple nonlinear loads, Example 7.7.

The harmonic analysis is carried out first without modeling the angles of the harmonics and the results of the current and voltage distortions are shown in Tables 7.6 and 7.7. The total THD$_I$ at the PCC is 38.34% and THD$_V$ is 7.76%.

Next model all the harmonic angles calculated in Tables 7.3–7.5. The current and voltage distortions are shown in Tables 7.8 and 7.9. The THD$_I$ is reduced by 23.48% and the THD$_V$ is reduced by 5.0%. Also note from these tables that the distortions at individual harmonics vary over large values, with angles modeled and without angles.

This example clearly demonstrates the misleading results that can be obtained when nonlinear loads of different types with varying waveforms are applied and harmonic angles are ignored.

Further modeling results are presented when all the nonlinear loads are replaced only with one type, say fluorescent lighting. The calculations with or without modeling the harmonic phase angles give same results (Tables 7.10 and 7.11).

Thus, even with multiple sources same results hold, provided their harmonic spectra are identical in amplitude and phase angle.

**TABLE 7.3   Harmonic Spectrum of a Voltage Source PWM
ASD, Harmonic Angles Referred to Fundamental
Frequency Load Flow Angle**

Harmonic Order	Magnitude	Phase Angle
1	100	−4
3	0.35	−172
5	61.0	−195
7	33.8	165
9	0.50	160
11	3.84	122
13	7.78	131
15	0.41	75
17	1.28	−36
19	1.60	103
21	0.35	26
23	1.10	−54
25	0.18	−51

**TABLE 7.4   Spectrum of Typical Switch-Mode Power
Supply, Harmonic Angles Referred to Fundamental
Frequency Load Flow Angle**

Harmonic	Magnitude	Angle	Harmonic	Magnitude	Angle
1	100	−4.1	14	0.1	−32
2	0.2	5	15	1.9	172
3	67	4	17	1.8	8
4	0.4	160	19	1.1	176
5	39	29	21	0.6	−14
6	0.4	8	23	0.8	173
7	13	−118	25	0.4	−41
8	0.3	96.3			
9	4.4	17			
11	5.3	−161			
12	0.1	106			
13	2.5	23			

7.4   MEASUREMENTS OF HARMONICS

It is reiterated that measurement of the harmonics is an accurate way to estimate the harmonic distributions in a network and start the harmonic analysis study. The measurements are also required for the following:

TABLE 7.5 Typical Harmonic Current Spectrum of Fluorescent Lighting with Electronic Ballasts, Harmonic Angles Referred to Fundamental Frequency Load Flow Angle

Harmonic Order	Magnitude	Angle (deg)
FUND	100	−4.1
2	0.2	−104
3	19.9	−144
5	7.4	−178
7	3.2	−159
9	2.4	−171
11	1.8	−129
13	0.8	−103
15	0.4	−93
17	0.1	−44
19	0.2	141
21	0.1	160
23	0.1	−154
27	0.1	161
32	0.1	−84

TABLE 7.6 Harmonic Currents as Percentage of Fundamental Current, PCC, Harmonic Angles not Modeled, Example 7.7

2	4	5	7	8	11	13	14	17	19	23	25	32
0.1	0.1	35.66	16.13	0.06	2.5	2.79	0.02	0.57	0.51	0.30	0.07	0.01

$\mathrm{THD}_I = 38.34\%$

TABLE 7.7 Harmonic Voltages as Percentage of Fundamental Voltage at PCC Harmonic Angles not Modeled, Example 7.7

2	4	5	7	8	11	13	14	17	19	23	25	32
0.01	0.01	6.39	4.05	0.02	0.99	1.31	0.01	0.35	0.35	0.26	0.06	0.01

$\mathrm{THD}_V = 7.76\%$

- Ascertaining that the harmonic injection into a utility system from PCC meets the requirements of the standards.
- A power system may not contain a source of harmonic, yet it may be impacted by harmonic infiltration from utility source or the consumers connected to a common utility service point.
- The measurements can be used to identify the resonant conditions.

TABLE 7.8   Harmonic Currents as Percentage of Fundamental Current, PCC, Harmonic Angles Modeled, Example 7.7

2	4	5	7	8	11	13	14	17	19	23	25	32
0.06	0.1	18.70	13.94	0.06	1.83	1.95	0.02	0.52	0.43	0.17	0.07	0.01

$\text{THD}_I = 23.48\%$

TABLE 7.9   Harmonic Voltages as Percentage of Fundamental Voltage at PCC, Harmonic Angles Modeled, Example 7.7

2	4	5	7	8	11	13	14	17	19	23	25	32
0.00	0.01	3.35	3.50	0.02	0.73	0.92	0.01	0.33	0.30	0.14	0.06	0.01

$\text{THD}_V = 5.0\%$

TABLE 7.10   Harmonic Currents as Percentage of Fundamental Current at PCC, with or without Harmonic Angles, All Harmonic Loads Are Fluorescent Lighting, Example 7.7

2	4	5	7	8	11	13	14	17	19	23	25	32
0.19	0.00	6.52	2.16	0.00	1.22	0.49	0.00	0.05	0.09	0.04	0.00	0.03

$\text{THD}_I = 7.15\%$

- After the mitigation strategies, such as active and passive filters are implemented, the measurements can validate the results of the study.
- Measurements can be used to identify the overloading of a capacitor bank or capacitive filter. Measurements can plot the entire spectrum of harmonics in passive filters, cables, transmission lines, transformers, and so on.
- Measurements of harmonic currents and voltages with their respective phase angles can help deriving the transfer impedance at a given location.

The PCC can be on either the high side or low side of the utility interconnecting transformer. It is also the metering point where the CTs and PTs for measurement will be available. The accuracy of these devices is discussed in a section to follow. The harmonic current measurements on the secondary windings can be transferred to the primary windings in the turns ratio of the transformer, but it is not true for harmonic voltage measurements. The third harmonics will be blocked by the delta winding of the transformer. Also transformer connection affects the phase angle of the harmonics. A transformer with primary and secondary windings both connected in wye-grounded connection will allow zero sequence currents to flow in the primary system. Unbalanced triplen harmonics are not zero sequence components. The harmonics currents are expressed as the percentage of average maximum demand to calculate TDD (Chapter 10). The phase angle of the harmonics should be included in the measurements – it provides a better picture of harmonic cancellation from various harmonic source types. All phase angles should be related to the same reference,

TABLE 7.11   Harmonic Voltages as Percentage of Fundamental Voltage at PCC, with or without Harmonic Angles, All Harmonic Loads Are Fluorescent Lighting, Example 7.7

2	4	5	7	8	11	13	14	17	19	23	25	32
0.01	0.00	1.18	0.66	0.00	0.49	0.23	0.00	0.03	0.07	0.03	0.00	0.04

$THD_V = 1.45.0\%$

commonly selected as the zero crossing of the fundamental frequency line-to-neutral voltage on phase a, Example 7.7.

7.4.1   Monitoring Duration

Monitoring duration should be long enough to capture the time stamp of the harmonics with varying processes. The time trend of the harmonics is required to be captured, from which a probability histogram can be developed. For facilities such as steel mills and arc furnaces that can have harmonic emissions, which vary from day to day, monitoring over longer durations is required. It is also important to evaluate the effect of different operating system conditions at harmonic levels:

- Effect of power factor capacitors and passive filters
- Effect of outage of a filter
- Effect of variations in the utility source, for example, the load may be supplied over redundant feeders each having different characteristics. Utility source impedance effects harmonic generation
- Effect of nearby consumers with significant harmonic generation
- Effect of variations of loads in the facility.

7.4.2   IEC Standard 6100 4-7

For purpose of measurements, IEC divides harmonics into three categories:

1. Slowly varying (quasi-stationary)
2. Fluctuating
3. Rapidly varying; short bursts of harmonics
   - Measurements up to 50th harmonic are recommended. The time intervals recommended are as follows:
     - T_{vs}: very short interval, 3 s
     - T_{sh}: Short interval, 10 min
     - T_L: Long interval, 1 h
     - T_D: 1 day
     - T_W: 1 week.

For thermal effects, the rms value of each harmonic and cumulative probabilities from 1% to 99% are calculated.

For instantaneous effects, the maximum value of each harmonic and a cumulative probability of 95% and 99% are considered.

The measurement instrument should be able to handle the category of harmonic being measured.

Category 1: Quasi-stationary harmonics can be measured with a rectangular window of 0.1–0.5 s, Chapter 2, with some gap between the windows.

Category 2: Hanning window can be used, Chapter 2.

Category 3: A rectangular window of 0.08 s can be used without a gap between successive windows.

7.4.3 Measurement of Interharmonics

If the measured waveform contains an interharmonic, then the measurement time interval kT, where T is the time period at fundamental frequency), should be

$$kT = mT_i \tag{7.18}$$

where T_i = time period of interharmonic.

That is, it should be the least common multiple of the periods of fundamental component and interharmonic component. If Eq. (7.18) is not satisfied, then the rms value of interharmonics and the powers associated with it are incorrectly measured. For $m = 20$, the interharmonic will be measured with a maximum error of $\pm 0.2\%$.

If at least one of the interharmonics of order h is an irrational number, then the observed waveform is not periodic. A waveform of two or more nonharmonically related frequencies may not be periodic. In such cases, kT should be infinitely large.

The monitoring equipment based on FFT will have errors. The application of proper window function and zero padding, Chapter 2, can improve the performance.

IEC recommends sampling interval of the waveform to 10 and 12 cycles for 50 Hz and 60 Hz systems, respectively, resulting in a set of spectra with 5 Hz resolution for harmonics and interharmonics. This may be simplified by summing the components between harmonics into one single interharmonic group. Figure 7.21 shows these groupings for 50 Hz (Ref. [7]).

$$X_{\text{IH}}^2 = \sum_{i=2}^{8} X_{10n+1}^2 \quad \text{for 50 Hz system} \tag{7.19}$$

$$X_{\text{IH}}^2 = \sum_{i=2}^{10} X_{12n+1}^2 \quad \text{for 60 Hz system} \tag{7.20}$$

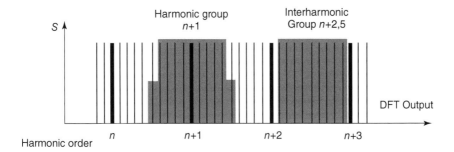

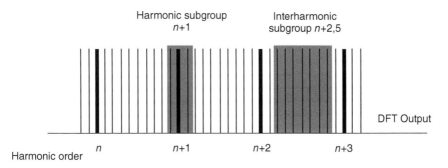

Figure 7.21   A histogram of voltage THD, presentation of harmonic measurement results.

7.5   MEASURING EQUIPMENT

The measuring equipments are as follows:

- Storage oscilloscope can be used and will visually indicate the resonance and distortion. The stored data can be later on transferred to a computer, Ref. [5].

- Spectrum analyzers display the power distribution of a signal as a function of frequency on a CRT or chart recorder.

- Harmonic analyzers or wave analyzers can measure the amplitude and phase angle and provide the line spectrum of the signal. The output can be recorded or monitored with analog or digital meters.

- Distortion analyzers will indicate THD directly. Digital measurements use two basic techniques. First by means of a digital filter, which is set up with proper bandwidth to capture small magnitude of harmonics when the fundamental current is large. And the second by means of the FFT techniques, which are powerful algorithms and with selection of proper window and bandwidth can eliminate picket fence and aliasing (Chapter 2).

- Microprocessor-based *online* meters are available, which will continuously monitor and store the harmonic spectra and distortion data with varying processes. This continuous measurement is helpful in ascertaining the time stamp of harmonics.

7.5.1 Specifications of Measuring Instruments

Accuracy The minimum accuracy requirements of measuring instruments are specified in IEEE 519[5]. The instrument must perform measurement of a steady-state harmonic component with an uncertainty no more than 5% of the permissible limit. For example, if 11th harmonic of 0.70% is to be measured in a 480 V system, it means a harmonic line-to-neutral voltage of 1.94 V. The instrument should have an uncertainty of less than $\pm(0.05)(1.94) = \pm0.097$ V. Thus, the accurate measurements of harmonics become a problem when their order increases and the amplitudes are correspondingly reduced. The ultimate accuracy results should consider not only the measuring instrument but also the transducers used for harmonic measurement, that is, overall measuring system.

Attenuation The second parameter is attenuation that dedicates instrument capability to separate harmonic components of different frequencies. Minimum attenuation limits for a frequency and time domain instrument are specified in Ref. [5] and are reproduced in Table 7.12. Almost all measurements can meet 60 dB (0.1% of fundamental). More expensive instruments can have 90 dB (0.00316%).

The load harmonics can be rapidly fluctuating, and averaging over a period of time can give a false picture of harmonic emission and distortion. If an average over a period of 3 s is required, then the response of the output meter should be identical to the first-order low-pass filter with a time constant of 1.5 ± 0.15s.

Bandwidth The bandwidth of the instrument will have a profound effect, especially when the harmonics are fluctuating. Instruments with a constant bandwidth for the entire range of frequencies must be used. The bandwidth should be 3 ± 0.5 Hz between -3 dB points with a minimum attenuation of 40 dB at a frequency of $f_h + 15$ Hz. When interharmonics are to be measured, a larger bandwidth will cause large positive errors.

TABLE 7.12 Minimum Required Attenuation (dB)

Injected Frequency (Hz)	Frequency Domain Instrument	Time Domain Instrument
60	0	0
30	50	60
120–720	30	50
720–1200	20	40
1200–2400	15	35

Source: Ref. [5].

7.5.2   Presentation of Measurement Results

The harmonic measured data are presented in the tabular form, as shown in the tables and waveforms in the time domain and spectra in the frequency domain in previous chapters. The results can also be presented as variation of harmonics with time, a time-trend plot.

The time-varying harmonics can be presented as a probability histogram (Fig. 7.22). This allows direct evaluation of the harmonic levels, bar height represents relative frequency of occurance.

Figure 7.23 shows inverse distribution curve. The data acquisition period T_D can be divided into m intervals. $mT = T_D$, then the mean value of current for each interval is

$$\sum_1^k \frac{I_{kh}}{k} \tag{7.21}$$

where over the subinterval T, k measurements were taken.

The mean square value is

$$\sum_1^k \frac{I_{kh}^2}{k} \tag{7.22}$$

The standard deviation is calculated from the equation:

$$I_h = \sqrt{I_{h,\max}^2 - I_{h,\min}^2} \tag{7.23}$$

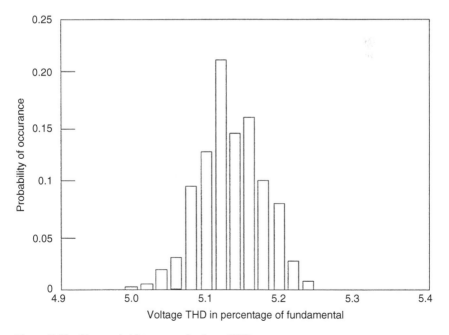

Figure 7.22   Harmonic histogram of voltage THD.

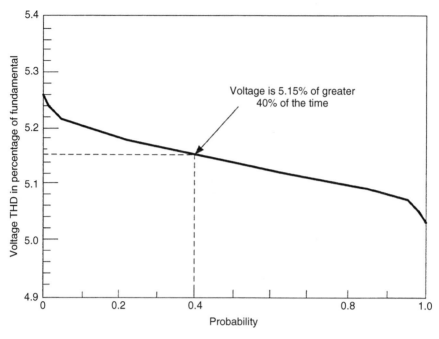

Figure 7.23 Harmonic distribution curve of voltage THD.

where

$I_{h,\max}$	=	maximum I_h over k measurements
$I_{h,\min}$	=	minimum I_h over k measurements

The harmonic measurements should be taken in each phase and these will show variations (Fig. 4.32 (a) and (b); see Section 7.7 for explanation of statistical terms).

7.6 TRANSDUCERS FOR HARMONIC MEASUREMENTS

Current Transformers The CT accuracies in ANSI/IEEE standards are specified at 60 Hz. According to Ref. [5], the CTs have accuracy in the range of 3% for frequencies up to 10 kHz. The accuracy is a function of the impedance of the CT and the external burden connected to it. Figure 7.24 shows CT ratio correction factor (RCF) with burden and without burden. The percentage error, RCF, as compared to the fundamental can be calculated (Ref. [8]):

$$RCF = 1 + (Z_s + Z_b)\left(\frac{Z_{cs}Z_e}{Z_{cs} + Z_e}\right) \qquad (7.24)$$

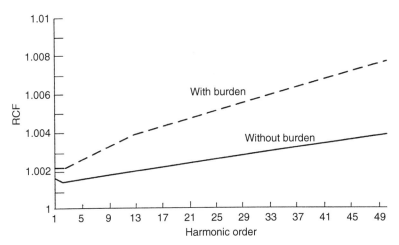

Figure 7.24   Ratio correction factor (RCF) of a CT with burden and without burden versus harmonics.

where

Z_s is the secondary winding resistance, Z_e is the excitation impedance of the CT, Z_{cs} is the secondary winding capacitance, and Z_b is the impedance of the CT burden.

Shielded conductors, coaxial cables, or triaxial cables are a must for accurate results. Proper grounding and shielding procedures are required to reduce pickup of parasitic voltages.

Hall effect probes, which are similar to clamp-on CTs, have generally not been used for harmonic measurements. Hall effect device allows DC as well as AC currents to be measured; the specified accuracy for commercial probes is 2–5% over a range of 500–1000 Hz.

Rogowski coils   Rogowski coils or Maxwell worms are devices where coils are wound on flexible plastic mandrels so that these can be used as clamp-on devices. Magnetic saturation is avoided when large currents up to 100 kA or DC currents are to be measured.

Voltage measurements   The magnetic voltage transformers are designed to operate at 60 Hz. Harmonic frequency resonance between winding inductances and capacitances can cause large ratio and phase errors (Fig. 7.25). For harmonics of frequency less than 5 kHz, the accuracy of most potential transformers is within 3%.

Capacitive voltage transformers cannot be used for harmonic measurements because typical resonant frequency peak appears at 200 Hz. Capacitive voltage dividers can be easily built. High-voltage bushings are provided with a capacitive tap for voltage measurements.

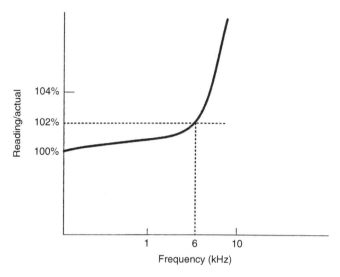

Figure 7.25   Potential transformer accuracy.

7.7   CHARACTERIZING MEASURED DATA

The accuracy of the measured data is important in several applications such as design of harmonic filters and imposition of stress on electrical equipment, and it is not as simple as may be made out from the earlier description. IEEE Task Force publication in two parts [9,10] provides valuable insight into the problem of measurement of nonstationary voltages and current waveforms, characterization of recorded data, harmonic summation, and cancellation in systems with multiple nonlinear loads and probabilistic harmonic power flow. It starts with two typical recorded data and then analyzing it. These recorded data from Ref. [9] at one of the sites (site A) are shown in Fig. 7.26(a) and (b).

The data representing site A shows a customer's 13.8 kV bus having rolling mill equipped with 12-pulse DC drives and tuned harmonic filters. After 2.5 h, the current and voltage distortions go down because rolling mill is not operating and the background distortion is captured. Site B measurements, not reproduced here, show that while the current distortion is high the corresponding voltage distortion is low because of *stiff system. This is a very important concept.*

The common techniques in harmonic estimating are based on DFT (Chapter 2), which gives accurate results provided:

- The signal is stationary, but practically it is not.
- The sampling frequency is greater than twice the frequency being measured, Nyquist theorem, Chapter 2.
- The number of periods sampled is an integer.
- The waveform does not contain frequencies that are noninteger multiples of fundamental frequency, that is, interharmonics.

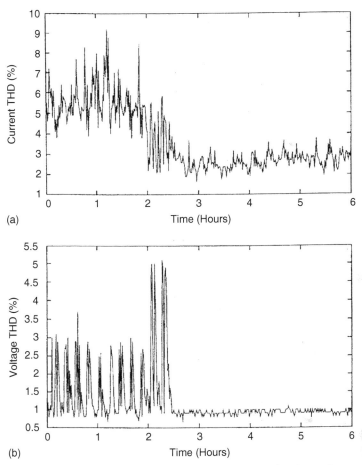

Figure 7.26   (a) Signature of current signal; (b) voltage signal, actual measurements. Source: Ref. [9].

When interharmonics are present, multiple periods need to be sampled, discussions in Section 7.4.3. Time variations of the individual harmonics are generated by windowed Fourier transformations or short-time Fourier transform. Each harmonic spectrum corresponds to the each window section of the continuous signal. Thus, different window sizes give different spectra. Chapter 2 discuses aliasing, leakage, and picket fence effects. Both leakage and picket fence effect can be mitigated by spectral windows.

Some recent approaches to improve the accuracy of measurements are as follows:

- Kalman filter-based analyzer [11]
- Self-synchronizing Kalman filter approach [12]
- A scheme based on Parseval's relation and energy concepts [13]

- A Fourier linear combiner using adaptive neural networks [14]

The irregularities in harmonic recorded data may fail to confirm to a coherent pattern, yet some patterns will show a deterministic component. In such cases, the signal can be expressed as a *deterministic component and random component.*

7.8   PROBABILISTIC CONCEPTS

We use probabilistic concepts in electrical engineering, not discussed in this book, Refs. [15–17] may be seen. To understand the presentation of harmonic data, the following brief introduction is provided:

The frequency distribution data can be represented by a histogram, frequency polygon, frequency curve, bar chart, and pie diagrams. The arithmetic mean for n numbers $x_1, x_2, \ldots, x_n$ is given by

$$x_m = \frac{\sum x}{n} \tag{7.25}$$

If x_1 occurs f_1 times, x_2 occurs f_2 times, $\ldots$, and f_n occurs n times, then the arithmetic mean is

$$x_m = \frac{\sum fx}{\sum f} \tag{7.26}$$

If a is the assumed arithmetic mean and d is deviation of the variate x from a, we can write

$$\frac{\sum fd}{\sum f} = \frac{\sum f(x - a)}{\sum f} = x_m - \frac{a \sum f}{\sum f}$$

$$x_m = a + \frac{\sum f(x - a)}{\sum f} = a + \frac{\sum fd}{\sum f} \tag{7.27}$$

Median is the measure of the central item of the distribution when it is arranged in ascending or descending order. For $n =$ odd, it is given by

$$M_d = \frac{n + 1}{2}\text{th} \quad \text{item} \tag{7.28}$$

For even frequency, there are two middle terms, and the mean of $n/2$ and $(n/2 + 1)$ terms gives the median.

Mode is defined as the size of the variable in a population that occurs most frequently:

$$\text{Mean} - \text{mode} = 3(\text{mean} - \text{median}) \tag{7.29}$$

Geometric mean is defined as

$$G = (x_1 \times x_2 \times \ldots \ldots \times x_n)^{1/n} \tag{7.30}$$

Mean and Standard Deviation   Average deviation or mean deviation is defined as the mean of the absolute values of the deviations of a given set of numbers from their arithmetic mean:

$$\text{Mean deviation} = \frac{\sum_{n=1}^{n} f_n |x_n - x_m|}{\sum f} \tag{7.31}$$

where x_1, x_2, $\ldots$, x_n is a set of numbers with frequencies f_1, f_2, $\ldots$, f_n. Standard deviation is defined as square root of the mean of the square of the deviation from the arithmetic mean:

$$\text{SD} = \sigma = \sqrt{\frac{\sum_{n=1}^{n} f_n (x_n - x_m)^2}{\sum f}} \tag{7.32}$$

The square of the SD, σ^2, is called variance. It is also called the second moment about the mean and denoted by μ_2. The coefficient of variance is given by

$$\frac{\sigma}{x_m} \times 100 \tag{7.33}$$

The standard deviation can be found from the following expression:

$$\sigma = \sqrt{\frac{\sum f d^2}{\sum f} - \left(\frac{\sum f d}{\sum f}\right)^2} \tag{7.34}$$

The statistical measures of signals in Fig 7.26 can be calculated using the above-mentioned equations. However, it cannot be concluded that these will follow Gaussian distribution. If the signal is totally random, then Gaussian distribution can be applied. When the signal contains a significant deterministic component, it will deviate considerably from Gaussian distribution. The signals in Fig. 7.26(a), current signal, are divided into two parts, deterministic and random, as depicted in Fig. 7.27(a) and (b).

Again some explanation is required. Gaussian distribution and Weibull distribution are commonly applied in insulation coordination and lightning phenomena. Briefly, Gaussian distribution is a continuous distribution and plays an important role in measurement statistics. It is derived as the limiting form of binomial distribution.

As the measurements increase, the curve most commonly fitted is a bell-shaped curve, called the normal or Gaussian probability density function (pdf):

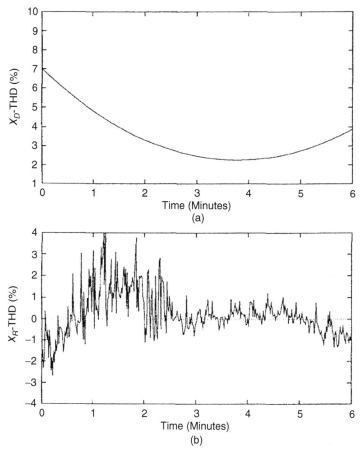

Figure 7.27   (a) and (b) Breakdown of current signal in Fig. 7.26(a) into definitive X_D and random X_R components.

$$p(x) = \frac{1}{\sigma\sqrt{2\pi}} e^{-\frac{1}{2}\left(\frac{x-\mu}{\sigma}\right)^2} \quad -\infty < x < \infty \tag{7.35}$$

where μ is the mean and σ is the SD.

The distribution is completely defined if the parameters σ and μ are known. The corresponding distribution function is given by

$$P(x) = P(X \leq x) = \frac{1}{\sigma\sqrt{2\pi}} \int_{-\infty}^{x} e^{-\frac{1}{2}\left(\frac{x-\mu}{\sigma}\right)^2} dx \tag{7.36}$$

The probability that x takes between values x_1 and x_2 is given by

$$P(x_1 < x < x_2) = \int_{x_1}^{x_2} \frac{1}{\sigma\sqrt{2\pi}} e^{-\frac{(x-\mu)^2}{2\sigma^2}} dx \tag{7.37}$$

If we substitute

$$z = \frac{x - \mu}{\sigma} \tag{7.38}$$

where z = standardized variable corresponding to x, then the mean of $z = 0$ and the variance is 1. In such a case, the density function becomes

$$f(z) = \frac{1}{\sqrt{2\pi}} e^{-\frac{z^2}{2}} \tag{7.39}$$

This is referred as *standard normal density function*. The corresponding distribution function is

$$P(z) = P(Z \leq z) = \frac{1}{\sqrt{2\pi}} \int_{-\infty}^{z} e^{-\frac{z^2}{2}} dz \tag{7.40}$$

A graph of the density function is shown in Fig. 7.28, and properties of Gaussian distribution are shown in many text books. Note that

- The curve is symmetrical about the y-axis. The mean, median, and mode coincide at the origin.
- The area of the curve is equal to the total number of observations.
- $f(z)$ decreases rapidly as z increases numerically. The curve extends on either side of the origin.
  In Fig. 7.28, the areas within 1–3 standard deviations of the mean are indicated.

A disadvantage of Gaussian distribution is that the time factor is totally lost with such statistical measurements; we cannot tell when the maximum distortion occurred, which is an important parameter.

7.8.1 Histogram and Probability Density Function

A histogram or pdf provides a picture of relative frequency of occurrence and shows total set of measurements that fall in various intervals. When scaled down so that the total area covered is equal to unity, the histogram becomes the exact pdf. A histogram may show irregularities due to deterministic component in signals.

Figures 7.29(a) and (b) are the pdf's related to recorded data in Fig. 7.26(a) and (b), respectively. The pdf of current contains multiple peaks. The pdf of voltage in Fig. 7.28(b) can be approximated by Rayleigh distribution [18]. Again the histograms provide total time duration, they hide the information when some events occurred in time whether it occurred in pulses or in continuous time.

7.8.2 Probability Distribution Function

The probability distribution function $P(x)$ is an integral of pdf (Eqs. (7.35) and (7.36)). Figure 7.23 shows an inverse distribution function while Fig. 7.30 shows the probability distribution function of voltage signal in Fig. 7.26(b). Again, similar comments apply to these distribution curves, though inverse distribution function will show the time duration for which the distortion is exceeded beyond a certain limits (Fig. 7.23).

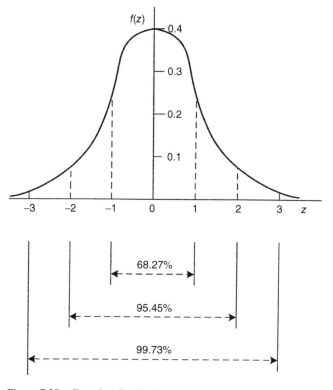

Figure 7.28   Gaussian distribution, see text.

7.8.3   Regression Methods: Least Square Estimation

We talked about deterministic and random parts of a signal. The deterministic component X_D can be extracted by fitting a polynomial function of a certain degree to the recorded measurements, using the *method of least squares*. Basically, it is a regression analysis to estimate one of the variables (dependent variable) from the other (independent variable). If y is estimated from x by some equation, it is referred to as regression. In other words, if the scatter diagram of two variables indicates some relation between these variables, then the dots will be concentrated around a curve. This curve is called the *curve of regression*. When the curve is a straight line, it is called *a line of regression*.

For some given data points, more than one curve may seem to fit. Intuitively, it will be hard to fit an appropriate curve in a scatter diagram and variation will exist.

Referring to Fig. 7.31, a measure of goodness for the appropriate fit can be described as

$$d_1^2 + d_2^2 + \ldots + d_n^2 = a \quad \text{minimum} \tag{7.41}$$

A curve meeting these criteria is said to fit the data in the *least square sense* and is called a least square regression curve, or simply a least square curve. The least square line imitating the points (x_1, y_1), ..., (x_n, y_n) has the equation:

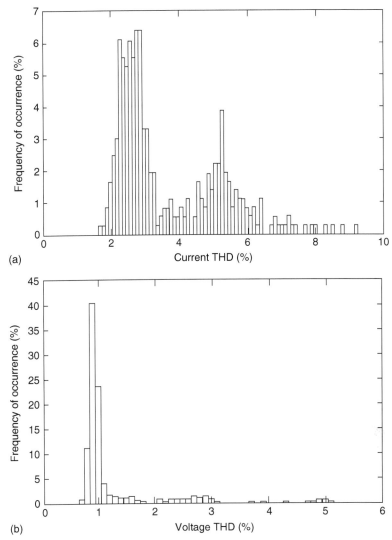

Figure 7.29   (a) Pdf of current signature in Fig. 7.26(a); (b) pdf of voltage signature in Fig. 7.26(b). Source: Ref. [9].

$$y = a + bx \tag{7.42}$$

The constants a and b are determined from solving simultaneous equations, which are called the normal equations for the least square line:

$$\sum y = an + b \sum x$$
$$\sum xy = a \sum x + b \sum x^2 \tag{7.43}$$

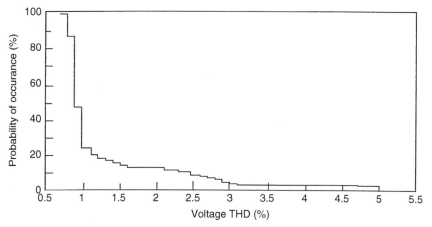

Figure 7.30 Probability distribution functions of voltage signal in Fig. 7.26(b). Source: Ref. [9].

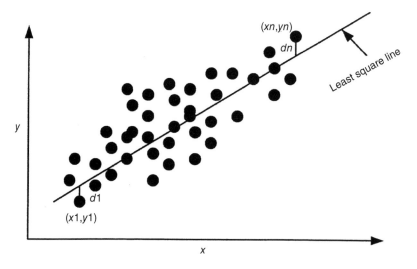

Figure 7.31 Least square line for some random data.

This gives

$$a = \frac{\left(\sum y\right)\left(\sum x^2\right) - \left(\sum x\right)\left(\sum xy\right)}{n\sum x^2 - \left(\sum x\right)^2}$$

$$b = \frac{n\sum xy - \left(\sum x\right)\left(\sum y\right)}{n\sum x^2 - \left(\sum x\right)^2} \tag{7.44}$$

Figure 7.31 shows a least square line plotted through some random measurements. Random measurements can take varied shapes and varied curves, such as parabola, or ellipse can be fitted. For further study on regression and multiple regression, Ref [19] may be seen.

The distribution of X_r approaches a Gaussian distribution, as the polynomial function representing X_D increases. For the data shown in Fig. 7.26(a) (current distortion signal), the following quadratic equation gives the best fit:

$$X_D(t) = 7 - 2.5t + 0.33t^2 \qquad (7.45)$$

where t is the time in hours. The distortion in Fig 7.27(a) is resolved into two parts, X_D and X_r, as shown in Fig 7.27(a) and (b). If a histogram of distortion in Fig. 7.27(b) is taken, it follows Gaussian distribution.

7.9   SUMMATION OF HARMONIC VECTORS WITH RANDOM ANGLES

The current harmonic amplitudes and phase angles vary. While the amplitudes of harmonics are commonly measured, the phase angles are not. Probabilistic concepts are provided in Ref. [20].

If we have two vectors V_{h1} and V_{h2} and different phase angles θ_{h1} and θ_{h2} then the angle θ_h between them can be written as

$$\theta_h = \theta_{h2} - \theta_{h1} = \cos^{-1} \frac{V_{hs}^2 - V_{h1}^2 - V_{h2}^2}{2 V_{h1} V_{h2}} \qquad (7.46)$$

The locus of V_{hs} is shown in Fig. 7.32.

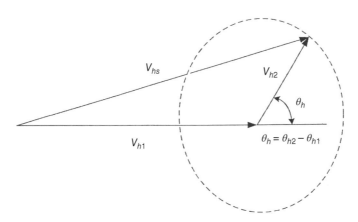

Figure 7.32   Summation of two vectors with random phase angles.

If phase angle difference is a uniform distribution around the circle, then the probability of PV_{hp} over PV_{hs} is defined as

$$PV_{hp} \geq PV_{hs} = \frac{1}{\pi} \cos^{-1} \frac{V_{hs}^2 - V_{h1}^2 - V_{h2}^2}{2V_{h1}V_{h2}} \qquad (7.47)$$

Accordingly probability of $V_{hs} > V_{h1}$ is

$$PV_{hs} \geq V_{h1} = \frac{1}{\pi} \cos^{-1} \frac{-V_{h2}}{2V_{h1}} \qquad (7.48)$$

Four cases are

(1) $|V_{h2}| \rightarrow 0, \quad P_{hs} \geq V_{h1} = 0.5$

(2) $|V_{h1}| = |V_{h2}| \quad P_{hs} \geq V_{h1} = 2/3$

(3) $V_{h\ min} = |V_{h1} - V_{h2}| \quad P_{hs} \geq V_{h\ min} = 1 \quad$ and $\quad P_{hs} = V_{h\ min} = 0$

(4) $V_{h\ max} = |V_{h1} + V_{h2}| \quad P_{hs} \geq V_{h\ max} = 0 \qquad (7.49)$

The uniform distribution of phase angle difference is

$$f(\theta) = \frac{1}{\pi}, \quad \theta \in [0, \pi] \qquad (7.50)$$

And their expected mean $E(\theta)$ is $\pi/2$, the standard deviation $\sigma(\theta)$ is $\pi/(2\sqrt{3})$. The statistically phase angle difference at negative standard deviation is

$$\theta_{-\sigma} = E(\theta) - \sigma(\theta) = \frac{\pi}{2}\left(1 - \frac{1}{\sqrt{3}}\right) = 38.04^o \qquad (7.51)$$

Then, the summation with known negative deviation becomes

$$V_{hs} = \sqrt{V_{h1}^2 + V_{h2}^2 + 2V_{h1}V_{h2}\cos 38.04^o}$$

$$= \sqrt{V_{h1}^2 + V_{h2}^2 + 1.575V_{h1}V_{h2}} \qquad (7.52)$$

Then the probability is

$$PV_{hp} \geq V_{hs} = \frac{38.04^o}{180^0} = 0.2113 \qquad (7.53)$$

IEC 61000-3-6 [21] gives the following summation equation:

$$V_{hs} = \sqrt[\lambda]{\sum i V_{hi}^\lambda} \qquad (7.54)$$

where $\lambda = 1$ for $h < 5$, $\lambda = 1.4$ for $5 \leq h \leq 10$, and $\lambda = 2$ for $h > 10$. These are exponents for summation of two vectors. For multisources, Eq. (7.54) is still valid but λ is different.

In terms of rectangular coordinates, we can write the amplitude of sum of random vectors as

$$Z = \sqrt{\left(\sum_{i=1}^{N} X_i\right)^2 + \left(\sum_{i=1}^{N} Y_i\right)^2} = \sqrt{S^2 + W^2} \qquad (7.55)$$

As S and W are random variables, Z is also a random variable described by its pdf, $f_z(z)$.

Then

$$f_z(z) = \int_0^{2\pi} f_{SW}(z \cos\theta, z \sin\theta)z \, d\theta \qquad (7.56)$$

Assuming a sufficiently large number of vectors, N, and applying central limit theorem (Section 7.10), the random variables S and W are normally distributed. If a correlation between S and W is neglected, the jpdf of S and W can be written as

$$f_{SW}(s, w) = \frac{e^{-\frac{(s^2+w^2)}{2\sigma^2}}}{2\pi\sigma^2} \qquad (7.57)$$

Substituting in Eq. (7.56),

$$f_z(z) = \frac{ze^{-\frac{z^2}{2\sigma^2}}}{\sigma^2}, \quad z > 0 \qquad (7.58)$$

For harmonic sources distributed in a stochastic network, the branch impedances and their angles are probabilistic quantities.

If we denote

φ_i, θ_{mi} as the angles associated with phasors I_i (branch current) and Z_{mi} (branch impedance), which are random variables and let $U_{mi} = |Z_{mi}||I_i|$ and $(\varphi_i + \theta_{mi}) = \varphi_{mi}$ and $D_{mi} = Z_{mi}I_i$, then pdf of D_{mi} is

$$f_{Dmi}(d_{mi}) = \int_{-\infty}^{\infty} (1/|e|)f_{zmiIi}(e, d_{mi}/e) \, de \qquad (7.59)$$

where $f_{zmiIi}(z_{mi}I_i)$ is pdf of the branch impedance magnitudes and harmonic currents at node i, and e is introduced as an auxiliary variable to facilitate determination of above equation.

The pdf of u_{mi} is given by

$$f_{umi}(u_{mi}) = (f_{Dmi}(u_{mi}) + f_{Dmi}(-u_{mi})), \quad u_{mi} > 0 \qquad (7.60)$$

The pdf of φ_{mi} is

$$f_{\varphi mi}(\varphi_{mi}) = \int_{-\infty}^{\infty} f_{\varphi mi}(\varphi_{mi} - \varphi_i, \varphi_i)d\varphi_i \tag{7.61}$$

Having found the pdf's of the variables in the above equations, the pdf of the sum of the probabilistic harmonic vectors can be found.

7.10 CENTRAL LIMIT THEOREM

We discussed the normal or Gaussian distribution in Section 7. There are also other distributions, for example, Poisson distribution (after Poisson in the early 19th century). If X is a random variable that takes the values 1, 2, ..., then the probability function of X is given by

$$f(x) = P(X = x) = \frac{\lambda^x e^{-\lambda}}{x!}, \quad x = 0, 1, 2, \dots \tag{7.62}$$

where λ is a positive coefficient. This distribution is called a Poisson distribution.

One wonders whether there is any other distribution in addition to binomial and Poisson that have normal distribution as the limiting case. The central limit theorem states that if X_1, X_2, ..., X_n be independent random variables that are identically distributed and have a finite mean and variance, and $S_n = X_1 + X_2 + X_3 + \dots X_n$, then

$$\lim_{n \to \infty} P\left(a \leq \frac{S_n - n\mu}{a\sqrt{n}} \leq b\right) = \frac{1}{2\pi} \int_a^b e^{-u^2/2}du \tag{7.63}$$

That is random variable

$$\frac{S_n - n\mu}{\sigma\sqrt{n}} \tag{7.64}$$

in which the standardized variable corresponding to S_n is asymptotically normal.

7.11 KALMAN FILTERING

The Kalman filter is recursive optimal estimator and has been used for harmonic measurements, suitable for online measurements (Ref. [12]). It requires redundant harmonic measurements, and DFT is used to get each harmonic spectrum measurements as an input to the estimator. It requires a state variable model for parameters to be estimated and a measurement equation that relates the discrete measurements to state variables parameters. The mathematics involved is discussed in many texts, a synopsis is provided here.

The state representation of signal with constant or time-varying magnitude can be written by expressing bus injection current as a function of time with reference rotating at ω as

$$s(t) = I \cos \omega t \cos \theta - I \sin \omega t \sin \theta \tag{7.65}$$

Let $x^R = I \cos \theta$, $x^I = I \sin \theta$. These include two components: one component is constant but unknown, and the other component may be time varying. These variables represent in-phase and quadrature phase components. The two state variables for the injection current can be expressed as

$$
\begin{vmatrix} x^R \\ x^I \end{vmatrix}_{k+1} = \begin{vmatrix} 1 & 0 \\ 0 & 1 \end{vmatrix} \begin{vmatrix} x^R \\ x^I \end{vmatrix}_k + \begin{vmatrix} w^R \\ w^I \end{vmatrix}_k \tag{7.66}
$$

where w^R and w^I allow state variables for random walk. The measurement equation will include the signal and noise and can be expressed as

$$
z_k = | \cos(\omega t_k) \quad -\sin(\omega t_k)| \begin{vmatrix} x_1^R \\ x_1^I \end{vmatrix}_k + v_k \tag{7.67}
$$

where v_k represents high-frequency noise.

Model 1:   For a power system with N buses that include harmonics, the noise free signal can be written as

$$
s(t) = \sum_{i}^{n} I_i(t) \cos(\omega t + \theta_i) \tag{7.68}
$$

where $I_i(t)$ represents ith harmonic current at time t and θ_i is the associated phase angle.

All bus currents are treated as state variables. The state equation and measurement equation can be written as

$$
\begin{vmatrix} x_1^R \\ x_1^I \\ x_2^R \\ x_2^I \\ \cdot \\ \cdot \\ x_n^R \\ x_n^I \end{vmatrix}_{k+1} = \begin{vmatrix} 1 & 0 & 0 & . & . & 0 \\ 0 & 1 & 0 & . & . & 0 \\ . & . & . & . & . & . \\ . & . & . & . & . & . \\ 0 & 0 & 0 & . & 1. & 0 \\ 0 & 0 & 0 & . & . & 1 \end{vmatrix} \begin{vmatrix} x_1^R \\ x_1^I \\ x_2^R \\ x_2^I \\ \cdot \\ \cdot \\ x_n^R \\ x_n^I \end{vmatrix}_k + \begin{vmatrix} w_1^R \\ w_1^I \\ w_2^R \\ w_2^I \\ \cdot \\ \cdot \\ w_n^R \\ w_n^I \end{vmatrix}_k \tag{7.69}
$$

And the measurement equation is

$$
z_k = H_k x_k + v_k = \begin{vmatrix} \cos(\omega k \Delta t) \\ -\sin(\omega k \Delta t) \\ . \\ \cos(n\omega \Delta t) \\ -\sin(n\omega \Delta t) \end{vmatrix}^t \begin{vmatrix} x_1^R \\ x_1^I \\ . \\ x_n^R \\ x_n^I \end{vmatrix} + v_k \tag{7.70}
$$

H_k is a time-varying vector.

Model 2:   If we consider state representation of a signal with time-varying magnitude using a stationary reference, then Eq. (7.66) becomes

$$
\begin{vmatrix} x_1^R \\ x_1^I \\ x_2^R \\ x_2^I \\ \cdot \\ \cdot \\ \cdot \\ x_n^R \\ x_n^I \end{vmatrix}_{k+1} = \begin{vmatrix} M_1 & & & \\ & M_2 & & \\ & & M_N & \end{vmatrix} \begin{vmatrix} x_1^R \\ x_1^I \\ x_2^R \\ x_2^I \\ \cdot \\ \cdot \\ \cdot \\ x_n^R \\ x_n^I \end{vmatrix}_{k} + \begin{vmatrix} w_1^R \\ w_1^I \\ w_2^R \\ w_2^I \\ \cdot \\ \cdot \\ \cdot \\ w_n^R \\ w_n^I \end{vmatrix}_{k} \tag{7.71}
$$

where the submatrices M_i are

$$
M_i = \begin{vmatrix} \cos(i\omega\Delta t) & -\sin(i\omega\Delta t) \\ -\sin(i\omega\Delta t) & \cos(i\omega\Delta t) \end{vmatrix} \tag{7.72}
$$

The measurement equation becomes

$$
z_k = H_k x_k + v_k = \begin{vmatrix} 1 \\ 0 \\ \cdot \\ 1 \\ 0 \end{vmatrix}^t \begin{vmatrix} x_1^R \\ x_1^I \\ \cdot \\ x_n^R \\ x_n^I \end{vmatrix} + v_k \tag{7.73}
$$

Having obtained the state equation of the signal and the measurement equation, the system covariance matrices for w_k and v_k are assumed as

$$
E[w_k w_k^t] = Q_k, \quad E[v_k v_k^t] = R_k \tag{7.74}
$$

The initial variable is assumed to be equal to 0:

$$
\hat{X}_{(0)} = 0 \tag{7.75}
$$

The initial covariance matrix is

$$
\hat{P}_0 = E[\hat{x} - \hat{x}_{(0)}] = E[(\hat{x})(\hat{x})^t] = \sigma \tag{7.76}
$$

The determination of initial covariance matrix depends on prior knowledge of likelihood of occurrence of harmonic sources and average load levels at certain buses.

The assumption that harmonic injections at distinct buses are correlated makes this matrix diagonal. The recursive steps are as follows:

Compute Kalman filter gain K_k

$$K_k = \widehat{P}_k H_k^t (H_k \widehat{P}_k H_k^t + R_k)^{-1} \tag{7.77}$$

Update with harmonic measurement z_k

$$\widehat{x}_k = \widehat{x}_k + K_k(z_k - H_k \widehat{x}_k) \tag{7.78}$$

Compute error covariance for updated estimate

$$P_k = (1 - K_k H_k)\widehat{P}_k \tag{7.79}$$

Proceed ahead:

$$\widehat{P}_{k+1} = \varphi_k P_k \varphi_k^t + Q_k$$
$$\widehat{X}_{k+1} = \varphi_k \widehat{X}_k \tag{7.80}$$

See reference [22] for further reading.

Although this chapter has provided estimation of harmonics through calculations and online measurements, from a practical point of view, it will be difficult for an application engineer to estimate correctly the harmonic spectra and their time stamp, especially in view of widely varying topologies of the power electronic system and their applications in the specific power systems. Generally, for the studies in the design stage of the project, the data supplied by vendors are more reliable, provided it is related to the Thévenin impedance at the point of application of the harmonic producing loads. For plants in operation, the online measurements over a period of time depending on the processes and nearby harmonic producing loads is the best choice.

REFERENCES

1. A. D. Graham and E. T. Schonholzer. "Line harmonics of converters with DC-motor loads," IEEE Trans Industry Applications, vol. 19, no.1, pp. 84–93, 1983.
2. J. C. Read. "The calculation of rectifier and inverter performance characteristics," JIEE, UK, pp. 495–509, August 1945.
3. M. Grötzbach and R. Redmann. "Line current harmonics of VSI-Fed adjustable-speed drives," IEEE Transactions on Industry Applications, vol. 36, pp. 683–690, 2000.
4. M. Grötzbach, R. Redmann. "Analytical predetermination of complex line current harmonics in controlled AC/DC converters," IEEE Transactions on Industry Applications, vol. 33, no. 3, pp. 601–611, 1997.
5. IEEE Standard 519. IEEE recommended practice and requirements for harmonic control in electrical power systems, 1992.

6. M. Grötzbach and B. Draxler, "Effect of DC ripple and commutation on the line harmonics of current controlled AC/DC converters," IEEE Transactions on Industry Applications, vol. 29, no. 3, pp. 997–1005, 1993.

7. E. W. Gunther, "Interharmonics recommended updates to IEEE 519," IEEE Power Engineering Society Summer Meeting, pp. 950–954, July 2002.

8. P. E. Sutherland. "Harmonic measurements in industrial power systems," IEEE Transactions on Industry Applications, vol. 31, no. 1, pp. 175–183, 1995.

9. Y. Baghzouz, R. F. Burch, A. Capasso, A. Cavallini, et al. "Time varying harmonics Part I-Characterizing harmonic data," IEEE Transactions on Power Delivery, vol. 13, no. 3, pp. 938–944, 1998.

10. Y. Baghzouz, R. F. Burch, A. Capasso, A. Cavallini, et al. "Time varying harmonics Part II-Harmonic simulation and propagation" IEEE Transactions on Plasma Science, vol. 17, no. 1, pp. 279–285, 2002.

11. A. A. Girgis, W.B. Chang and E.B. Makram, "A digital recursive measurement scheme for on-line tracking of power system harmonics," IEEE Transactions on Power Delivery, vol. 6, no. 3, pp. 1153–1160, 1991.

12. I. Kamwa, R. Grondin and D. McNabb, "On-line tracking of changing harmonics in stressed power transmission systems-Part II: Applications to Hydro-Quebec network," in conference record, IEEE PES Winter meeting, Baltimore, MD, 1996.

13. C. S. Moo, Y.N. Chang and P. P. Mok, "A digital measuring scheme for time –varying transient harmonics," in conference record, IEEE PES Summer Meeting, 1994.

14. P. K. Dash, S. K. Patnaik, A. C. Liew, and S. Rahman, "An adaptive linear combiner for on-line tracking of power system harmonics" in conference record, IEEE PES, Winter Meeting, Baltimore, MD, 1996.

15. F. M. Dekking, C. Kraaikamp, H. P. Loup, and L. E. Meester, A Modern Introduction to Probability and Statistics, Springer Texts in Statistics, Springer, NY, 2007.

16. J. L. Devore, Probability and Statistics for Engineering and Sciences, Brooks/Cole, Boston, MA, 2011.

17. M. R. Spiegel, J.J. Schiller, and R. A. Srinivasan, Theory and Problems of Probability and Statistics, Second Edition, Schaum's Outline Series, McGraw Hill, New York, 2000.

18. G. R. Cooper and C.D. McGillem, Probabilistic Methods of Signal and System Analysis, Oxford University Press, NY, 1999.

19. F. Scheid, Theory and Problems of Numerical Analysis, Second Edition, Schaum Outline Series, McGraw Hill, New York, 1988.

20. Y. Xiao, X. Yang, "Harmonic summation and assessment based on probability distribution," IEEE Transactions on Power Delivery, vol. 27, no.2, pp. 1030–1032, 2012.

21. IEC Standard 61000-3-6, Electromagnetic Compatibility (EMC) Part 3: Limits-Section 6: Assessment of Emission Limits for Distorting Loads in MV and HV Power Systems, Second Edition, 2008.

22. R. G. Brown, D. Y. C. Hwang. Introduction to Random Signal Analysis and Kalman Filtering, John Wiley, New York, 1992.

CHAPTER *8*

EFFECTS OF HARMONICS

Harmonics have deleterious effects on electrical equipment. These can be itemized as follows [1]:

1. Capacitor bank failure because of reactive power overload, resonance, and harmonic amplification. Nuisance fuse operation.

2. Excessive losses, heating, harmonic torques, and oscillations in induction and synchronous machines, which may give rise to torsional stresses.

3. Increase in negative sequence current loading of synchronous generators, endangering the rotor circuit and windings.

4. Generation of harmonic fluxes and increase in flux density in transformers, eddy current heating, and consequent derating.

5. Overvoltages and excessive currents in the power system, resulting from resonance.

6. Derating of cables due to additional eddy current heating and skin effect losses.

7. Inductive interference with telecommunication circuits.

8. Signal interference in solid-state and microprocessor-controlled systems.

9. Relay malfunction.

10. Interference with ripple control and power line carrier systems, causing misoperation of the systems, which accomplish remote switching, load control, and metering.

11. Unstable operation of firing circuits based on zero-voltage crossing detection and latching.

12. Interference with large motor controllers and power plant excitation systems.

13. Possibility of subsynchronous resonance described in Chapter 5.

14. Flicker described in Chapter 5.

Nonlinear loads in the presence of capacitors can bring about a resonant condition with one of the load-generated harmonics, where none existed before. This also has following additional effects:

- Increase the transient inrush current of the transformers and prolong its decay rate [2].

Power System Harmonics and Passive Filter Designs, First Edition. J.C. Das.
© 2015 The Institute of Electrical and Electronics Engineers, Inc. Published 2015 by John Wiley & Sons, Inc.

- Increase the duty on the switching devices.
- Possibility of part-winding resonance exists if the predominant frequency of the transient coincides with natural frequency of the transformer.

8.1 ROTATING MACHINES

8.1.1 Induction Motors

Harmonics will produce elastic deformation, that is, shaft deflection, parasitic torques, vibration noise, additional heating, and lower the efficiency of rotating machines.

The movement of the harmonics is with or against the direction of the fundamental. Criterion of forward or reverse rotation is established from $h = 6m \pm 1$, where h is the order of harmonic and m is any integer. If $h = 6m \pm 1$, the rotation is in the forward direction, but at $1/h$ speed. Thus, 7th, 13th, 19th, ... harmonics rotate in a direction same as the fundamental.

If $h = 6m - 1$, the harmonic rotates in the reverse direction to the fundamental. Thus, 5th, 11th, 17th, ... are the reverse rotating harmonics. Harmonics of the order $2, 5, 8, 11, 14, \dots$ are the negative sequence harmonics (Chapter 1).

The magnitude of harmonic current in a three-phase induction motor may be calculated from the expression

$$I_h = \frac{V_h}{h\omega_0 L_{lh}} \tag{8.1}$$

where I_h is the hth harmonic current, V_h is the hth harmonic voltage, and L_{lh} is the stator and rotor leakage inductance at hth harmonic referred to the stator. The effective inductance tends to decrease as h increases. Approximately L_{lh} is equal to L_1 (stator leakage reactance), the minimum value when internal bar inductance is negligible. An accurate model of the harmonic impedance of an induction motor is derived in Chapter 12.

With certain assumptions, the harmonic losses can be defined as [1]

$$\frac{P_h}{P_{RL}} = k \sum_{h=5}^{h=\infty} \frac{V_h^2}{h^{3/2} V_1^2} \tag{8.2}$$

where

$$k = \frac{(T_s/T_R)E}{(1 - S_R)(1 - E)} \tag{8.3}$$

P_h is the harmonic loss, P_{RL} is the loss at the rated point with sinusoidal supply, T_s is the starting torque, T_R is the rated torque, S_R is the slip, and E is the efficiency. In NEMA class C motors [3], k can range up to 25 or higher.

A motor distortion index (MDI) is defined as

$$\text{MDI} = \frac{1}{V_1} \left(\sum_{h=5}^{h=\infty} \frac{V_h^2}{h^{3/2}} \right)^{1/2} \tag{8.4}$$

The use of Eq. (8.4) permits a convenient comparison of different motor designs, but it will not evaluate localized heating. A similar ratio can be derived for rotor heating only. The motors with large deep bars or double cage will have the highest harmonic heating.

In a detailed analysis, the effect of the harmonics on motor losses should consider the subdivision of losses into windage and friction, stator copper loss, core loss, rotor copper loss, and stray loss in the core and conductors and the effect of harmonics on each of these components. The effective rotor and stator leakage inductance decreases and the resistance increases with frequency. The equivalent circuits of the induction motors for positive and negative sequences are shown in Chapter 12 and the effect of negative sequence currents will be more pronounced at higher frequencies. The rotor resistance may increase four to six times the DC value while leakage reactance may reduce to a fraction of the fundamental frequency value. The stator copper loss increases in proportion to the square of the total harmonic current plus an additional increase due to skin effect on resistance at higher frequencies. Harmonics contribute to magnetic saturation, and the effect of distorted voltage on core losses may not be ignored. Major loss components influenced by harmonics are stator and rotor copper losses and stray losses. A harmonic factor of 11% gives approximately 25% derating of general-purpose motors.

Figure 8.1 from NEMA (North American Manufacturers Association) [3] shows derating factor with respect to harmonic voltage factor, which is another name for the voltage distortion factor.

8.1.2   Torque Derating

When the motor is operated below rated speed, torque derating occurs because of reduced cooling, and NEMA provides the derating curves. In case of inverter operation, the torque reduces due to additional temperature rise on account of harmonic losses and also due to voltage–frequency characteristics of some inverters. When determining derating the thermal reserve of the motor is important, and the derating factor at rated frequency may vary from 0% to 20%. NEMA [3] states that there is no established method for determining the derating curve of a particular motor. The preferred method is to test representative samples of the motor designs under load

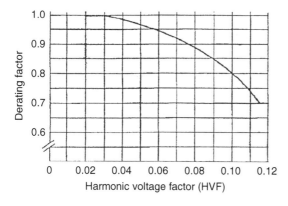

Figure 8.1   Proposed derating curve due to harmonic voltage factor (HVF) same as THDv, all machines. Source: NEMA Part 30 [3].

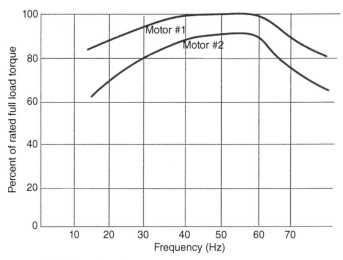

Figure 8.2 Examples of torque derating of NEMA motors when used with inverters. Source: NEMA, Part 30 [3].

while operating from a representative prototype of the inverter design and measure the temperature rise of the windings. Figure 8.2 is reproduced from NEMA [3]. For operation above 90 Hz at a required hp level, it may be necessary to utilize a motor with a greater 60-Hz rating. Motors of 1.15 service factors are a better choice. The derating curves in Fig. 8.2 are not specific; two curves are provided and labeled "design #1" and "design #2." This indicates that torque derating under same operating conditions is dictated by motor design.

8.1.3 Pulsating Fields and Dynamic Stresses

In a synchronous machine, the frequency induced in the rotor is net rotational difference between fundamental frequency and the harmonic frequency. Fifth harmonic rotates in reverse with respect to stator and with respect to the rotor, the induced frequency is that of sixth harmonic. Similarly, the forward rotating seventh harmonic with respect to stator produces sixth harmonic in the rotor. The interaction of these fields produces a pulsating torque at 360 Hz and results in the oscillations of the shaft. Similarly, the harmonic pair 11 and 13 produces a rotor harmonic of 12th. If the frequency of the mechanical resonance exists close to these harmonics during starting, large mechanical forces can occur.

The same phenomena occur in induction motors. Considering slip of the induction motors, the positive sequence harmonics, $h = 1, 4, 7, 10, 13, \ldots$, produce a torque of $(h - 1 + s)\omega$ in the direction of rotation, and the negative sequence harmonics, $h = 2, 5, 8, 11, 14, \ldots$, produce a torque of $-(h + 1 - s)\omega$ opposite to that of rotation. Here, s is the slip of the induction motor.

It is possible that harmonic torques are magnified due to certain combinations of stator and rotor slots, and cage rotors are more prone to circulation of harmonic currents as compared to the wound rotors (Chapter 3).

The zero sequence harmonics ($h = 3, 6, \ldots$) do not produce a net flux density. These produce ohmic losses.

All parasitic fields produce noise and vibrations. The harmonic fluxes superimposed upon the main flux may cause tooth saturation, and zigzag leakage can generate unbalanced magnetic pull, which moves around the rotor. As a result, the rotor shaft can deflect and run through a critical resonant speed amplifying the torque pulsations.

Torque ripples may exist at various frequencies. If the inverter is a six-pulse type, then a sixth harmonic torque ripple is created, which would vary from 36 to 360 Hz, when the motor is operated over the frequency range of 6–60 Hz. At low speeds, such torque ripple may be apparent as observable oscillations of the shaft speed or as torque and speed pulsations, usually termed *cogging* (Chapter 3). It is also possible that some speeds within the operating range may correspond to natural mechanical frequencies of the load or support structure. At such frequencies amplification can occur, giving rise to large dynamic stresses. Operation other than momentary, that is, during starting, should be avoided at these speeds.

The oscillating torques in synchronous generators can simulate the turbine generator into complex coupled mode of vibration that result in torsional oscillations of rotor elements and flexing of turbine buckets. If the frequency of a harmonic coincides with the torsional frequency of the turbine generator, it can be amplified by the rotor oscillation.

A documented case of failure of a large generator is described in Ref. [4]. A control loop within a SVC unit in a nearby steel mill resulted in modulation of 60 Hz waveform. This created upper and lower sidebands, producing 55 and 65 Hz current components. The reverse-phase rotation manifested itself as a 115 Hz stimulating frequency on the rotor, which drifted between 114 and 118 Hz. This excited sixth-mode natural frequency of the rotor shaft, creating large torsional stresses (see Chapter 5).

8.2   EFFECT OF NEGATIVE SEQUENCE CURRENTS ON SYNCHRONOUS GENERATORS

Synchronous generators have both a continuous and short-time unbalanced current capabilities, which are shown in Tables 8.1 and 8.2 [5,6]. These capabilities are based on 120 Hz negative sequence currents induced in the rotor due to continuous unbalance or unbalance under fault conditions. In the absence of harmonics and impedance asymmetries (i.e., nontransposition of transmission lines), it is a standard requirement that the synchronous generators should be able to supply some unbalance currents. When these capabilities are exploited for harmonic loading, the variations in loss intensity at different harmonics versus 120 Hz should be considered. The following expression can be used for equivalent heating effects of harmonics translated into negative sequence currents:

$$I_{2,\text{equiv}} = \left[\left(\frac{6f}{120} \right)^{1/2} (K_{5,7})(I_5 + I_7)^2 + \left(\frac{12f}{120} \right)^{1/2} (K_{11}, K_{13})(I_{11} + I_{13})^2 + \ldots \right]^{1/2}$$

(8.5)

TABLE 8.1 Requirements of Unbalanced Faults on Synchronous Machines

Type of Synchronous Machine	Permissible $I_2^2 t$
Salient pole generators	40
Synchronous condensers	30
Cylindrical rotor generators	
Indirectly cooled	30
Directly cooled (0–800 MVA)	10
Directly cooled (801–1600 MVA)	10–(0.00625) (MVA–800)*

[a] Thus, for a 1600-MVA generator, $I_2^2 t = 5$.

TABLE 8.2 Continuous Unbalance Current Capability of Generators

Type of Generator and Rating	Permissible I_2
Salient pole with connected amortisseur windings	10
Salient pole with nonconnected amortisseur windings	5
Cylindrical rotor, indirectly cooled	10
Cylindrical rotor, directly cooled to	
960 MVA	8
961–1200 MVA	6
1201–1500 MVA	5

where $K_{5,7}$, $K_{11,13}$, ... are correction factors to convert from maximum rotor surface loss intensity to average loss intensity [7], and these can be read from Fig. 8.3, $f =$ fundamental frequency and I_5, $I_7 =$ harmonic current in pu values.

Example 8.1:

(a) Consider a synchronous generator, with continuous unbalance capability of 0.10 pu (Table 8.2). It is subjected to fifth and seventh harmonic loadings of 0.07 and 0.06 pu, respectively. Is the unbalance capability exceeded? From Fig. 8.3 and harmonic ratio $0.06/0.07 = 0.857$, $K_{5,7} = 0.4$. From Eq. (8.5),

$$I_{2\text{equiv}} = \left[\sqrt{3}\,(0.4)\,(0.07 + 0.06)^2 \right]^{1/2} = 0.108$$

The continuous negative sequence capability is exceeded. If we simply sum up the harmonics, we get 13%. Thus, calculation with simple summation is inaccurate.

(b) A harmonic analysis study shows that a generator of 13.8 kV, 100 MVA, absorbs the following harmonic currents in percentage of the generator's full-load

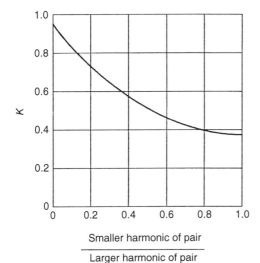

Smaller harmonic of pair
⎯⎯⎯⎯⎯⎯⎯⎯⎯⎯⎯⎯⎯⎯
Larger harmonic of pair

Figure 8.3   Ratio K for average loss to maximum loss based on harmonic pair. Source: Ref. [7].

current: 5th harmonic = 40%, 7th harmonic = 30%, 11th harmonic = 18%, 13th harmonic = 10%, 17th harmonic = 5%, and 19th harmonic = 2%. The generator has $I_2^2 T = 30$. The negative sequence relay for the protection of the generator is set at 25 for conservatism. Find the operating time of the relay. Here, only the negative sequence harmonics need to be considered:

$$[(0.4)^2 + (0.18)^2 + (0.05)^2]t = 25$$

The relay will trip in 128.3s. The generator capability to withstand the negative sequence harmonics is 153s. Harmonic loading of generators must be carefully calculated and synchronous generators must be protected for it.

8.3   INSULATION STRESSES

The high-frequency operation of modern pulse width modulation (PWM) converters with IGBTs is discussed in Chapter 4. It subjects the motors to high dv/dt. It has an adverse effect on the motor insulation and contributes to the motor-bearing currents and shaft voltages. The rise time of the voltage pulse at the motor terminals influences the voltage stresses on the motor windings. As the rise time of the voltage becomes higher, the motor windings behave like a network of capacitive elements in series. The first coils of the phase windings are subject to overvoltages, which can give rise to ringing. There has been a documented increase in the insulation failure rate caused by turn-to-turn shorts or phase-to-ground faults due to high dv/dt stress [8]. The common remedies are to provide inverter grade insulation or to add filters. The soft switching slows initial rate of rise as discussed in Chapter 4.

NEMA [3] has established limitations on voltage rise for general-purpose NEMA design A and B induction motors and definite-purpose inverter-fed motors. Windings designed for definite-purpose inverter grade motors use magnet wires with increase build, and these polyester-based wires exhibit higher breakdown strength. A motor experiencing rise time shorter than 100 ns and surges higher than 3.7 pu has a risk of stator winding problems [3].

- The stator winding insulation system of general-purpose motors rated $\leq 600\,\text{V}$ shall withstand $V_{peak} = 1\,\text{kV}$ and rise time $\geq 2\,\mu\text{s}$, and for motors rated $> 600\,\text{V}$ these limits are $V_{peak} \leq 2.5\,\text{pu}$ and rise time $\geq 1\,\mu\text{s}$.

- For definite-purpose inverter-fed motors with base voltage rating $\leq 600\,\text{V}$, $V_{peak} \leq 1600\,\text{V}$ and the rise time is $\leq 0.1\,\mu\text{s}$. For motors with base voltage rating $> 600\,\text{V}$, $V_{peak} \leq 2.5\,\text{pu}$ and rise time is $\leq 0.1\,\mu\text{s}$. V_{peak} is of single amplitude and 1 pu is the peak of the line-to-ground voltage at the maximum operating speed point.

The derating due to harmonic factor, effect on motor torque, starting current, and power factor are described in NEMA [3] (also see Refs [9–12]).

8.3.1 Common-Mode Voltages

The motor windings can be exposed to higher than normal voltages due to neutral shift and common-mode voltages [13], and in some current source inverters it can be as high as 3.3 times the crest of the nominal sinusoidal line-to-ground voltage. The generation of common-mode voltages can be described with reference to a three-phase six-pulse bridge rectifier circuit (Fig. 8.4). Only two phases conduct at a time, and the DC plus and negative voltages to the midpoint are shown. These voltages do not add to zero and the midpoint oscillates at three times the AC supply system frequency. The DC positive and negative buses have common-mode voltages, and its magnitude changes with the firing angle. The peak of the voltage is approximately $0.5V_{In}$, where V_{In} is the peak line-to-neutral point input voltage.

The operation of the output bridge creates a common-mode voltage by exactly the same mechanism as the input bridge does, where the back EMF of the motor is analogous to the line voltage. Thus, the worst-case condition for the common-mode voltage is no-load, full-speed operation, as the phase-back angle is 90° for both the converters, and the motor voltage is essentially equal to the line voltage. The sum of both common-mode voltages is approximately V_{In} at six times the input frequency. As the input and output frequencies are generally different, the motor experiences a waveform with beat frequencies of both input and output frequencies and there will be instances when two times the rated voltage is experienced.

The grounding systems can be designed so that the motor insulation system is not stressed beyond its design level [13].

Harmonics also impose higher dielectric stresses on the insulation of other electrical apparatuses. Harmonic overvoltages can lead to corona, void formation, and degradation.

The calculation of the surge voltages at the motor terminals can be done by assuming that the motor stator windings and interconnecting cable have certain

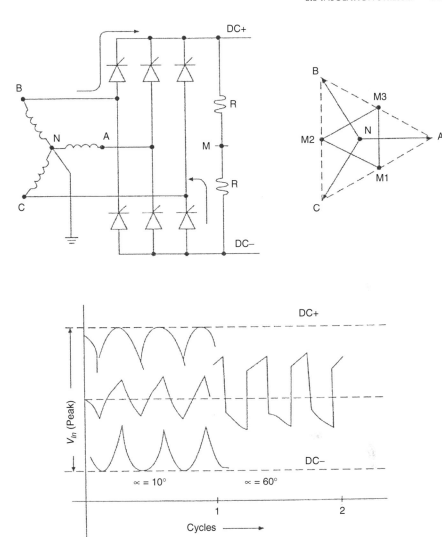

Figure 8.4   Generation of common-mode voltage in a six-pulse converter, see text.

surge impedance and using transmission line traveling wave theory. Reference [14] describes a wide-band 50–10 MHz monitor. (Fourier analysis of a 50-ns rise time of surge will show frequencies up to 6 MHz.)

8.3.2   Bearing Currents and Shaft Voltages

Shaft voltages are caused due to asymmetries in the windings, slotting, eccentricity, and key ways, and shaft currents can flow due to inductive or capacitive couplings. These will cause currents to flow through the bearings. PWM drives can produce increased shaft currents due to high-frequency currents produced by common-mode currents – this is an inductive effect. Also at high frequencies, capacitive currents depend on the first few turns of stator windings. A motor that is symmetrical at fundamental frequency can become asymmetrical at higher frequencies. The degree of damage caused is dependent on many factors, including the quality of bearings. Sometimes, semiconducting grease is used to provide a path for the electrical currents, but it degrades the life of the bearings. If shaft voltages higher than 300 mV peak occur, the motor should be equipped with insulated bearings.

Both the bearings, at drive end and opposite drive end (ODE), must be insulted. The capacitive coupling currents flow to the ground. If only one bearing is insulted, all the current will flow through the uninsulated bearing resulting in rapid failure. Also mechanical load must be insulted. If it is not possible to insulate the mechanical load, a shaft grounding brush should be added to provide a low-impedance path to ground. Figure 8.5(a) shows inductive circulating currents, and Fig. 8.5(b) illustrates capacitive coupled current flow. PWM drives using BJT (bipolar junction transistors) and IGBTs can cause electric discharge machining (EDM) currents. PWM inverters excite capacitive coupling between the stator windings, the rotor, and the stator frame. This common-mode current does not circulate but travels to ground (Fig. 8.5(b)). The

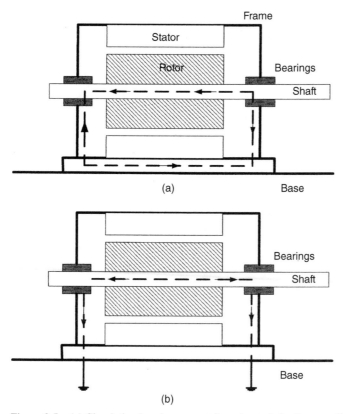

Figure 8.5   (a) Circulating bearing current flow due to inductive coupling; (b) capacitive current flow to ground.

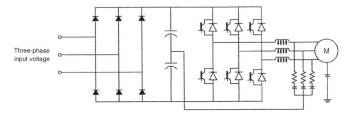

Figure 8.6   Output filters to avoid differential-mode and common-mode high dv/dt.

capacitance between the windings and stator frame is typically 30–100 times higher than the capacitance between the stator windings and the rotor.

High-frequency currents can travel on the surface of the grounding conductor. The motor mounting base should be welded to the mechanical load base for effective grounding. Sometimes, the bearings are left uninsulated and a shaft grounding brush is added. The stator frame should be effectively grounded for high frequencies. This can be ensured by welding the motor base frame to the driven mechanical load base frame.

The common-mode noise caused by PWM is 10 times or more compared to a sine wave. The addition of a filter at motor terminals can reduce differential mode and common mode dv/dt at motor terminals and also induced shaft voltage and leakage current to ground. Figure 8.6 shows an output filter to reduce differential mode and common mode dv/dt at the motor terminals [15–18]. Figure 8.7(a) depicts that the rms value of the noise-canceled stator current increases as the fault develops; the bearing is at 50% load level. Figure 8.7(b) shows that the vibrations increase as the fault develops [18].

8.3.3   Effect of Cable Type and Length

When the motor is connected through long cables, the high dv/dt pulses generated by PWM inverters give rise to traveling wave phenomena on the cables, resulting in reinforcement of incident and reflected waves due to impedance discontinuity at the motor terminals. The voltages can reach twice the inverter output voltage. The cable-to-motor impedance ratio and cable run length are important factors with respect to reflection coefficients (Chapter 12). An analogy can be drawn with long transmission lines and traveling wave phenomena. The incident traveling wave is reflected at the motor terminals, and reinforcement of incident and reflected waves occurs. Owing to dielectric losses and cable resistance, damped ringing occurs as the wave is reflected from one end of the cable to the other. The ringing frequency is a function of cable length and wave propagation velocity and is of the order of 50 kHz to 2 MHz [14].

An approximate check for voltage doubling possibility can be made by the following calculation:

$$L_c = \frac{vt_r}{2} \tag{8.6}$$

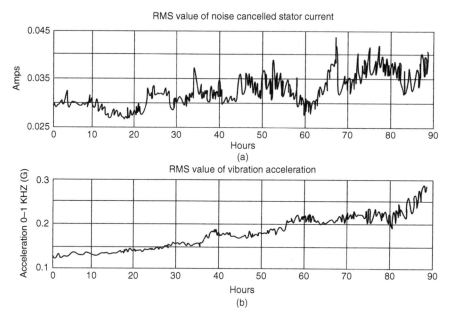

Figure 8.7   (a) Rms value of the noise-canceled stator current increases as the fault develops; the bearing is at 50% load level; (b) increase in vibrations as the fault develops. Source: Ref. [18].

where

L_c is the critical length of the cable

v is the velocity of propagation in the cable, which may be taken as 50% of the speed of light $= 150\,\text{m}/\mu\text{s}$

t_r is the rise time of the pulse in microseconds.

For the fastest IGBT, $t_r = 0.1\,\mu\text{s}$. This gives a critical cable length of 7.5 m, and for the slowest pulse rise of $4.0\,\mu\text{s}$, $L_c = 360\,\text{m}$.

Figure 8.8 is constructed on this basis; also see Table 8.3.

A first-order RC filter (consisting of a resistance and capacitance in series connected to ground at the motor terminals) can be added at the motor terminals to limit the overvoltages [11]. If we limit the reflected wave at the motor terminals to no more than 0.2 times the incident wave, then the terminal voltage at the motor is limited to 20% above inverter DC-link voltage. Make $R =$ surge impedance of the cable $= Z$, and the value of C is given by the expression:

$$C = \frac{l_c C_c}{0.22314} \tag{8.7}$$

where l_c is the length of the cable in feet and C_c its capacitance per foot (Fig. 8.9(a)).

The type of cables between motor and the drive system is important. Reference [14] reports test results of various types of cables with respect to concerns of the following:

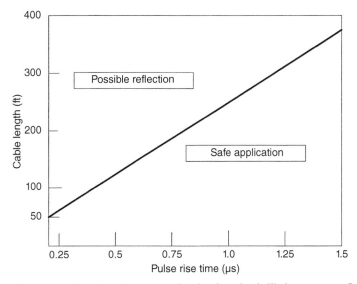

Figure 8.8   Cable length versus pulse rise time that is likely to cause reflection. Source: Ref. [14].

TABLE 8.3   Minimum Cable Length and PWM Rise Time

PWM Rise Time (μs)	Minimum Cable Length (ft)
0.1	19
0.5	97
1.0	195
2.0	390
3.0	585
4.0	780

- Currents in the National Electric Code (NEC) equipment grounding circuits
- Common-mode current
- Motor frame voltage to PE (Protective Earth, an IEC terminology) ground at the motor
- Cross talk between adjacent motor circuits.

It recommends that electrical shielding is properly connected to ground and a symmetrical six conductor (three ground conductors), and continuous corrugated aluminum armor-type sheath, NEC (National Electric Code) cable-type MC-metal-clad, should be used (see Fig. 8.9(b) and (c) for termination at the motor).

By adding output filters, the cable charging current as well as dielectric stresses on the motor insulation can be reduced. The common filter types are as follows:

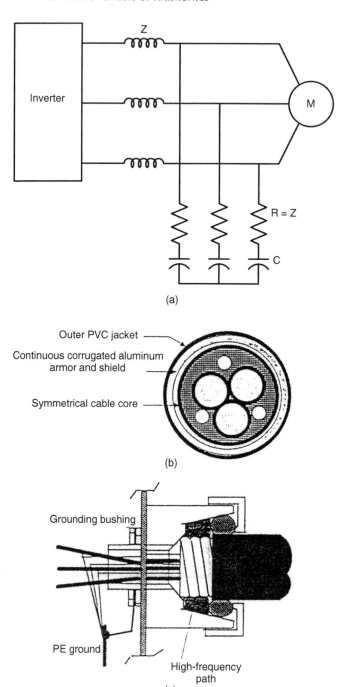

(a)

(b)

(c)

Figure 8.9 (a) Output filter at motor terminals to prevent reflections, (b) cable construction for ASDs, and (c) cable termination.

- output line inductors
- output limit filter
- sine-wave filter
- motor termination filter.

An output inductor reduces dv/dt at the inverter and the motor. The ringing and overshoot may also be reduced. Output limit filters may consist of laminated core inductor or ferrite core inductors. A sine-wave filter is a conventional low-pass filter formed from an output inductor, capacitor, and a damping resistor. Motor termination filters are first-order resistor/capacitor filters (Chapter 16).

8.4 TRANSFORMERS

A transformer supplying nonlinear load may have to be derated. Harmonics effect transformer losses and eddy current loss density. The upper limit of the current distortion factor is 5% of the load current, and the transformer should be able to withstand 5% overvoltage at rated load and 10% at no load. The harmonic currents in the applied voltage should not exceed these limits.

In addition to derating due to harmonic currents and induced eddy current loss, a drive system transformer may be subjected to severe current cycling and load demand depending on the drive system.

8.4.1 Losses in a Transformer

The linear model of a two-winding transformer has been described in Section 3.1.1, and its equivalent circuit and phasor diagram were developed in Figs. 3.1 and 3.2. The expressions for hysteresis and eddy current loss are given in Eqs. (3.4) and (3.5).

On a simplified basis, the transformer positive sequence or negative sequence model is given by its percentage reactance specified by the manufacturer, generally on the transformer natural cooled (ONAN, Oil Natural Air Natural) MVA rating base. This reactance remains fairly constant and is obtained by a short-circuit test on the transformer. The magnetizing circuit components are obtained by an open-circuit test.

The hysteresis and eddy current losses constitute no-load losses, and these can be ascertained by open circuit test. The test is conducted with the secondary winding open-circuited and rated voltage applied to the primary winding. For high-voltage transformers, the secondary winding may be excited and the primary winding opened. At constant applied frequency, B_m is directly proportional to applied voltage, and the core loss is approximately proportional to B_m^2. The magnetizing current rises steeply at low flux densities, then more slowly as iron reaches its maximum permeability, and thereafter again steeply, as saturation sets in.

From Fig. 3.1(a), the open circuit admittance is

$$Y_{OC} = g_m - jb_m \tag{8.8}$$

This neglects the small voltage drop across r_1 and x_1. Then,

$$g_m = \frac{P_0}{V_1^2} \tag{8.9}$$

where P_0 is the measured power and V_1 is the applied voltage. Also,

$$b_m = \frac{Q_0}{V_1^2} = \sqrt{\frac{S_0^2 - P_0^2}{V_1^2}} \tag{8.10}$$

where P_0, Q_0, and S_0 are the measured active power, reactive power, and volt-amperes on open circuit. Note that the exciting voltage E_1 (Fig. 3.2(a)) is not equal to V_1 due to the voltage drop that no-load current produces through r_1 and x_1. Corrections can be made for this drop. When the secondary does not have any load and is open circuited, some small amount of current flows through r_1, which gives some copper loss, this can be accounted for in the calculations.

The short-circuit test is conducted at the rated current of the winding, which is shorted and a reduced voltage is applied to the other winding to circulate a full-rated current:

$$P_{sc} = I_{sc}^2 R_1 = I_{sc}^2(r_1 + n^2 r_2) \tag{8.11}$$

where P_{sc} is the measured active power on short-circuit and is the representative of copper loss, and I_{sc} is the short-circuit current:

$$Q_{sc} = I_{sc}^2 X_1 = I_{sc}^2(x_1 + n^2 x_2) \tag{8.12}$$

Thus, the losses in a transformer are discussed in the following sections.

Fixed Losses These consist of eddy current and hysteresis loss (core loss and dielectric losses), given by no-load test and corrected for $I^2 R$ no-load loss. The hysteresis loss accounts for 75–80% of the core loss.

Direct Loss Copper loss in primary and secondary windings, which will depend on the load current and its power factor.

Stray Load Loss The stray load loss includes eddy current losses in conductors and other parts of the transformers, such as tank walls and constructional parts.

Cooling System Losses These losses account for ventilation fans for forced cooling and forced oil pumps. A manufacturer may specify the losses as shown in Table 8.4.

Note that the stray loss is not separately specified in Table 8.4, but it can be calculated. *It is essential to obtain the actual transformer loss data before calculations for derating of the transformer for nonlinear loads can proceed.*

Figure 8.10 from Ref. [19] shows the electromagnetic field produced by the current in a core-type transformer. Each metallic conductor has induced voltages that

TABLE 8.4   Transformer Test Data

Test Parameter	Test Results	Remarks
Winding resistance	Measured between H1–H2, H1–H3, and H2–H3 Measured between X1–X2, X1–X3, and X2–X3	Variation between measurements, temperature of measurement, and average values are indicated. The measurements are provided at all taps
Core loss	Test voltage and frequency, average voltage, amperes, and watts	Supplied at 90%, 100%, and 110% of the voltage at rated output
Winding loss (copper loss plus stray load loss)	Average amperes, watts, rms voltage, frequency, test temperature, and voltage at rated MVA	Supplied at various primary tap voltages and corresponding impedances of the transformer corrected to 75 °C

produce eddy currents. Eddy current loss is dissipated in the form of heat. This eddy current loss can be divided into two parts: that occurring in the winding called "eddy current loss," and the portion outside the windings is called the "other stray loss." Inner winding of a core-type transformer has higher loss because electromagnetic flux has tendency to fringe toward the low-reluctance path of core leg. Also highest eddy current losses occur in end conductors of the inner winding, as this region has highest radial electromagnetic flux density. The IEEE Standard Ref. [19] makes simplifying assumptions with respect to relative proportions of eddy current loss in the inner and outer windings for calculations of transformer derating.

8.4.2   Derating of Transformers Supplying Nonlinear Loads

The following calculations are based on Ref. [19].

On the basis of the discussions in Section 8.4.1, the total transformer loss P_{LL} is

$$P_{LL} = P + P_{EC} + P_{OSL} \tag{8.13}$$

where P is $I^2 R$ loss. Here, the stray load losses is subdivided into the losses in the windings, and losses in the nonwinding components of the transformer, that is, core clamps, structures, and tank. P_{EC} is the winding eddy current loss and P_{OSL} is other stray loss.

Effects of Harmonic Current on Losses   *The winding eddy current loss P_{EC} for* power frequency and the frequencies associated with harmonics tend to be proportional to the square of the current and approximately proportional to the square of

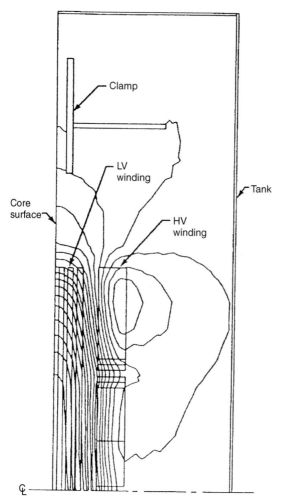

Figure 8.10 Electromagnetic flux in a core-type transformer on loading, see text. Source: Ref. [19].

the frequency. This characteristic causes excessive winding loss and hence abnormal temperature rise.

The other stray load loss will also increase proportional to the square of the current. However, this will not increase proportional to the square of the frequency. The studies show that eddy current loss in bus bars, connections, and structural parts increase by a harmonic exponential factor of 0.8 or less. The effects of these losses vary depending on the type of transformer. The temperature rise in these nonwinding parts will not be critical for dry-type transformers, but must be accounted for in the liquid-immersed transformers.

For the liquid-immersed transformers, the top oil rise θ_{TO} will increase as the total load losses increase due to harmonic loading. Unlike dry-type transformers, where P_{OSL} is ignored, it must be considered for oil-immersed transformers as it impacts the top oil temperature.

The DC component of the load current and harmonics are not considered in Ref. [19], but it may increase the magnetizing current (Fig. 3.2 component I_m) and audible noise substantially.

Equation (8.13) can be written in terms of pu:

$$P_{\text{LL-R(pu)}} = 1 + P_{\text{EC-R(pu)}} + P_{\text{OSC-R(pu)}} \tag{8.14}$$

If the rms value of the current including harmonics is the same as the fundamental current, I^2R loss will be maintained the same. If the rms value due to harmonics increases, so does the I^2R loss:

$$I_{\text{(pu)}} = \left[\sum_{h=1}^{h=\text{max}} \left(I_{h\text{(pu)}} \right)^2 \right]^{1/2} \tag{8.15}$$

Here, h is the harmonic order, $h = \text{max}$ is the highest significant harmonic order, and $I_h(\text{pu})$ is the per-unit rms current at harmonic h.

The eddy current loss P_{EC} is assumed to vary in proportion to the square of the electromagnetic field strength. Square of the harmonic current or the square of the harmonic number may be considered to be representative of it. Owing to skin effect, the electromagnetic flux may not penetrate conductors at high frequencies. The leakage flux has its maximum concentration between interfaces of the two windings:

$$P_{\text{EC(pu)}} = P_{\text{EC-R(pu)}} \sum_{h=1}^{h=\text{max}} I_{h\text{(pu)}}^2 h^2 \tag{8.16}$$

where $P_{\text{EC-R(pu)}}$ is the winding eddy current loss under rated conditions and $I_h(\text{pu})$ is the per-unit rms current at harmonic h. To facilitate actual field measurements, define winding eddy current loss at measured current and the power frequency by another term $P_{\text{EC-O}}$. Then, we can write the equation:

$$P_{\text{EC}} = P_{\text{EC-O}} \times \frac{\displaystyle\sum_{h=1}^{h=h_{\text{max}}} I_h^2 h^2}{I^2} = P_{\text{EC-O}} \times \frac{\displaystyle\sum_{h=1}^{h=h_{\text{max}}} I_h^2 h^2}{\displaystyle\sum_{h=1}^{h=h_{\text{max}}} I_h^2} \tag{8.17}$$

where I is the rms load current.

8.4.3   Harmonic Loss Factor for Winding Eddy Currents

Define harmonic loss factor F_{HL} for the windings, which represents effective heating as a result of harmonic load currents as a ratio: $P_{\text{EC}}/P_{\text{EC-O}}$ in Eq. (8.17).

$$F_{\text{HL}} = \frac{\displaystyle\sum_{h=1}^{h=h_{\text{max}}} \left[\frac{I_h}{I} \right]^2 h^2}{\displaystyle\sum_{h=1}^{h=h_{\text{max}}} \left[\frac{I_h}{I} \right]^2} \tag{8.18}$$

In the above equation I, *the rms load current* can be substituted with I_1, where I_1 is the rms fundamental load current:

$$F_{\text{HL}} = \frac{\sum_{h=1}^{h=h_{\max}} \left[\frac{I_h}{I_1}\right]^2 h^2}{\sum_{h=1}^{h=h_{\max}} \left[\frac{I_h}{I_1}\right]^2} \tag{8.19}$$

It can be shown that *whether we normalize with respect to I or I_1*, the F_{HL} calculation gives the same results.

Example 8.2: Consider a fundamental current of 1500 A, 5th harmonic 300 A, 7th harmonic 200 A, 11th harmonic 80 A, and 13th harmonic 50 A. Higher order harmonics are ignored. Calculate F_{HL} using Eqs (8.18) and (8.19).

To use Eq. (8.18), first normalize to rms load current.
Calculate rms load current:

$$I = \sqrt{1500^2 + 300^2 + 200^2 + 80^2 + 50^2} = 1545.61 \text{ A}$$

Table 8.5 shows the calculations, note the ratio I_h/I and further calculations.

$$F_{\text{HL}} = \frac{3.206332}{1.00} = 3.206$$

To use Eq. (8.19), normalize to rms *fundamental* load current.
Table 8.6 shows the calculation steps.

$$F_{\text{HL}} = \frac{3.40307}{1.061735} = 3.2052$$

This confirms the result within errors of calculations.

TABLE 8.5   Harmonic Distribution Normalized to rms Load Current

h	I_h	I_h/I	$(I_h/I)^2$	h^2	$(I_h/I)^2 h^2$
1	1500	0.97049	0.9418	1	0.9418
5	300	0.1941	0.03767	25	0.94185
7	200	0.1294	0.0167	49	0.82046
11	80	0.052	0.00267	121	0.32416
13	50	0.0323	0.0011	169	0.17685
Σ			1.0		3.206

TABLE 8.6   Harmonic Distribution Normalized to rms Fundamental Load Current

h	I_h	I_h/I_1	$(I_h/I_1)^2$	h^2	$(I_h/I_1)^2 h^2$
1	1500	1	1.0	1	1.0
5	300	0.2	0.04	25	1.0
7	200	0.1333	0.0178	49	0.8722
11	80	0.0533	0.00284	121	0.3436
13	50	0.0333	0.00111	169	0.1876
Σ			1.061		3.4034

8.4.4   Harmonic Loss Factor for Other Stray Loss

Dry-Type Transformers   The heating due to other stray losses is not a consideration for dry-type transformers as the heat generated is dissipated by the cooling air.

Liquid-Filled Transformers   For the liquid-filled transformers, the other stray loss cannot be ignored:

$$P_{\text{OSL}} = P_{\text{OSL-R}} \times \sum_{h=1}^{h=\max} \left(\frac{I_h}{I_1}\right)^2 h^{0.8} \tag{8.20}$$

where $P_{\text{OSL-R}}$ is the other stray load loss under rated conditions.

The harmonic loss factor for the other stray load loss can be written similar to Eq. (8.18):

$$F_{\text{HL-STR}} = \frac{\displaystyle\sum_{h=1}^{h=h_{\max}} \left[\frac{I_h}{I}\right]^2 h^{0.8}}{\displaystyle\sum_{h=1}^{h=h_{\max}} \left[\frac{I_h}{I}\right]^2} \tag{8.21}$$

Again I can be replaced with I_1 as in Eq. (8.19) and the same results will be obtained.

8.4.5   Calculations for Dry-Type Transformers

With $P_{\text{OSL}} = 0$, all the stray losses are assumed to occur in the windings. P_{LL} can be written as

$$P_{\text{LL(pu)}} = \sum_{h=1}^{h=\max} I_{\text{(pu)}}^2 \times (1 + F_{\text{HL}} P_{\text{EC-R(pu)}})\text{pu} \tag{8.22}$$

To adjust the per-unit loss density in the individual windings, the effect of F_{HL} must be known on each winding. The per-unit value of the nonsinusoidal current for the dry-type transformers, which will make the result of Eq. (8.22) equal to the design value of the loss density in the highest loss region for rated frequency and for rated current, is given by the following equation:

$$I_{\text{max(pu)}} = \left[\frac{P_{\text{LL-R(pu)}}}{1 + [F_{\text{HL}} \times P_{\text{EC-R(pu)}}]} \right]^{1/2} \tag{8.23}$$

The calculations of $P_{\text{EC-R}}$ can be made from the transformer test data. The maximum eddy current loss density is assumed 400% of the average value for that winding. The division of eddy current loss between the windings is the following:

- A total of 60% in inner winding and 40% in outer winding in all transformers having a self-cooled rating of < 1000 A regardless of turns ratio.
- A total of 60% in inner winding and 40% in outer winding in all transformers for all transformers having turns ratio of 4:1 or less.
- A total of 70% in the inner winding and 30% in outer winding for all transformers having turns ratio > 4 : 1 and also having one or more windings with a maximum self-cooled rating of > 1000 A.
- The eddy current loss distribution within each winding is assumed nonuniform.

The stray loss component of the load loss is calculated by the following expression:

$$P_{\text{TSL-R}} = \text{Total load loss} - \text{copper loss}$$

$$= P_{\text{LL}} - K[I_{(1-R)}^2 R_1 + I_{(2-R)}^2 R_2] \tag{8.24}$$

In this expression for copper loss, R_1 and R_2 are the resistances measured at the winding terminals (i.e., H_1 and H_2 or X_1 and X_2) and should not be confused with winding resistances of each phase. $K = 1$ for single-phase transformers and 1.5 for the three-phase transformers.

A total of 67% of the stray loss is assumed to be winding eddy losses for dry-type transformers:

$$P_{\text{EC-R}} = 0.67 P_{\text{TSL-R}} \tag{8.25}$$

A total of 33% of the total stray loss is assumed to be winding eddy losses for liquid immersed.

$$P_{\text{EC-R}} = 0.33 P_{\text{TSL-R}} \tag{8.26}$$

The other stray losses are given by

$$P_{\text{OSL-R}} = P_{\text{TSL-R}} - P_{\text{EC-R}} \tag{8.27}$$

As the low-voltage winding is the inner winding, maximum $P_{\text{EC-R}}$ is given by

$$\text{max } P_{\text{EC-R(pu)}} = \frac{K_1 P_{\text{EC-R}}}{K(I_{(2-R)})^2 R_2} \text{pu} \tag{8.28}$$

where K_1 is the division of eddy current loss in the inner winding, equal to 0.6 or 0.7 multiplied by the maximum eddy current loss density of 4.0 per unit, that is, 2.4 or 2.8 depending on the transformer turns ratio and the current rating and K has already been defined depending the number of phases.

Example 8.3: A delta–wye connected dry-type isolation transformer of 13.8–2.4 kV, 3000 kVA, serves nonlinear load with the following current spectrum in the percentage of fundamental frequency current I_1.

$$I_1 = 1, I_5 = 0.20, I_7 = 0.125, I_{11} = 0.084, I_{13} = 0.07, I_{17} = 0.05, I_{19} = 0.04,$$

$$I_{23} = 0.03, I_{25} = 0.025$$

Calculate if the transformer will be overloaded due to harmonic current spectrum. The following data are supplied by the manufacturer: $R_1 = 1.052\ \Omega$, $R_2 = 0.0159\ \Omega$, and total load loss $= 39,000\,\text{W}$ at $75\,°C$ when supplying rated full-load current.

The rated fundamental frequency current of the transformer is 721.7 A. The primary 13.8 kV winding is connected in delta; thus, fundamental frequency line current at 13.8 kV $= 125.5\,\text{A}$. The copper loss in the windings at the fundamental frequency is

$$1.5[(1.052)(125.5)^2 + (0.0159)(721.71)^2] = 37,229.3\,\text{W}$$

From the given loss data:

$$P_{\text{TSL-R}} = 39,000 - 37,229.3 = 1770.7\,\text{W}$$

As the transformer's turns ratio exceeds 4:1 and the secondary winding current is $< 1000\,\text{A}$,
the winding eddy loss is then

$$P_{\text{EC-R}} = 1770.7 \times 0.67 = 1186.4\,\text{W}$$

The transformer has a ratio of $> 4 : 1$, but secondary current is $< 1000\,\text{A}$. Then, from Eq. (8.28), the maximum $P_{\text{EC-Rmax}}$ is

$$P_{\text{EC-Rmax}} = \frac{2.4 \times 1186.4}{1.5(0.0159)(721.7)^2} = 0.2254\,\text{pu}$$

Table 8.7 is constructed based on the transformer fundamental current of 721.7 A as the base current, the steps of calculation are obvious. Then,

$$F_{\text{HL}} = 6.6149/1.0732 = 6.164$$

$$P_{\text{LL(pu)}} = 1.0732 \times (1 + 0.2254 \times 6.164) = 2.5643$$

TABLE 8.7 Calculations for Derating of a Dry-Type Transformer due to Harmonic Loads

h	I_h/I_1	$(I_h/I_1)^2$	h^2	$(I_h/I_1)^2 h^2$
1	2	3	4	5
1	1	1.0	1.0	1.0
5	0.2	0.04	25	1.00
7	0.125	0.0156	49	0.766
11	0.084	0.0071	121	0.854
13	0.07	0.0049	169	0.8281
17	0.05	0.0025	289	0.7225
19	0.04	0.0016	361	0.5776
23	0.03	0.0009	529	0.4761
25	0.025	0.00063	625	0.3906
Σ		1.0732		6.6149

Then from Eq. (8.23), the maximum permissible value of nonsinusoidal current is

$$I_{\max(pu)} = \sqrt{\frac{1.2254}{1 + 6.164 \times 0.2254}} = 0.716\,pu$$

Thus, the transformer capability with the given nonsinusoidal load current harmonic composition is approximately $721.7 \times 0.716 = 516.74$ A.

8.4.6 Calculations for Liquid-Filled Transformers

The liquid-filled transformers are similar to dry-type transformers except that effects of all stray losses are considered. For self-cooled OA (ONAN) mode, the top oil temperature rise above ambient is given by

$$\theta_{TO} = \theta_{TO\text{-}R}\left[\frac{P_{LL} + P_{NL}}{P_{LL\text{-}R} + P_{NL}}\right]^{0.8} \,^\circ C \tag{8.29}$$

where $\theta_{TO\text{-}R}$ is the top oil temperature rise over ambient underrated conditions. P_{NL} is the no-load loss, and P_{LL} is the load loss under rated conditions. Also,

$$P_{LL} = P + F_{HL}P_{EC} + F_{HL\text{-}STR}P_{OSL} \tag{8.30}$$

where $F_{HL\text{-}STR}$ is the harmonic loss factor for the other stray loss, and P is the I^2R portion of the load loss.

The winding hottest spot conductor rise is given by

$$\theta_g = \theta_{g\text{-}R}\left(\frac{1 + F_{HL}P_{EC\text{-}R(pu)}}{1 + P_{EC\text{-}R(pu)}}\right)^{0.8} \tag{8.31}$$

where θ_g and $\theta_{g\text{-R}}$ are the hottest spot conductor rise over top oil under harmonic loading and underrated conditions, respectively.

For liquid-immersed transformers, Eq. (8.31) becomes

$$\theta_{g1} = \theta_{g1\text{-R}} \left(\frac{1 + 2.4 F_{\text{HL}} P_{\text{EC-R(pu)}}}{1 + 2.4 P_{\text{EC-R(pu)}}} \right)^{0.8} \,°\text{C} \tag{8.32}$$

or

$$\theta_{g1} = \theta_{g1\text{-R}} \left(\frac{1 + 2.8 F_{\text{HL}} P_{\text{EC-R(pu)}}}{1 + 2.8 P_{\text{EC-R(pu)}}} \right)^{0.8} \,°\text{C} \tag{8.33}$$

where θ_{g1} is the hottest spot HV conductor rise over top oil temperature and $\theta_{g1\text{-R}}$ is the hottest spot HV conductor rise over top oil underrated conditions.

Example 8.4: A 13.8–4.16 kV, 2500 kVA, delta–wye connected three-phase transformer ONAN is subjected to the following harmonic current spectrum in per unit, calculated in terms of transformer full-load current at 4.16 kV = 346.97A:

$\quad I_1 = 1, I_5 = 0.65, I_7 = 0.40, I_{11} = 0.07, I_{13} = 0.05, I_{17} = 0.04, I_{19} = 0.025.$
The top oil temperature rise and winding hottest spot conductor rise are required to be calculated. The vendor supplied data are

$$\text{No load loss} = 5000\,\text{W}$$

$$R_1 = 1.01 \ \Omega$$

$$R_2 = 0.032 \ \Omega$$

Total load loss = 25, 600W at full load, 75 °C.
The full-load copper loss is

$$1.5[(1.01)(104.6)^2 + (0.032)(346.97)^2] = 22, 354.5\,\text{W}$$

Then

$$P_{\text{TSL-R}} = 25, 600 - 22, 354.5 = 3245.5$$

$$P_{\text{EC-R}} = 3245.5 \times 0.33 = 1071.0\,\text{W}$$

$$P_{\text{OSL-R}} = P_{\text{TSL-R}} - P_{\text{EC-R}} = 3245.5 - 1071.0 = 2174.5\,\text{W}$$

Assume the following temperature rises, which are normal for 55 °C rated transformers.

$$\text{HV and LV windings average} = 55\,°\text{C}$$

$$\text{Top oil rise} = 55\,°\text{C}$$

$$\text{Hottest spot conductor rise} = 65\,°\text{C}$$

TABLE 8.8 Calculations for Derating of a Liquid-Filled Transformer due to Harmonic Loads

h	I_h/I_1	$(I_h/I_1)^2$	h^2	$(I_h/I_1)^2 h^2$	$h^{0.8}$	$(I_h/I_1)^2 h^2$
1	2	3	4	5		5
1	1	1.0	1.0	1.0	1.0	1.0
5	0.65	0.422	25	10.56	3.62	1.60
7	0.40	0.160	49	7.84	4.74	0.758
11	0.07	0.0049	121	0.5929	6.81	0.0334
13	0.05	0.0025	169	0.4225	7.78	0.0194
17	0.04	0.0016	289	0.4624	9.65	0.0154
19	0.025	0.000625	361	0.2256	10.54	0.0066
Σ		1.592		21.10		3.433

TABLE 8.9 Loss Calculation, Liquid-Immersed Transformer, Example 8.4

Type of Loss	Rated Losses (W)	Load Losses (W)	Harmonic Multiplier	Corrected Losses (W)
No-load	5000	5000		5000
Copper	22,354.5	22,577.5		22,577.5
Winding eddy	1071.0	1082.06	13.254	14,337.0
Other stray	2174.2	2195.9	2.156	4734.4
Total	30,600			46,649

From Table 8.8, the harmonic loss factor is 13.254. Harmonic loss factor for the other stray loss is the summation of column 7 divided by the summation of column 3 = 3.433/1.592 = 2.156.

If we consider that the transformer is loaded only to 80% of its rating, the losses must be adjusted to actual loading conditions. These calculations are in Table 8.9.

From Table 8.8, the rms current is given by the square root of the third-column summation, that is, $\sqrt{1.592} = 1.2618$. The total losses must be corrected to reflect rms current and also load factor. Therefore, correction for the rms current is

$$P_{LL} = I^2_{rms-pu} L^2_f = 1.2618^2 \times 0.8^2 = 1.01$$

Correct the losses using harmonic multipliers calculated earlier, as shown in Table 8.9.

Thus, the total losses are now = 46, 649 W.

Then, the top oil temperature from Eq. (8.29) is

$$\theta_{TO} = 55 \times \left(\frac{46,649}{30,600} \right)^{0.8} = 77.06\,^\circ C$$

The rated inner winding losses corrected for rms current are

$$1.01 \times 1.5(0.032)(346.97)^2 = 5836.4\,\text{W}$$

Also loss under rated condition = 5778.6 W.

The low-voltage winding currents are less than 1000 A. A total of 60% of the winding eddy loss is assumed to occur in the LV winding, and the maximum eddy loss at the hottest region is assumed to be four times average eddy loss. The hottest spot conductor rise over top oil temperature is calculated using Eq. (8.31).

$$\theta_g = (65 - 55) \times \left(\frac{5836.4 + 14,337 \times 2.4}{5778.6 + 1071 \times 2.4} \right)^{0.8} = 35.19\,^\circ\text{C}$$

Then, the hottest spot temperature is $77.06 + 35.19 = 112.25\,^\circ\text{C}$.

This exceeds the maximum of $65\,^\circ\text{C}$ by $47.25\,^\circ\text{C}$. The transformer on continuous operation will be seriously damaged at 80% load factor.

If the transformer is loaded to 50%, then

$$\text{Total losses} = 20,523\,\text{W}$$

$$\theta_{TO} = 55 \times \left(\frac{20,523}{30,600} \right)^{0.8} = 39.9\,^\circ\text{C}$$

and

$$\theta_g = (65 - 55) \times \left(\frac{2323 + 5707 \times 2.4}{5778.6 + 1071 \times 2.4} \right)^{0.8} = 16.84\,^\circ\text{C}$$

Then the hot spot temperature is $56.74\,^\circ\text{C}$. This is acceptable.

Thus, the transformer is capable of supplying approximately 56.7% of its rated load. This high derating is due to high harmonic content in the spectrum. This demonstrates that liquid-filled or air-cooled transformers cannot be loaded to the rated fundamental frequency rating when supplying nonlinear loads, and the derating must be calculated in each case.

8.4.7 UL K Factor of Transformers

The UL (Underwriter's Laboratories) standards [20, 21] specify transformer derating (K factors) when carrying nonsinusoidal loads. The UL definition of K factor is given by the expression

$$K\text{-factor} = \sum_{h=1}^{h=h_{\max}} I^2_{h\text{-pu}} h^2 \qquad (8.34)$$

The UL definition of K factor is based on using the transformer rated current in the calculation of per-unit current in the earlier equation:

$$K\text{-factor} = \frac{1}{I_1^2} \sum_{h=1}^{h=h_{\max}} I^2_{h\text{-pu}} h^2 \qquad (8.35)$$

This can be shown to be equal to

$$K\text{-factor} = \left(\frac{\displaystyle\sum_{h=1}^{h=\max} I_h^2}{I_1^2} \right) F_{\text{HL}} \tag{8.36}$$

The harmonic loss factor is a function of the harmonic current distribution and independent of its relative magnitude. UL K factor is dependent on both the magnitude and distribution of the harmonic current. For a new transformer with harmonic currents specified as per unit of the rated transformer secondary currents, the K factor and harmonic loss factor will have the same numerical values. That is, the K factor is equal to the harmonic factor only when the square root of the sum of harmonic currents squared equals the rated secondary current of the transformer.

There is marked difference between the derating in ANSI/IEEE recommendations and UL K factor. If a transformer has 3% eddy current loss and a K factor of 5, then the eddy current loss increases by $3 \times 5 = 15\%$. The UL derating ignores eddy current loss gradient. Figure 8.11 shows higher derating with the ANSI method of calculation for a six-pulse harmonic current spectrum, assuming a theoretical magnitude of harmonics of $1/h$, for a K factor load of approximately 9.

Example 8.5: Calculate the UL K factor for the transformer loading harmonic spectrum of Example 8.1.

From Table 8.7, the UL K factor $= 6.164/1.07 = 5.76$ K factor of 13 is usually required for a transformer that has to supply 100% electronic load that may consist of SMP supplies.

8.4.8   Inrush Current of Transformers

If the capacitors and transformers are switched together, the inrush currents can increase and a resonance at a certain frequency can occur (also see Refs [22–25]).

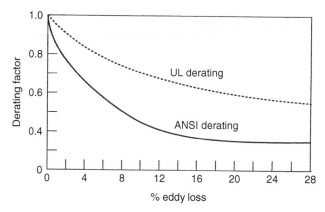

Figure 8.11   Relative derating of transformers ANSI method and UL K factor.

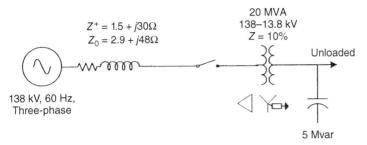

Figure 8.12   System configuration of a 20-MVA transformer with 5-Mvar capacitor bank on the secondary windings switched together.

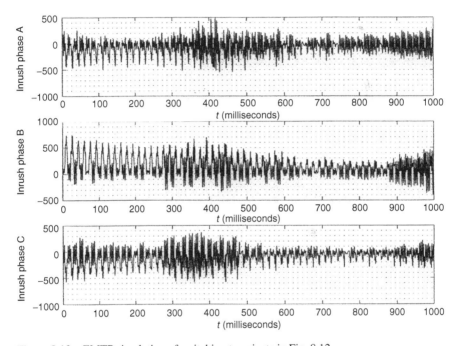

Figure 8.13   EMTP simulation of switching transients in Fig. 8.12.

Example 8.6:   Figure 8.12 illustrates that a 20-MVA transformer is switched together with a 5-Mvar capacitor bank on the secondary side of the transformer at no-load. An EMTP simulation of the inrush current in this configuration is shown in Fig. 8.13. The resulting inrush transients do not decay much even after 1 s. Usually, the inrush currents of the transformers last for 0.1–0.2 s (Fig. 3.8(b)).

8.5   CABLES

A nonsinusoidal current in a conductor causes additional losses. The AC conductor resistance is changed due to skin and proximity effects. Both these effects are

TABLE 8.10   Ratio of AC/DC Resistance

Conductor Size (KCMIL or AWG)	5- to 15-kV Nonleaded Shielded Power Cable, Three Single Concentric Conductors in Same Metallic Conduit	
	Copper	Aluminum
1000	1.36	1.17
900	1.30	1.14
800	1.24	1.11
750	1.22	1.10
700	1.19	1.09
600	1.14	1.07
500	1.10	1.05
400	1.07	1.03
350	1.05	1.03
300	1.04	1.02
250	1.03	1.01
4/0	1.02	1.01
3/0	1.01	< 1%
2/0	1.01	< 1%

dependent on frequency, conductor size, cable construction, and spacing. Even at 60 Hz, the AC resistance of conductors is higher than the DC resistance (Table 8.10). With harmonic currents, these effects are more pronounced. The AC resistance is given by

$$\frac{R_{AC}}{R_{DC}} = 1 + Y_{cs} + Y_{cp} \tag{8.37}$$

where Y_{cs} is due to conductor resistance resulting from skin effect, and Y_{cp} is due to proximity effect. The skin effect is an AC phenomenon, where the current density throughout the conductor cross section is not uniform and the current tends to flow more densely near the outer surface of the conductor than toward the center. This is because an AC flux results in induced EMFs, which are greater at the center than at the circumference, so that potential difference tends to establish currents that oppose the main current at the center and assist it at the circumference. The result is that the current is forced to the outside, reducing the effective area of the conductor. The effect is utilized in high-ampacity hollow conductors and tubular bus bars to save material costs. The skin effect is given by [26]

$$Y_{cs} = F(x_s) \tag{8.38}$$

where Y_{cs} is due to skin effect losses in the conductor and $F(x_s)$ is the skin effect function:

$$x_s = 0.875\sqrt{f\frac{k_s}{R_{DC}}} \tag{8.39}$$

where the factor k_s depends on the conductor construction.

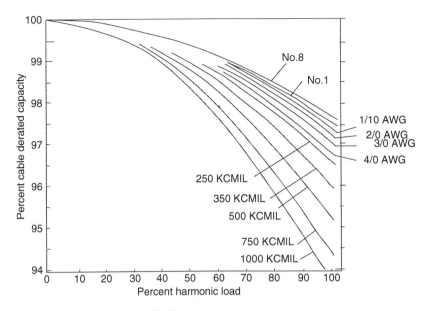

Figure 8.14   Derating of cables for a six-pulse current source converter.

The proximity effect occurs because of the distortion of current distribution between two conductors in close proximity. This causes concentration of current in parts of the conductors or bus bars closest to each other (currents flowing in forward and return paths). The expressions and graphs for calculating the proximity effect are given in Ref. [26]. The increased resistance of conductors due to harmonic currents can be calculated, and the derated capacity is

$$\frac{1}{1 + \sum_{h=1}^{h=\max} I_h^2 R_h} \tag{8.40}$$

where I_h is the harmonic current and R_h is the ratio of the resistance at that harmonic with respect to the conductor DC resistance.

For a typical six-pulse harmonic spectrum, the derating is approximately 3–6% depending on the cable size, Fig. 8.14 [27]. This treatment does not consider harmonic resonance conditions. Cables involved in harmonic resonance can be subjected to high overvoltages and corona and possible damage.

8.6   CAPACITORS

The main effect of harmonics on capacitors is that a resonance condition can occur with one of the load-generated harmonics. See Chapters 9 and 13 for further discussion and placement of capacitors in the power systems.

8.7 VOLTAGE NOTCHING

The commutation notches are shown in Fig. 4.10. These occur due to commutation as discussed in Chapter 4. At the bus bars to which the commutating bridge is connected, the voltage will be reduced to zero. The normal high short-circuit currents do not flow during the brief period of commutation. Figure 8.15 shows the zero crossing and oscillations in greater detail and illustrates that the voltage waveform has more than one zero crossing. Much electronic equipment uses crossover of the waveform to detect frequency or generate a reference for the firing angle. Such items of equipment are timers, domestic clocks, and UPS (uninterruptible power supply) systems. Many controllers that use an inverse cosine type of control circuit will be susceptible to zero crossing. Automatic voltage regulators of generators are another example. The control of firing angle and hence the DC voltage and excitation will be affected. All drives using rectifier front end can be affected and converter disruption can occur. Line harmonics seem to affect the accuracy and operation of magnetic devices and peripheral equipment also. Also note the oscillations that are caused by the inductive–capacitive couplings. The capacitances may be those of power factor correction capacitors or cables. The frequency range of this oscillation is from 5 to 50 kHz. This can couple to the communication circuits through coupling of power line with the telephone line (mutual inductance that may not be identical in three phases). If the firing angle is large, close to the peak of the voltage wave, as shown in Fig. 8.15, the system can be subjected to overvoltage and insulation stresses. The high dv/dt can impact the insulation systems, failure of solid-state devices, and electromagnetic interference (EMI).

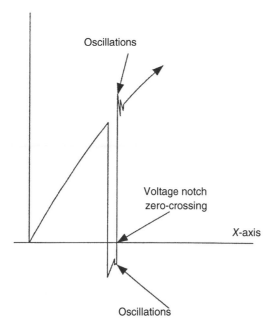

Figure 8.15   Zero crossing of voltage notch toward the peak of the voltage wave at higher firing angle with oscillations.

The IEEE limitations on voltage notching and an example of calculation are in Chapter 10.

8.8   EMI (ELECTROMAGNETIC INTERFERENCE)

Disturbances generated by switching devices such as BJTs (bipolar junction transistors), IGBTs (insulated gate bipolar transistors), and high-frequency PWM modulation systems, and voltage notching due to converters, generate high-frequency switching harmonics. Also, a short rise and fall time of 0.5 μs or less occurs due to the commutation action of the switches. This generates sufficient energy levels in the radio frequency range in the form of damped oscillations between 10 kHz and 1 GHz. The following approximate classification of electromagnetic disturbance by frequency can be defined:

Below 60 Hz: subharmonic

60 Hz to 2 kHz: harmonics

16–20 kHz: acoustic noise

20–150 kHz: range between acoustic and radio frequency disturbance

150 kHz to 30 MHz: conducted radio frequency disturbance

30 MHz to 1 GHz: radiated disturbance

A section of the electromagnetic spectrum with respect to frequencies of interest is shown in Fig. 8.16. The high-frequency disturbances are referred to as EMI. The radiated form is propagated in free space as electromagnetic waves, and the conducted form is transmitted through the power lines, especially at distribution level. Conducted EMI is much higher than radiated noise. Two modes of conducted EMI are recognized:

- Symmetrical or differential mode
- Asymmetrical or common mode.

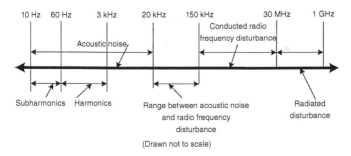

Figure 8.16   EMI frequency spectrum.

The symmetrical mode occurs between two conductors that form a conventional return circuit, and common-mode propagation takes place between a group of conductors and ground or other group of conductors.

The EMI issues can be aggravated due to higher voltages, number of drive systems in the same location, solidly grounded systems, motor leads greater than 100 ft, PLC digital communications, and poor grounding practices.

The ungrounded systems are practically extinct because of phenomena of arcing grounds and the high transient voltages these can generate due to capacitive–inductive couplings [25]. Yet, from drive systems point of view, an ungrounded transformer neutral breaks the return path of the CM current back to the ASD input. Thus, CM noise substantially reduces current in ground grid. A solidly grounded system will complete a transient CM noise current conductive return path from ASD output to the ground grid and back to ASD. An HRG (high-resistance grounding) system, which is more popular in the industry for medium-voltage and low-voltage systems, does significantly reduce the peak CM ground current and provide additional circuit damping.

The conducted EMI propagates through wires, and the relevant propagation mechanisms between wires include capacitive and inductive couplings, Fig. 8.17 [28]. The radiation phenomena is related to loops and wires performing as antennas.

The main category of EMI investigations can be divided into the following:

- generation of EMI;
- EMI coupling mechanisms;
- mitigation;
- effects on equipment;
- modeling of EMI;
- EMI standards and test methods.

Reference [29] discusses these aspects for modern PWM drives and provides a list of 53 references for further reading. The common-mode noise is most difficult to mitigate. The parasitic capacitances of the heat sink and transformer windings excited by high dv/dt of switching devices cause common-mode currents to flow. The differential-mode noise is caused by the voltage induced across equivalent series inductance and resistance of input and output voltage capacitors. Frequency spectrum of common-mode voltage and differential-mode voltage of six-step PWM waveform is shown in Fig. 8.18(a) and (b) [29]. The switching frequency is $f_c = 500\,\text{Hz}$ and pulse rise time = 200 ns. The spectrum is normalized to DC bus value. EMI components centered at drive f_c and harmonics of f_c are seen.

The noise at frequencies around a few kilohertzs can interfere with audiovisual equipment and electronic clocks. Shielding and proper filtering are the preventive measures. Lightning surges, arcing type of faults, and operation of circuit breakers and fuses also produce EMI. The use of shielded cables, grounding, and filter designs are some techniques for limiting EMI. For effective EMI filter designs, the phase currents and neutral currents must be decomposed into differential mode and common mode. The mitigation techniques are as follows:

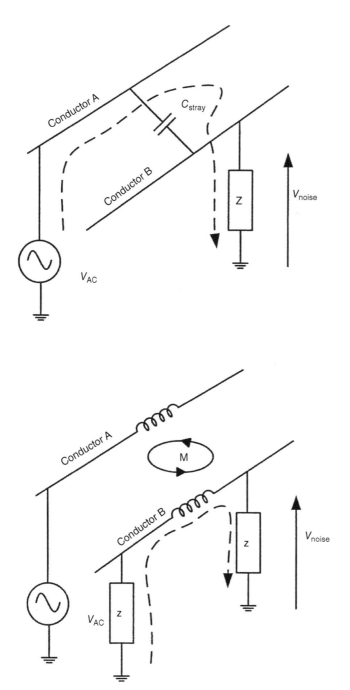

Figure 8.17   Noise generation in conductors due to capacitive and inductive couplings.
Source: Ref. [28].

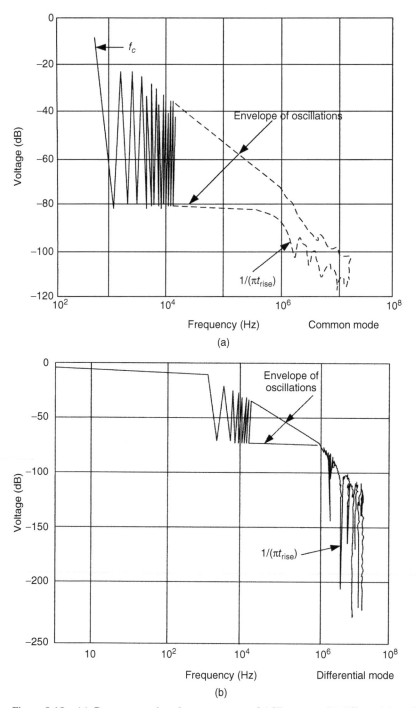

Figure 8.18 (a) Common-mode voltage spectrum of ASD output; (b) differential-mode output voltage spectrum. Source: Ref. [29].

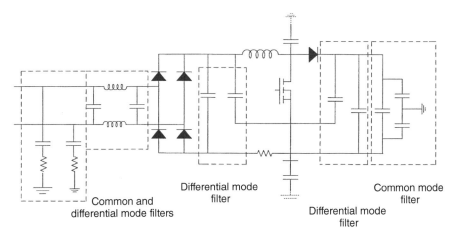

Figure 8.19 Common-mode and differential-mode filters for a boost PFC (power factor controller). Source: Ref. [28].

- proper low- and high-frequency grounding;
- shielded cables, shielding noise away from sensitive equipment;
- attenuating the noise source; and
- capturing noise and returning it to the source.

Figure 8.19 from Ref. [28] shows common-mode and differential-mode filters for a PFC.

8.8.1 FCC Regulations

The FCC standard [1, pt. 15, subpt J] was written for computers that interface with other equipment. The definition of computing device is an electronic device, which generates or uses timing signals at a clock rate >10 kHz. It defines radiated (30–1000 MHz) and conducted interference (0.45–30 MHz). The main problem is conducted interference. With respect to drive systems, if it does not use an oscillator > 10 kHz, it is exempt from FCC regulations.

8.9   OVERLOADING OF NEUTRAL

Figure 4.19 shows that the line current of switched-mode power supplies flows in pulses. Also, the PBM technique gives rise to neutral currents. At low level of current, the pulses are nonoverlapping in a three-phase system, that is, only one phase of a three-phase system carries current at any one time. The only return path is through the neutral, and as such the neutral can carry the summed currents of the three phases (Fig. 8.20). Then,

$$I_{\text{phase}} = (1.0 + 0.7^2)^{1/2} = 1.22$$

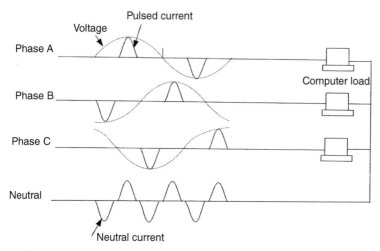

Figure 8.20   Summation of neutral current in a three-phase four-wire system with single-phase nonlinear loads.

$$I_{\text{neutral}} = (0.7 + 0.7 + 0.7) = 2.1$$

$$\frac{I_{\text{neutral}}}{I_{\text{phase}}} = 1.72 \tag{8.41}$$

Therefore, the rms value of current in the neutral is 172% of the line current. As the load increases, the pulses in the neutral overlap and the neutral current as a percentage of the line current reduce. The third harmonic is the major contributor to the neutral current; other triplen harmonics have insignificant contributions. A minimum of 33% third harmonic is required to produce 100% neutral current in a balanced wye system.

An approximate expression for calculating the neutral rms current is

$$I_{\text{rms,neutral}} = 3\sqrt{\frac{0.5P_{\text{nl}}^2}{1 + 0.5P_{\text{nl}}^2}} I_{\text{rms,phase}} \tag{8.42}$$

This is based on the assumption that circuit loading is balanced, and that nonlinear loads are a fraction P_{nl} of the total load and the load current has a third harmonic component of 70% of fundamental. Figure 8.21 shows neutral current as the percentage of electronic load.

The NEC (National Electric Code), published by NFPA – National Fire Protection Association [30], recommends that, where the major portion of loads consists of nonlinear loads, the neutral shall be considered as a current-carrying conductor. Usually, a reduced neutral cross section as compared to phase conductors is specified in NEC. In some installations, the neutral current may exceed the maximum phase current. The single-phase branch circuits can be run with a separate neutral for each phase rather than using multiwire branch circuit with a shared neutral.

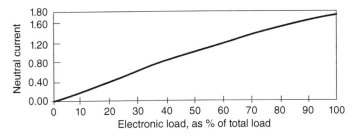

Figure 8.21   Neutral current as a percentage of electronic load.

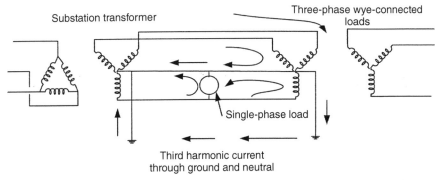

Figure 8.22   Flow of third harmonic current through ground and neutral due to single-phase loading.

Figure 8.22 shows that the neutral currents flowing through ground can cause problem at the substation, where these must return to neutral connection. High-stray electrical fields can occur near the substation and the ground fault relays can misoperate. Another concern is that the substation transformer can overheat due to the flow of neutral currents. In core-type transformers, the return path of the zero sequence currents is through the tank walls, which can cause hot spots.

A zigzag or delta–wye transformer connected on the three-phase, four-wire wye-connected transformer serving pulsed power supply loads will act as zero sequence traps (Chapter 15).

8.10   PROTECTIVE RELAYS AND METERS

Harmonics may result in possible relay nuisance operation. Relays that depend on crest voltage and/or current or voltage zeros are affected by harmonic distortion on the wave. The excessive third harmonic zero sequence current can cause ground relays to false trip. A Canadian study [31] documents the following effects:

1. Relays exhibited a tendency to operate slower and/or with higher pickup values rather than to operate faster and/or with lower pickup values.

2. Static underfrequency relays were susceptible to substantial changes in the operating characteristics.

3. In most cases, the changes in the operating characteristics were small over a moderate range of distortion during normal operation.

4. Depending on the manufacturer, the overcurrent and overvoltage relays exhibited various changes in the operating characteristics.

5. Depending on harmonic content, the operating torque of the relays could be reversed.

6. Operating time could vary widely as a function of the frequency mix in the metered quantity.

7. Balanced beam distance relays could exhibit both underreach and overreach.

8. Harmonics could impair the operation of high-speed differential relays.

9. The impedance relays are set for the fundamental frequency impedance. Presence of harmonics under fault conditions could cause measurement errors.

Harmonic levels of 10–20% are generally required to cause problems with relay operation. These levels are much higher than what will be tolerated in the power systems. Reference [31] states the impossibility of completely defining relay responses because of the variety of relays in use and the variations in the nature of distortions that can occur, even if the discussions is limited to 6-pulse and 12-pulse converters. Not only the harmonic magnitudes and predominant harmonic orders vary, but relative phase angels can also vary. The relays will respond differently to two wave shapes that have the same characteristic magnitudes but different phase angles relative to the fundamental.

A specific study of the effect of nonsinusoidal current waveforms on electromechanical and solid-state relays is included in Ref. [32]. This investigates an induction pattern relay and a solid-sate relay of a specific type of one manufacturer. The operation was simulated with third harmonic current of 33.33%, fifth harmonic 20%, and seventh harmonic 14.3%. The induction pattern relay set at 1 A pickup, time dial of 1, operated 54% faster at 1.3 times the pickup current, 15.4% faster at three times the pickup current , 4.4% faster at six times the pickup current, and 3.5% faster at eight times the pickup current with harmonic currents. There was some difference with respect to operating times when equivalent rms current for complex wave = 1.0823 A was applied. Trip timings of the solid-state relay varied much with 50 Hz plus third, fifth, and seventh harmonic currents and equivalent rms current.

Table 8.11 is from Ref. [33] and shows the test results of various relay types.

8.10.1 Modern MMPR (Multifunction Microprocessor-Based Relays)

Modern multifunction microprocessor-based relays (MMPRs) use filters for current and voltage waveforms. These may utilize various measuring techniques – digital sampling, digital filtering, asynchronous sampling, and rms measurements. A microprocessor relay using a digital filter is immune to the effect of harmonics because it extracts the fundamental from the waveform. Rms measurements with asynchronous

TABLE 8.11   Test Results – Effect of Harmonics on Operation of Protective Relays

Measuring Unit	Approximate Pickup for Single Frequency Inputs	Pickup with Mixed Frequency Inputs
Clapper	$\alpha\sqrt{f}$ (may chatter)	Slightly higher RMS pickup
Induction disk (overcurrent)	Increases with frequency	Essentially unchanged
Induction disk (phase unbalance)	15% lower at 120, 180 Hz; 15–30% higher up to 540 Hz	Essentially unchanged
Negative sequence overcurrent	$\alpha\sqrt{f}$	Essentially unchanged
Cylinder (directional)	αf	Slightly higher
Cylinder (bus differential)	αf^2	Slightly higher
Cylinder (undervoltage)	$\alpha f^{0.85}$	Same fundamental pickup
Polar (transformer differential)	$\alpha f^{0.25}$	Slightly higher
Polar (harmonic restraint)	Blocks at second, decreases beyond second	Blocks at second
Microprocessor (asynchronous sampling)	No change	RMS responsive

sampling measure the rms value of the input current directly and use it as an operating quantity. This approach does not use a digital filter and waveform is sampled sufficient number of times to account for the higher frequencies. Each sample is squared and summed, and the average of the square root of the sum over the samples per cycle is taken as the rms value. This puts a high computational burden, and the processing time is limited by the number of samples required per cycle.

A relevant observation is that the electromechanical relays are no longer being applied in the industry. These have many limitations from the protective relaying consideration, for example, fixed time–current characteristics, high maintenance and calibration costs, and downtime because of moving parts. Contrary to that, microprocessor-based relays have programmable characteristics, fault diagnostics, communication, and metering functions.

8.10.2   Metering and Instrumentation

Metering and instrumentation are affected by harmonics. Close to resonance, higher harmonic voltages may cause appreciable errors. A 20% fifth harmonic content can produce a 10–15% error in a two-element, three-phase watt transducer. The error due to harmonics can be positive, negative, or smaller with third harmonics depending on the type of the meter. An electromechanical kilowatt-hour meter will read high with consumer-generated ASD harmonics. The presence of harmonics reduces the reading on power factor meters.

The solid-state instruments will measure true power irrespective of the waveforms. Modern rms sensing voltmeters and ammeters are practically immune to the

CHAPTER 8   EFFECTS OF HARMONICS

waveform distortion, as long as the harmonics are within operating bandwidth of the instrument. Average and peak reading meters calibrated in rms are not suitable in the presence of harmonics.

8.11   CIRCUIT BREAKERS AND FUSES

Harmonic components can affect the current interruption capability of circuit breakers. The high di/dt at current zero can make interruption process more difficult. Figure 4.7 for a six-pulse converter shows that the current zero is extended and di/dt at current zero is very high. There are no definite standards in the industry for derating. One method is to arrive at the maximum di/dt of the breaker, based on interrupting rating at fundamental frequency, and then translate it into maximum harmonic levels assuming that the harmonic in question is in phase with the fundamental [31].

A typical reduction in current-carrying capacity of molded case circuit breakers, when in use at high-frequency sine-wave currents, is shown in Table 8.12 based on the published data of one manufacturer. The effect on larger breakers can be more severe as the phenomena are also related to skin effect and proximity effect.

Harmonics will reduce the current-carrying capacity of fuses. Also, the time–current characteristics can be altered, and the melting time will change. Harmonics also affect the interrupting rating of the fuses. Excessive transient overvoltages from current limiting fuses, forcing current to zero before a natural zero crossing, may be generated. This can cause surge arrester operation and capacitor failures.

Circuit breakers for capacitor switching should be "definite purpose" circuit breakers (Chapter 11).

8.12   TELEPHONE INFLUENCE FACTOR

We discussed telephone influence factor (TIF) in Chapter 1 (see Eqs (1.21) through (1.24)). Harmonic currents and voltages can produce electrical and magnetic fields that will impair the performance of communication systems. Due to proximity, there will be inductive coupling with the communication systems. Relative *weights* have been established by tests for the various harmonic frequencies that indicate

TABLE 8.12   Reduction in Current-Carrying Capability (%) of Molded Case Circuit Breakers, 40 °C Temperature Rise

Breaker Size	Sinusoidal Current	
	300 Hz	420 Hz
70 A	9	11
225 A	14	20

On the basis of published data of one manufacturer.

disturbance to voice frequency communication. This is based on disturbance produced by the injection of a signal of the harmonic frequency relative to that produced by a 1-kHz signal similarly injected. The TIF weighting factor is a combination of C-message weighting characteristics, which account for relative interference effects of various frequencies in the voice band, and a capacitor, which produces weighting that is directly proportional to the frequency to account for an assumed coupling function [31]. It is dimensionless quantity that is indicative of the waveform and not the amplitude.

The term balanced is used when signals included in Eq. (1.21) are only positive or negative sequence. When zero sequence signals are included in Eq. (1.21), the term residual is used:

$$TIF = \sqrt{TIF_r^2 + TIF_b^2} \qquad (8.43)$$

For the TIF current or voltage, we can write

$$\frac{\sqrt{\sum X_f^2 W_f^2}}{X_t} \qquad (8.44)$$

where X_f is single frequency current or voltage and X_t is total voltage or current.

The TIF weighting function that reflects C-message weighting and coupling normalized to 1 kHz is given by Eq. (1.24) repeated here:

$$W_f = 5P_f f \qquad (8.45)$$

where P_f is the C-message weighting at frequency f, under considerations. For example, TIF weighting at 1 kHz is

$$W_f = (5)(1)(1000) = 5000 \qquad (8.46)$$

because C-message attenuation is unity.

Therefore, C-message index is defined as

$$C_I = \frac{\sqrt{\sum_{h=1}^{h=max} c_h^2 I_h^2}}{I_{rms}}$$

$$C_V = \frac{\sqrt{\sum_{h=1}^{h=max} c_h^2 V_h^2}}{V_{rms}} \qquad (8.47)$$

where c_h is the weighting factor divided by five times the harmonic order h.

The single-frequency TIF values are listed in Table 8.13 and graphically in Fig. 8.23. The weighting is high in the frequency range 2–3.5 kHz in which human hearing is most sensitive.

TABLE 8.13   1960 Single Frequency TIF Values

FREQ	TIF	FREQ	TIF	FREQ	TIF	FREQ	TIF
60	0.5	1020	5100	1860	7820	3000	9670
180	30	1080	5400	1980	8330	3180	8740
300	225	1140	5630	2100	8830	3300	8090
360	400	1260	6050	2160	9080	3540	6730
420	650	1380	6370	2220	9330	3660	6130
540	1320	1440	6560	2340	9840	3900	4400
660	2260	1500	6680	2460	10,340	4020	3700
720	2760	1620	6970	2580	10,600	4260	2750
780	3360	1740	7320	2820	10,210	4380	2190
900	4350	1800	7570	2940	9820	5000	840
1000	5000						

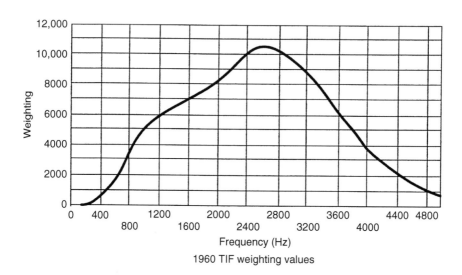

Figure 8.23   1960 TIF weighting values. Source: Ref. [31].

Telephone interference is often expressed as a product of current and TIF, that is, IT product where I is the rms current in amperes. Alternatively, it is expressed as a product of voltage and TIF weighting, where the voltage is in kilovolts, that is, kV-T product:

$$IT = \text{TIF} \times I_{rms} \qquad VT = \text{TIF} \times V_{rms} \tag{8.48}$$

Table 8.14 [31] gives balanced IT guidelines for converter installations. The values are for circuits with an exposure between overhead systems, both power and telephone. Within an industrial plant or commercial building, the interference between power cables and twisted-pair telephone cables is low, and interference

TABLE 8.14   Balanced IT Guidelines for Converter Installations Tie Lines

Category	Description	IT
I	Levels unlikely to cause interference	Up to 10,000
II	Levels that might cause interference	10,000–50,000
III	Levels that probably will cause interference	>50,000

is not normally encountered. Telephone circuits are particularly susceptible to the influence of ground return currents.

Example 8.7:   A large industrial system is served from a dedicated 115 kV bus, also designated PCC (point of common coupling, see Chapter 10). The three-phase symmetrical short-circuit current at 115 kV is 30 kA, and the load demand is 50 MVA = 251 A. The harmonic injection at the PCC from the plant loads and the calculation of TIF is shown in Table 8.15. The higher order harmonics are of importance as the weighting is maximum at approximately 47th harmonic. Also even order and triplen harmonics (not shown in Table 8.15), which may be present, should not be ignored. The calculations in Table 8.15 show an IT of 57,879.4, which is > 50,000 and interference is likely to occur. The TIF is 230.03. IT product is high beyond category III (Table 8.14).

8.12.1   Psophometric Weighting

The psophometric weighting by CCITT (International Telecommunication Union, Geneva) is extensively used in Europe. There is a slight difference in the C-message weighting and psophometric weighting.

The level of interference is described in terms of telephone form factor (TFF), which ignores the geometric configuration of the coupling:

$$\text{TFF} = \sqrt{\sum_{h=1}^{h_{\max}} \left(\frac{U_h F_h}{U} \right)^2} \tag{8.49}$$

where

$$F_h = \frac{p_h h f}{800} \tag{8.50}$$

and p_h is the psophometric weighting factor, f is the fundamental frequency = 50 Hz, U_h is the component at harmonic h of disturbing voltage, and U is the rms voltage. The required limit of TFF is typically 1%.

TABLE 8.15 Calculation of TIF, Example 8.7

Harmonic Order/ Frequency	Harmonic Current in Amperes	TIF Value from Table 8.13	$X_f W_f$	$(X_f W_f)^2$
1/60	251	0.5	126	15,876
5/300	10	225	2250	5,062,500
7/420	9	650	5850	34,222,500
11/660	6	2260	13,560	183,873,600
13/780	5	3360	16,800	282,240,000
17/1020	4	5100	20,400	416,160,000
19/1140	4.2	5630	23,646	550,559,296
21/1260	4.0	6050	24,200	585,640,000
23/1380	2.2	6370	14,014	196,392,196
25/1500	1.9	6680	12,692	161,086,864
29/1740	1.8	7320	13,176	173,606,976
31/1860	1.8	7820	14,076	198,133,776
35/2100	1.3	8830	11,479	131,767,441
37/2220	1.2	9330	11,196	125,350,416
41/2460	1.2	10,340	12,408	153,958,464
43/2580	1.0	10,600	10,600	112,360,000
47/2820	1.0	10,210	10,210	104,244,100
49/2940	1.0	9820	9820	96,432,400

$IT = \sqrt{\sum (X_f W_f)^2} = 57,879.4$

$X_t = 251.62$

$TIF = 230.03$

CCITT recommends that total psophometric weighted noise on a telephone circuit has an emf of less than 1 mV; the telephone circuit is terminated in its characteristics impedance of 600 ohms. The psophometric weighted noise across terminating resistance should be 0.5 mV. The CCITT also defines equivalent disturbing current:

$$I_p = \left(\frac{1}{p800}\right) \sqrt{\sum_f (h_f p_f I_f)^2} \tag{8.51}$$

where I_f is the component of frequency f of the current causing disturbance, P_f is the psophometric weighting factor at frequency f, and h_f is a function of frequency and considers type of coupling between lines. By definition, $h_{800} = 1$.

The interference can be reduced by

- phase multiplication, Chapter 6;
- shielding and twisted-pair conductors can be used to minimize the influence of ground return currents;

- reactance of utility and converter transformers contribute to the commutating reactance of the converter, which will cause IT product and kVT product at the line terminals of converter to increase rapidly with phase retard angle;
- series and shunt filters can be applied.

The effect of harmonics can be felt at a distance from their point of generation. Sometimes, it eludes the intuition, till a rigorous study is conducted. Reference [31] details some abnormal conditions of harmonic problems. These are natural resonance of transmission lines, overexcitation of transformers, and harmonic resonance in the zero sequence circuits, as further discussed in this book.

REFERENCES

1. IEEE. A report prepared by Load Characteristics Task Force. "The effects of power system harmonics on power system equipment and loads," IEEE Transactions, vol. PAS 104, pp. 2555–2561, 1985.
2. J.F. Witte, F.P. DeCesaro, and S.R. Mendis. "Damaging long term overvoltages on industrial capacitor banks due to transformer energization inrush currents," IEEE Transactions of Industry Applications, vol. 30, no. 4, pp. 1107–1115, 1994.
3. NEMA. Motors and Generators, Parts 30 and 31, 1993. Standard MG-1.
4. IEEE. Working Group J5 of Rotating Machinery Protection subcommittee, Power System Relaying Committee. "The impact of large steel mill loads on power generating units," IEEE Transactions of Power Delivery, vol. 15, pp. 24–30, 2000.
5. ANSI. Synchronous generators, synchronous motors and synchronous machines in general, 1995. Standard C50.1.
6. ANSI. American standard requirements for cylindrical rotor synchronous generators, 1965. Standard C50.13.
7. M.D. Ross and J.W. Batchelor. "Operation of non-salient-pole type generators supplying a rectifier load," AIEE Transactions, vol. 62, pp. 667–670, 1943.
8. A.H. Bonnett. "Available insulation systems for PWM inverter fed motors," IEEE Industry Applications Magazine, no. 4, pp. 15–26, 1998.
9. G. Stone, S. Campbell, and S. Tetreault. "Inverter-fed drives: which stators are at risk?," IEEE Industry Applications Magazine, vol. 6, no. 5, pp. 17–22, 2000.
10. M. Hodowanec. "Proper application of motors operated on adjustable frequency control," IEEE Industry Applications Magazine, vol. 6, no. 5, pp. 41–46, 2000.
11. A. van Jouanne, P. Enjeti, and W. Gray. "Application issues for PWM Adjustable speed AC motor drives," IEEE Industry Applications Magazine, vol. 2, no. 5, pp. 10–18, 1996.
12. S. Bell and J. Sung. "Will your motor insulation survive a new adjustable frequency drive," IEEE Transactions of Industry Applications, vol. 33, pp. 1307–1311, 1997.
13. J.C. Das and R.H. Osman. "Grounding of AC and DC low-voltage and medium-voltage drive systems," IEEE Transactions on Industry Applications, vol. 34, pp. 205–216, 1998.
14. J.M. Bentley and P.J. Link. "Evaluation of motor power cables for PWM AC drives," IEEE Trans Industrial Applications, vol. 33, pp. 342–358, 1997.
15. D. Macdonald and W. Gary. "PWM drive related bearing failures," IEEE Industry Applications Magazine, vol. 5, no. 4, pp. 41–47, 1999.
16. P.J. Link. "Minimizing electric bearing currents in ASD systems," IEEE Industry Applications Magazine, vol. 5, no. 4, pp. 55–65, 1999.
17. J. Erdman, R. Kerman, D. Schlegel, and G. Skibinski. "Effect of PWM inverters on AC motor bearing currents and shaft voltages," IEEE Transactions of Industry Applications, vol. 32, pp. 250–259, 1996.
18. W. Zhou, B. Lu, T.G. Habetler, and R.G. Harley. "Incipient bearing fault detection via motor stator noise cancellation using Wiener filter," IEEE Transactions of Industry Applications, vol. 45, no. 4, pp. 1309–1316, 2009.

19. IEEE Standard C57.110. IEEE recommended practice for establishing liquid-filled and dry-type power and distribution transformer capability when supplying nonsinusoidal load currents, 2008.
20. UL. Dry-type general purpose and power transformers, 1994. Standard UL 1561.
21. UL. Transformers distribution, dry-type over 600V, 1994. Standard UL 1562.
22. R.S. Bayless, J.D. Selmen, D.E. Traux, and W.E. Reid. "Capacitor switching and transformer transients," IEEE Transactions PWRD, vol. 3, no. 1, pp. 349–357, 1988.
23. J.C. Das. "Analysis and control of large shunt capacitor bank switching transients," IEEE Transactions of Industry Applications, vol. 41, no. 6, pp. 1444–1451, 2005.
24. J.C. Das. "Surge transference through transformers," IEEE Industry Applications Magazine, vol. 9, no. 5, pp. 24–32, 2003.
25. J.C. Das. Transients in Electrical Systems, McGraw Hill, New York, 2011.
26. J.H. Neher and M.H. McGrath. "The calculation of the temperature rise and load capability of cable systems," AIEE Transactions, PAS, vol. 76, pp. 752–764, 1957.
27. A. Harnandani. "Calculations of cable ampacities including the effects of harmonics," IEEE Industry Applications Magazine, vol. 4, pp. 42–51, 1998.
28. L. Rossetto, P. Tenti, and A. Zuccato. "Electromagnetic compatibility issues in industrial equipment," IEEE Industry Application Magazine, vol. 5, no. 6, pp. 34–46, 1999.
29. G.L. Skibinski, J. Kerkman, and D. Schlegel. "EMI emissions of modern PWM ac drives," IEEE Industry Application Magazine, vol. 5, no. 6, pp. 47–80, 1999.
30. NFPA. National Electric Code 2009. NFPA 70.
31. IEEE Standard 519. IEEE Recommended Practice and Requirements for Harmonic Control in Electrical Systems, 1992.
32. P.M. Donohue and S. Islam. "The effect of non-sinusoidal current waveforms on electromechanical and solid state overcurrent relay operation," IEEE Transactions of Industry Applications, vol. 46, no. 6, pp. 2127–2133, 2010.
33. W.A. Elmore, C.A. Kramer, and S. Zocholl. "Effect of waveform distortion on protective relays," IEEE Transactions of Industry Applications, vol. 29, no. 2, pp. 404–411, 1993.

CHAPTER 9

HARMONIC RESONANCE

The harmonic resonance is an important factor affecting the system harmonic levels. A load-generated harmonic can be magnified many times due to harmonic resonance. A harmonic resonance without application of shunt capacitor banks can practically be ignored in industrial systems, but not so in the distribution and transmission systems. Distribution system frequency response is affected by the shunt capacitances and the system inductances. A number of small capacitor banks can be applied in a distribution system, and these will give rise to multiple resonant frequencies. The transmission systems have complex frequency response and cable, line, and transformer capacitances must be modeled. It is common to apply large capacitor banks and static var compensators (SVCs) on transmission systems, and when these are switched, the frequency response characteristic changes.

The harmonic resonance in a power system cannot be tolerated and must be avoided. The magnified harmonics will have serious effects on equipment heating, harmonic torque generation, nuisance operation of protective devices, derating of electrical equipment, damage to the shunt capacitors due to overloading, and can precipitate shutdowns.

9.1   TWO-PORT NETWORKS

In harmonic analysis, we derive frequency response of a network, which means its performance over a range of frequencies of interest. This frequency response can spot the resonance in the system. To display the frequency response, a plot of the impedance magnitude and also its angle versus frequency are constructed. A representation of a two-port network is shown in Fig. 9.1(a) and (b). The output can be terminated in impedance, it can be open-circuited or short-circuited.

We define the following parameters with respect to two-port network.

$$Z_{in} = \frac{1}{Y_{in}} = H_z = \frac{V_1}{I_1}$$

$$H_v = \frac{V_2}{V_1} = \frac{Z'}{Z_1 + Z'}$$

$$V_2 = \frac{Z'}{Z_1 + Z'} V_1 \qquad (9.1)$$

Power System Harmonics and Passive Filter Designs, First Edition. J.C. Das.

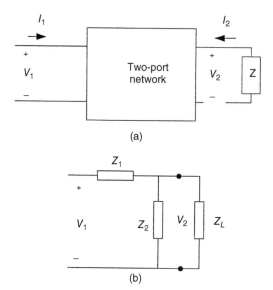

(a)

(b)

Figure 9.1 (a) Two-port networks; (b) Generalized representation.

TABLE 9.1 Transfer Functions – Two-Port Networks

Function→ Output↓	$H_z = V_1/I_1$	$H_v = V_2/V_1$	$H_i = I_2/I_1$	$H_vH_z = V_2/I_1$	$H_i/H_z = I_2/V_1$
Connected to load, Z_L	$Z_1 + Z'$	$\dfrac{Z'}{Z_1 + Z'}$	$-\dfrac{Z_2}{Z_2 + Z_L}$	Z'	$\dfrac{-Z'}{Z_L(Z_1 + Z')}$
Open-circuited	$Z_1 + Z_2,$	$\dfrac{Z_2}{Z_1 + Z_2}$	0	Z_2	0
Short-circuited	Z_1	0	-1	0	$-1/Z_1$
Units	Ohms	Dimensionless	Dimensionless	Ohms	Siemens

where

$$Z' = \frac{Z_2 Z_L}{(Z_2 + Z_L)}$$

$$H_i = \frac{I_2}{I_1}$$

$$Z_{tr} = \frac{V_2}{I_1} \, or \, \frac{V_1}{I_2} \tag{9.2}$$

Z_{in} and Z_{tr} are the input and transfer impedances and H functions are known as *network* or *transfer* functions. These are complex functions of the real variable of frequency (see Table 9.1 for transfer functions).

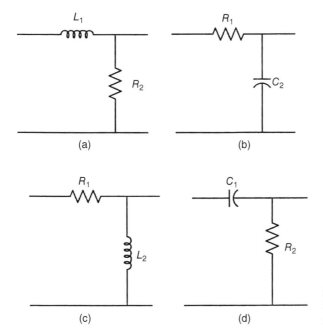

Figure 9.2 (a) and (b) Low-pass circuits and (c) and (d) high-pass circuits.

9.1.1 High-Pass and Low-Pass Circuits

High-pass and low-pass circuits are commonly referred and are shown in Fig. 9.2. Circuits in Fig. 9.2(a) and (b) are low-pass circuits, while those in Fig. 9.2(c) and (d) are high-pass circuits.

- If the magnitude of H_v decreases as frequency increases, it is called *high-frequency roll-off* and the circuit is a low-pass circuit.
- If the magnitude of H_v decreases as the frequency decreases, it is a high-pass circuit. It will have *low-frequency roll-off*.

Let us examine the high-pass circuit of Fig. 9.2(c):

$$H_z = R_1 + j\omega L_2 \tag{9.3}$$

Dividing by R_1

$$\frac{H_z}{R_1} = 1 + j\frac{\omega}{\omega_x} < \tan^{-1}\left(\frac{\omega}{\omega_x}\right) \tag{9.4}$$

where

$$\omega_x = \frac{R_1}{L_2} \tag{9.5}$$

   If we plot the function H_z/R_1, its magnitude and phase angle θ_H graphs as shown in Fig. 9.3(a) and (b) result. The magnitude will approach infinity with increasing frequency and phase angle will approach 90°.

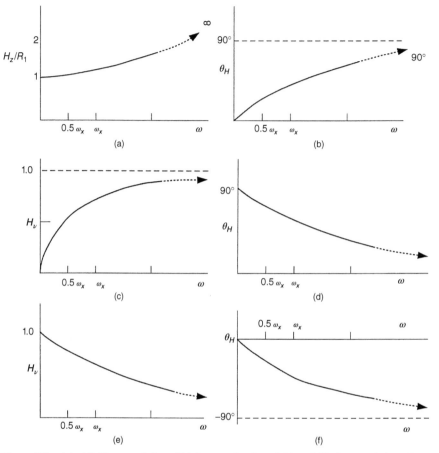

Figure 9.3 (a)–(d) Characteristics of high-pass circuit and (e) and (f) characteristics of low-pass circuit.

We can also write

$$H_v = \frac{j\omega L_2}{R_1 + j\omega L_2} = \frac{1}{1 - j(\omega_x/\omega)} \tag{9.6}$$

and

$$\theta_H = \tan^{-1}\left(\frac{\omega_x}{\omega}\right) \tag{9.7}$$

Figure 9.3(c) and (d) shows these plots. The transfer function approaches unity at high frequency, and the output voltage is the same as the input voltage; hence, the description, "low-frequency roll-off" and high-pass circuit.

For the low-pass circuit in Fig. 9.2(a)

$$H_v = \frac{j\omega R_2}{R_2 + j\omega L_1} = \frac{1}{1 + j(\omega/\omega_x)} \tag{9.8}$$

and

$$\theta_H = \tan^{-1}\left(-\frac{\omega}{\omega_x}\right) \qquad (9.9)$$

Figure 9.3(e) and (f) shows these plots for low-pass circuit.

9.1.2 Half-Power Frequency

In a network having capacitors or inductors or both, we define half-power frequency ω_x, as the frequency at which

$$|H(\omega)| = 0.707|H(\omega_x)| \qquad (9.10)$$

This frequency may not always correspond to 50% power. There are two half-power frequencies, one above and the other below the peak frequency, and these are called upper and lower half-power frequencies.

The separation between the upper and lower half-power frequencies gives the *bandwidth*. These concepts are utilized in the following discussions of series and parallel resonances.

9.2 RESONANCE IN SERIES AND PARALLEL RLC CIRCUITS

Amplification of the voltages and currents can occur in RLC series and parallel resonant circuits.

9.2.1 Series RLC Circuit

The impedance of a series circuit, Fig. 9.4(a), is

$$z = R + j\left(\omega L - \frac{1}{\omega C}\right) \qquad (9.11)$$

At a certain frequency, say f_0, z is minimum when

$$\omega L - \frac{1}{\omega C} = 0 \quad \text{or} \quad \omega = \omega_0 = \frac{1}{\sqrt{LC}} \qquad (9.12)$$

The value of the current at resonance is

$$I_r = \frac{V_1}{R} = \frac{V_2}{R} \qquad (9.13)$$

The current is limited only by the resistance, and if it is small, current can be high.

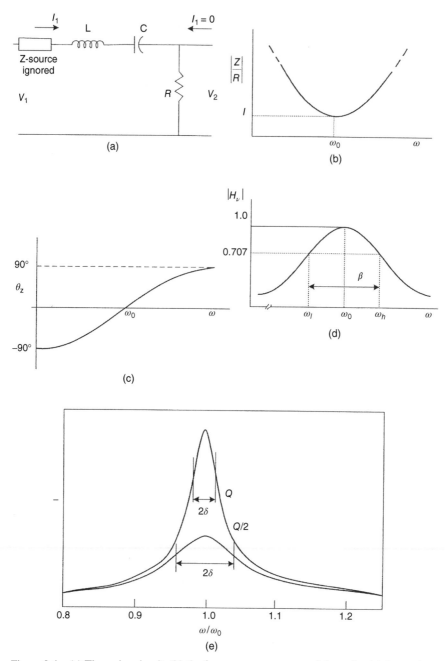

Figure 9.4 (a) The series circuit, (b) the frequency response, modulus only, (c) the angle, (d) half-power points and frequencies, (e) performance of RLC circuit as a function of Q and resonance in terms of Q.

At resonance the voltage across inductor is

$$V_L = \frac{j\omega_0 LV}{R} = jQV \tag{9.14}$$

Likewise, the voltage across the capacitor is

$$V_c = -jQV \tag{9.15}$$

These voltages are equal in magnitude and opposite in sign. The voltages developed in L and C are larger than the system voltage by factor Q, and Q can be large for sharpness of tuning.

Figure 9.4(b) shows the frequency response, the capacitive reactance; inversely proportional to frequency is higher at low frequencies, while the inductive reactance directly proportional to frequency is higher at higher frequencies. Thus, the reactance is capacitive and angle of Z negative below f_0 and above f_0 the circuit is inductive and angle Z is positive (Fig. 9.4(c)). The voltage transfer function $H_v = V_2/V_1 = R/Z$ is shown in Fig. 9.4(d). This curve is reciprocal of Fig. 9.4(b). The half-power frequencies are expressed as

$$\omega_h = \frac{R}{2L} + \sqrt{\left(\frac{R}{2L}\right)^2 + \frac{1}{LC}} = \omega_0 \left(\sqrt{1 + \frac{1}{4Q_0^2}} + \frac{1}{2Q_0}\right)$$

$$\omega_l = -\frac{R}{2L} + \sqrt{\left(\frac{R}{2L}\right)^2 + \frac{1}{LC}} = \omega_0 \left(\sqrt{1 + \frac{1}{4Q_0^2}} - \frac{1}{2Q_0}\right) \tag{9.16}$$

The bandwidth β shown in Fig. 9.4(d) is given by

$$\beta = \frac{R}{L} = \frac{\omega_0}{Q_0}(= \omega_h - \omega_l)$$

or

$$Q_0 = \frac{\omega_0 L}{R} \tag{9.17}$$

The quality factor Q is defined as

$$Q_0 = 2\pi \left(\frac{\text{maximum energy stored}}{\text{energy dissipated per cycle}}\right) \tag{9.18}$$

Q factor is called the *figure of merit*, given by the ratio of maximum energy stored per cycle and energy dissipated per cycle multiplied by (2π). The energy stored in a reactor is $0.5LI_m^2$ and dissipated in its series resistance R is $RI_m^2/2$. Here, I_m is the peak current. Then, Q is

$$Q = \frac{2\pi LI_m^2}{RI_m^2/f} = \frac{\omega L}{R}, \text{ as before} \tag{9.19}$$

A capacitor has some losses and is represented by an ideal capacitor in parallel with a high resistance. Then, Q for a capacitor becomes

$$Q = \frac{2\pi C V_m^2}{V_m^2 / Rf} = \omega CR \tag{9.20}$$

where V_m is the maximum voltage across the parallel combination of resistor and capacitor. The Q is, therefore, a frequency-dependent function as R does not vary with the frequency. As R of a capacitor is very large in resonant circuits, Q of the inductor is the controlling factor. Q of a resonant circuit implies the measured value of the circuit. For a series circuit product of 1/2 power frequencies is:

$$\omega_h \omega_i = \frac{1}{LC} = \omega_0^2 \tag{9.21}$$

Resonant circuits are frequency selective; it is desired that the circuit responds to a narrowband of frequencies and should have no response to the other frequencies. Close to resonance, we can write the circuit impedance of a series circuit as

$$Z = R + j\sqrt{\frac{L}{C}}\left(\omega\sqrt{LC} - \frac{1}{\omega\sqrt{LC}}\right) \tag{9.22}$$

$$Z = R + j\sqrt{\frac{L}{C}}\left(\frac{\omega}{\omega_0} - \frac{\omega_0}{\omega}\right)$$
$$= R\left[1 + jQ\left(\frac{\omega}{\omega_0} - \frac{\omega_0}{\omega}\right)\right] \tag{9.23}$$

Introduce a new variable δ, defined as

$$\delta = \left(\frac{\omega}{\omega_0} - \frac{\omega_0}{\omega}\right) \quad \text{or} \quad \frac{\omega}{\omega_0} = 1 + \delta \tag{9.24}$$

δ is the fractional deviation of the actual frequency from the resonant frequency. Then,

$$Z = R\left[1 + jQ\left(1 + \delta - \frac{1}{1+\delta}\right)\right]$$
$$(1+\delta)^{-1} \approx 1 - \delta + \delta^2 \tag{9.25}$$

Thus,
$$Z \approx R[1 + jQ\delta(2 - \delta)]$$

This gives the impedance of the series resonant circuit for small deviations from the resonant frequency.

How well a series resonant circuit will perform as a frequency selector is illustrated in Fig. 9.4(e). (A circuit of high Q gives more selective curve.) The bandwidth can be defined in cycles, at the frequency at which the power in the circuit is one-half the maximum power.

At half-power

$$\frac{P}{2} = \frac{I_r^2 R}{2} \tag{9.26}$$

where I_r is the current at resonance in peak value.

Then,

$$\frac{I_r}{\sqrt{2}} = \frac{1}{\sqrt{2}}\frac{V}{R} = \frac{V}{\sqrt{R^2 + X^2}} \tag{9.27}$$

That is,

$$R = X \tag{9.28}$$

The resistance and reactance are equal at half-power frequencies and may be found from Eq. (9.25):

$$1 = Q\delta_{1/2}(2 - \delta_{1/2}) \tag{9.29}$$

For most selective circuits, value of $\delta_{1/2}$ at half-power frequency is small with respect to factor 2:

$$2Q\delta_{1/2} = 1 \tag{9.30}$$

The frequency deviation of each half-power point from resonance will be $\delta(1/2)$, and therefore deviation between two half-power frequencies will be $2(\delta(1/2))$. Then, bandwidth in cycles is

$$\beta = 2\delta_{1/2}f_0 = \frac{f_0}{Q} \tag{9.31}$$

Here, Q is for the complete series circuit, inductor capacitor, and source. Note that in Fig. 9.4(a), source impedance is not shown. The total impedance of the series circuit is the *total* resistance, and this means that series circuit should be connected to voltage source of low resistance.

9.2.2 Parallel RLC Circuit

In a parallel RLC circuit, Fig. 9.5(a), the admittance is high at resonance frequency.

$$y = \frac{1}{R} + \frac{1}{j\omega L} + j\omega C \tag{9.32}$$

Thus, the resonant condition is

$$-\frac{1}{\omega L} + \omega C = 0 \quad \text{or} \quad \omega = \omega_\alpha = \frac{1}{\sqrt{LC}} \tag{9.33}$$

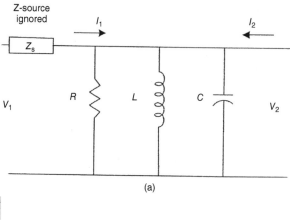

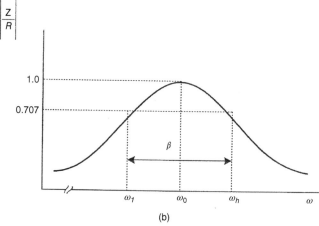

Figure 9.5   (a) Parallel RLC circuit, (b) the frequency response, modulus only.

Compare Eq. (9.33) for parallel resonance with Eq. (9.12) for the series circuit. The magnitude Z/R is plotted in Fig. 9.5(b). Half-power frequencies are indicated in the plot. The bandwidth is given by

$$\beta = \frac{\omega_a}{Q_a} \tag{9.34}$$

where the quality factor for the parallel circuit is given by

$$Q_a = \frac{R}{\omega_a L} = \omega_a RC = R\sqrt{\frac{C}{L}} \tag{9.35}$$

A practical inductor will have some series resistance and a capacitor has some power losses.

Example 9.1: Derive the expressions for Q factor for the series and parallel RLC circuits.

Series Circuit

The instantaneous stored energy in the system is

$$W_s = \frac{1}{2}Li^2 + \frac{q_C^2}{2C}$$

For a maximum,

$$\frac{dW_s}{dt} = Li\frac{di}{dt} + \frac{q_C}{C}\frac{dq}{dt}$$

$$= i(v_L + v_C) = 0$$

For $i = 0$, v_C is maximum, as voltage lags current by $90°$,

$$W_{s,max} = \frac{Q_{max}^2}{2C} = \frac{1}{2}CV_{max}^2 = \frac{I_{max}^2}{2C\omega^2} \quad \omega \le \omega_0$$

For the second condition,

$$v_L + v_C = 0, \text{ the current is maximum}$$

$$W_{s,max} = \frac{1}{2}LI_{max}^2 \quad \omega \ge \omega_0$$

The energy dissipated in the resistor is

$$W_r = \frac{I_{max}^2 R\pi}{\omega}$$

Therefore,

$$Q = 1/\omega CR \quad \omega \le \omega_0$$

$$= \omega L/R \quad \omega \ge \omega_0 \tag{9.36}$$

Parallel Circuit

The instantaneous stored energy in the system is

$$W_s = \frac{1}{2}Li_L^2 + \frac{q_C^2}{2C}$$

For a maximum,

$$\frac{dW_s}{dt} = Li_L\frac{di}{dt} + \frac{q_C}{C}i_C$$

$$= v(i_L + i_C) = 0$$

For $v = 0$, $q_C = 0$:

$$i_L = \pm I_{L,\max} = \pm \frac{V_{\max}}{\omega L}$$

This gives

$$W_{s,\max} = \frac{1}{2} \frac{V_{\max}^2}{L\omega^2}$$

For

$$(i_L + i_C) = 0, \, i_L = i_C = 0, \, q_C = \pm CV_{\max}$$

This gives

$$W_{s,\max} = \frac{1}{2} CV_{\max}^2$$

Therefore,

$$Q = \frac{R}{L\omega} \quad \omega \le \omega_0$$

$$= \omega CR \quad \omega \ge \omega_0 \tag{9.37}$$

Consider an RLC circuit with $R = 10 \ \Omega$, $L = 25$ mH, and $C = 0.5 \ \mu\text{F}$, then

$$\omega_0 = \frac{1}{\sqrt{0.025 \times 0.5 \times 10^{-6}}} = 8944.3 \, \text{rad/s}$$

$$Q = \frac{\omega_0 L}{R} = 22.4$$

$$Q = \frac{1}{\omega_0 CR} = 22.4$$

$$Q = \frac{\omega_0}{\beta} = 22.4$$

A physical explanation of the Q factor can be stated as the efficiency with which the energy is stored. Capacitors and inductors are energy storage elements.

Example 9.2: This example is an EMTP simulation of a series and parallel RLC circuit to clarify the relations between these circuits. Consider a resistance of 0.01 ohms, reactance of 0.1 mH, and a capacitor of 2.814 μF. The inductance and capacitance are chosen so that these have equal impedances at 300 Hz, fifth harmonic. A 300-Hz voltage of 480 V is applied to the series and parallel circuits. Figure 9.6(a) and (b) shows the connection diagram with all component values for the series and parallel circuits.

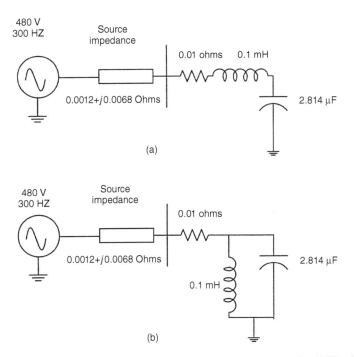

Figure 9.6   (a) Series resonance circuit, (b) a "tank" circuit for EMTP simulation.

In the series circuit (Fig. 9.6(a))

$$j\omega L - \left(\frac{1}{j\omega C}\right) = 0$$

The source current is high and is given by the resistance and source impedance of the circuit. A plot of this current is shown in Fig. 9.7. The current drawn is 13.5-kA rms symmetrical. The voltage developed across the capacitor and inductor reach high values, 2695 V, rms line-to-line. (In EMTP simulations, the voltage or current is shown line-to-neutral in peak.) These voltages cancel each other (Fig. 9.8).

In the circuit of Fig. 9.6(b), the current drawn from the source is small = 4 A rms (Fig. 9.9). However, the current through the capacitor and the reactor is high, 2572 A rms (Fig. 9.10). These currents circulate in the parallel circuit of inductor and capacitor.

9.3   PRACTICAL LC TANK CIRCUIT

The resistance of the inductor was ignored in parallel resonant circuit (Fig. 9.5(a)). Although the capacitor may be treated as an ideal capacitor, the inductor must be modeled with some practical value of resistance. The circuit of Fig. 9.11(a) is called

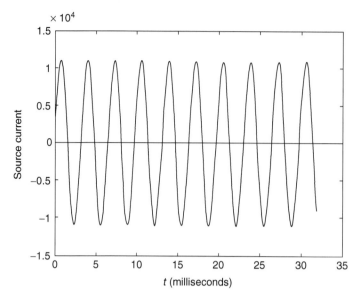

Figure 9.7 Source current in series resonant circuit, EMTP simulation.

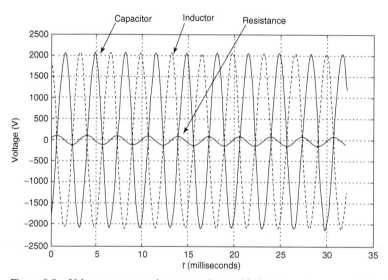

Figure 9.8 Voltages across resistor, capacitor and inductor; series resonant circuit, EMTP simulation.

a *tank circuit*. Consider that it is excited by a source of zero resistance. Then, its admittance can be written as

$$Y = \frac{R}{R^2 + \omega^2 L^2} - j\left(\frac{\omega L}{R^2 + \omega^2 L^2} - \omega C\right) \tag{9.38}$$

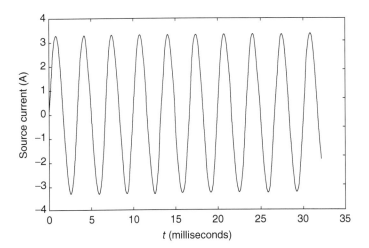

Figure 9.9   Source current parallel tank circuit, EMTP simulation.

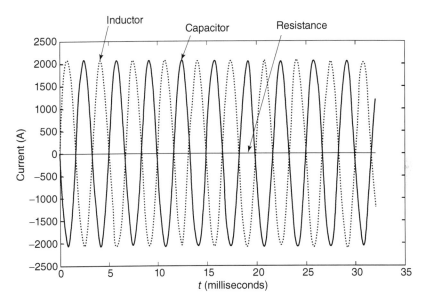

Figure 9.10   Current in inductor, capacitor and resistor, tank circuit, EMTP simulation.

For resonance, the circuit must have a unity power factor, that is,

$$\left(\frac{\omega_0 L}{R^2 + \omega_0^2 L^2} - \omega_0 C \right) = 0 \tag{9.39}$$

This shows that the resonance occurs at

$$\omega_0 = \sqrt{\frac{1}{LC} - \frac{R^2}{L^2}} \tag{9.40}$$

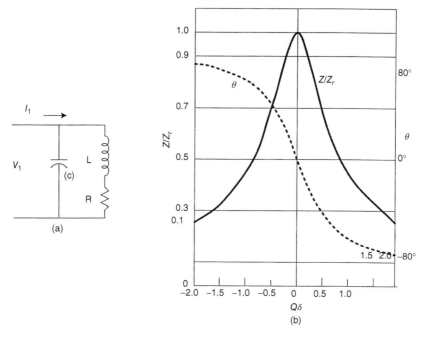

Figure 9.11 (a) Tank resonant circuit, (b) resonance curve as a function of $Q\delta$.

This means that resonance cannot occur if

$$\frac{R^2}{L^2} > \frac{1}{LC} \tag{9.41}$$

In contrast, a series circuit can be resonant at all values of resistance, only Q factor and sharpness of tuning will change.

Equation (9.40) can be written as

$$\omega_0 = \sqrt{\frac{1}{LC}}\sqrt{1 - \frac{R^2 C}{L}} = \sqrt{\frac{1}{LC}}\sqrt{1 - \frac{1}{Q^2}} \tag{9.42}$$

Thus, it differs from the series circuit by the factor

$$\sqrt{1 - \frac{1}{Q^2}} \tag{9.43}$$

If Q is large, say > 10, the error is less than 1%. Also resonance is not possible with $Q < 1$. From Eq. (9.42)

$$\omega_0^2 LC = 1 - \frac{1}{Q^2}$$

$$X_L = X_C\left(1 - \frac{1}{Q^2}\right) \tag{9.44}$$

Thus, unlike series resonant circuit, inductive and capacitive branches are not exactly equal.

The impedance variation with frequency can be examined similar to series circuit. We can write the admittance in Eq. (9.38) as

$$Y = \frac{R\left(1 - \frac{j\omega L}{R} + \frac{\omega^2 C L^2}{R} + \omega CR\right)}{R^2 + \omega^2 L^2} \tag{9.45}$$

Writing

$$\frac{\omega L}{R} = \frac{\omega_0 L}{R}\frac{\omega}{\omega_0} = Q(1 + \delta)$$

$$\omega^2 LC = \omega_0^2 LC(1 + \delta)^2 = \left(1 - \frac{1}{Q^2}\right)(1 + \delta)^2 \tag{9.46}$$

and substituting in Eq. (9.45)

$$Y = \frac{1 - jQ(1 + \delta)\left\{1 - \left(1 - \frac{1}{Q^2}\right)(1 + \delta)^2\left(1 - \frac{1}{Q^2(1+\delta)^2}\right)\right\}}{R[1 + Q^2(1 + \delta)^2]} \tag{9.47}$$

This can be simplified assuming $Q > 10$

$$Y = \frac{1 - jQ(1 + \delta)\{1 - (1 + \delta)^2\}}{RQ^2(1 + \delta)^2} \tag{9.48}$$

or

$$Z = \frac{RQ^2(1 + \delta)^2}{1 + jQ\delta(1 + \delta)(2 + \delta)} \tag{9.49}$$

At resonance $\delta = 0$ and therefore

$$Z_r \approx R(1 + Q^2) \approx RQ^2 \tag{9.50}$$

Then

$$\frac{Z}{Z_r} = \frac{(1 + \delta)^2}{1 + jQ\delta(1 + \delta)(2 + \delta)} \tag{9.51}$$

Considering that close to resonance $\delta \ll 1$

$$\frac{Z}{Z_r} = \frac{1}{1 + j2Q\delta} = A < \theta° \tag{9.52}$$

A plot is shown in Fig. 9.11(b), in terms of circuit Q and small variation in δ and angle θ. For negative values of δ or for frequencies below resonance, the circuit is inductive, while for positive values of δ, it is capacitive.

The maximum possible impedance of the circuit does not occur at unity power factor. The square of the admittance is

$$|Y|^2 = \frac{1 - 2\omega^2 LC + \omega^2 C^2 (R^2 + \omega^2 L^2)}{R^2 + \omega^2 L^2} \qquad (9.53)$$

This can be minimized with respect to ω by

$$\frac{d|Y|^2}{d\omega} = 0 \qquad (9.54)$$

This gives

$$\omega = \left[\frac{1}{LC} \sqrt{1 + \frac{2CR^2}{L} - \frac{R^2}{L^2}} \right]^{1/2} \qquad (9.55)$$

Therefore, the maximum impedance as the frequency is varied does not occur at unity power factor. If Q is large, then it will reduce to that condition.

9.4   REACTANCE CURVES

Using reactance curves is an easy method to visualize resonance conditions. We can use reactance or susceptance values. On the basis of the definition of these elements, the plots are as follows:

- *Linear Positive*: inductive reactance and capacitive susceptance
- *Hyperbolic Negative*: capacitive reactance and inductive susceptance
- The reciprocal of a positive linear relation is a negative hyperbola and vice versa. When resistance is small, the reactance and susceptance curves can be combined, giving composite resonance curves.

For a series resonant circuit, Fig. 9.12 illustrates curves of X_L and X_C and the added curve X_T. This gives the circuit performance as frequency is varied. It shows zero reactance at resonant point, a capacitive reactance at frequencies below resonance, and an inductive reactance at frequencies above resonance.

For the tank circuit, susceptance curves are shown in Fig. 9.13. The reciprocal of b_T is plotted as X_T, giving a curve of reactance versus frequency. At resonance point, the reactance, theoretically, goes to infinity.

Example 9.3: Plot the reactance curves of the circuit shown in Fig. 9.14(a).

The circuit can be considered in parts and the curve plotted for each part and then combined. The L_2 and C_2 series branch circuit is depicted in Fig. 9.14(b) as X_2. The reciprocal of X_2, that is, b_2 is plotted in Fig. 9.14(c) and added algebraically to b_{C_1}, as the two branches are in parallel. Then, the total susceptance of the two branches b_{12} results (Fig. 9.14(c)). The reciprocal of b_{12}, that is, X_{12}, is plotted in Fig. 9.14(d). To X_{12} is added X_{L_1}, a series element, giving total reactance curve X_T, which gives idealized performance of the circuit.

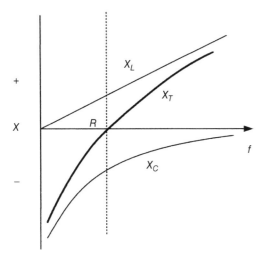

Figure 9.12   Reactance curves of a series LC circuit.

In general, a circuit will have series and parallel resonances not greater than the number of meshes +1.

9.5   FOSTER'S NETWORKS

The concept of poles and zeros is widely applied in control systems. In a resonant network, at a series resonance the reactance is zero (*zeros*), while in a parallel network the reactance is theoretically infinite (*poles*). Locations of poles and zeros, thus, specify the complete network. Foster's reactance theorem states that input impedance is completely defined by the location of its poles and zeros, and by its value at a nonzero, nonpole, and frequency.

The input impedance of a reactive network can be expressed as

$$Z_{in} = \frac{\Delta}{\Delta_{11}} \tag{9.56}$$

In circuit theory using matrices, an impedance matrix of an electrical circuit can be written, which consists of self and mutual impedances between circuit elements. The determinant of the Z matrix is Δ, and Δ_{11} is obtained with the first row and first column deleted [1–4].

Both Δ and Δ_{11} are polynomials in ω. Then the input impedance can be written as

$$Z_{in} = S \frac{a_0 + a_2\omega^2 + a_4\omega^4 + \cdots + a_{2m}\omega^{2m}}{b_0 + b_2\omega^2 + b_4\omega^4 + \cdots + b_{2m}\omega^{2m}} \tag{9.57}$$

The coefficient S is

$$S = \pm j\omega K \quad \text{or} \quad = \pm \frac{K}{j\omega} \tag{9.58}$$

K is a scale factor

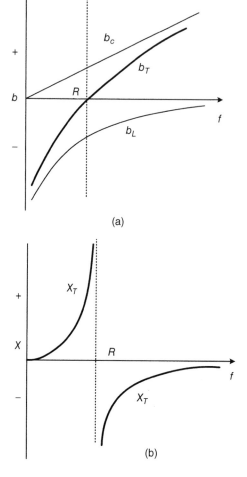

(a)

(b)

Figure 9.13   (a) Reactance curves of a parallel LC circuit, (b) total reactance X_T.

In general, Δ will involve even powers of ω and the denominator will involve odd powers or vice versa. The numerator and denominator differ by one in degree of variable ω, as determined by the position of ω in S. The highest power is $2m$, where m is the order of network determinant Δ.

By factoring,

$$Z_{in} = S \frac{(\omega^2 - \omega_1^2)(\omega^2 - \omega_3^2)(\omega^2 - \omega_5^2) \ldots (\omega^2 - \omega_{2m-1}^2)}{(\omega^2 - \omega_2^2)(\omega^2 - \omega_4^2)(\omega^2 - \omega_6^2) \ldots (\omega^2 - \omega_{2m}^2)} \tag{9.59}$$

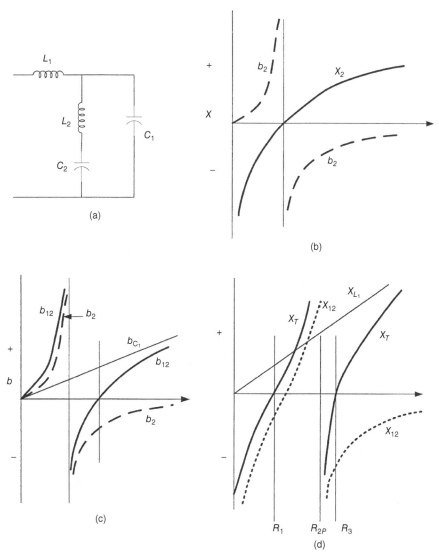

Figure 9.14   (a) A circuit configuration for plotting reactance curves, (b)–(d) step-by step procedure for plotting the final curves by breaking circuit in (a) in parts and successive graphic summation (Example 9.3).

Roots of Δ contribute zeros and roots of Δ_{11} contribute poles. This means that at

$$\omega = \omega_1, \omega_3, \omega_5, \ \ldots$$

impedance function has a zero, and all these frequencies are resonant angular frequencies. Similarly, at

$$\omega = \omega_2, \omega_4, \omega_6 \ \ldots$$

impedance function is infinite and has a pole. Because ω is present in S, the value $\omega = 0$ may be either a pole or zero. As ω becomes infinite, the impedance may approach infinity or zero. Thus, Z_{in} may occur in any of the four forms.

Equation (9.59) can be expanded by use of partial fractions:

$$Z_{in} = j\omega H \left(1 + \frac{A_0}{\omega^2} + \frac{A_2}{\omega^2 - \omega_2^2} + \cdots + \frac{A_{2m-2}}{\omega^2 - \omega_{2m-2}^2} \right) \tag{9.60}$$

where $S = -j\omega H$.

It may be realized that the first term is a representative of a series inductance, the second of a series capacitor, and the succeeding terms an inductor and capacitor in parallel. The coefficients A_k can be evaluated by equating Eqs (9.59) and (9.60) and multiplying by ω_2.

$$\frac{(\omega^2 - \omega_1^2)(\omega^2 - \omega_3^2)(\omega^2 - \omega_5^2) \cdots (\omega^2 - \omega_{2m-1}^2)}{(\omega^2 - \omega_2^2)(\omega^2 - \omega_4^2)(\omega^2 - \omega_6^2) \cdots (\omega^2 - \omega_{2m}^2)}$$

$$= \omega^2 + \frac{A_0}{1} + \frac{A_2\omega^2}{\omega^2 - \omega_2^2} + \cdots + \frac{A_{2m-2}\omega^2}{\omega^2 - \omega_{2m-2}^2} \tag{9.61}$$

If $\omega = 0$, A_0 is known

Then multiplication with $(\omega^2 - \omega_k^2)$ and making $\omega = \omega_k$ will give other coefficients.

The slope of the reactance function is always positive. At a pole, the function must change sign and pass through zero before reaching another pole; this means that zeros and poles alternate. This is the separation property of the poles and zeros. In Fig. 9.14(d) we have zero, then a pole, and then again zero.

9.6   HARMONIC RESONANCE

Harmonic resonance occurs at a certain harmonic or harmonics. We call a resonance as "harmonic resonance" when it takes place at one of the nonlinear load-generated harmonics. When shunt capacitors are connected for power factor improvement, voltage support, or reactive power compensation, these act in parallel with the system impedance.

This means that the system impedance considered inductive (with some resistance in series) is in parallel with a capacitor. As discussed in previous sections, resonance occurs at a frequency where the inductive and capacitive reactances are equal; the impedance of the combination is infinite for a lossless system. The impedance angle changes abruptly as the resonant frequency is crossed. The inductive impedance of the power source and distribution (utility, transformers, generators, and motors) as seen from the point of application of the capacitors equals the capacitive reactance of the power capacitors at the resonant frequency.

$$j2\pi f_0 L = \frac{1}{j2\pi f_0 C} \tag{9.62}$$

where f_n is the resonant frequency. Figure 9.15 is a rehash of Fig. 9.13 and shows that the resonance occurs at fifth harmonic, which is commonly generated by the electronic loads. The resonance point depends on relative values of the inductance and capacitance. If the resonant frequency is 300 Hz, $5j\omega L = 1/(5j\omega C)$, where ω pertains to the fundamental frequency (see Example 9.2). When excited at the resonant frequency, a harmonic current magnification occurs in the parallel-tuned circuit, though the exciting input current is small, as illustrated in Example 9.2. This magnification can be many times the exciting current and can even exceed the fundamental frequency current. It overloads the capacitors and may result in nuisance fuse operation, severe amplification of the harmonic currents resulting in waveform distortions, which has consequent deleterious effects on the power system components.

Resonance with one of the load-generated harmonics is the major effect of harmonics in the presence of power capacitors. This condition has to be avoided in any application of power capacitors. More conveniently, the resonant condition in the power system can be expressed as

$$h = \frac{f_0}{f} = \sqrt{\frac{kVA_{sc}}{kvar_c}} \tag{9.63}$$

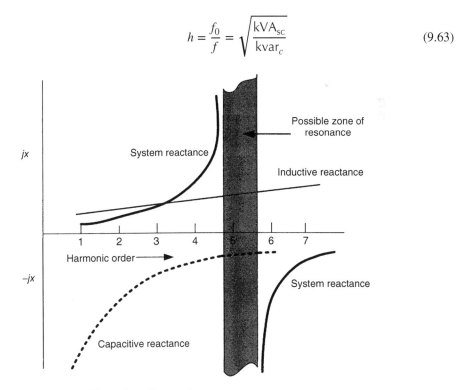

Figure 9.15   Illustration of harmonic resonance.

where h is the order of harmonics, f_n is the resonant frequency, f is the fundamental frequency. kVA_{sc} is the short-circuit duty at the point of application of the shunt capacitors, and $kvar_c$ is the shunt power capacitor rating in kvar.

Consider that the short-circuit level at the point of application of power capacitors is 500 MVA. Resonance at 5th, 7th, 11th, 13th, ... harmonics will occur for a shunt capacitor size of 20, 10.2, 4.13, 2.95, ... Mvar, respectively. The smaller the size of the capacitor, the higher is the resonant frequency.

The short-circuit level in a system is not a fixed entity. It varies with the operating conditions. In an industrial plant, these variations may be more pronounced than in the utility systems, as a plant generator or part of the plant-rotating loads may be out of service, depending on the variations in processes and operation. The resonant frequency in the system will *float around.*

The lower order harmonics are more troublesome from a resonance point of view. As the order of harmonics increases, their magnitude reduces. Sometimes, a harmonic analysis study is limited to 25th or 29th harmonic only; however, possible resonances at higher frequencies may be missed.

It can be concluded that

- the resonant frequency will swing around depending on the changes in the system impedance, for example, switching a tie-circuit on or off, operation at reduced load. Some sections of the capacitors may be switched out altering the resonant frequency;
- an expansion or reorganization of the distribution system may bring out a resonant condition where none existed before;
- even if the capacitors in a system are sized to escape current resonant condition, immunity from future resonant conditions cannot be guaranteed owing to system changes, for example, increase in the short-circuit level of the utility system (also see Ref. [5]).

Example 9.4:   In a series resonant circuit $X_c = 300$ ohms, and resonance occurs at 10th harmonic. Find the value of the reactor and plot impedance of the series circuit as the frequency varies from 60 to 600 Hz. Consider a Q factor of 50.

At 600 Hz:

$$X_c = X_L$$

$X_c = 300$ ohms at fundamental frequency, Therefore, at 600 Hz, it is equal to 30 ohms.

Therefore, X_L at 600 Hz = 30 ohms. At fundamental frequency, it is equal to 3 ohms. The Q factor (at fundamental frequency) is 50. Therefore, $R = 0.06$ ohms. At resonance, the impedance is equal to $R = 0.06$ ohms. The plot of the impedance of the series circuit is therefore shown in Fig. 9.16.

We will repeat this example for a parallel resonant circuit. Applying Eqs (9.16) and (9.17), the calculated impedance plot is shown in Fig. 9.17.

Example 9.5:   Calculate half-power frequencies in Example 9.4. What is the significance of these frequencies?

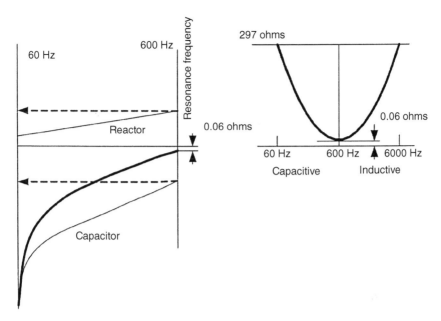

Figure 9.16   Plot of resonance, series RLC circuit (Example 9.4).

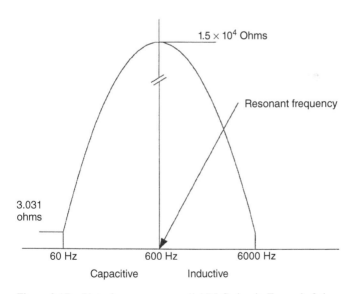

Figure 9.17   Plot of resonance, parallel RLC circuit, Example 9.4.

We considered that the specified Q is at fundamental frequency. Therefore, at the resonant frequency, $Q_0 = 500$. Using Eq. (9.16), the frequencies are not much different from the resonant frequency of 600 Hz.

The half-power frequencies determine the sharpness of tuning. Here, the resistance is low.

If we calculate for lower Q values, with $Q = 30$ or higher, the frequencies are not much different from the resonant frequency. If we consider Q at resonance of say 10 only, then the frequencies can be calculated approximately 6.24% around the resonant frequency taken as unity.

This shows the impact on sharpness of tuning with change in resistance. Practically, the fundamental frequency loss on selection of a certain Q becomes important in harmonic filter designs. A lower value of Q can give high fundamental frequency loss and much heat generation. See Chapter 15, which also provided equations for optimized Q for filters.

9.7 HARMONIC RESONANCE IN A DISTRIBUTION SYSTEM

The techniques used for the harmonic load flow study are discussed in Chapter 14. An elementary tool to accurately ascertain the resonant frequency of the system in the presence of capacitors and nonlinear loads is to run a frequency scan on a digital computer (Chapter 14). Fundamentally, the frequency is applied in incremental steps, say at 2-Hz increment, for the range of harmonics to be studied to calculate resonance frequency accurately (a larger step reduces computer time but gives a 50% tolerance band with respect to frequency step used.

System component models used for power frequency applications, for example, transformers, generators, reactors, and motors, are modified for higher frequencies (Chapter 12).

Consider the distribution system shown in Fig. 9.18. A 5-MVA transformer supplies a six-pulse drive system (three-phase fully controlled bridge circuit, Chapter 1) load of 5 MVA. There is a utility tie transformer of 30 MVA, and also a 35.3-MW generator operating in synchronism with the utility source and a 6-Mvar capacitor bank at the 13.8-kV bus. The spectrum of the harmonic injection at bus 2 is shown in Table 9.2.

Much alike fundamental frequency load flow calculations, all the system impedances must be accurately modeled, furthermore, these impedances are modified at each incremental frequency chosen for the frequency scan.

A frequency scan of bus 2, both impedance modulus and angle, is shown in Figs 9.19 and 9.20, respectively. A resonance occurs close to 11th harmonic. Figure 9.19 shows that the angle abruptly changes at the resonant frequency. The calculated resonant frequency is 672 Hz. Figure 9.21 shows the voltage spectrum of 13.8-kV bus 1 and 4.16-kV bus 2. This shows that the 11th harmonic voltage at 4.16-kV bus 2 is 17.2%, while at 13.8-kV bus 1, it is 11.8%. Figure 9.22 shows distorted waveform of the voltage at these two buses for one cycle. Figure 9.23 depicts the spectrum of current flow through the capacitor, the 11th harmonic current is 130% of the fundamental current at 13.8 kV (the fundamental frequency current is not shown). The injected 11th harmonic current, a current at 4.16 kV bus 2, is 45.1 A based on the harmonic spectrum shown in Table 9.2.

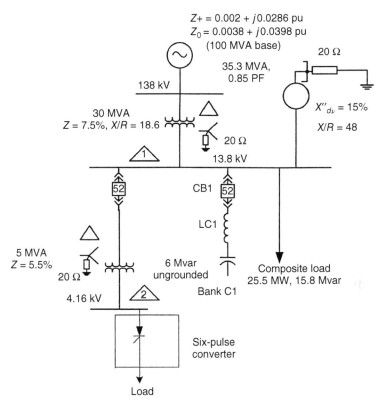

Figure 9.18   A distribution system for the study of resonance.

TABLE 9.2   Harmonic Emission, Six-Pulse Converter, Example 9.7

h	5	7	11	13	17	19	23	25	29	31
%	18	13	6.5	4.8	2.8	1.5	0.5	0.4	0.3	0.2

Thus, there is severe magnification at the 11th harmonic current. We will illustrate in the chapters to follow how these large waveform distortions are mitigated with the application of passive filters.

9.8   ELUSIVENESS OF RESONANCE PROBLEMS

The resonance problem in power system is a potential problem and a serious one. There are many documented cases of resonance, component failures, and plant shutdowns. It may not surface immediately and may appear at certain operating condition of the power system. It may sometimes "come" and "go" without attracting immediate attention, as it may be a "partial" resonance – thereby implying that the resonant

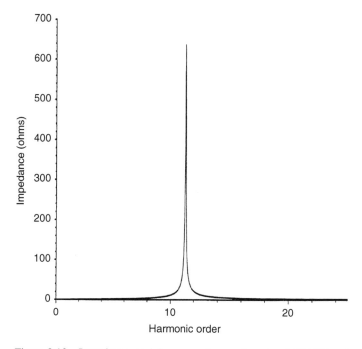

Figure 9.19   Impedance modulus versus harmonic order, 4.16-kV bus 2.

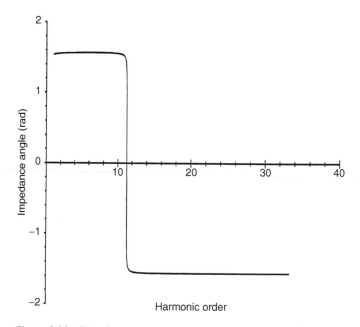

Figure 9.20   Impedance angle versus harmonic order, 4.16-kV bus 2.

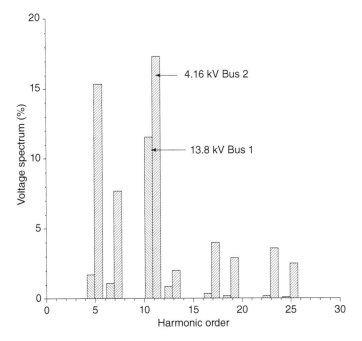

Figure 9.21   Voltage spectrum, 13.8-kV bus 1 and 4.16-kV bus 2.

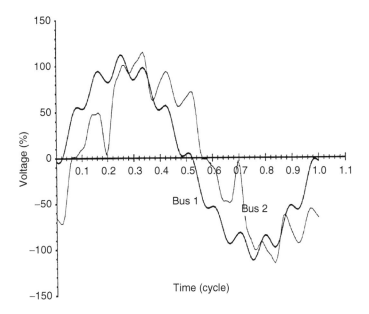

Figure 9.22   Distorted voltage waveforms 13.8-kV bus 1 and 4.16-kV bus 2.

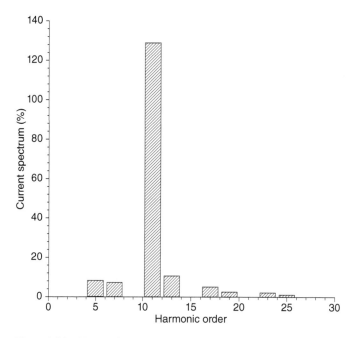

Figure 9.23 Harmonic current spectrum through 6-Mvar capacitor bank.

impedance did not exactly produce $Q\delta = 0$ situation. Thus, most of the time, a resonant condition is "elusive." It may require long-term online measurements to establish the disturbing element in the system. References [6,7] provide some documented cases.

Harmonic distortions may come and go with time of the day. The patterns can be recognized and correlated with particular type of disturbing load, such as mass transit systems, rolling mills, and arc furnaces. Arc-lighting harmonics can be recognized from dusk to dawn patterns.

Figure 9.24 is an illustration of such a harmonic distortion caused by the load pattern of a consumer. Note the correlation of the harmonic distortion with the load pattern [5].

9.9 RESONANCE DUE TO SINGLE-TUNED FILTERS

Chapter 15 discusses the various harmonic filter types. Single-tuned filters are commonly applied, and these are very efficient is shunting the desired harmonic from the power system. However, it is shown in Chapter 15 that the harmonic resonance is not eliminated, but the harmonic resonance frequency shifts to a lower value with respect to the tuned frequency of the ST filter. Figure 9.25 illustrates this. In this figure,

f_1 is the tuned frequency of the ST filter. The filter will shunt out the desired harmonic close to the frequency at which it is tuned.

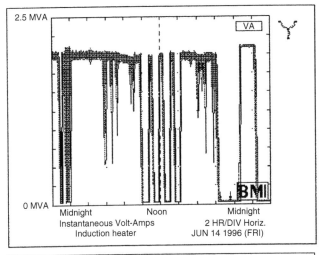

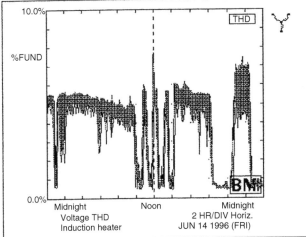

Figure 9.24 Voltage distortion versus time on a distribution system, and the load profile of a particular customer that is causing harmonic distortion. Source: Ref. [5].

f_2 is the original resonant frequency, without application of an ST filter, Chapter 15.

f_3 is the shifted resonant frequency due to ST filter.

If the ST filter is tuned to shunt out the fifth harmonic, its shifted resonance frequency may be the third or fourth harmonic or something in between these two values, which can be calculated. We have examined that interharmonics and even harmonics can be present, and these can be magnified by the shifted resonant frequencies.

It is common to use more than one ST filter. A number of ST filters tuned to a different frequency, say 5th, 7th, and 11th, may be required. Each of these ST filters will create a resonant frequency below its tuned frequency.

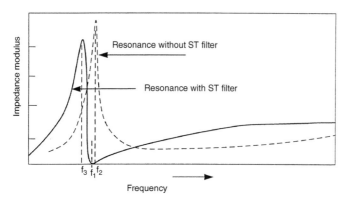

Figure 9.25   Shifted resonance frequency, ST filter.

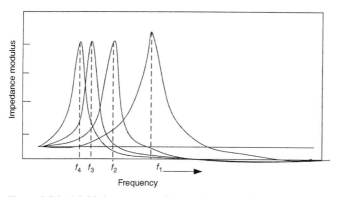

Figure 9.26   Multiple resonance frequencies, capacitor banks switched in sequence to maintain load power factor.

The placement of shifted resonance frequencies in ST filter designs is one important consideration (Chapter 15).

9.10   SWITCHED CAPACITORS FOR POWER FACTOR IMPROVEMENT

For the power factor, improvement switched capacitors are normally applied at low voltage or medium voltage. As the load increases and power factor drops, a capacitor is switched in. In the presence of nonlinear loads, this will give rise to multiple resonant frequencies. Consider a four-step switching. With reference to Fig. 9.26 when the first capacitor bank is switched, it gives rise to resonant frequency f_1. As the load increases and the power factor drops, the second capacitor bank is switched and the resonant frequency shifts to f_2. Similarly, when the third capacitor bank is switched, frequency shifts to f_3 and, finally, to f_4 when the last capacitor is switched in. These frequencies will vary with the size of capacitor bank being switched in each step.

Here, the objective is to keep power factor within certain operating limits as the load varies.

This gives rise to two problems:

- Transients due to back-to-back switching and duties on the switching devices (Chapter 11).
- With shifting resonant frequencies, multiple resonances can occur as the frequency shifts from f_1 to f_4. These will be difficult to control.

If each switched capacitor is turned into a single-tuned filter; tuned to the same frequency, then there will be only one resonant frequency to deal with. With the proper design of the ST filters, the resonance at a load-generated harmonic can be escaped. There is one more consideration, that is, that ST filters tuned to the same frequency may drive each other due to component tolerances. Sometimes, a slight purposeful detuning is adopted in ST parallel filter designs.

9.10.1 Nearby Harmonic Loads

A utility source may serve more than one customer from the same source. A consumer may design a filter to take care of his harmonic producing loads, and yet there can be problems because of adjacent harmonic producing loads. *The effects of harmonics can be present at considerable distances from their origins.*

The disturbing loads of the other consumers, especially on the same close by service, can make the filter designs ineffective and even overload these filters. The presence of capacitors along with filters can create more distortion and degrade the performance of filter, sometimes creating additional resonance frequencies. Monitoring of each customer loads and the distortion these are producing becomes important.

9.11 SECONDARY RESONANCE

When there are secondary circuits that have resonant frequencies close to the switched capacitor bank, the initial surge can trigger oscillations in the secondary circuits that are much larger than the switched circuit. The ratio of these frequencies is given by

$$\frac{f_c}{f_m} = \sqrt{\frac{L_m C_m}{L_s C_s}} \tag{9.64}$$

where f_c is the coupled frequency, f_m is the main circuit switching frequency, L_s and C_s are the inductance and capacitance in the secondary circuit, and L_m and C_m are the inductance and capacitance in the main circuit. Figure 9.27(a) shows the circuit diagram, and Fig. 9.27(b) shows the amplification of transient voltage in multiple capacitor circuits [8–10]. The amplification effect is greater when the natural frequencies of the two circuits are almost identical. Damping ratios of the primary and coupled circuits will affect the degree of interaction between the two circuits.

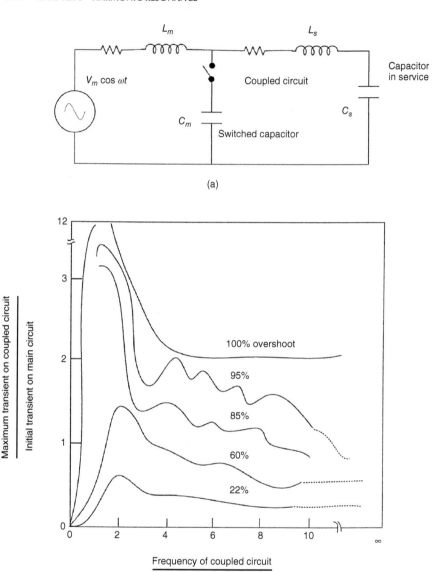

Figure 9.27 (a) Circuit configuration to illustrate secondary resonance, (b) overvoltages on switching depending on coupled circuit parameters.

The switched capacitor can be on in the utility system while the secondary capacitor in service can be at a user's electrical system. This overvoltage can cause damage to the secondary capacitors, result in nuisance tripping of ASDs, and damage to the sensitive equipment.

The maximum switching surges can be expected if

$$L_m C_m = C_s (L_s - L_m) \tag{9.65}$$

The angular frequencies of the circuit are

$$\beta_1 = \left(\frac{\alpha}{2} - \sqrt{\frac{\alpha^2}{4} - \beta} \right)$$

$$\beta_2 = \left(\frac{\alpha}{2} + \sqrt{\frac{\alpha^2}{4} - \beta} \right) \tag{9.66}$$

where

$$\alpha = \frac{1}{L_m C_m} + \frac{1}{L_s C_s} + \frac{1}{L_s C_m}$$

$$\beta = \frac{1}{L_m C_m L_s C_s} \tag{9.67}$$

The voltage across capacitor C_s is

$$V_s = V_m \left[1 - \frac{\beta_2^2 \cos \beta_1 t - \beta_1^2 \cos \beta_2 t}{\beta_2^2 - \beta_1^2} \right] \tag{9.68}$$

And the maximum voltage is

$$V_{s,\text{max}} = V_m \frac{\beta_2^2 + \beta_1^2}{\beta_2^2 - \beta_1^2} \tag{9.69}$$

Example 9.6: This example is an EMTP simulation of a secondary resonance. Figure 9.18 is modified to show a 200-kvar capacitor on the secondary side of the 2-MVA transformer, as depicted in Fig. 9.28. This capacitor remains in service, while 6-Mvar capacitor at 13.8-kV bus is switched. The simulation in Fig. 9.29 shows that a peak voltage of 1220 V line-to-neutral is developed across the secondary of 480-V transformer. This overvoltage is 3.1 times the rated system voltage. The calculations show that $f_c/f_m \approx 3.8$. Figure 9.30 shows the transient current that flows through the 200-kvar capacitor that has a peak value of 2200 A, approximately 6.5 times the full-load current of 200 kvar capacitor.

When capacitors are applied at multivoltage level in a distribution system, apart from the overvoltages due to resonance illustrated earlier, these can also prolong the decay of the transients.

It is, therefore, best to apply capacitors at one voltage level only. If these are applied at multivoltage levels, a rigorous switching transient analysis to predetermine the resonance points and eliminate them is required, which is generally not undertaken due to complexity of modeling.

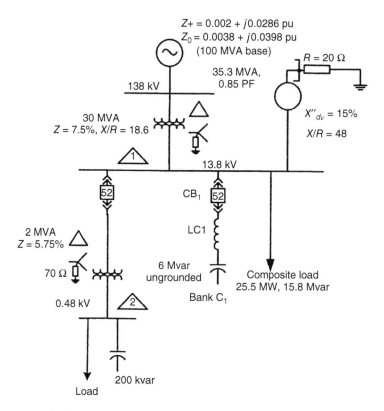

Figure 9.28 A system configuration for study of secondary resonance, EMTP simulation.

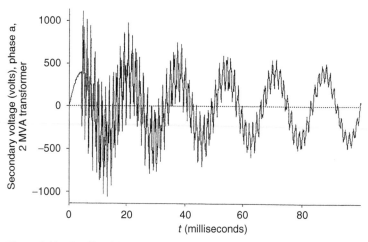

Figure 9.29 Profile of 2-MVA transformer secondary (480-V) transient voltage on switching of 6-Mvar, 13.8-kV capacitor bank, EMTP simulation.

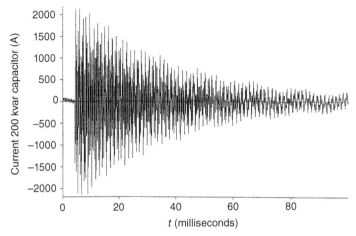

Figure 9.30   Profile of transient current through 200-kvar capacitor bank on switching of 6-Mvar, 13.8-kV capacitor bank, EMTP simulation.

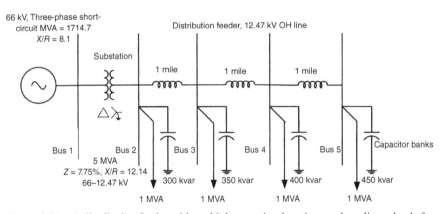

Figure 9.31   A distribution feeder with multiple capacitor locations and nonlinear loads for the study of harmonic resonance.

9.12   MULTIPLE RESONANCES IN A DISTRIBUTION FEEDER

Multiple resonances can occur in a distribution feeder due to location of capacitor banks. This is illustrated with following example:

Example 9.7:   Figure 9.31 shows a distribution feeder circuit, emanating from a substation, a 66–12.47 kV, 5-MVA transformer. A 12.47-kV overhead line takes off from the secondary side of the substation transformer and serves nonlinear consumer loads, fluorescent lighting, switch-mode power supplies, and ASDs. Their spectra are

TABLE 9.3   Parameters of Distribution Feeder, Ohms per 1000 ft or Siemens/1000 ft

Parameter	Calculated	Parameter	Calculated
R_1	0.099945	R_0	0.189825
X_1	0.104826	X_0	0.608527
Y_1	0.0000015	Y_0	0.000006

TABLE 9.4   Resonant Frequencies and Impedance Modulus

Bus ID	Z, Modulus in Ohms	Harmonic	Frequency
2	92.84	5.833	350
	87.1	24.133	1448
3	128.47	5.833	350
	34.92	24.2	1452
4	158.55	5.833	350
	2.68	24.1	1446
5	176.42	5.833	350
	58.42	24.34	1460

modeled as shown in Chapter 4. To support the voltages in the system, shunt capacitors of the ratings as shown are applied at the load points. With adjustment of 5-MVA transformer off-load taps to provide 2.5% voltage boost to the 12.47-kV secondary windings, fundamental frequency load flow shows that approximately rated operating voltages can be maintained at the loads.

The 12.47-kV line conductors of AAC 4/0 are spaced 1.2 ft in flat formation at a height of 25 ft from the ground level, and the distribution line carries a 3/#10 AWG steel ground wire. The soil resistivity is 100 ohms/m. With these line parameters, the line constants are calculated using a computer routine and are shown in Table 9.3.

A frequency scan shows resonant frequencies as illustrated in Table 9.4. The phase angle and impedance modulus plots are depicted in Figs 9.32 and 9.33, respectively. These show multiple harmonic resonances, and these do not occur at integer multiple of the fundamental frequency.

9.13   PART-WINDING RESONANCE IN TRANSFORMER WINDINGS

Part-winding resonance in transformers occurs mostly due to switching surges, current chopping in circuit breakers (low inductive currents), and VFTs (very fast transients) in gas-insulated substations. Failure of four autotransformers of 500 kV and

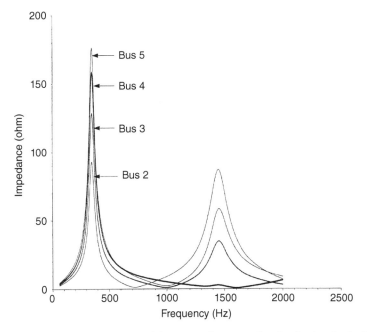

Figure 9.32   Impedance modulus versus frequency for distribution feeder in Fig. 9.31.

765 kV systems of American Electric Power in 1968 and 1971 led to the investigations of winding resonance phenomena [11–13]. When an exciting oscillating overvoltage arises due to line or cable switching or faults and coincides with one of the fundamental frequencies of part of a winding in the transformer, high overvoltages and dielectric stresses can occur. Low-amplitude and high-oscillatory switching transients cannot be suppressed by surge arrestors, but can couple to the transformer windings.

The locations of poles and zeros of an impedance circuit are already discussed. Terminal resonance can be defined to occur at maximum current and minimum impedance, it is also called a series resonance. Here, the reactive component of the terminal impedance is zero. The terminal antiresonance is defined as the minimum current and maximum impedance. The internal resonance escalates the voltage and leads to possible insulation failure. *The terminal resonance and internal resonance may not necessarily bear a direct relation.* A part-winding resonance may significantly influence transient oscillations of a major part of the transformer winding, but its effect may not show up in the terminal impedance plot. A *terminal* reactance and resistance plot are shown in Fig. 9.34.

Generally, the first natural resonant frequency of the transformer is above 5 kHz. Leaving aside VFTs, the resonant frequencies in core-type transformer lie from 5 kHz to few hundred kilohertzs. The resonant frequencies may not vary much between transformers of different manufacturers and cannot be altered much, though efforts have been directed in this direction too. Generally, efforts are made to avoid network conditions, which can produce oscillating voltages. (In-phase switching and resistance switching are discussed in Chapter 12.) In the design stage, a lumped

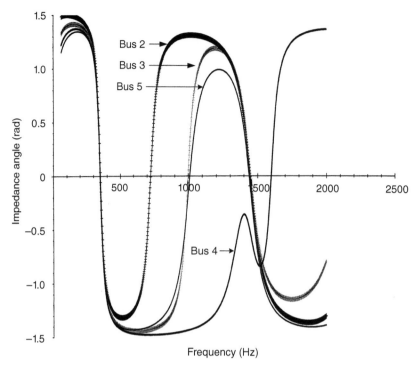

Figure 9.33 Impedance angle versus frequency for distribution feeder in Fig. 9.31.

equivalent circuit of winding inductances and capacitance can be analyzed; for the transformers in operation, the frequency response can be measured.

Under steady-state conditions, it is unlikely that harmonics (generally under 3 kHz) will produce a part-winding resonant condition. However, on switching a capacitor bank, high inrush currents and frequencies occur. IEEE limits on back-to-back switching of a shunt capacitor bank with "definite-purpose circuit breakers" (see Chapter 11) are the following:

- Indoor circuit breakers, K factor = 1 to 38 kV = 2 to 4.2 kHz
- Outdoor circuit breakers rated 72.5 kV and below including circuit breakers in GISs (gas-insulted substations): 3.360–6.8 kHz
- Outdoor circuit breakers rated 123–550 kV including circuit breakers in GIS: 4.25 kHz

An EMTP simulation of switching transients on back-to-back switching of capacitor bank is shown in Fig. 11.26. Figure 9.35 shows fast Fourier transform (FFT) of harmonic contents of voltage on switching of a 13.8-kV, 9-Mvar isolated capacitor bank. The current FFT follows approximately the same pattern. The frequencies are limited to approximately 3.5 kHz and should not be of concern. However, higher frequencies can occur on switching operation in interconnected systems. The probability of exciting part-winding resonance due to switching operations of capacitor banks is small, but cannot be ruled out.

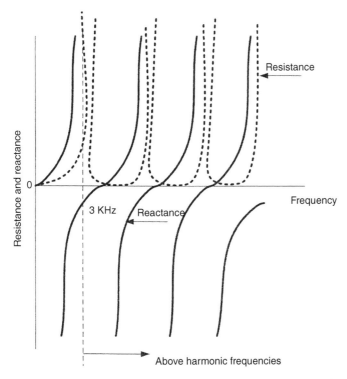

Figure 9.34   A transformer terminal resistance and reactance over frequency.

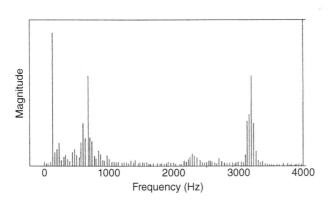

Figure 9.35   FFT of switching overvoltages of a shunt capacitor.

9.14   COMPOSITE RESONANCE

The impedance of AC system interacts through the converter characteristics to present entirely different impedance to the DC side, which gives rise to resonance

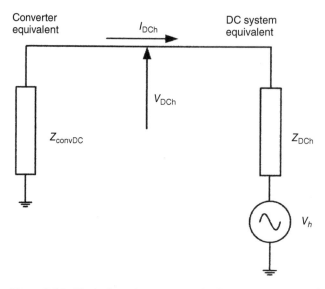

Figure 9.36   Equivalent of converter and DC system. Source: Ref. [16].

frequencies. These frequencies depend on all three parameters, namely, (1) AC system impedance, (2) DC system impedance, and (3) switching of converter.

The composite resonance can be conceived as a matrix quantity [14]. Any single harmonic current flowing into composite impedance produces a multitude of voltage harmonics. Amplification factors have been used to isolate resonance frequencies, and these factors can be defined as transfer functions from a fictitious voltage source placed in series with the converter to the voltage across the DC filter.

Referring to Fig. 9.36, the composite frequencies are determined by selecting a point near the converter, say the DC terminals of the converter. The equivalent impedances looking each way (into the DC system and looking into the converter) are added. The resulting impedance will indicate a resonance when the reactive components cancel each other and are equal to zero.

A composite resonance frequency can be excited by a system transient or by an unbalance in the converter components and controls. An equivalent series resonance circuit can be derived to match the composite resonance at the resonant frequency, from which the amplifying factor can be derived. An RLC network is a second-order system and from elementary differential equation, the time solution of the current is

$$i(t) = e^{-\sigma t}(A_1 \cos \omega_r t + A_2 \sin \omega_r t) \tag{9.70}$$

where

$$\sigma = \frac{R}{2L} \tag{9.71}$$

is the damping factor and A_1 and A_2 are calculated from initial conditions. Figure 9.37 shows the magnification of second harmonic based on bus capacitance. At some

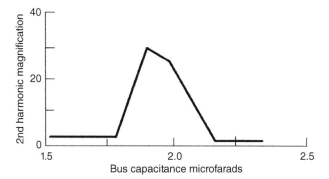

Figure 9.37   Amplification of second harmonic with respect to bus capacitance.

value, the inductive and capacitive components are canceling each other or have approximately the same values.

A test case is CIGRE benchmark model of HVDC link [15]. Three cases of composite circuit series impedances are discussed. The model is designed to represent parallel resonance on AC side and a series resonance at fundamental frequency on the DC side. The rectifier DC terminals are chosen as the point where the system impedances are added. Two examples consider constant control gains at the rectifier, with slightly negative and positive damping factors. A third case is the where a high-pass filter is placed in parallel with the PI control path, to increase the damping factor at composite resonance, without affecting transient response at lower frequencies. Figure 9.38 shows resistive and reactive components of the series composite resonance circuit for cases 1 and 3 [16].

9.15   RESONANCE IN TRANSMISSION LINES

The transmission lines have natural resonance frequencies determined by their length. The input impedance can be zero (series resonance) or very large (parallel resonance) at the natural resonant frequencies. It may happen that the series resonant frequency is close to the dominant harmonic generated by converters, harmonic resonance can occur, and this can lead to severe telephone interference.

The solutions to control this situation are not so straightforward, for example, a series blocking filter and shunt filters can be applied or the length of the line can be changed. The discussions are continued in Chapter 12 (also see References [17,18]).

9.16   ZERO SEQUENCE RESONANCE

A system configuration for possible zero sequence resonance is shown in Fig. 9.39. If the generator neutrals are grounded through reactors or generators are connected in step-up configuration *with wye-connected windings on the generator side* and these wye-connected windings are solidly grounded or grounded through neutral reactors,

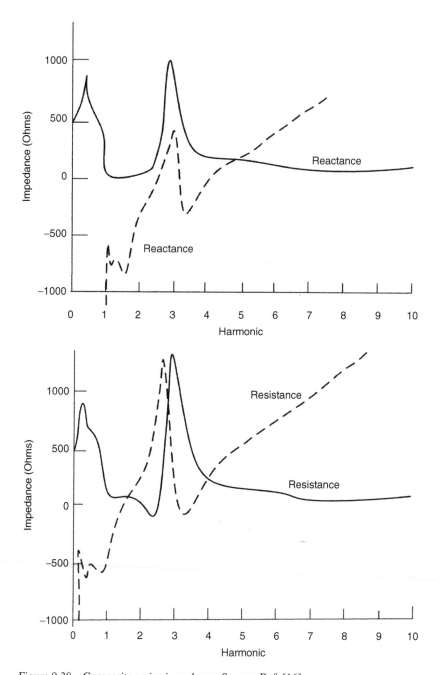

Figure 9.38   Composite series impedance. Source: Ref. [16].

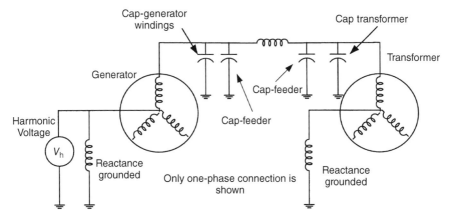

Figure 9.39   A circuit for possible zero sequence resonance.

zero sequence harmonic resonance can occur. First, it may be remarked that as per industrial grounding practice in the United States, generators and transformer neutrals are rarely grounded through reactors. The generator zero sequence harmonic voltages are discussed in Chapter 7. These voltages can act similar to harmonic voltage sources and are connected in series combination of generator, transformer, feeder and neutral grounding reactances, and a capacitive reactance (generator windings, transformer interwinding, and winding and bushing capacitances). If these two happen to be of same magnitude at a harmonic frequency, a large harmonic current will flow in the loop. It can cause unusual problems of step and touch voltage, and false operation of ground fault relays. Note that commonly, the generators are connected through step-up transformers with generator-side windings in delta connection, which breaks the resonant loop.

9.17   FACTORS AFFECTING HARMONIC RESONANCE

Some factors impacting harmonic resonance are the following:

- Synchronous and asynchronous machines and loads in the power system will absorb some of the generated harmonics and change the resonance points. Their correct modeling is an important factor.

- The harmonic impedance of the utility source must be ascertained and accounted for. It is not merely given by the three-phase, short-circuit current, when harmonic sources are present.

- The shunt power capacitors are not recommended to be applied in the presence of load-generated harmonics. These must be turned into harmonic filters after a careful study and applied at an appropriate location in the power system.

- Secondary resonance can occur when the shunt capacitors are applied at multi-voltage level in a distribution system. This is an important consideration. When a capacitor bank on the high side of distribution is switched, overvoltages of the order of four to five times can occur on the capacitor bank, which is in service at a lower voltage.

- The load resistance plays an important role in the system resonance (Fig. 9.5(b)). The impedance modulus and sharpness of the tuning of the ST filters vary with resistance.

- The motor loads should be appropriately modeled (Chapter 14). These appear primarily inductive at harmonic frequencies.

- The presence of single-phase loads must be considered.

- The harmonic mitigation and passive filter designs should be properly applied. Application of single-tuned or band-pass filter does not eliminate harmonic resonance, but merely shifts the resonant frequency. A proper choice of passive filter type has to be exercised (Chapter 15).

- Harmonic analysis for transmission systems requires rigorous modeling. Also the transmission systems undergo changes and the limitation of the computer models and practical conditions apply (Chapter 14).

- Resonant conditions may not be experienced all the time. A resonant condition can vanish with a system change and vice versa.

- Online measurements over a period of time are required to capture a resonant condition.

- Subharmonic resonance is discussed in Chapter 5.

REFERENCES

1. J.O. Bird. Electrical Circuit Theory and Technology, Butterworth Heinmann, Oxford, UK, 1997.
2. R.M. Foster. "A reactance theorem," Bell Systems Technical Journal, pp. 259–267, April 1924.
3. W.R. LePage and S. Sealy. General Network Analysis, McGraw-Hill, New York, 1952.
4. M.B. Reed. Alternating Current Circuits, Harper, New York, 1948.
5. IEEE P519.1/D9a. IEEE Guide for Applying Harmonic Limits on Power Systems, Unapproved Draft, 2004.
6. P.C. Buddingh. "Even harmonic resonance-an unusual problem," IEEE Transactions of Industry Applications, vol. 39, no. 4, pp. 1181–1186, 2003.
7. G. Lemieux. "Power system harmonic resonance-a documented case," IEEE Transactions on Industry Applications, vol. 26, no. 3, pp. 483–488, 1990.
8. J. Zaborszky and J.W. Rittenhouse. "Fundamental aspects of some switching overvoltages in power systems," AIEE Transactions on Power and Systems, vol. PAS-81, issue 3, pp. 822–830, 1962.
9. A. Kalyuzhny, S. Zissu and D. Shein. "Analytical study of voltage magnification transients due to capacitor switching," IEEE Transactions on Power Delivery, vol. 24, no. 2, pp. 797–805, 2009.
10. J.C. Das. "Analysis and control of large–shunt-capacitor-bank switching transients," IEEE Transactions on Industry Applications, vol. 41, no. 6, pp. 1444–1451, 2005.
11. H.B. Margolis, J.D. Phelps, A.A. Carlomagno and A.J. McElroy, "Experience with part-winding resonance in EHV autotransformers; diagnosis and corrective measures," IEEE Transactions on Power Apparatus and Systems, vol. PAS-94, no. 4, pp. 1294–1300, 1975.
12. A.J. McElroy. "On significance of recent transformer failures involving winding resonance," IEEE Transactions on Power Apparatus and Systems, vol. PAS-94, no. 4, pp. 1301–1316, 1975.
13. P.A. Abetti. "Transformer models for determination of transient voltages," AIEEE Transactions, vol. 72, pp. 468-480, 1953.

14. S.R. Naidu and R.H. Lasseter. "A study of composite resonance in AC/DC converters," IEEE Transactions on Power Delivery, vol. 18, no. 3, pp. 1060–1065, 2003.

15. M. Szechtman, T. Weiss and C.V. Thio. "First benchmark model for HVDC control studies," Electra, vol. 135, pp. 55–75, 1991.

16. A. R. Wood, J. Arrillaga, "Composite resonance; a circuit approach to the waveform distortion dynamics of an HVDC converter," IEEE Transactions on Power Delivery, vol. 10, no. 4, pp. 1882–1888, 1995

17. T.E. Shea. Transmission Networks and Wave Filters, Nostrand, New York 1929.

18. L. Wanhammer. Analog Filters using MATLAB, Springer, New York, 2009.

CHAPTER 10

HARMONIC DISTORTION LIMITS ACCORDING TO STANDARDS

10.1 STANDARDS FOR LIMITATION OF HARMONICS

The deleterious effects of harmonics and the varied nature of harmonics and interharmonics generation have been described in the previous chapters. Many countries have enacted their standards for harmonic limitations. IEC Standard Series 61000 provides internationally accepted information for the control of harmonics and interharmonics. This consists of number of standards, see Table 10.1. EN standards represent European Norms approved by CENELEC [1]. The European standardization bodies are the following:

CEN, Comité Europeén de Normalization

CENLEC, Comité Europeén de Normalization Electro-technique

ETSI, European Telecommunication Standard Institute.

These organizations establish standards for all member states. EN standards are published in EU official journal. The CE markings and labeling are somewhat like UL labeling in the United States. The standards establish emission requirements such as harmonics, voltage fluctuations, radio frequency (RF) disturbance, and also immunity requirements.

10.1.1 IEC Standards

IEC Series 61000 is a vast collection of standards running into many parts. For example:

- 61000-1 has 6, 1-1, 1-2, 1-3, 1-4, 1-5, and 1-6
- 61000-2 has 14 parts
- 61000-3 has 12 parts
- 61000-4 has 12 parts and so on.

Power System Harmonics and Passive Filter Designs, First Edition. J.C. Das.
© 2015 The Institute of Electrical and Electronics Engineers, Inc. Published 2015 by John Wiley & Sons, Inc.

TABLE 10.1 Some Important IEC Series 61000 (Electromagnetic Compatibility) Standards

Standard Number	Description	Year of Publication
1-1, Ed. 1.0	General – Section 1: Applications and interpretations of fundamental definitions and terms	1992
1-4, Ed. 1.0	General: Historical rationale for the limitation of power frequency conducted harmonic current emissions from the equipment in the frequency range up to 2 kHz	2005
1-6, Ed. 1.0	General: Guide to the assessment of measurement uncertainty	2012
2-1, Ed. 1	Part 2, Environment, Section 1: Description of the environment. Electromagnetic environment for low frequency conducted disturbances and signaling in public power supply systems	1990
2-2, Ed. 2	Part 2-2, Environment, Section 1: Description of the environment. Compatibility levels for low frequency conducted disturbances and signaling in public low-voltage power supply systems	2002
2-12, Ed. 1.0	Part 2-2, Environment, Section 1: Description of the environment. Compatibility levels for low frequency conducted disturbances and signaling in public medium-voltage power supply systems	2003
3-2, Ed. 3.2	Part 3-2: Limits for harmonic current emissions (equipment input current less than or equal to 16 A per phase)	2009
3-4, Ed. 1.0	Part 3-4: Limits for harmonic current emissions in low-voltage power supply systems for equipment with rated current > 16 A	1998
3-6, Ed. 2.0	Part 3-6: Limits – Assessment of emission limits for the connection of distorting installations to MV, HV and EHV power systems	2008
3-7, Ed. 2.0	Part 3-7: Limits – Assessment of emission limits for the connection of fluctuating installations to MV, HV and EHV power systems	2008
3-12, Ed. 2.0	Part 3-12: Limits for harmonic currents produced by equipment connected to public low-voltage systems with input currents > 16 A and less than or equal to 75 A	2011
3-14, Ed. 1.0	Part 3-14: Assessment of emission limits for harmonics, interharmonics, voltage fluctuations and unbalance for the connections of disturbing installations to LV power systems	2011
4-7, Ed. 2.1	Part 4-7: Testing and Measurement Techniques – General guide on harmonics and interharmonics measurements and instrumentation for power supply systems and equipment connected thereto	2009
4-13, Ed. 1.1	Part 4-13: Testing and Measurement Techniques: Harmonics and interharmonics including mains signaling at ac power port, low-frequency immunity tests	2009
4-15, Ed. 2	Part 4-15: Testing and Measurement Techniques: Flickermeter – Functional and design specifications	2010

There are approximately 184 publications for series 61000. Table 10.1 details some important standards from harmonic emission point of view.

10.1.2   IEEE Standard 519

In North America, the harmonic limits described in IEEE 519 [2] prevail. Also these limits are accepted in other countries too. We discussed planning and compatibility levels in Chapter 5, IEC concept. In the various studies and examples contained in this book, the limits specified in IEEE 519 are followed without any further lowering of these limits, irrespective of the power system under considerations and the study. Note that IEEE 519 indirectly allows higher current distortion limits from large consumers, as the short-circuit levels in their systems will, *generally*, be higher. Although this may not always be true.

10.2   IEEE 519 HARMONIC CURRENT AND VOLTAGE LIMITS

The standard provides recommended harmonic indices:

- Depth of notches, total notch area, and distortion of bus voltage distorted by commutation notches, low-voltage systems
- Individual and total current distortion
- Individual and total voltage distortion.

The standard realizes that the harmonic effects differ substantially depending on the characteristics of equipment being affected, and the severity of harmonic effects cannot be perfectly correlated to a few simple indices. Harmonic characteristics of utility circuit seen from point of common coupling (PCC) are not accurately known. Good engineering judgments are required on a case-by-case basis, and the recommendations in standards *in no way override such judgment.*

Also, it is acknowledged that strict adherence to harmonic limits will not always prevent problems, particularly when the limits are approached.

The harmonic indices, current, and voltage are defined at the PCC [2]. The concept of PCC is explained further in Section 10.3. The following points are of importance when interpreting the harmonic limits specified in IEEE 519 [2].

- High-voltage DC (HVDC) systems and static var compensators (SVCs) owned and operated by the utility are excluded from the definition of the PCC. Harmonic measurements are recommended at PCC. It is assumed that the system is characterized by the short-circuit impedance, and that the effect of capacitors is neglected. The recommended current distortion limits are concerned with total demand distortion (TDD), which is defined as the total root sum square harmonic current distortion as a percentage of maximum demand load current (15 or 30 minutes demand; see Eq. (1.19)).
- The limits for the current distortion that a consumer must adhere to are shown in Tables 10.2 and 10.3. The ratio I_{sc}/I_L is the ratio of the short-circuit current

TABLE 10.2   Current Distortion Limits for General Distribution Systems (120 V–69 kV)

	Maximum Harmonic Current Distortion in % of Fundamental Harmonic Order (Odd Harmonics)[a]					
I_{sc}/I_L [b]	<11	$11 \leq h < 17$	$17 \leq h < 23$	$23 \leq h < 35$	$35 \leq h$	TDD
< 20[c]	4.0	2.0	1.5	0.6	0.3	5.0
20–50	7.0	3.5	2.5	1.0	0.5	8.0
50–100	10.0	4.5	4.0	1.5	0.7	12.0
100–1000	12.0	5.5	5.0	2.0	1.0	15.0
>1000	15.0	7.0	6.0	2.5	1.4	20.0

Current distortions that occur in a DC offset, for example, half-wave converters are not allowed.

For general subtransmission systems (69,001 V through 161,000 V), the limits are 50% of the limits shown in Table 10.2.

[a]Even harmonics are limited to 25% of the odd harmonic limits above.

[b]I_{lc} = maximum short-circuit current at PCC; I_L = maximum load current (fundamental frequency) at PCC.

[c]All power generation equipment is limited to these values of current distortion regardless of I_{sc}/I_L.

Source: Ref. [2].

TABLE 10.3   Current Distortion Limits for General Transmission Systems (> 161 kV)

	Dispersed Generation and Cogeneration Maximum Harmonic Current Distortion in % of Fundamental Harmonic Order (Odd Harmonics)[a]					
I_{sc}/I_L [b]	<11	$11 \leq h < 17$	$17 \leq h < 23$	$23 \leq h < 35$	$35 \leq h$	TDD
<50[c]	2.0	1.0	0.75	0.3	0.15	2.5
>50	3.0	1.5	1.15	0.45	0.22	3.75

Current distortions that occur in a DC offset, for example, half-wave converters are not allowed.

[a]Even harmonics are limited to 25% of the odd harmonic limits above.

[b]I_{sc} = Maximum short-circuit current at PCC; I_L = maximum load current (fundamental frequency) at PCC.

[c]All power generation equipment is limited to these values of current distortion regardless of I_{sc}/I_L.

Source: Ref. [2].

available at the PCC to the maximum fundamental frequency current and is calculated on the basis of the average maximum demand for the preceding 12 months. As the size of user load decreases with respect to the size of the system (size of the system is rather vague, here meant by the stiffness of the power system given by the three-phase symmetrical short-circuit current at the PCC), the percentage of the harmonic current that the user is allowed to inject into the utility system increases.

- To calculate the ratio I_{sc}/I_L, it is implied that a three-phase, short-circuit current calculations should be carried out followed by the load flow calculation. Here, note that the intention is not to calculate the short-circuit duties on the switching devices, such as circuit breakers and fuses. Usually, the short-circuit calculation procedures detailed in ANSI/IEEE standards are followed without postmultiplying factors for the circuit breaker duties. It is not only establishing the current demand at PCC, but also fundamental frequency load flow should establish that the operating voltages at all power system buses are within acceptable limits. This will require (1) optimum tap settings on transformers, (2) providing reactive power compensating equipment, (3) automatic voltage regulators, (4) under load tap changing transformers (ULTCs), and so on. The short-circuit calculations and fundamental frequency load flow calculations are not discussed in this book. References for short circuit and load flow are separately listed at the end of this chapter.

- Tables 10.2 and 10.3 are applicable to six-pulse rectifiers. For higher pulse numbers, the limits of the characteristic harmonics are increased by a factor of $\sqrt{P/6}$, *provided* that the amplitudes of noncharacteristic harmonics are less than 25% of the limits specified in the tables.

- An overriding article is provided that the transformer connecting the user to the utility system should not be subjected to harmonic currents in excess of 5% of the transformer-rated current. When this requirement is not met, a higher rated transformer should be provided. The derating of transformers serving nonlinear loads has been discussed in Chapter 8 with solved examples of calculations.

- The injected harmonic currents may create resonance with the utility's system, and a consumer must insure that harmful series and parallel resonances are not occurring. The utility's source impedance may have a number of resonant frequencies, see Chapter 14, and it is necessary to model the utility's system in greater detail for harmonic analysis calculations.

- The limit of harmonic current injection does not limit a user's choice of converters or selection of harmonic producing equipment technology. Neither does it lay down limits on the harmonic emission from the nonlinear or electronic equipment. It is left to the user how he would adhere to the limits of harmonic current injection, whether by choice of an alternative technology, use of passive or active filters, or any other harmonic mitigating device. This is in contrast to the IEC, which has laid out the maximum emission limits from the equipment, for example, IEC 61000-3-2 [3].

- The current distortion limits assume that there will be some diversity between harmonic currents injected by different consumers (see Example 6.1). This diversity can be with respect to time, phase angle of harmonics, and harmonic spectra. Table 10.4 from this standard shows the basis of current limits.

- The objectives of current limits are to limit the maximum individual frequency voltage harmonic to 3%.

- An important qualification is that TDD is for maximum demand load current, 15 or 30 minutes demand. The limits specified are to be used for the worst-case condition for normal operation, conditions lasting for more than 1 hour. For shorter periods during startups or unusual conditions, the limits may be exceeded by 50%.

TABLE 10.4 Basis for Harmonic Current Limits

SCR at PCC	Maximum Individual Frequency Voltage Harmonic	Related Assumption
10	2.5–3.0%	Dedicated system
20	2.0–2.5%	1–2 large customers
50	1.0–1.5%	A few relatively large customers
100	0.5–1.0%	5–20 medium-size customers
1000	0.05–0.10%	Many small customers

Source: Ref. [2].

- Ideally, the harmonic distortion caused by a single consumer should be limited to an acceptable level *at any point in the power system*. And the entire power system should be operated without substantial distortion anywhere in the system. The objectives of the current limits are to limit the maximum individual frequency, voltage harmonic to 3% of the fundamental, and the voltage THD to 5% for systems without a major parallel resonance at one of the injected harmonic frequencies.

- The idea of ensuring that the harmonic limits are met throughout a system may not be achieved in practice. The intent of the IEEE limits applies to the PCC only. Downstream, the user equipment may be able to tolerate higher harmonic distortion limits and operate satisfactorily. IEEE Draft Guide [4] for applying harmonic limits on power systems is based on IEEE 519.

10.3 POINT OF COMMON COUPLING (PCC)

PCC is the point of metering of power supply from the utility to a consumer, or any point as long as both the consumer and utility can either access the point for direct measurement of the harmonic indices meaningful to both or can estimate the harmonic indices at the point of interference (POI) through mutually agreeable methods. *The PCC is also the point where another consumer can be served from the same system.* The PCC could be located at the primary or secondary of the supply transformer, depending whether or not multiple customers are supplied from the transformer. A utility may supply more than one consumer from the secondary of a transformer.

As stated in Chapter 1, the delta transformer windings are a sink to the zero sequence currents and third harmonics. Thus, if the primary of the transformer is declared as the PCC, these harmonics will not enter into the evaluations.

The utilities may accept higher harmonic limits from the customers than those specified in the standards. Consider that in a consumer's installation, all the TDD limits, TDD total, and on individual harmonics are met, except that 11th harmonic is 5% higher than IEEE 519 limits. This situation may be acceptable to the utility.

10.4 APPLYING IEEE 519 HARMONIC DISTORTION LIMITS

Harmonics from small consumers with a limited amount of disturbing load may not require detailed analysis. A technique developed in IEC Standard 6100-3-6 [5] uses weighting factors.

The weighted disturbing power is calculated as follows:

$$S_{Dw} = \sum_i (S_{Di} \times W_i) \qquad (10.1)$$

For all the disturbing loads in the facility, where S_{Di} is the power rating of an individual disturbing load in kilovolt-ampere and W_i is the weighting factor.

If

$$S_{Dw}/S_{sc} < 0.1\% \qquad (10.2)$$

where S_{sc} is the short-circuit capacity at the PCC in kilovolt-ampere and Eq. (10.2) is satisfied, then automatic acceptance occurs, without detailed analysis. This criterion is proposed in IEC Standard 61000-3-6. Weighting factors are given in Table 10.5.

Even a simpler evaluation is advocated in IEC 6100-3-6 [5]. If the majority of the load is one of the types in first three rows of Table 10.5, a detailed analysis is only necessary if the nonlinear load is more than 5% of the facility load.

For the other types of loads with current distortions less than 50%, the nonlinear load can be as high as 10%. The waveforms of the disturbing loads are shown in Fig. 10.1.

A harmonic analysis study is required if the customer is planning to provide capacitors for the power factor improvement.

TABLE 10.5 Weighting Factors

Type of Load	Current Distortion	Weighting Factor
Single-phase power supply	80%, high third	2.5
Semiconverters, see Chapter 1	High second, third, and fourth at partial load	2.5
Six-pulse converter, capacitive smoothing, no series reactor	80%	2.0
Six-pulse converter, capacitive smoothing, series inductance >3%, or DC drive	40%	1.0
Six-pulse converter, with large reactor for current smoothing	28%	0.8
12-pulse converter	15%	0.5
AC voltage regulator	Varies with firing angle	0.7
Fluorescent lighting	20%	0.5

Source: Ref. [4].

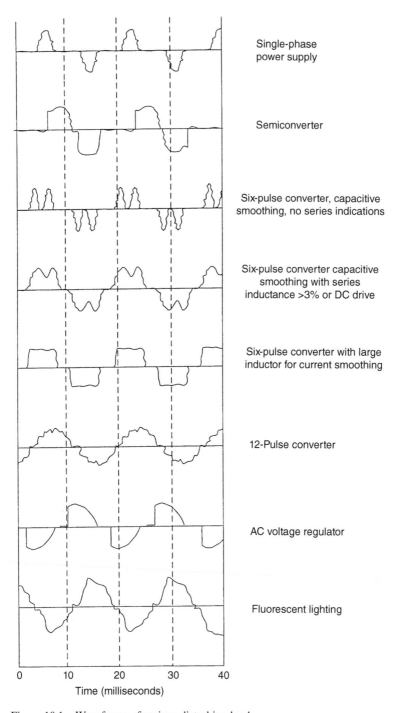

Single-phase
power supply

Semiconverter

Six-pulse converter, capacitive
smoothing, no series indications

Six-pulse converter capacitive
smoothing with series
inductance >3% or DC drive

Six-pulse converter with large
inductor for current smoothing

12-Pulse converter

AC voltage regulator

Fluorescent lighting

Figure 10.1   Waveforms of various disturbing loads.

As an application of this table, for a 13.8-kV system, with short-circuit level of 1000 MVA at the PCC, a 2-MVA, 12-pulse load is acceptable.

10.5   TIME VARYING CHARACTERISTICS OF HARMONICS

We alluded to time variation of the harmonic distortion in Fig. 9.24, corresponding to load pattern of a consumer. Also probabilistic concepts are discussed in Chapter 7. IEEE 519 provides for probabilistic applications of harmonic distortion limits (Fig. 10.2). The steady-state harmonic levels are compared to measured level that is not exceeded 95% of the time (95% probability point). This is consistent with the compatibility level of IEC 61000-2-2 [6]. IEEE 519 also states that higher harmonic limits up to 150% are acceptable for short period of times, 1 hour per day, that is, approximately 4% of the time. This limitation is consistent with design limits of 95% probability level not being exceeded. The higher limit can be compared with the measured harmonic level that is not exceeded 99% of the time (99% probability level).

It is rather rare that the harmonic limits will behave similar to a rectangular pulse of long duration; practically, this may consist of a number of pulses of shorter duration. These two profiles will have different impact on motor or transformer heating (Fig. 10.3).

This requires that the harmonics are measured over a considerable period of time (Chapter 7). The three factors are the following:

- The total duration of harmonic bursts is the summation of all the time intervals in which the measured levels exceed a particular level.

- The maximum duration of a single burst is the longest time interval in which the measured levels exceed a particular level.

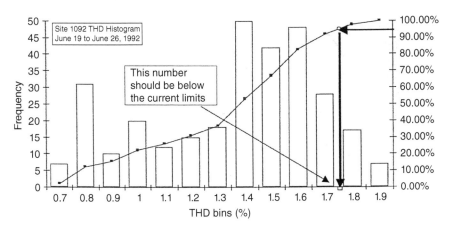

Figure 10.2   Probability plot illustrating variable nature of harmonic levels. Source: Ref. [4].

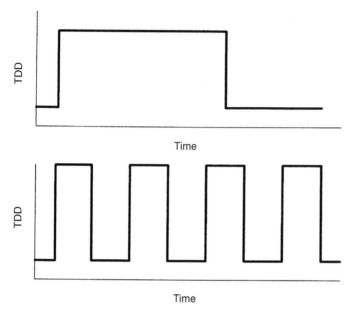

Figure 10.3   Two different harmonic emission trends, which will have a different heating impact on equipment.

- The measurements should account for presence of capacitors, harmonic filters, outage of a filter bank, effect of alternate sources from the utility with different short-circuit levels, and so on.

Both the parameters in bulleted items 1 and 2 above are considered. Figure 10.4 shows a time duration plot versus TDD. The 6% TDD is exceeded for 8 minutes. The probability of a specified harmonic being exceeded can be evaluated with histograms and cumulative probability density and distribution plots (Chapter 7).

As the harmonic levels are continuously changing, a single sample cannot be used to evaluate compliance with harmonic limits. Harmonic producing loads may be swathed in or out, or these may operate for a specific duration during the day depending on the processes.

Table 10.6 provides magnitude/duration limits that can be used to evaluate time-varying nature of harmonics. It is understood that the actual variations may not exactly follow the patterns shown in this table.

A flow chart of recommended procedure for harmonic limit evaluation from Ref. [4] is shown in Fig. 10.5.

10.6   IEC HARMONIC CURRENT EMISSION LIMITS

The IEC Standards 61000-3-2 and 61000-2-2 [3,6] define limits on harmonic current injected into a public distribution network by nonlinear appliances with an input current less than or equal to 16 A (at 220 V). This classifies such appliances into four

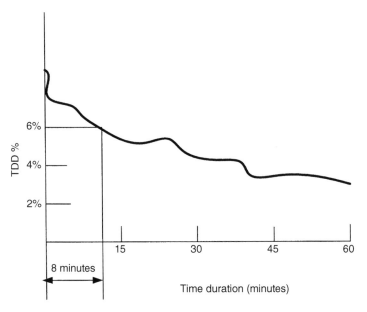

Figure 10.4   Plot of TDD versus duration.

TABLE 10.6   Short-Duration Harmonic Limits Based on a 24-Hour Measurement Period

Acceptable Harmonic Distortion Level (Individual or TDD)	Maximum Duration of Single Harmonic Burst $T_{maximum}$	Total Duration of All Harmonic Bursts T_{Total}
3.0× (design limits)	1 second $< T_{maximum} <$ 5 seconds	15 seconds $< T_{Total} <$ 60 seconds
2.0× (design limits)	5 seconds $< T_{maximum} <$ 10 minutes	60 seconds $< T_{Total} <$ 40 minutes
1.5× (design limits)	10 minutes $< T_{maximum} <$ 30 minutes	40 minutes $< T_{Total} <$ 120 minutes
1.0× (design limits)	30 minutes $< T_{maximum}$	120 minutes $< T_{Total}$

Source: Ref. [4].

classes (A–D). For each class, harmonic current emission limits are established up to the 39th harmonic. The classes are the following:

Class A: General-purpose loads. These are balanced three-phase loads, the line currents differing by no more 20%, and all other equipment except as described in the following categories.

Class B: Portable appliances and tools.

Class C: Lighting equipment, with the exception of light dimmers (class A) and self-ballasted lamps having class D wave shapes.

Class D: Appliances with input current with an assigned special wave shape and an active power input $P < 600\,\text{W}$ measured according to the method illustrated in the standard. This special wave shape is shown in Fig. 10.6.

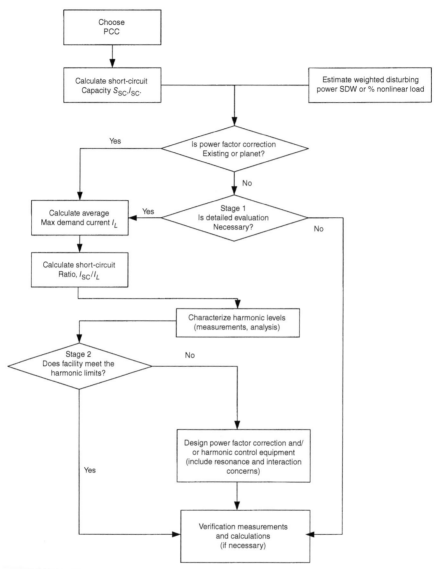

Figure 10.5 Flow chart showing general procedure for evaluating harmonic limits. Source: Ref. [4].

The standard lays down methods for testing individual appliance harmonic emissions in the regulated frequency range. The harmonic limits were developed to limit the impact of these loads on the overall system. A system impedance of $R = 0.4$ ohms and $X = 0.25$ ohms (50 Hz) is used to evaluate the impact of these loads on local voltage distortion levels.

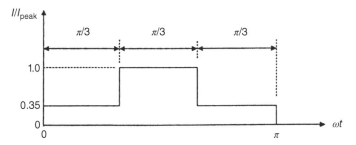

Figure 10.6   IEC class D equipment wave shape. Each half-cycle of input current is within the envelope for at least 95% of the time.

TABLE 10.7   IEC 61000-3-2 Limits for Class A Equipment (General-Purpose Loads)

Harmonic Order	Odd Harmonics Maximum Permissible Current in Amperes	Harmonic Order	Even Harmonics Maximum Permissible Current in Amperes
3	2.3	2	1.08
5	1.14	4	0.43
7	0.77	6	0.3
9	0.4	8–40	$0.23(8/n)$
11	0.33		
13	0.21		
15–39	$0.15(15/n)$		

TABLE 10.8   IEC 61000-3-2 Limits for Class C Equipment (Lighting)

Harmonic order	Maximum Value Expressed as a Percentage of the Fundamental Input Current of Luminaries
2	2.0%
3	$30\% \times PF$
5	10%
7	7%
9	5%
11–39	3%

Tables 10.7 through 10.9 show the limits for class A, C, and D equipments. Limits for class B equipment are 1.5 times of that given for class A equipment. The electronic equipment falls in class D. Figure 10.7 shows these limits expressed as a percentage of full-load current for different load power levels [4].

TABLE 10.9   Harmonic Emission Limits for Class D Appliances

Harmonic Order	Maximum Admissible Harmonic Current (mA/W) 75 W $< P <$ 600 W	Maximum Admissible Harmonic Current (A) $P >$ 600 W
3	3.4	2.3
5	1.9	1.14
7	1.0	0.77
9	0.5	0.4
11	0.35	0.33
$13 \leq n \leq 39$	$3.85/n$	$0.15 \, (15/n)$

Source: Ref. [3].

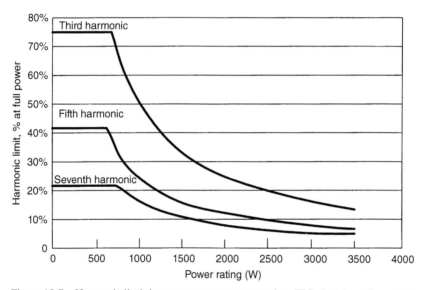

Figure 10.7   Harmonic limit in percent versus power rating, IEC class D equipment. Source: Ref. [4].

10.7   VOLTAGE QUALITY

10.7.1   IEEE 519

Although a user can inject only a certain amount of harmonic current into the utility system as discussed in the previous sections, the utilities and power producers must meet requirements of a certain voltage quality to the consumers. The recommended voltage distortion limits are given in Table 10.10 [2]. The index used is THD (total harmonic voltage distortion) as a percentage of nominal fundamental frequency voltage. The limits are system design values for *worst case* for normal operation, conditions lasting for more than 1 hour. For shorter periods, during startups or unusual

TABLE 10.10   Harmonic Voltage Limits for Power Producers (Public Utilities or Cogenerators)

	Harmonic Distortion in % at PCC		
	<69kV	>69–161 kV	>161 kV
Maximum for individual harmonic	3.0	1.5	1.0
Total harmonic distortion (THD)%	5.0	2.5	1.5

High-voltage systems can have up to 2.0% THD where the cause is an HVDC terminal that will attenuate by the time it is tapped for a user.

Source: Ref. [2].

conditions, the limits can be exceeded by 50%. If the limits are exceeded, harmonic mitigation through use of filters or stiffening of the system through parallel feeders is recommended.

The limits of distortion on most medium-voltage systems are 5%. The consumer may accept a higher voltage distortion depending on the sensitivity of his loads to voltage distortions. Note that the IEEE 519 does not lay down any limits of voltage distortion at the PCC *which a consumer must adhere to.* The utility company has to maintain a certain voltage quality at the PCC. Example of a filter design in Chapter 16 shows that for weak power systems, the voltage distortions at PCC are a bigger problem. The consumer nonlinear load can seriously distort the voltage at PCC.

10.7.2   IEC Voltage Distortion Limits

IEC 6100-2-2 [6] provides compatibility levels for various types of power quality characteristics. The compatibility level for the distortion on low-voltage systems is 8%.

Table 10.11 gives the voltage harmonic distortion limits in public low-voltage networks (IEC-61000-2-2). These limits are the same as in IEC 61000-2-4 for class 2. For class 3, as per this standard, the limits are shown in Table 10.12.

The THD (voltage) is ≤8% for class 2 and ≤10% for class 3.

Class 2: PCC and IPC (in-plant point of coupling) in industrial environment, in general

Class 3: Applies to IPCs in industrial environment

10.7.3   Limits on Interharmonics

IEC 61000-4-15 [7] has established a method of measurement of harmonics and interharmonics utilizing a 10- or 12-cycle window for 50 and 60 Hz systems. This results in a spectrum with 5 Hz resolution. The 5 Hz bins are combined to produce various groupings and components for which limits and guidelines can be referenced. The IEC limits the interharmonic voltage distortion to 0.2% for the frequency range from DC to 2 kHz, see also Chapter 7.

TABLE 10.11 IEC 61200-2-2 Harmonic Voltage Limits in Public Low-Voltage Network. Also IEC 61000-2-4. Harmonic Voltage Limits in Industrial Plants, Class 2

Odd Harmonics		Even Harmonics		Triplen Harmonics	
h	$\%Vh$	h	$\%Vh$	h	$\%Vh$
5	6	2	2	3	5
7	5	4	1	9	1.5
11	3.5	6	0.5	15	0.3
13	3	8	0.5	≥ 21	0.2
17	2	10	0.5		
19	1.5	≥ 12	0.2		
23	1.5				
25	1.5				
≥ 29	x				

$x = 0.2 + 12.5/h$, for $h = 29.31, 35$, and 37; $V_h = 0.63, 0.60\%, 0.56\%$, and 0.54%.

TABLE 10.12 IEC 61200-2-4 Harmonic Voltage Limits, Class 3

Odd Harmonics		Even Harmonics		Triplen Harmonics	
h	$\%Vh$	h	$\%Vh$	h	$\%Vh$
5	8	2	3	3	6
7	7	4	1.5	9	2.5
11	5	≥ 6	1	15	2
13	4.5			21	1.75
17	4			≥ 27	1
19	4				
23	3.5				
25	3.5				
≥ 29	y				

$y = \sqrt[5]{11/h}$

For $h = 29, 31, 35$, and 37; $V_h = 3.1, 3.0\%, 2.8\%$, and 2.7%.

There are yet no limits of interharmonics assigned in IEEE 519-1992 [2]. The following recommendations are from Ref. [8]. *Note that these are not the recommendations of the Working group of IEEE 519 standard, but the ideas of a single author.*

- Limit of 0.2% for frequencies less than 140 Hz to address flicker of incandescent lamps and fluorescent lamps.
- Limit individual interharmonics component distortion to less than 1% above 140 Hz up to some frequency, yet to be determined, to protect low-frequency PLC (power line carrier), address sensitivity to light flicker within 8 Hz of harmonic frequencies, and account for resonances created by harmonic filters.

- For higher frequencies, limit interharmonics voltage component and total distortion to some percentage related to the proposed frequency-dependent harmonic voltage limits to protect high-frequency PLC and filter resonances. Alternatively, define a linear limit curve with increasing slope to recognize the reduced impact on light flicker with increasing frequency.

It may be necessary to impose severe restrictions on interharmonics if there are concerns of torsional interactions at nearby generating plants. In other cases, interharmonics need not be treated differently from the integer harmonics.

Example 10.1: The harmonic current and voltage spectrum of a 12-pulse LCI inverter, operating at $\alpha = 15°$, is presented in first two columns of Table 10.13. The demand current is 1200 A, which is also the inverter maximum operating current. The available short-circuit current at the PCC is 12 kA. Calculate the distortion limits. Are these exceeded?

Table 10.13 is extended to show permissible limits of the TDD at each of the harmonics, and individual and total permissible current distortions are calculated. From Table 10.2 for $I_{sc}/I_L < 20$ (actual $I_{sc}/I_L = 10$). The noncharacteristic harmonics are reduced by 25%, and the characteristic harmonics are multiplied by a factor of the square root of p over 6, that is, $\sqrt{2}$.

TABLE 10.13 Calculations of TDD, 12-Pulse Converter, Example 10.1

h	$I_h(A)$	Harmonic Distortion (%)	IEEE Limits (%)	$V_h(V)$
5	24.32	2.027	1.0	20.86
7	13.43	1.119	1.0	16.13
11	65.42	5.452	2.82	123.46
13	39.69	3.331	2.82	88.52
17	1.87	0.156	0.375	5.45
19	0.97	0.081	0.375	3.16
23	8.45	0.704	0.846	33.34
25	8.54	0.711	0.846	36.62
29	0.98	0.081	0.15	4.87
31	0.76	0.063	0.15	4.04
35	5.02	0.418	0.423	30.14
37	3.27	0.273	0.423	20.75
41	0.23	0.019	0.075	1.62
43	0.23	0.019	0.075	1.70
47	3.45	0.287	0.423	27.81
49	2.58	0.215	0.423	21.69
TDD		6.887%	5%	4.08%

Table 10.13 shows that current distortion limits on a number of harmonics are exceeded. Also, the TDD is greater than the permissible limits. The total THD (voltage) is below the 5% limit, but the distortion at the 11th harmonic is 3.32%, which exceeds the maximum permissible limit of 3% on an individual voltage harmonic.

In this example, the distortion limits exceed the permissible limits, though a 12-pulse converter is used, and the base load equals the converter load, that is, the entire load is nonlinear. Generally, the harmonic producing loads will be some percentage of the total load demand, and this will reduce TDD as it is calculated on the basis of total fundamental frequency demand current.

10.8 COMMUTATION NOTCHES

Commutation notches are shown in Fig. 10.8(a). As we examined in Chapter 4, commutation produces two primary notches per cycle plus four secondary notches of lesser magnitude due to notch reflection from other leg of the bridge. The full-wave

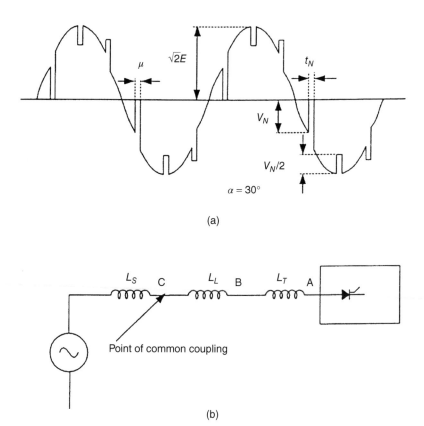

(a)

(b)

Figure 10.8 (a) Notching in current source converters with DC-link reactors and (b) the notch area varies at various points shown in this figure.

TABLE 10.14   Low-Voltage System Classification and Distortion Limits

	Special Applications	General Systems	Dedicated Systems
Notch depth	10%	20%	50%
THD (voltage)	3%	5%	10%
Notch area (A_n)	16,400	22,800	36,500

The notch area for voltages other than 480 V is given by multiplication by a factor $V/480$. Notch area is given in volts-microseconds at rated voltage and current. Special applications include hospitals and airports. A dedicated system is exclusively dedicated to the converter load.

Source: Ref. [2].

diode bridge with capacitor load operates in discontinues mode and does not produce notches. For low-voltage systems, the notch depth, the total notch area of the line-to-line voltage at the PCC, and THD should be limited as shown in Table 10.14. The notch area is given by

$$A_N = V_N t_N \tag{10.3}$$

where A_N is the notch area in volt-microseconds, V_N is the depth of the notch in volts, line-to-line (L-L) of the deeper notch in the group, and t_N is the width of the notch in microseconds.

Consider the equivalent circuit of Fig. 10.8(b) and define the following inductances:

- L_t is the inductance of the drive transformer
- L_L is the reactance of the feeder line
- L_s is the inductance of the source

The notch depth depends on where we look into the system. The primary voltage goes to zero at the converter terminals, point A in Fig. 10.8(b), and the depth of the notch is the maximum. At B, the depth of the notch in per unit of the notch depth at A is given by

$$\frac{L_s + L_L}{L_s + L_t + L_L} \tag{10.4}$$

If point C in Fig. 10.8(b) is defined as the PCC, then the depth of the notch at the PCC in per unit with respect to notch depth at the converter is

$$\frac{L_s}{L_L + L_s + L_t} \tag{10.5}$$

For actual values, multiply Eq. (10.4) or (10.5) with e, where e is the instantaneous voltage (L-L) prior to the notch. That is, $V_N = e$ multiplied by Eq. (10.4) or (10.5).

The impedances in the converter circuit are acting a sort of potential divider. The width of the notch is given by the expression:

$$t_N = \frac{2(L_L + L_t + L_s)I_d}{e} \tag{10.6}$$

The area of the notch at the converter terminals is given by

$$A_N = 2I_d(L_L + L_t + L_s)$$

A relationship between line notching and distortion factor is given by

$$V_h = \sqrt{\frac{2V_N^2 t_N + 4(V_N/2)^2 t_N}{1/f}} = \sqrt{3V_N^2 t_N f} \qquad (10.7)$$

where f is the power system fundamental frequency. If a factor ρ equal to the ratio of total inductance to common system inductance is written as

$$\rho = \frac{L_L + L_t + L_s}{L_L} \qquad (10.8)$$

then

$$V_{\mathrm{NMAX}} = \frac{\sqrt{2}E_L}{\rho} \qquad (10.9)$$

and

$$\mathrm{THD}_{\mathrm{MAX}} = 100\sqrt{\frac{3\sqrt{2}.10^{-6}A_N f}{\rho E_L}} \qquad (10.10)$$

In Eq. (5.7), the two deeper and four less deep notches per cycle, Fig. 10.8(a), are considered (also see [2]).

Example 10.2:   Consider a system configuration as shown in Fig. 10.9(a). It is required to calculate the notch depth and notch area at PCC.

The inductances throughout the system are calculated and are shown in Fig. 10.9(b). The source inductance (reflected at 480 V), based on the given short-circuit data, is 0.85 µH. It is a stiff system at 13.8 kV, and the source reactance is small.

A 1-MVA transformer reactance referred to the 480-V side, based on the X/R ratio, $X_t = 5.638\% = 0.1299$ ohms. This gives an inductance of 34.4 µH at 480 V. The feeder inductance is 13.8 µH. The notch depth at PCC as a percentage of depth at the converter $= (0.85 + 34.4)/(0.85 + 34.4 + 13.8) = 71.9\%$. Referring to Table 10.14, this exceeds the limits even for a dedicated system.

Consider that the converter is supplying a motor load of 500 hp at 460 V, and the DC current is continuous and equal to 735 A.

Then the DC current of the converter is $(735)/0.85 = 864.70$ A.

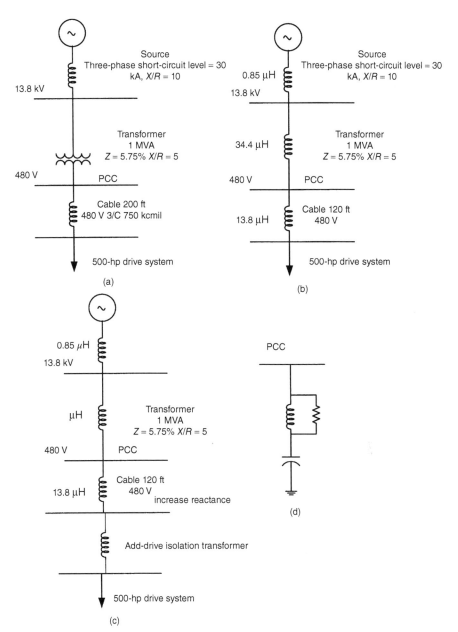

Figure 10.9   (a) A system configuration for calculation of notch area, (b) equivalent circuit showing reactance values, (c) circuit with addition of drive isolating transformer, and (d) use of a high-pass filter to reduce notch area.

TABLE 10.15 Harmonics for Calculations of Notching, CSI and VSI Converters, Example 10.2

Harmonic Order	CSI	VSI
1	78	78
5	16	45
7	10	28
11	4	6.3
13	3	5.7
17	2	4.3

Notch area at converter is equal to

$$(0.85 + 34.4 + 13.8) \times 864.70 = 84,826 \, \text{V} \, \mu\text{s} \text{ and notch area at PCC}$$

$$= 60,990 \, \text{V} \, \mu\text{s}$$

Notch width at converter and at PCC is the same:

$$= \frac{84,826}{\sqrt{2} \times 480}$$

Thus, notch width $= 125 \, \mu\text{s}$.

To calculate percentage distortion, the harmonic emission (truncated at 17th harmonic for this example) is given in Table 10.15. Note that this is in terms of DC output current. If X_L is the line impedance (at fundamental frequency), then the expression $I_{DC}[\sum (hI_h X_L)^2]^{1/2} = I_{DC} X_L [\sum hI_h^2]^{1/2}$ can be calculated.

Then for CSI

$$X_L I_{DC}[(5 \times 0.16)^2 + (7 \times 0.1)^2 + (11 \times 0.04)^2 + (13 \times 0.03)^2 + (17 \times 0.02)^2]^{1/2}$$

$$= 1.29 X_L I_{DC}$$

In terms of converter full load:

$$I_{DC} = 572LI_{conv}$$

Therefore, percentage of distortion at converter is

$$\frac{100(572)(0.85 + 34.4 + 13.8)864.70}{480} = 5.05\%$$

And at PCC

$$\frac{100(572)(0.85 + 34.4)864.70}{480} = 3.60\%$$

Using Eq. (10.10) maximum distortion $= 11.2\%$ at PCC

A similar calculation can be conducted for diode bridge with capacitor load; the distortion will be much higher.

The example shows that the calculated values are much above the permissible values. The standard requirements can be met if the cable inductance is increased or drive isolating transformer is provided. This means that in this example, the source-plus transformer inductance should be much lower than the cable-plus drive transformer inductance. In the earlier example, no drive isolation transformer is considered. However, note that the distortion at PCC increases with increasing reactance, and this may have to be reduced for controlling the distortion.

This example suggests that problems of notching can be mitigated by the following:

- Decreasing the source impedance behind the PCC bus.
- Increasing the impedance ahead of (on the load side) of the PCC bus. Isolation transformers and line reactors are commonly used and serve the same purpose (Fig. 10.9(c)).
- Providing second-order high-pass filters is another option, which can also provide reactive power compensation and improve the voltage profile on low-power factor loads (Chapter 8; see Fig. 10.9(d)). The high-pass filter will provide a low-impedance source of commutating current to reduce notching.

A case of distribution system oscillations due to notching is illustrated in Ref. [4]. A 25-kV bus, which serves drive system loads and is connected to 144-kV system, experiences oscillations close to 60th harmonic and failure of surge capacitors.

10.9 APPLYING LIMITS TO PRACTICAL POWER SYSTEMS

A manufacturer may specify that the harmonic emissions from his equipment meet IEEE 519 requirements, without additional filters or harmonic mitigation devices. Three factors need to be considered:

- The I_{sc}/I_r ratio to which the harmonic emission limits relate? If the limits are specified at a ratio higher than what is actual ratio in the power system, the IEEE limits at PCC may not be met.
- Are the specified limits applicable when the harmonic producing equipment is the only load in the system? The TDD is based on the total load in the system: linear and nonlinear. This mix will give variable results at PCC.
- In some cases where the I_s/I_r ratio is much lower than 20 (minimum at which the harmonic limits are specified in IEEE 519), higher distortion at PCC can occur, see case study in Chapter 16.

Assuming that the specified limits pertain to the actual I_{sc}/I_r ratio of the system and there are some other linear loads in the system, it may be assumed that distortion limits at PCC will be within IEEE limits, provided there is no amplification of harmonics in the power system under consideration. This means that there are no capacitors in the system. Yet, the capacitance of cables and OH lines cannot be ignored. Even if there are no capacitors in the system, resonance at higher frequencies can occur.

Thus, the assumption that no further study is required when the manufacturer confirms harmonic emission within IEEE limits has to be made with caution. Owing to much variations in electrical power systems, a harmonic analysis study is warranted with correct modeling taking into account the operation of the system.

REFERENCES

1. EN 50160. Voltage characteristics of electricity supplied by public distribution systems, Brussels, 1994.
2. IEEE Standard 519. IEEE recommended practice and requirements for harmonic control in electrical systems, 1992.
3. IEC 61000-3-2. See Table 10-1.
4. IEEE Draft P519.1/D9a. Guide for applying harmonic limits on power systems, 2004.
5. IEC 61000-3-6. See Table 10-1.
6. IEC 61000-2-2. See Table 10-1.
7. IEC 61000-4-15. See Table 10-1
8. E.W. Gunther, "Interharmonics recommended updates to IEEE 519," IEEE Power Engineering Society Summer Meeting, pp. 950–954, 2002.

References for Load Flow and Short-Circuit Calculations

9. J.C. Das. Power System Analysis-Short-Circuit Load Flow and Harmonics, Second Edition, CRC Press, Boca Raton, FL, 2011.
10. IEEE Standard 399. IEEE recommended practice for power system analysis, 1990.
11. ANSI/IEEE Standard C37.010. Application guide for AC high-voltage circuit breakers rated on a symmetrical current basis, 1999 (R-2005).
12. IEEE Standard 551 (Violet Book). IEEE recommended practice for calculating short-circuit currents in industrial and commercial power systems, 2006.
13. IEEE Standard C37.013. IEEE standard for AC high-voltage generator circuit breakers rated on a symmetrical current basis, 1997.
14. ANSI/IEEE. Standard C37.13. Standard for low-voltage AC power Circuit Breakers used in Enclosures, 2008.
15. W.F. Tinney and C.E. Hart. "Power flow solution by Newton's method," Transactions of IEEE, PAS 86, pp. 1449–1456, 1967.
16. B. Stott and O. Alsac. "Fast decoupled load flow," Transactions of IEEE, PAS 93, pp. 859–869, 1974.
17. N.M. Peterson and W.S. Meyer. "Automatic adjustment of transformer and phase-shifter taps in the Newton power flow," Transactions of IEEE, vol. 90, pp. 103–108, 1971.
18. M.S. Sachdev and T.K.P. Medicheria. "A second order load flow technique," IEEE Transactions of PAS, PAS 96, pp. 189–195, 1977.
19. W. Stagg and A.H. El-Abiad. Computer Methods in Power System Analysis, McGraw-Hill, New York, 1968.
20. F.J. Hubert and D.R. Hayes. "A rapid digital computer solution for power system network load flow," IEEE Transactions of Power and Systems, pp. 90, pp. 934–940, 1971.
21. H.E. Brown, G.K. Carter, H.H. Happ and C.E. Person. "Power flow solution by impedance iterative methods," IEEE Transactions of Power and Systems 2, pp. 1–10, 1963.
22. H.E. Brown. Solution of Large Networks by Matrix Methods, New York, John Wiley, 1972.

23. A.R. Bergen and V. Vittal. Power System Analysis, Second Edition, Prentice Hall, New Jersey, 1999.
24. J.J. Granger and W.D. Stevenson. Power System Analysis, McGraw-Hill, New York, 1994.
25. L. Powell. Power System Load Flow Analysis, McGraw-Hill, New York, 2004.
26. X.-F. Wang, Y. Song and M. Irving. Modern Power System Analysis, Springer, New York, 2008.
27. D.P. Kothari and I.J. Nagrath. Power System Engineering, Second Edition, Tata McGraw-Hill, New Delhi, 2008.

CHAPTER 11

APPLICATION OF SHUNT CAPACITOR BANKS

11.1 SHUNT CAPACITOR BANKS

The term capacitor bank signifies a *complete assembly*, which may consist of a number of series and parallel groups of power capacitors, individual capacitor element fuses or group fusing, surge arresters, interlocked grounding switches, protection relays, neutral unbalance protection, primary switching devices, and so on. It may be a collection of components assembled at the operating site or may include one or more pieces of factory-assembled equipment.

The power capacitors can be used in series or shunt connections. The series capacitors are used in high-voltage (HV) transmission lines for series compensation to improve the power-handling capabilities (Chapter 5). Also series capacitors can be used in filters (Chapter 15).

The shunt capacitors can be applied to an electrical system for multiple tasks in *one single application*.

11.1.1 Power Factor Improvement

These can be switched with induction motors, can be provided at substations with load-dependent switching, or can be connected to the buses at medium- and low-voltage levels for power factor improvement, which also reduces system losses.

$$\% \text{ power loss} = 100 \left(\frac{\text{original PF}}{\text{improved PF}} \right)^2 \tag{11.1}$$

$$\% \text{ loss reduction} = 100 \left[1 - \left(\frac{\text{original PF}}{\text{improved PF}} \right)^2 \right] \tag{11.2}$$

If power factor is improved from 0.8 lagging to unity, there is 36% loss reduction. Equation (11.2) is applicable for lagging power factor. The improvement of

Power System Harmonics and Passive Filter Designs, First Edition. J.C. Das.

induction motor power factor through capacitors has certain limitations, as discussed in Section 11.11.

The utilities penalize a consumer for poor power factor. There may be many types of tariffs, but invariably the demand charge is based on kilovolt-ampere plus fixed or sliding charges for the energy consumed. The harmonic filters can admirably serve the dual purpose of harmonic mitigation and simultaneously improve the power factor.

The cost of power factor improvement equipment is balanced with cost of energy saving, and many computer programs will provide an optimum level of power factor improvement:

$$PF = \sqrt{1 - \left(\frac{C}{S}\right)^2} \tag{11.3}$$

where C is the cost per kvar of capacitor bank, S = cost per kvar of the system equipment, and PF is the optimum power factor.

Consider a 1000-kW load:

- kvar required to improve power factor from 0.7 to 0.8 = 270
- kvar required to improve power factor from 0.8 to 0.9 = 366
- kvar required to improve power factor from 0.9 to unity = 484.

An improvement in power factor above 0.95 is not economical and normally not applied.

11.1.2 Voltage Support

Shunt capacitors will provide voltage support to a bus in case of failure of a tie line or sudden dip in the voltage. An elementary calculation is as follows:

Consider the power flow on a mainly inductive tie line (Fig. 11.1). The load demand is shown as $P + jQ$, the series admittance $Y_{sr} = g_{sr} + b_{sr}$, and $Z = R_{sr} + jX_{sr}$. The power flow equation from the source bus (an infinite bus) is given by

$$P + jQ = V_r e^{-j\theta}[(V_s - V_r e^{j\theta})(g_{sr} + jb_{sr})]$$

$$= [(V_s V_r \cos\theta - V_r^2)g_{sr} + V_s V_r b_{sr} \sin\theta]$$

$$+ j[(V_s V_r \cos\theta - V_r^2)b_{sr} - V_s V_r g_{sr} \sin\theta] \tag{11.4}$$

Figure 11.1 Power flow through a short transmission line.

If resistance is neglected,

$$P \doteq V_s V_r b_{sr} \sin \theta \tag{11.5}$$

$$Q \doteq (V_s V_r \cos \theta - V_r^2) b_{sr} \tag{11.6}$$

If the receiving end load changes by a factor $\Delta P + \Delta Q$, then

$$\Delta P = (V_s b_{sr} \sin \theta) \Delta V + (V_s V_r b_{sr} \cos \theta) \Delta \theta \tag{11.7}$$

and

$$\Delta Q = (V_s \cos \theta - 2V_r) b_{sr} \Delta V - (V_s V_r b_{sr} \sin \theta) \Delta \theta \tag{11.8}$$

where ΔV_r is the scalar change in voltage V_r and $\Delta \theta$ is the change in angular displacement. If θ is eliminated from Eq. (11.4) and resistance is neglected, a dynamic voltage equation of the system is obtained as follows:

$$V_r^4 + V_r^2 (2QX_{sr} - V_s^2) + X_{sr}^2 (P^2 + Q^2) = 0 \tag{11.9}$$

From Eq. (11.6) as θ is usually small:

$$\frac{\Delta Q}{\Delta V} = \frac{V_s - 2V_r}{X_{sr}} \tag{11.10}$$

If the three phases of the line connector are short circuited at the receiving end, the receiving end short-circuit current is

$$I_r = \frac{V_s}{X_{sr}} \tag{11.11}$$

This assumes that the resistance is much smaller than the reactance. At no-load $V_r = V_s$, therefore

$$\frac{\partial Q}{\partial V} = -\frac{V_r}{X_{sr}} = -\frac{V_s}{X_{sr}} \tag{11.12}$$

Thus,

$$\left| \frac{\partial Q}{\partial V} \right| = \text{short-circuit current} \tag{11.13}$$

Alternatively, we could say that

$$\frac{\Delta V_s}{V} \approx \frac{\Delta V_r}{V} = \frac{\Delta Q}{S_{sc}} \tag{11.14}$$

where S_{sc} is the short-circuit level of the system. This means that the voltage regulation is equal to the ratio of the reactive power change to the short-circuit level. This gives the obvious result that the receiving end voltage falls with the decrease in

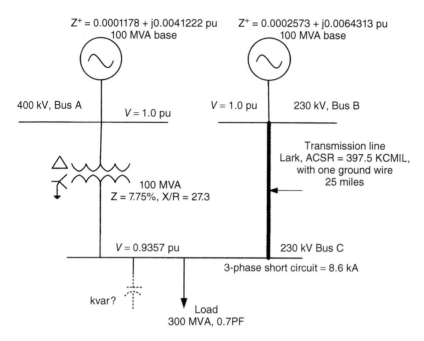

Figure 11.2 To illustrate reactive power compensation is a power system (Example 11.1).

system short-circuit capacity, or increase in system reactance. A stiffer system tends to uphold the receiving end voltage.

Example 11.1: Consider the system of Fig. 11.2. Bus C has two sources of power: one transformed from the 400-kV bus A and the other connected through a transmission line to 230-kV bus B. These sources run in parallel at 230-kV bus C. The voltages at buses A and B are maintained equal to the rated voltage. A certain load demand at bus C dips the voltage by 6.47%. The short-circuit level at bus C is 8.3 kA rms symmetrical. What is the reactive power compensation required at bus C to bring the voltage to its rated value?

An approximate solution is given by Eq. (11.14), which can be written as

$$\Delta V \approx \frac{I_c}{I_{sc}} = \frac{Q_C}{S_c} \tag{11.15}$$

where ΔV is the per-unit voltage rise, I_c is the capacitor current, I_{sc} is the short-circuit current, Q_c is the capacitor three-phase kvar, and S_c is the system short-circuit kilovolt-ampere. The voltage at bus $C = 0.9$ pu, that is, a voltage dip of 23 kV. Therefore, 213.5 Mvar of reactive power compensation is required.

When the system is unloaded, the voltage at bus C will rise by more than 230 kV due to off-loading the system and Ferranti effect of transmission line (see Chapter 12). Equation (11.15) is approximate as resistance is neglected. The voltage regulation of power system under varying loads is one major consideration of operation.

Devices that act on voltage dip and voltage rise such as ULTC, SVC, STATCOM, and so on are required. A load flow program will provide reactive power compensation requirements in a power system.

The voltage rise is also approximately given by the expression:

$$\Delta V \approx \frac{C_{\text{kvar}}(\%Z_t)}{T_{\text{kVA}}} \tag{11.16}$$

where C_{kvar} is the kvar of the capacitor, $\%Z_t$ is the transformer % impedance, and T_{kVA} is the transformer kilovolt-ampere. If a 1-Mvar capacitor bank is switched on the secondary side of a 1.0-MVA transformer of 5.5% impedance, the voltage will rise by 5.5%. This is the steady-state voltage rise, ignoring the switching transients.

Conversely, overvoltage on switching a large capacitor bank can also become of concern. Such overvoltages should be limited to safe levels. Some capacitors in a system may be kept continuously in service, while the others may use voltage, current, power factor, or kvar-dependent switching controls.

11.1.3   Improve Active Power-Handling Capability

As the electrical equipment is rated on kilovolt-ampere basis, provision of capacitors to meet a certain reactive power requirement will increase active power-handling capability of the equipment.

11.1.4   Application of Capacitors

Capacitors are an essential component of

- passive and active filters,
- electronic converters, multilevel converters, switch-mode power supplies, chopper circuits, and so on,
- applications in transmission, subtransmission, and distribution systems as compensating devices,
- SVCs (static var compensators),
- TCRs (thyristor-controlled reactors),
- TSCs (thyristor-switched capacitors),
- energy storage systems,
- snubber circuits,
- FACT controllers,
- surge suppression equipment.

We discussed some of these applications in earlier chapters of the book; we are primarily interested in shunt power capacitors used in passive filters for harmonic mitigations.

11.2 LOCATION OF SHUNT CAPACITORS

In industrial and commercial distribution systems, there is much freedom in locating the power capacitors when there are no nonlinear loads:

- At low-voltage substations
- Switched with low-voltage or medium-voltage motors
- At any medium- or high-voltage bus in a power distribution system
- At multivoltage levels.

The natural resonant frequency of most industrial power systems is much higher than that of load-generated harmonics, because the stray capacitances are small. Thus, a resonance will occur at a much higher frequency beyond the harmonics generated by nonlinear loads. A qualification is required, and that in today's proliferation of electronic equipment, there is no absolutely harmonic-free environment. Fluorescent lighting, household appliances, and switch-mode power supplies commonly used for computers and copy machines, battery chargers, and uninterruptible power supply systems create harmonics.

Capacitors may be located in an industrial plant or close to such a plant that has significant harmonic producing loads. This location is very possibly subject to harmonic resonance and should be avoided or the capacitors can be used as harmonic filters.

On a subtransmission system where the capacitors may be located close to or far from the harmonic producing loads, the propagation of harmonics through the interconnecting systems needs to be studied. The linear loads served from a common feeder, which also serves nonlinear loads of some other consumers, may become susceptible to harmonic distortion.

There have been two approaches:

- One approch is to consider capacitor placement from a reactive power consideration and then study the harmonic effects.
- The second is to study the fundamental frequency voltages, reactive power, and harmonic effects simultaneously. A consumer system that does not have harmonic producing loads can be subject to harmonic pollution due to harmonic loads of other consumers in the system.

The optimum location of capacitors in a distribution system is a fairly complex algorithm [1,2]. The two main criteria are voltage profiles and system losses. Correcting the voltage profile will require capacitors to be placed toward the end of the feeders, while emphasis on loss reduction will result in capacitors being placed near load centers. An automation strategy based on intelligent customer meters, which monitor the voltage at consumer locations and communicate this information to the utility, can be implemented [2].

A capacitor placement algorithm based on loss reduction and energy savings may use dynamic programming concepts [3]. We can define the objective functions as

- peak power loss reduction,
- energy loss reduction,
- voltage and harmonic control,
- capacitor cost.

The variables to be solved are fixed and switched capacitors, and their number, size, location, and switched time. Certain assumptions in this optimization process are the loading conditions of the feeders, type of feeder load, and capacitor size based on the available standard ratings and voltages.

The HV transmission lines and long length of power cables have resonant frequency profiles of their own (Chapter 12), and the capacitors are generally turned into suitable filter types.

11.3 RATINGS OF CAPACITORS

The capacitors can be severely overloaded due to harmonics, especially under resonant conditions and can even be damaged. The capacitors are capable of continuous operation under contingency system and capacitor bank conditions, provided that none of the following is exceeded [4–7].

- 110% of the rated rms voltage should not be exceeded. If harmonics are present, this means

$$V_{rms} \leq 1.1 = \left[\sum_{h=1}^{h=h_{max}} V_h^2 \right]^{1/2} \tag{11.17}$$

- The crest should not exceed $1.2 \times \sqrt{2}$ times the rated rms voltage, including harmonics but excluding transients:

$$V_{crest} \leq 1.2 \times \sqrt{2} \sum_{h=1}^{h=h_{max}} V_h \tag{11.18}$$

- The rms current should not exceed 135% of nominal rms current based on rated kvar and rated voltage, including fundamental and harmonic currents:

$$I_{rms} \leq 1.35 = \left[\sum_{h=1}^{h=h_{max}} I_h^2 \right]^{1/2} \tag{11.19}$$

- The kvar loading should not exceed 135% of rated kvar, that is, if harmonics are present:

$$\text{kvar}_{\text{pu}} \leq 1.35 = \left[\sum_{h=1}^{h=h_{\text{max}}} (V_h I_h) \right] \tag{11.20}$$

The limitation of 135% of rated kvar in IEEE Standard 18 is based on dielectric heating at fundamental frequency and on thermal stability test. The dielectric loss can be written as

$$L_{\text{Dielectric}} = fCV^2 \tag{11.21}$$

where f is the rated frequency and C is the rated capacitance.

Capacitors should be manufactured in accordance with UL 810-2008 [6] and protected with a UL-listed/recognized device. The capacitors which do not contain a UL-listed device should be protected with external current-limiting fuses or other protection devices which are UL listed.

11.3.1   Testing

IEEE 18 [5] specifies that terminal to case test for the internal insulation of indoor capacitors should be performed at 3-kV rms (capacitors rated 300 V or less) or 5-kV rms (capacitors rated 301 V to 1199 V) for 10 seconds. The terminal-to-terminal test should be of 10 seconds at twice the rated rms voltage (AC test) or 4.3 times the rated rms voltage (DC test).

11.3.2   Discharge Resistors

Each capacitor rated 600 V or less should be provided with discharge resistors to reduce the residual voltage from peak of rated to less than 50 V within 1 minute of deenergization and 5 minutes for capacitor elements rated higher than 600 V.

11.3.3   Unbalances

Unbalances within a capacitor bank due to capacitor element failures and or individual fuse operations result in overvoltages on capacitor units that remain in service. Typically, voltage is allowed to increase by 10% before unbalance detection removes capacitor banks from service. In a filter bank, failure of a capacitor element will cause detuning.

The limitations of the capacitor bank loadings become of importance in the design of capacitor filters.

11.3.4   Short-Duration Overvoltage Capability

The overvoltage capability of the capacitor banks is shown in Fig. 11.3 [4]. This shows that the overvoltage capability even for short durations is limited to 2.2 pu. A capacitor tested according to IEEE Standard 18 [5] will withstand a combined total

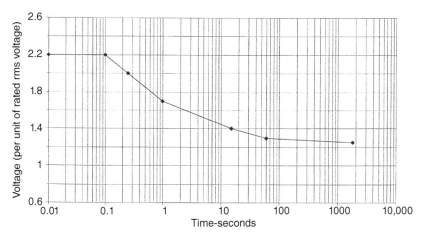

Figure 11.3   Maximum contingency power frequency overvoltage capability of capacitor units. Source: Ref. [4].

of 300 applications of power frequency terminal-to-terminal overvoltages without superimposed transients or harmonic content, the magnitude and duration are shown in this figure. This capability is shown for a capacitor unit and not for a capacitor bank. For a capacitor bank, the unbalance in voltages should be allowed for. The x-axis of this figure indicates the maximum time for a *single* event. This means two continuous events at 1.73 per unit are more severe than two separate events of 1 second each with intervening time to allow localized cooling and absorption of corona gasses. The damage is occurring more rapidly at the end of the event than at the beginning. This is so because overvoltages cause localized damaging discharges in the capacitor unit, and trapped charges plus gas generation in the first half-cycle make the discharge more severe in the second half-cycle. For frequently switched capacitors, current and voltage peaks must be held to lower values. Figure 11.4 shows the overvoltage capability versus number of transients per year [4]. A capacitor may reasonably be expected to withstand transient overvoltages in this figure. Note that the capacitor can tolerate peak voltage of 3.5 times its crest (peak) rated voltage, provided the transients are no more than 4 in 1 year. (Figure 11.4 is in terms of per unit of rated voltage.)

A capacitor can tolerate an overvoltage of 10% continuously; however, this capability should be utilized with care in any harmonic filter design, there are number of other considerations to select the voltage rating for capacitors in filters.

For transient events, generally less than 1/2 cycle of power frequency, such as capacitor bank switching, circuit breakers restrikes, and so on, the voltage rating is calculated from

$$V_r = \frac{V_{tr}}{\sqrt{2}k} \tag{11.22}$$

where V_{tr} is the peak transient voltage, V_r is the rated rms voltage of the capacitor, and k factor is read from Fig. 11.4. If there are 100 transient events in a year,

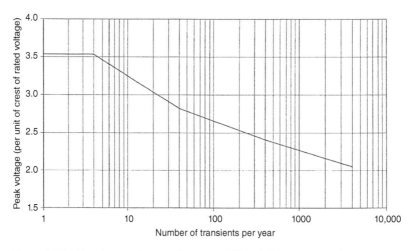

Figure 11.4   Transient peak overvoltage capability of the capacitor units.

$k = 2.65$. For a 13.8-kV rated rms voltage of the capacitor:$V_{tr} = 13.8 \times \sqrt{2} \times 2.65 = 51.70$ kV peak.

This implies that the capacitors can withstand 100 surges in a year of magnitude approximately 2.65, provided the transient voltage is limited to 51.70 kV peak. Higher voltage surges will be applicable if the frequency of surges is decreased.

For dynamic events generally lasting for a few fundamental cycles to several seconds, such as transformer energization and bus fault clearance, the voltage rating of the capacitor is calculated from the equation:

$$V_r = \frac{V_d}{\sqrt{2}} \tag{11.23}$$

where V_d is the voltage across the capacitor reached during dynamic event. The power frequency short-time overvoltage capability is specified in Fig. 11.3. A capacitor unit can withstand 2.2 pu of rated voltage for 0.01–0.1 seconds, and at about 15 seconds the overvoltage withstand capability is reduced to 1.4 pu.

Power system studies, for example, EMTP switching transient studies, are required to establish the transient and dynamic overvoltages in a system. The rated voltage selected should be the highest of the voltages given by Eqs (11.22) and (11.23). On a continuous basis, an operating voltage of 8–10% higher than the nominal system voltage is selected. This reduces the reactive capability at the voltage of application as the square of the voltages.

11.3.5   Transient Overcurrent Capability

The capacitor unit is also expected to withstand transient currents inherent in the operation of power systems, which include infrequent high lightning currents and discharge currents due to nearby faults. For frequent back-to-back switching (Section 11.9), the peak capacitor unit current should be held to a value lower than that shown

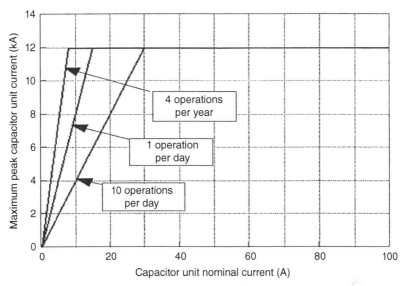

Figure 11.5   Transient current capability of capacitor units for regularly occurring transients. Source: Ref. [4].

in Fig. 11.5. The capacitor bank current is the number of capacitor unit current multiplied by the number of capacitor units or strings in parallel. In this figure, the curves are based on straight lines from origin to 12 kA, and the slopes are 1500 times the nominal current for 4 operations per year, 800 times the nominal current for 1 operation per day, and 400 times the nominal current for 10 operations per day.

Example 11.2:   Calculate the capacitor bank loading for the resonant condition illustrated in Section 9.7.

We can read the harmonic voltages and currents from the simulations shown in Figs. 9.21 and 9.23 and construct Table 11.1.

Then from Eqs (11.17)–(11.20)

$$I_{rms} = 163\%$$

$$V_{rms} = 1.032 \text{ pu}$$

$$kvar_{pu} = 1.256$$

The rms current exceeds the limit of 135%.

TABLE 11.1   Harmonic Currents and Voltages for Calculation of Capacitor Bank Loading

Harmonic	5th	7th	11th	13th	17th	19th	23rd	25th
V_h (%)	1.8	1	11	1.4	0.5	0.2	0.1	0.1
I_h (%)	8	7.8	128	8	6	4	3.6	2

Example 11.3: A capacitor bank is to be constructed for operation at 13.8-kV nominal voltage using 400-kvar capacitor units. Select the rated voltage of a unit capacitor to form a wye-ungrounded capacitor bank based on the following:

- The operating voltage can be 5% higher due to load regulation.
- The phase unbalance system detection is arranged to shutdown the capacitor bank in case the voltage rises by 10%.
- For control of the power factor, 400 switching operations per year are required, the switching transient overvoltage is calculated as 25-kV peak.
- The transient inrush current for back-to-back switching is limited to 6 kA.

As the capacitor units are to be connected in wye-ungrounded configuration, the rated line-to-neutral voltage is 7967.7 V. Although the power capacitors have the continuous overvoltage capability of 10%, and this capability can be utilized to keep the bank in service if one of the parallel fuses operates, it is desirable to reserve this margin for conservatism and for deterioration over a period of time. Adding 15% voltage margin, capacitor-unit voltage becomes 9162.9 V.

Table 11.2 is constructed based on the standard sizes of capacitor units that are commercially available. Although we calculated a voltage rating of 9162, a total of

TABLE 11.2   Standard Ratings of Single-Phase High-Voltage Capacitor Units, Film/Foil Type, Class IIIB Combustible Fluid, Non-PCB Dielectric

Unit Voltage	BIL kV	50 kvar	100 kvar	150 kvar	200 kvar	300 kvar	400 kvar
2400	75	Yes	Yes	Yes	Yes	No	No
2770	75					Yes	No
4160	75					Yes	Yes
4800	75					Yes	Yes
6640	95	Yes	Yes	Yes	Yes	Yes	Yes
7200	95						
7620	95						
7960	95						
8320	95						
9960	95	Yes	Yes	Yes	Yes	Yes	Yes
12,470	95						
13,280	95						
13,800	95						
14,400	95						
19,920	125	No	Yes, single bushing only	Yes, single bushing only	Yes, single bushing only	Yes, single bushing only	Yes, single bushing only
21,600	125						

19,920 and 21,600 volts rated capacitor units available in single bushing designs only.

9 V, the next standard capacitor unit of 9.54 kV, can be selected. For 400 operation, $k = 2.4$, and from Eq. (11.22) using 9.54 kV capacitor unit, the $V_{tr} = 32.37$ kV. Therefore, the number of switching operations should be reduced. The dynamic voltage V_d is not stated, but the rated current of the capacitor unit string, considering five units in parallel per string, 400-kvar, 9.54-kV rated voltage when connected to 13.8-kV source voltage in wye configuration is equal to 174 A. The inrush current is limited to 6-kA peak. Therefore, the switching overcurrent capability is acceptable.

11.4   SHUNT CAPACITOR BANK ARRANGEMENTS

Formation of shunt capacitor banks from small to large sizes and at various voltages is required for harmonic filter designs and reactive power compensation. The shunt capacitors can be connected in a variety of three-phase connections, which depend on the best utilization of the standard voltage ratings, fusing, and protective relaying. To meet certain kvar and voltage requirements, the banks are formed from standard unit power capacitors available in certain ratings and voltages and are connected in series and parallel groups. Table 11.3 [7] shows the number of series groups for wye-connected capacitor banks required for line operating voltages from 12.47 to 500 kV.

There are three main considerations.

- On failure of a capacitor-can, the voltage on remaining parallel-connected capacitors should be controlled to no more than 10% voltage rise.

- Only a certain number of capacitors in parallel (total of 3200 kvar) can be used when expulsion type fuses are used (Section 11.5.2).

- Failure of a unit causes detuning, shifts the resonant frequency, and may increase TDD if the capacitor bank is to be used in a filter application.

- Selection of rated voltage is of much importance (see Example 11.3 and further discussions in Chapter 15).

11.4.1   Formation of a 500-kV Capacitor Bank

As an example, for three-phase 500-kV application, consider that a wye-connected capacitor bank to provide 200 Mvar is required. From voltage considerations alone, line-to-neutral voltage is = 288.68 kV. A total of 14 series strings of 21.6-kV rated capacitors will give 302.4 kV, or 38 strings of 7.62-kV rated capacitors will give 289.6 kV. To limit the number of strings, use 14 strings of 21.6-kV capacitor-cans (elements) in a wye-connected formation as shown in Fig. 11.6(a). Table 11.4 shows the number of elements to be connected in parallel to limit the overvoltage on the remaining units to no more than 10%. From this table, a minimum of 11 elements/strings are required for wye connection.

A 200-Mvar three-phase rating is required. This means that each series group should have 4.76 Mvar [200/(3)(14)]. This gives a unit size of 476.2 kvar, which is not even commercially available. Also as stated in Section 11.5.2, the limitation with

TABLE 11.3 Number of Series Groups in Y-Connected Capacitor Banks

V_{ll} (kV)	V_{ln} (kV)	21.6	19.92	14.4	13.8	13.28	12.47	9.96	9.54	8.32	7.96	7.62	7.2	6.64
							Available Capacitor Voltage kV Per Unit							
500.0	288.7	14	15	20	21	22		29	30	35	36	38		
345.0	199.2		10			15	16	20	21	24	25	27		
230.0	132.8					10			14	16	17	18		20
161.0	92.9					7							13	14
138.0	79.7		4	6	6	6		8			10		11	12
115.0	66.4					5			7	8	9	9		10
69.0	39.8		2		3	3		4			5			6
46.0	26.56					2								4
34.5	19.92		1					2						3
24.9	14.4			1									2	
23.9	13.8				1									
23.0	13.28					1								
14.4	8.32									1	1			
13.8	7.96										1			
13.2	7.62											1		
12.47	7.2												1	

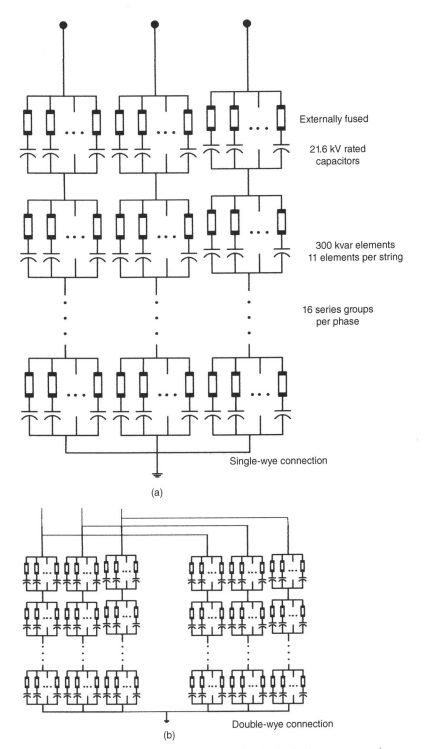

Externally fused

21.6 kV rated
capacitors

300 kvar elements
11 elements per string

16 series groups
per phase

Single-wye connection

(a)

Double-wye connection

(b)

Figure 11.6   (a) Formation of a 500-kV capacitor bank, single-wye connection,
(b) double-wye connection.

TABLE 11.4 Minimum Number of Units in Parallel per Series Group to Limit Voltage on Remaining Units to 110% with One Unit Out

Number of Series Groups	Grounded Y or Δ	Ungrounded Y	Double Y, Equal Sections
1	–	4	2
2	6	8	7
3	8	9	8
4	9	10	9
5	9	10	10
6	10	10	10
7	10	10	10
8	10	11	10
9	10	11	10
10	10	11	11
11	10	11	11
12 and over	11	11	11

use of expulsion fuses is that no more than 3200 kvar can be connected in parallel. Thus, the required size of 200 Mvar cannot be achieved in single wye connection.

Continuing with the formation of the bank, only 300-kvar 11 units in parallel can be used per series string. This gives an Mvar output of 46.2 Mvar per phase or 138.6 Mvar in total. This will be at a rated voltage of 523.75 kV. At 500 kV, the available output will be 126.7-Mvar, single wye-connected bank.

The 14 groups in series give a rated voltage of 302.4 kV and line-to-line voltage of 523.7 kV. This is higher from the rated voltage of 500 kV by only 4.75%, and this overvoltage margin is not adequate. With 16 series groups, line-to-line voltage is 598.5 and the Mvar rating at 500 kV = 109.3 Mvar.

It is necessary to use double-wye connection as shown in Fig. 11.6(b), then the total Mvar = 218.6 at 500 kV, which should be acceptable.

11.5 FUSING

The capacitor banks can be externally fused, internally fused, or fuseless.

Each phase of each filter step should be protected by fuses. Fuses should be current limiting, rated for the fault current at its location, UL listed, class J or T.

11.5.1 Externally Fused

The classification for external fusing is the following:

- Group fusing
- Individual fusing

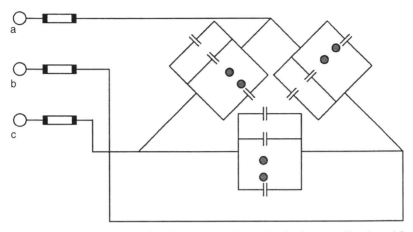

Figure 11.7   Group fusing of a delta-connected capacitor bank, generally adopted for low-voltage systems.

Group Fusing   In group fusing, more than one capacitor unit can be protected with a fuse. Low-voltage capacitors generally connected in delta configuration are group fused with fuses in phase conductors only (Fig. 11.7). Group protection is also applied to pole-mounted capacitor racks. The maximum size of capacitor bank for group fusing is limited by the maximum size of the available fuse, current limiting, or T and K expulsion-type links. The expulsion-type fuses are generally used for outdoor rack-mounted capacitor banks.

Individual Fusing   A bank using individual capacitor unit fusing has smaller sized fuses, sharing a small portion of the total transient current. On operation of a fuse, the bank need not be shut down, the unbalance and overvoltages on the remaining units in service can be calculated, and the operation can be sustained. Individual fusing offers increased protection against case rupture in all cases, and the use of smaller rated fuse link will usually result in faster clearing times.

The fuse is sized with the following considerations:

- Maximum continuous current, including harmonics.
- Switching inrush current should not operate the fuse. This should include back-to-back switching (Section 11.9). Even for "definite-purpose circuit breakers," the isolated and back-to-back switching currents are limited according to standards [8].
- Restrike current; this considers restrikes in the switching device during opening, not discussed.
- Lightning surge current and discharge current into a failing capacitor unit.
- The overvoltage duration that will occur on the capacitors remaining in service is of much concern.
- The available short circuit and its clearing time at the point of application should be compared with the manufacturer's published data on short circuit withstand

on capacitor units. As the capacitor enclosures are grounded, a phase-to-ground fault can still occur, even if the capacitors are connected in ungrounded configuration.

- In an ungrounded wye-connected bank, the maximum short-circuit current is only three times the full-load current, and a group fuse selection to meet all the criteria and fast clearance time becomes difficult.

- For externally fused banks, fuses should be fast to coordinate with the fast unbalance relay settings, but should not operate during switching or external faults.

- The most important consideration is that individual capacitor-can fusing is selected to protect the rupture/current withstand rating of the can. The fuse must operate to interrupt the maximum power frequency fault current.

- The maximum clearing time curve of the fuse and the case rupture curve are plotted together. The case rupture characteristics vary with the design and size of the capacitor element, and these data should be obtained from a manufacturer. The curves are sometimes plotted for 10% and 50% probability boundary. The probability of case rupture can be defined as opening of the case from a mere cracked seam or bushing seal to a violent bursting of the case. Within safe zone, no greater damage than a slight swelling of the case will occur, though a case rupture is possible for low levels of short-circuit current flowing for extended period of time. Figure 11.8 shows a typical case rupture curve from Ref. [4]. Many utilities may not accept capacitor units, which exhibit probability tank rupture curves.

- Definite tank rupture curves indicate that effectively there is no chance of capacitor-can rupture if a fuse or protective device coordinates with the curve. This may be termed as zero probability curve.

- A proper fuse protection will decrease the probability of case rupture, but not eliminate it. Each unit contains many series sections of individual capacitor elements or rolls. Gas can be generated at each failure site, pressurizing the unit. The capacitor cans can be provided with individual unit pressure switch, which can be wired to trip the capacitor bank.

11.5.2 Expulsion-Type Fuses

There is a limit to the number of parallel elements connected in a group when expulsion-type fuses are used for individual capacitor-can protection. The energy liberated and fed into the fault when a fuse operates is required to be limited, depending on the fuse characteristics (Fig. 11.9). The energy release may be as follows:

$$E = 2.64 \text{ J per KVAC rated voltage} \qquad (11.24)$$

$$E = 2.64(1.10)^2 \text{ J per KVAC } 110\% \text{ voltage} \qquad (11.25)$$

$$E = 2.64(1.20)^2 \text{ J per KVAC } 120\% \text{ voltage} \qquad (11.26)$$

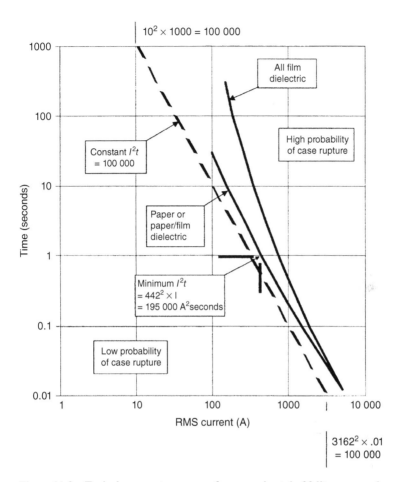

Figure 11.8   Typical case rupture curves for approximately 30 liters case volume.

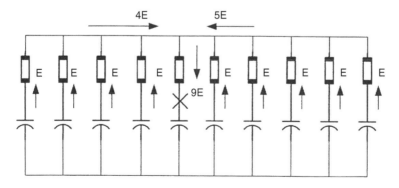

Figure 11.9   Energy fed into a fault from parallel capacitor units.

TABLE 11.5   Short-Circuit Current Withstand with Expulsion Fuses

Capacitor Unit Size (kvar)	Maximum Fault Current for Expulsion-Type Fuses (A)
50	4000
100	5000
150	6000
200	6000
300–400	7000

Usually, capacitor elements totaling up to 3100 kvar can be connected in parallel when expulsion fuses are used. This limit can be exceeded if capacitor elements are fused with current-limiting fuses (generally limited to indoor metal-enclosed installations).

Manufacturers publish data of the maximum fault currents for expulsion fuses for their capacitor unit sizes (see Table 11.5).

When short-circuit currents exceed these values, current-limiting fuses can be used. The current-limiting fuses are of no value if the available fault current is at such a level that it takes more than 0.5 cycles to melt the fuse. This implies that the fuse should operate in its current-limiting zone and the fault current should be above the "threshold" current (see Refs [9,10] for operation of current-limiting fuses).

Example 11.4: The case rupture curve of a 300-kvar capacitor element, 50% and 10% case rupture probability curves as supplied by a manufacturer are shown in Fig. 11.10. The time–current characteristics of K-type fuse links are in Fig. 11.11. The 100K fuse characteristics and the 300-kvar capacitor unit probability damage curves are plotted together in Fig. 11.10. Note that the link protects both the 50% and 10% probability curves, except that on a long-time basis 10% probability curve is not protected. If the characteristics of a higher link size of 140K are plotted, it will not even protect 50% probability curve.

Whether it is group fusing or the individual fusing, the I^2t energy of the surge current must be compared to the fuse melt I^2t. The fuse should not melt on transient currents. The duration of the transients is no more than a few cycles, and during this short time there is no heat transfer away from the fuse element. The heating value of the transient current can be compared with the I^2t capability of the fuse.

11.5.3   Internally Fused

The interior of a capacitor unit consists of a number of parallel- and series-connected group of elements. One fuse is connected in series with each capacitor element (Fig. 11.12). On a puncture or short circuit in a capacitor element, the current through the fuse increases in proportion to the number of units in parallel. This will melt the fuse in a short time of few milliseconds. The capacitor units are designed with a large number of elements in parallel, and the voltage rise on the remaining units in

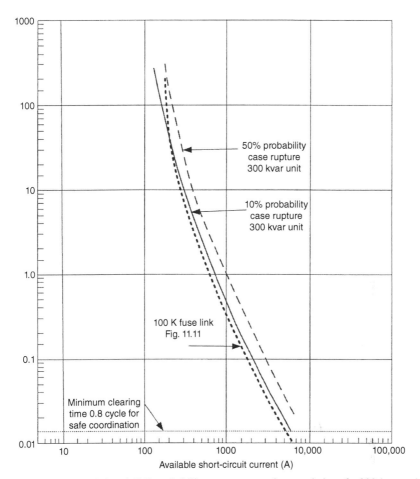

Figure 11.10   50% and 10% probability case rupture characteristics of a 300-kvar unit, showing protection with 100K T-link expulsion fuse.

parallel is controlled. The operation of a fuse results in decrease in the capacitance of the unit. Figure 11.12 shows 14 parallel elements, three series groups, and two blown fuses in one group.

Generally, the series groups are reduced and the number of elements in parallel increased compared to externally fused banks. Internally fused banks have two bushing units, in order to isolate the parallel group of capacitor units in one string from the parallel group in the adjacent string. The basic insulation level (BIL) of the bushings and internal insulation of the capacitor units are determined by the maximum voltage to the rack. The rack potential is established by connecting the rack to one series strings of capacitor units.

The settings on unbalance protection consider the capability of internal fuses, transient overvoltage capability of elements, and consequences of failure to the case

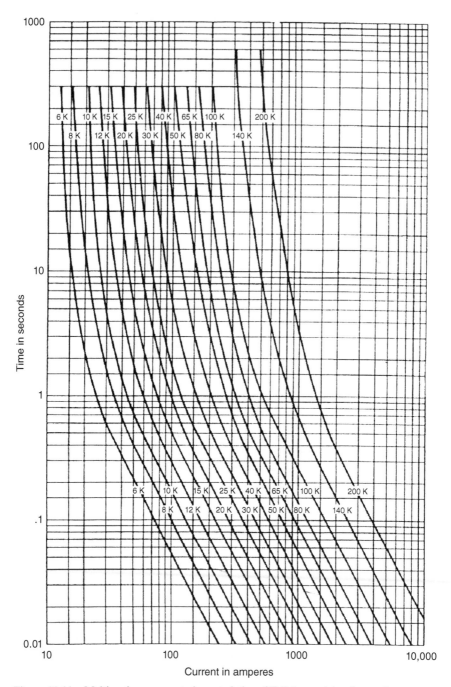

Figure 11.11 Melting time–current characteristics of T-link expulsion fuses of a manufacturer.

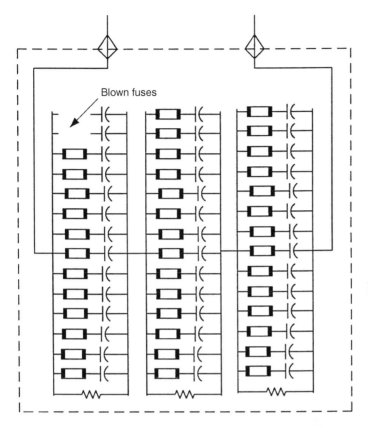

Figure 11.12   Internally fused capacitor bank.

or the failure of an internal fuse. The overvoltages to the healthy capacitor units should be limited to 110% of the rated voltage.

11.5.4   Fuseless

The fuseless capacitor banks have an arrangement that does not use fuses, but it is not simply the elimination of fuses. When an all-film capacitor develops a dielectric short, the film burns away resulting in a weld between the foils – it creates a low-resistance short, which does not generate large amount of heat or gas, as long as the current is limited through the dielectric short, allowing the capacitor to continue in operation indefinitely. These designs are not used with paper/film capacitors because dielectric shorts are more likely to have localized heating and gassing. The desired three-phase kvar is achieved by putting series strings of capacitor units in parallel. The sum of the individual capacitor unit voltages in a string should equal or exceed the normal phase-to-ground or phase-to-neutral voltage of the capacitor bank. A more sensitive unbalance protection and isolation is required.

Fuseless capacitor banks should have two bushings. For system voltages less than 35 kV, the failure of a single element, where there are 10 elements or less in a

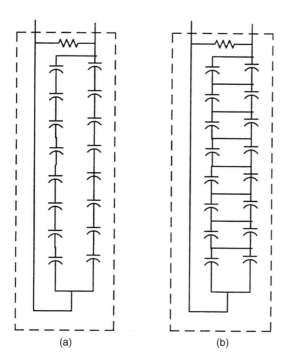

(a)   (b)

Figure 11.13   Fuseless capacitor bank.

series string, may result in excessive voltage on remaining elements requiring imme-diate shutdown of the bank. Figure 11.13 shows schematic of fuseless capacitor bank. Figure 11.13(a) shows a series-connected capacitor unit, while Fig. 11.13(b) shows a parallel-connected unit. The series unit shows two series groups having 8 elements, a total of 16 elements. If one element fails, the voltage across remaining elements is 15/16. The series groups can be increased. The parallel connection illustrated eight series groups of two elements each.

Externally fused capacitor banks are most popular in the United States. The external fuses have tell-tale signs that indicate that the fuse has operated. The original capacitance of the bank can be reestablished by replacing the fuse, which is of importance for harmonic filter designs. The change in the capacitance values of a harmonic filter is not desirable; it will alter the tuned and resonant frequencies in a single-tuned harmonic filter designs; in fact, close tolerance components are used.

11.6   CONNECTIONS OF BANKS

The capacitor banks may be connected in

- ungrounded wye connection,
- ungrounded double-wye neutrals,

- grounded double-wye neutrals,
- delta connection,
- H-bridge connection for large banks.

These connections are shown in Fig. 11.14. Delta connection is common for low-voltage application with one-series group rated for line-to-line voltage. A wye-ungrounded group can be formed with one group per phase when the required operating voltage corresponds to standard capacitor unit rating. The wye neutral is left ungrounded.

11.6.1   Grounded and Ungrounded Banks

For impedance grounded or ungrounded systems, only ungrounded-wye or delta capacitor bank configurations should be used. The medium-voltage industrial systems at 23, 13.6, 4.16, and 2.4 kV are invariably resistance grounded; thus ungrounded connections are applicable for these systems.

For effectively grounded systems, there is a choice of ungrounded or grounded-wye banks. Grounded Y neutrals and multiple series groups are common for voltages above 34.5 kV. Multiple series groups limit the fault current. Advantages of grounded-wye arrangements are the following:

- Grounded capacitors provide a low-impedance path for lightning surge currents and give some protection from surge voltages; however, third-harmonic currents can circulate and these may overheat the system grounding impedances.
- Initial cost of capacitor bank may be lower as the neutral has not to be insulated from ground for full-system BIL.
- Capacitor switch recovery voltages are reduced.

    However, the grounded banks have the following disadvantages:

- High inrush currents may occur in station grounds and structures, which may cause instrumentation problems.
- Grounded neutrals may draw zero sequence currents and cause telephone interference problems.
- A short circuit in one phase of the capacitor bank results in system line-to-ground fault with high magnitude of fault currents (for effectively grounded systems).
- Owing to high-fault currents, in capacitor banks with one-series group only, current-limiting fuses will be required.

In the case of an ungrounded multiple series group wye-connected bank, third-harmonic currents do not flow, but the entire bank including the neutral should be insulated for the line voltage. Double-wye banks and multiple series groups are used when a capacitor bank becomes too large for the 3100 kvar per group for the expulsion type of fuses.

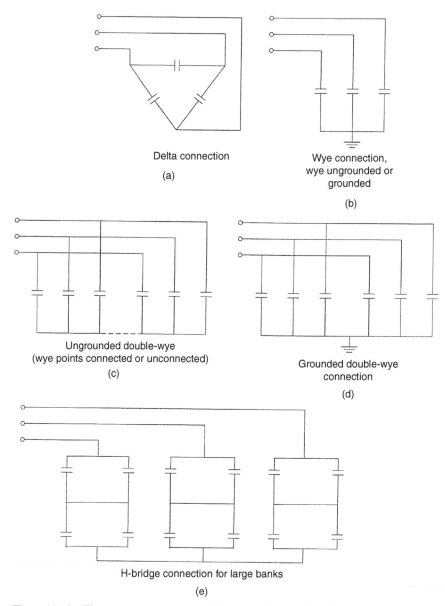

Delta connection

(a)

Wye connection,
wye ungrounded or
grounded

(b)

Ungrounded double-wye
(wye points connected or unconnected)

(c)

Grounded double-wye
connection

(d)

H-bridge connection for large banks

(e)

Figure 11.14   Three-phase connections of shunt capacitor banks. (a) Delta connection, (b) wye connection, it can be grounded or ungrounded, (c) double-wye connection, ungrounded, wye points connected or unconnected, (d) grounded double-wye connection, and (e) H-bridge connection for large banks.

11.6.2   Grounding Grid Designs

The design of grounding grids and the connection of neutral points of capacitor banks are of importance. Two methods of grounding are

- single-point grounding,
- Peninsula grounding.

With single-point grounding, the neutrals of capacitor banks of a given voltage are all connected through an insulted cable, which is connected to the grid at one point. There will be substantial voltage of the order of tens of kilovolts between the ends of the neutral bus and single-point ground during switching. The use of shielded cable will help reduce the voltage stress. In the event of a fault, high-frequency currents can flow back into power system via the substation ground grid.

With peninsula grounding, the grounding grid is built under capacitor banks and bus work in a form resembling a series of peninsulas. In this arrangement, one or more ground conductors may be carried under the capacitor rack of each phase of each group and tied to main station ground at one point at the edge of capacitor area. All capacitor neutral connections are made to this isolated peninsula grounding grid conductors [7]. The capacitor bank and the associated current transformer and voltage transformer potential will rise during capacitor switching, but the transients in the rest of the system will be reduced. Also with peninsula grounding, all equipments at the neutral ends tend to rise to the same potential and differential voltages can be avoided.

11.7   UNBALANCE DETECTION

An external fuse should be sized to meet a number of requirements as stated earlier, yet a failure of a fuse in a capacitor bank involving more than one series group per phase and in all ungrounded capacitor banks irrespective of the series groups will increase the voltage on the capacitors remaining in service. This magnitude can be calculated from the following expressions:

For grounded-wye, double-wye or delta-connected or double-wye grounded banks:

$$\%V = \frac{100S}{1 + \left(1 - \frac{\%R}{100}\right)(S-1)} \tag{11.27}$$

where S is the number of series groups > 1, R is the percentage of capacitor units removed from one series group, and $\%V$ is the percent of nominal voltage on the remaining units in affected series group. For example, a series group has five capacitors in parallel and there are three series groups per phase ($S = 3$), then failure of a fuse in one unit means 20% outage, ($R = 20$) and $\%V = 115\%$. This is graphically illustrated in Fig. 11.15.

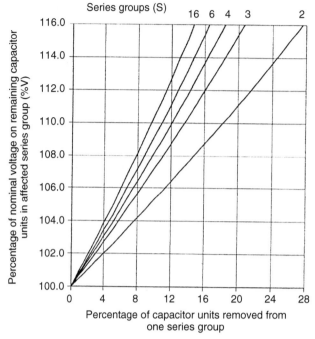

Figure 11.15 Grounded-wye connected, delta, or grounded-double-wye connected capacitor bank: voltage on remaining capacitor units in series group versus percentage of capacitor units removed from series group.

For ungrounded-wye or ungrounded double-wye banks, neutrals isolated, the following expression is applicable:

$$\%V = \frac{100S}{S\left(1 - \frac{\%R}{100}\right) + \frac{2}{3}\left(\frac{\%R}{100}\right)} \qquad (11.28)$$

For the same configuration as for grounded bank, $\%V$ will be 118.4%. This is illustrated in Fig. 11.16.

For ungrounded double-wye connected bank, neutral tied together, the equation is

$$\%V = \frac{100S}{S\left(1 - \frac{\%R}{100}\right) + \frac{5}{6}\left(\frac{\%R}{100}\right)} \qquad (11.29)$$

This relation is graphically shown in Figure 11.17. Reference [7] provides further details on unbalance calculations.

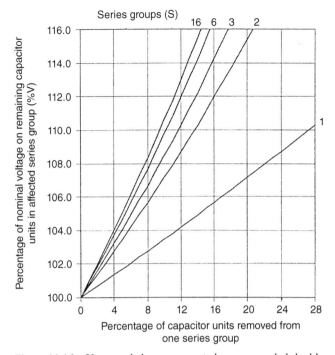

Figure 11.16   Ungrounded-wye connected or ungrounded-double–wye connected (neutrals isolated) capacitor bank: voltage on remaining capacitor units in series group versus percentage of capacitor units removed from series group.

11.7.1   Detuning due to Fuse Failure

The failure of an element results in detuning in case of a filter bank. Generally, the voltage rise is limited to a maximum of 10% and sometimes even lower. A bank can be designed so that on loss of a single element, the capacitor bank continues to operate in service and tripped immediately on loss of two elements in the same group. The protection schemes to detect unbalance are not discussed; however, with modern microprocessor-based relays, it is possible to detect an unbalance as low as 1–2% on failure of a capacitor element in banks of parallel and series groups.

11.8   DESTABILIZING EFFECT OF CAPACITOR BANKS

The var output of shunt capacitors is given by the expression:

$$\text{var}s = \frac{V^2}{X_c} = 2\pi f C V^2 \qquad (11.30)$$

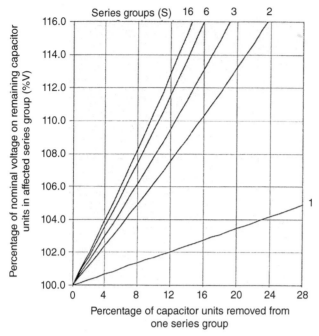

Figure 11.17   Ungrounded-double-wye connected capacitor bank, neutrals tied together with capacitor bank: voltage on remaining capacitor units in series group versus percentage of capacitor units removed from series group.

The shunt capacitors present a constant impedance type of load, and the capacitive power output varies with the square of the voltage:

$$Q_{V,\text{new}} = Q_{V,\text{rated}} \left(\frac{V_{\text{new}}}{V_{\text{rated}}} \right)^2 \tag{11.31}$$

Consider that the voltage dips by 10%, then the reactive power output will reduce by 81% of the rated reactive power. Compare this with a synchronous generator. If the voltage dips, the generator excitation and voltage regulator act so that generator supplies more reactive power on a short-time basis. This behavior of shunt capacitor is called the *destabilizing* effect compared to the operation of a synchronous generator, which is called the *stabilizing* effect in terms of reactive power output.

Power capacitors are designed for operation at the system-rated frequency. When these are operated at frequencies other than the rated, both the reactive power and the capacitive currents are reduced:

$$Q_{\text{OP}} = \left(\frac{f_{\text{OP}}}{f_R} \right) Q_R$$

$$I_{\text{OP}} = I_{\text{NOM}} \left(\frac{f_{\text{OP}}}{f_R} \right) \tag{11.32}$$

where

Q_{OP} = operating reactive power of the capacitor

Q_R = rated reactive power of the capacitor

I_{OP} = the capacitor current at applied frequency and rated voltage

I_{NOM} = the capacitor current at rated voltage and frequency

f_R = rated frequency of the capacitor

Example 11.5: Form a capacitor bank of 15 Mvar, operating voltage 44 kV in wye configuration from unit capacitor sizes in Table 11.2. Expulsion fuses are to be used for fusing of individual capacitor units.

The line-to-line voltage is 44 kV and the line-to-neutral voltage is 25.40 kV. If an overvoltage factor of 10% is considered, it gives 27.94 kV. Referring to Table 11.3, 14.4-kV elements can be selected and provided in two series groups. The bank's size is 15 Mvar. This means a rating of 5 Mvar per phase is required. As there are two series groups, each group can be rated for 2.5 Mvar. Then,

$(2.5)\left(\frac{28.8}{25.40}\right)^2 = 3.214$ Mvar. As 400-kvar units are being used, 8.03 units per phase will be required. If we use 8 units, the effective Mvar at operating voltage of 44 kV will be 14.93 instead of desired 15 Mvar. Check that the bank thus formed meets the requirements in Table 4.4 for the number of units in Y, and also that expulsion fuses can be used. It is obvious that when a certain bank size is to be formed, it will be slightly lower or higher than the required Mvar and the study must account for the actual bank size.

11.9   SWITCHING TRANSIENTS OF CAPACITOR BANKS

Historically, capacitor-switching transients have caused problems that have been studied in the existing literature. During the period from the late 1970s to 1980s, switching of capacitor banks in transmission systems caused high phase-to-phase voltages on transformers and magnification of transients at consumer-end distribution capacitors. Problems with switchgear restrikes caused even higher transients. Problems were common in industrial distributions with capacitors and DC drive systems, and the advent of pulse width modulated (PWM) inverters created a whole new concern of capacitor switching.

On connecting to a power source, a capacitor is a sudden short circuit because the voltage across the capacitor cannot change suddenly. The voltage of the bus to which the capacitor is connected will dip severely. This voltage dip and the transient step change are a function of the source impedance behind the bus. The voltage will then recover through a high-frequency oscillation. In the initial oscillation, the transient voltage can approach 2 per unit of the bus voltage. The initial step change and the subsequent oscillations are important. As these are propagated in the distribution system, these can couple across transformers and can be magnified. Transformer failures have been documented [11]. Surge arresters and surge capacitors can limit these transferred overvoltages and also reduce their frequency [12,13]. In a part-winding resonance, the predominant frequency of the transient can coincide with a natural frequency of the transformer. Secondary resonance, described in Chapter 8, is a potential problem, and the sensitive loads connected in the distribution system may trip.

The inrush current and frequency on capacitor current switching can be calculated by the solution of the following differential equation:

$$iR + L\frac{di}{dt} + \int \frac{idt}{C} = E_m \sin \omega t \tag{11.33}$$

The solution to this differential equation is discussed in many texts and is of the form:

$$i = A \sin(\omega t + \alpha) + Be^{-Rt/2L} \sin(\omega_0 t - \beta) \tag{11.34}$$

where

$$\omega_0 = \sqrt{\frac{1}{LC} - \frac{R^2}{4L^2}} \tag{11.35}$$

The first term is a forced oscillation, which in fact is the steady-state current, and the second term represents a free oscillation and has a damping component $e^{-Rt/2L}$. Its frequency is given by $\omega_0/2\pi$. Resistance can be neglected, and this simplifies the solution. The maximum inrush current is given at an instant of switching when $t = \sqrt{(LC)}$. For the purpose of evaluation of switching duties of circuit breakers, the maximum inrush current on switching an isolated bank is

$$i_{max,peak} = \frac{\sqrt{2}E_{LL}}{\sqrt{3}}\sqrt{\frac{C_{eq}}{L_{eq}}} = 1.300\sqrt{\frac{kvar_c}{L_{eq}}} \tag{11.36}$$

where $i_{max,peak}$ is the peak inrush current in amperes, without damping, E_{LL} is the line-to-line voltage in volts, C_{eq} is the equivalent capacitance in farads, and L_{eq} is the equivalent inductance in henrys, and $kvar_c$ is the capacitor being switched. (All inductances in the switching circuit including that of cables and buses must be considered.) The frequency of the inrush switching current f is given by

$$f = \frac{1}{2\pi\sqrt{L_{eq}C_{eq}}} \tag{11.37}$$

The voltage across the capacitor, which will also be the bus voltage in Laplace transform, is given by

$$V_c(s) = \frac{i(s)}{sC} = \frac{V}{LC}\left[\frac{1}{s\left(s^2 + 1/T^2\right)}\right] \tag{11.38}$$

Resolve into partial fractions:

$$V_c(s) = \frac{V}{LC(1/T^2)} \left[\frac{1}{s} - \frac{s}{(s^2 + 1/T^2)} \right] \qquad (11.39)$$

Taking inverse transform:

$$V_c = V \left(1 - \cos \frac{1}{\sqrt{LC}} t \right) = V(1 - \cos \omega_0 t) \qquad (11.40)$$

Thus, the maximum voltage occurs at

$$\omega_0 t = \pi \qquad (11.41)$$

On back-to-back switching (i.e., switching a dead bank on the same bus to which an energized bank is connected) can result in high-current magnitudes and frequencies as the interconnecting inductance between the two banks on the same bus will be small.

The switching inrush current and frequency can be calculated from

$$i_{max,peak} = \frac{\sqrt{2}E_{LL}}{\sqrt{3}} \sqrt{\frac{C_1 C_2}{(C_1 + C_2)L_m}} \qquad (11.42)$$

And the frequency of the inrush current is

$$f = \frac{1}{2\pi \sqrt{L_m \frac{C_1 C_2}{(C_1 + C_2)}}} \qquad (11.43)$$

where C_1 and C_2 are the sizes of the capacitors in farad and L_m is the inductance between them in henry. Apart from stresses on the circuit breaker, the high-frequency transient can stress the other equipment too.

Figure 11.18 [14,15] shows the statistical analysis of the overvoltages on capacitors C_2 and C_3 of single-tuned filters for furnace installation when switched together. The overvoltage is a function of ratio C_2/C_3, and the ratio of system short-circuit level MVA$_{sc}$ and effective Mvar of capacitors Mvar$_c$, given by $(V^2 h^2)/(X_c(1 - h^2))$. Arc furnace installation requires careful studies due to transients:

- Restrikes in circuit breakers
- Switching of transformers
- Faults on the buses
- Energizing the harmonic filters.

These can create overvoltages on buses and in harmonic filters [15].

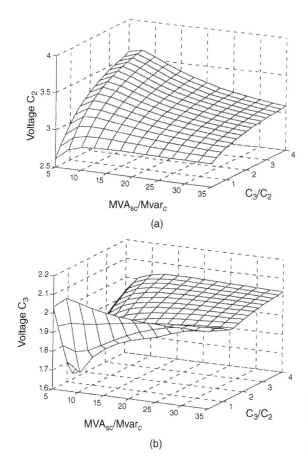

Figure 11.18 (a and b) Overvoltages on capacitors of single-tuned filters, arc furnace installation.

11.10 CONTROL OF SWITCHING TRANSIENTS

The capacitor-switching transients can be controlled by the following:

- Series inrush current-limiting reactors. However, care has to be exercised that when the series reactor is used solely for inrush current limitations, it may form a single-tuned filter and give rise to uncalled-for harmonic resonance
- Resistance switching
- Point-of-wave switching (synchronous breakers)
- Application of surge arresters
- Dividing the capacitor bank into smaller size banks. The smaller the size of the capacitor bank being switched, the lesser is the transient
- Avoiding application of capacitors at multivoltage levels to eliminate possibilities of secondary resonance, MOVs can be applied at lower voltage buses
- Converting the capacitor banks to capacitor filters because filter reactor reduce the inrush current amplitude and its frequency, though may prolong the decay

of the transient. This is one effective way to mitigate transients, eliminate harmonic resonance, and control harmonic distortion. Active filters can be used depending on the size of the capacitive compensation and harmonic mitigation required for a particular system

- Providing current-limiting reactors and chokes, which is a must for back-to-back switching and to have acceptable capacitor switching duties on circuit breakers and switching devices
- Considering steady-state voltage rise due to application of capacitors. The transformer taps may have to be adjusted.

11.10.1   Resistance Switching

In AC current interruption technology, the use of switching resistors in high-voltage breakers is well implemented to reduce the overvoltages and frequency of TRV. In medium-voltage cubical type or metal-clad circuit breakers, as the switching resistors are not integral to the breakers, two breakers can be used, as shown in Fig. 11.19, to preinsert the resistor for a short duration of four to six cycles.

Figure 11.20 illustrates a basic circuit of resistance switching. A resistor r is provided in parallel with the breaker pole and R, L, and C are the system parameters on the source side of the break. Consider the current loops in this figure. Following equations can be written as

$$u_n = iR + L\frac{di}{dt} + \frac{1}{C}\int i_c dt \tag{11.44}$$

$$\frac{1}{C}\int i_c dt = i_r r \tag{11.45}$$

$$i = i_r + i_c \tag{11.46}$$

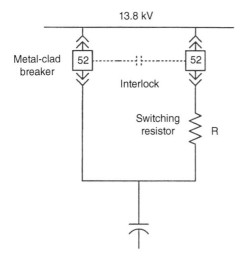

Figure 11.19   A circuit for resistance switching using two interlocked medium-voltage metal-clad circuit breakers.

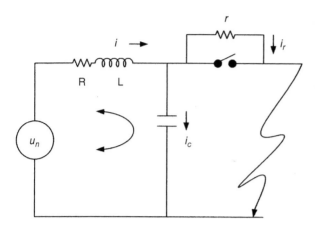

Figure 11.20 Circuit configuration for resistor switching; calculation of switching resistor.

This gives

$$\frac{d^2 i_r}{dt^2} + \left(\frac{R}{L} + \frac{1}{rC}\right)\frac{di_r}{dt} + \left(\frac{1}{LC} + \frac{R}{rLC}\right)i_r = 0 \qquad (11.47)$$

The frequency of the transient is given by

$$f_n = \frac{1}{2\pi}\sqrt{\frac{1}{LC} - \frac{1}{4}\left(\frac{R}{L} - \frac{1}{rC}\right)^2} \qquad (11.48)$$

In power systems, R is $<< L$. If a parallel resistor across the contacts of value $r < \frac{1}{2}\sqrt{L/C}$ is provided, the frequency reduces to zero. The value of r at which frequency reduces to zero is called the critical damping resistor. The critical resistance can be evaluated in terms of the system short-circuit current, I_{sc}

$$r = \frac{1}{2}\sqrt{\frac{u}{I_{sc}\omega C}} \qquad (11.49)$$

With properly selected preinsertion resistors, the switching transient can be totally eliminated except a minor hash in the first cycle.

11.10.2 Point-of-Wave Switching or Synchronous Operation

In all the above-mentioned examples, the transients are calculated with the switch closed at the peak of the voltage wave. A breaker can be designed to open or close with reference to the system voltage sensing and zero crossing. The switching device must have enough dielectric strength to withstand system voltage till its contacts touch. The consistency of closing within ±0.5 ms is possible. Grounded capacitor banks are closed with three successive phase-to-ground voltages reaching zero, for example, 60° separations. Ungrounded banks are controlled by closing the first two phases at a phase-to-phase voltage of zero, and then delaying the third phase 90° when phase-to-ground voltage is zero.

11.11   SWITCHING CAPACITORS WITH MOTORS

The switching of capacitors with motors in the presence of harmonic producing loads can give rise to resonance and impact filter design. As some motors may be out of service, and so their power factor improvement capacitors, the resonant frequency can vary over wide limits, which will be troublesome to analyze.

Power factor improvement capacitors are commonly applied in industrial power systems and may raise the operating power factor of the motors. The power factor of low-speed induction motors decreases due to leakage reactance of stator overhang windings. Generally, the induction motors do not operate at full load, which further lowers the operating power factor. Even though the power factor of the motor varies significantly with load, its reactive power requirement does not change much. Thus, with the application of power factor improvement capacitor, the motor power factor from no-load to full load will not vary much. The capacitors may be (1) connected directly to the motor terminals, (2) the capacitor and motor are switched as a unit, and (3) the capacitor is switched independently of the motor contactor through a separate switching breaker interlocked with the motor starting breaker.

Switching of the capacitors and motor as a unit can result into problems due to

- presence of harmonic currents,

- overvoltages due to self-excitation,

- excessive transient torques and inrush currents due to out-of-phase closing.

Overvoltages due to self-excitation are important for proper application of capacitors with induction motors.

The magnetizing current of the induction motors varies with the motor design. Premium high-efficiency motors operate less saturated than the previous U or T frame designs [16]. The motor and capacitor combination in parallel will circulate a current between motor and capacitor corresponding to their terminal voltage. In this manner, the network is said to *self-excite*. This is shown in Fig. 11.21(a) and (b). The same size of capacitor applied to a standard and high-efficiency motor has different results because of the motor magnetizing characteristics. In the case of high-efficiency design motor, it raises the terminal voltage to 700 V.

A capacitor size can be selected for the power factor improvement based on the following equation:

$$\mathrm{kvar}_c \leq \sqrt{3} I_0 \sin \phi_0 \qquad (11.50)$$

where kvar_c is the maximum capacitor that can be applied, I_o is the motor no-load current, and ϕ_0 is the power factor angle. This means that the capacitor size does not exceed the motor no-load reactive kvar. However, this no-load excitation motor data is not readily available, and the size of capacitor application for a specific motor design should be based on the manufacturer's recommendations. The capacitors should not be directly connected to the motor terminals when

- solid-state starters are used,

- open transition starting methods are applied,

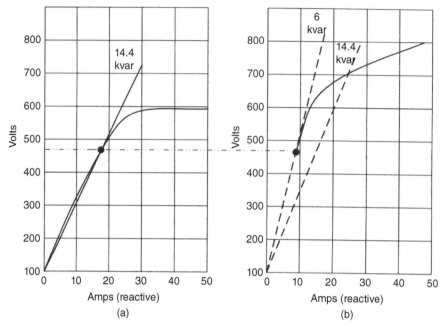

Figure 11.21 Self-excitation of motors with switched capacitors. (a) Motor of standard design and (b) motor of high-efficiency design.

- the motor is meant for repeated switching, jogging, inching, or plugging,
- a reversing or multispeed motor is used.

A 250-kvar capacitor is switched with a 2000-hp, 2.3-kV induction motor connected to a 2.5-MVA, 13.8–2.4-kV transformer of 6% impedance; the 13.8-kV, three-phase, short-circuit level is 29-kA rms symmetrical. The inertia constant $H = 0.5$. The inrush current profile in the capacitor is shown in Fig. 11.22, which is approximately eight times the rated current. This will increase the motor stating current and torque transients.

It is best to avoid power factor improvement capacitors switched with motors in presence of harmonic producing loads.

11.12   SWITCHING DEVICES

With respect to the application of circuit breakers for capacitance switching, consider (1) the type of application that is overhead line, cable or capacitor bank, or shunt capacitor filter switching, (2) power frequency and system grounding, and (3) the presence of single or two-phase to ground faults.

ANSI ratings [8] distinguish between general-purpose and *definite-purpose* circuit breakers, and there is a vast difference between their capabilities for capacitance current switching. Definite-purpose breakers may have different constructional features, that is, a heavy-duty closing and tripping mechanism

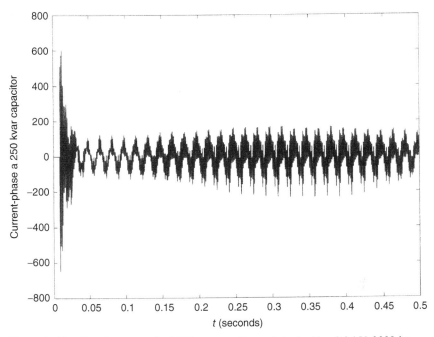

Figure 11.22   Inrush current of a 250 kvar capacitor switched with a 2.3-kV, 2000-hp, induction motor.

and testing requirements. Table 11.6 [8] shows the capacitance current switching ratings. General-purpose breakers do not have any back-to-back capacitance current-switching capabilities. A 121-kV general-purpose circuit breaker of rated current 2-kA and 63-kA symmetrical short circuit has overhead line charging current or isolated capacitor switching current capability of 50 A rms and no back-to-back switching capability.

Rated transient inrush current is the highest magnitude, which the circuit breaker will be required to close at any voltage up to the rated maximum voltage and will be measured by the system and unmodified by the breaker. Rated transient inrush frequency is the highest natural frequency, which the circuit breaker is required to close at 100% of its rated back-to-back shunt capacitor or cable switching current.

The following definitions are applicable:

1. Rated open wire line charging current is the highest line charging current that the circuit breaker is required to switch at any voltage up to the rated voltage.

2. Rated isolated cable charging and isolated shunt capacitor bank switching current is the highest isolated cable or shunt capacitor current that the breaker is required to switch at any voltage up to the rated voltage.

3. The cable circuits and switched capacitor bank are considered isolated if the rate of change of transient inrush current, di/dt, does not exceed the maximum rate of change of symmetrical interrupting capability of the circuit breaker at

TABLE 11.6 Preferred Capacitance Current Switching Rating for Outdoor Circuit Breakers 121 kV and Above, Including Circuit Breakers Applied in Gas-Insulated Substations

Rated Maximum Voltage (kV)	Rated Short-Circuit Current at Rated Maximum Voltage (kA, rms)	Rated Continuous Current at 60 Hz (A, rms)	General-Purpose Circuit Breakers Rated Overhead Line Current (A, rms)	General-Purpose Circuit Breakers, Rated Isolated Current (A, rms)	Overhead Line Current (A, rms)	Rated Isolated Current (A, rms)	Back-to-Back Switching Current (A, rms)	Inrush Current Peak current (kA)	Frequency (Hz)
123	31.5	1200, 2000	50	50	160	315	315	16	4250
123	40	1600, 2000, 3000	50	50	160	315	315	16	4250
123	63	2000, 3000	50	50	160	315	315	16	4250
145	31.5	1200, 2000	80	80	160	315	315	16	4250
145	40	1600, 2000, 3000	80	80	160	315	315	16	4250
145	63	2000, 3000	80	80	160	315	315	16	4250
145	80	2000, 3000	80	80	160	315	315	16	4250
170	31.5	1600, 2000	100	100	160	400	400	20	4250
170	40	2000, 3000	100	100	160	400	400	20	4250
170	50	2000, 3000	100	100	160	400	400	20	4250
170	63	2000, 3000	100	100	160	400	400	20	4250

The columns Overhead Line Current, Rated Isolated Current, Back-to-Back Switching Current, Peak current, and Frequency fall under the heading: Definite-Purpose Breakers Rated Capacitance Switching Current Shunt Capacitor Bank or Cable.

245	31.5	1600, 2000, 3000	160	160	200	400	400	20	4250
245	40	2000, 3000	160	160	200	400	400	20	4250
245	50	2000, 3000	160	160	200	400	400	20	4250
245	63	2000, 3000	160	160	200	400	400	20	4250
362	40	2000, 3000	250	250	315	500	500	25	4250
362	63	2000, 3000	250	250	315	500	500	25	4250
550	40	2000, 3000	400	400	500	500	500	25	4250
550	63	3000, 4000	400	400	500	500	500	25	4250
800	40	2000, 3000	500	500	500	500	–	–	–
800	63	3000, 4000	500	500	500	500	–	–	–

Source: Ref. [8].

the applied voltage [8]:

$$\left(\frac{di}{dt}\right)_{max} = \sqrt{2\omega}\left[\frac{\text{rated maximum voltage}}{\text{operating voltage}}\right]I \qquad (11.51)$$

where I is the rated short-circuit current in amperes.

4. Cable circuits and shunt capacitor banks are considered switched back-to-back if the highest rate of change of inrush current on closing exceeds that for which the cable or shunt capacitor can be considered isolated.

The oscillatory current on back-to-back switching is limited only by the impedance of the capacitor bank and the circuit between the energized bank and the switched bank.

The 2005 revision to IEEE standard [17,18] and also IEEE standard [19], classify the breakers for capacitance switching as classes C_0, C_1, and C_2, as well as mechanical endurance class M_1 or M_2 are assigned. These classes have specific type testing duties, coordinated as per standards, and is an attempt to harmonize with IEC standards [20].

Class C_1 circuit breaker is acceptable for medium-voltage circuit breakers and for circuit breakers applied for infrequent switching of transmission lines and cables. Class C_2 is recommended for frequent switching of transmission lines and cables [17]. An important consideration is the transient overvoltages that may be generated by restrikes during opening operation. The effect of these transients will be local as well as remote.

Example 11.6:   A 6-Mvar capacitor bank at 13.8-kV, ungrounded-wye connected has a source impedance of $0.02655 + j0.236$ ohms at the source. Calculate the peak inrush current and its frequency. Does it meet the requirement of switching an isolated bank? The 13.8-kV definite-purpose circuit breaker for switching has the following ANSI/IEEE ratings:

Rated current = 1200 A, maximum peak inrush current = 18 kA, inrush current frequency = 2.4 kHz, short-circuit interrupting rating of breaker 40 kA [8].

From Eq. (11.36),

$$I_{peak} = \frac{\sqrt{2}}{\sqrt{3}} \times 13.8 \times 10^3 \sqrt{\frac{83.6 \times 10^{-6}}{(0.236)/2\pi f}} = 4.1 \text{ kA}$$

And its frequency is

$$f_{inrush} = \frac{1}{2\pi\sqrt{83.6 \times 10^{-6}(0.236/2\pi f)}} = 695 \text{ Hz}$$

The maximum rate of change of current is

$$2\pi(695)(4100) \times 10^{-6} = 17.91 \text{ A}/\mu s$$

From Eq. (11.51), the breaker di/dt is

$$2\pi(60)\sqrt{2}(40 \times 10^3) \times 10^{-6} = 21.32 \text{ A}/\mu s$$

This is greater than 17.91 A/µs; the capacitor bank can be considered isolated.

Figure 11.23 shows an EMTP simulation of the switching transient current in the three phases. The switch is closed when the voltage wave is passing through zero in phase *a*, which gives the maximum current in this phase. Figure 11.24 depicts the bus voltage profile. The decay depends on system resistance.

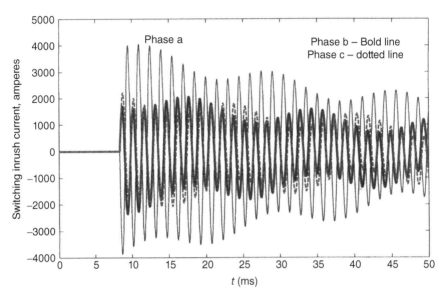

Figure 11.23   EMTP simulation of the inrush current of a 13.8-kV, 6-Mvar bank.

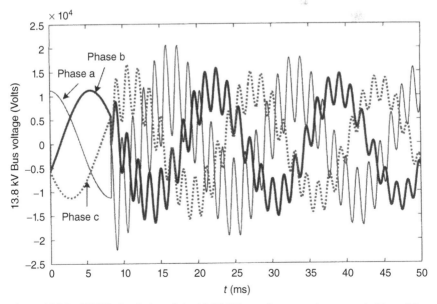

Figure 11.24   EMTP simulation of the 13.8-kV bus voltage transient on switching of the 6-Mvar capacitor bank.

Example 11.7: In Fig. 9.18, a 6-Mvar capacitor bank is shown. Consider that a similar capacitor bank of 6 Mvar is connected to the same bus. One of the capacitor bank is in service while the other is switched on, a back-to-back switching operation. The inductances of the buses, cables, and capacitor banks must be carefully calculated and a circuit diagram for study of back-to-back switching can be constructed as shown in Fig. 11.25. Note that capacitor banks also have some inductance. Once all the inductances have been calculated, the back-to-back switching transients can be estimated. Note that supply source reactance will have a minimum impact on the transients; in hand calculations, it can be ignored.

The total inductance between switched banks is 16.9 μH, using Eqs (11.42) and (11.43), the inrush current is 17.7 kA and its frequency is 5.88 kHz. This exceeds the capabilities of "definite-purpose" breaker. The back-to-back switching of capacitors connected to the same bus through small lengths of cables is not acceptable. Additional reactance is required to be introduced. Figure 11.26 is an EMTP simulation of the back-to-back switching, which confirms the results of hand calculations.

The protection, transients, and switching devices discussed earlier form an introduction. An interested reader may see Refs [21–26].

11.12.1 High Voltages on CT Secondaries

When a number of capacitors are connected to a bus and a fault occurs on a feeder circuit connected to this bus, all the connected capacitors will discharge into the fault.

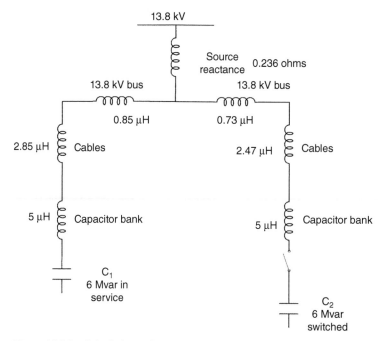

Figure 11.25   Calculations of reactances on back-to-back switching of 6-Mvar capacitor banks.

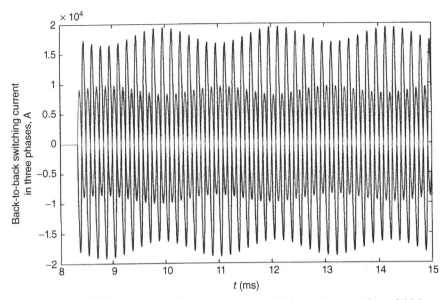

Figure 11.26   EMTP simulation of the back-to-back switching current transient of 6-Mvar capacitor banks.

The BCT (bushing current transformer) secondary voltage may reach high values. This secondary voltage can be estimated from

$$
\left(\frac{1}{\text{BCT ratio}} \right) (\text{crest transient current}) \times
$$

$$
\left((\text{relay burden}) \left(\frac{\text{transient frequency}}{\text{system frequency}} \right) \right) \tag{11.52}
$$

These high voltages are of concern for the health of the CT, and the CT saturation alone may not limit these voltages [7].

Example 11.8: A system configuration at 44 kV is considered, which is provided with 15 Mvar of capacitor bank for power factor improvement. A three-phase fault occurs on a feeder connected to this bus, some distance away from the bus. The total circuit inductance from the installation of the capacitor bank to the fault point is small, consisting of a section of bus length, connections to the capacitor bank, and 20' of cable from the bus to the point of fault. The total calculated inductance is 75 μH. The CT has a ratio of 600/5. The CT secondary burden, which is sum of the CT secondary resistance, CT leads, and the protective relay burden is $1 + j1$ ohms. Calculate the voltage across the CT secondary.

The capacitance of the 15-Mvar bank is 20.40 μF.

Then, the frequency of oscillation is

$$f = \frac{10^6}{2\pi\sqrt{75 \times 20.40}} = 4068\,\text{Hz}$$

The maximum inrush current is

$$\sqrt{\frac{2}{3}} \times 10^3 \times 44\sqrt{\frac{20.4}{75}} = 18.73\,\text{kA}$$

Therefore, secondary current = 156 A at 4068 Hz.
Then from Eq. (11.52), a maximum CT secondary voltage is

$$\sqrt{(156)^2 + 156^2(67.8)^2} = 10.58\,\text{kV}$$

If the ratio of the CT is 1200/5, this reduces to approximately 5.29 kV, still too high. Surge suppressors are provided across CT secondary windings as shown in Fig. 11.27.

11.13 SWITCHING CONTROLS

Capacitors can be switched in certain discrete steps and do not provide a stepless control of voltage. If a large capacitor bank is continuously left in service, it may create

- Unacceptable overvoltages

- Excess reactive power can be generated, which overcompensates the lagging reactive power load demand and makes the power factor leading. It will reverse the reactive power flow in the utility source interconnection and may result in nuisance operation of protective devices, apart from overvoltages

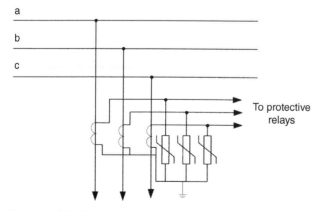

Figure 11.27   Surge arresters on the secondary side of current transformers to limit overvoltages on a capacitor discharge current.

- The maximum size of a capacitor filter that can be switched requires careful considerations, with respect to switching transients.

Often, it is necessary to switch capacitors in some predetermined steps in certain sizes. The switching control methods include the following:

- Power factor control
- Reactive power or active power sensing
- Load current-dependent control
- Voltage-dependent control.

The pros and cons of these switching strategies are not discussed. Figure 11.28 shows a two-step sequential reactive power switching control to maintain voltage within a certain band. As the reactive power demand increases and the voltage falls (shown linearly in Fig. 11.28 for simplicity), the first bank is switched at A, which compensates the reactive power and suddenly raises the voltage. The second step switching is, similarly, implemented at point B. On reducing demand, the banks are taken out from service at A′ and B′. A time delay is associated with switching in either direction to override transients.

Example 11.9:   A 100-MVA load at 0.8 power factor and 13.8 kV is supplied through a short transmission line of 0.943 per unit reactance (100-MVA base). Size a capacitor bank at load terminals to limit the voltage drop to 2% at full load. Ascertain number of switching steps and the capacitor sizes to maintain the load voltage within a band of ±2% of rated voltage as the load varies from zero to full load.

100 MVA at 0.8 power factor at 13.8 kV gives a current of 4183 A < −36.87°.

V_r is unknown. First consider that $V_r = V_s$ (i.e., the receiving end voltage is equal to sending end voltage).

The line reactance = 0.1 pu = 0.1794 ohms.

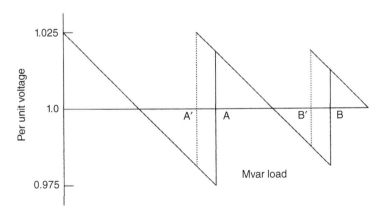

Figure 11.28   Reactive power-dependent switching controls of capacitors to maintain voltage within a certain plus minus band.

Then, voltage drop through the line is

$$I_r Z = (4184)(0.8 - j0.6)(j0.1794) = 4504 + j600$$

This gives receiving end voltage of

$$7967.7 - 4504 - j600 = 3464 < -9.8°$$

The load is considered as a constant impedance load; therefore, the load impedance is

$$Z_l = \frac{13800}{\sqrt{3} \times 4183 < -36.8°} = 1.90 < 36.8°$$

Therefore, in the second iteration, the load current is

$$\frac{3464 < -9.8°}{1.90 < 36.8°} = 1269 - j1346$$

The receiving end voltage is

$$7967.7 - (1269 - j1346)(j0.1794) = 7726.2 - j227$$

The voltage at receiving end can be solved like a load flow problem, and after a couple of iterations it is
$0.941 < -4.3°$ pu. The load supplied is 71 MW and 53.1 Mvar.
The current is

$$\frac{7489 < -4.3°}{1.90 < 36.8°} = 2967 - j2600 \text{ A}$$

Sending end power $= V_s I_s^* = (7967.7)(2967 - j2600) = 23.71 \times 10^6 + j20.73 \times 10^6$ per phase.

Approximately, 10 Mvar is lost in the transmission line.

It is a short line; therefore, the line susceptance is neglected. At no load, the receiving end voltage is equal to sending end voltage. At full load, it dips by 5.9%. The capacitor size to limit the voltage dip to 2% is approximately given by equation:

$$|V_{raised}| = |V_{available}| + \left[\frac{X_{th}}{|V_{available}|}\right] I_c \tag{11.53}$$

where X_{th} is the Thevenin reactance of the line. Here, $X_{th} = 0.1794$ ohms.

This gives $Q_c = 38.9$ Mvar. If a capacitor bank of this rating is switched at the maximum load, the voltage will improve to 0.98 pu. But this will not meet the requirement of maintaining the receiving end voltage within $\pm 2\%$ band from zero to full load.

When the line is carrying a load of 27 MW, the voltage dip will be 2%. If we switch a capacitor bank of 26.4 Mvar at this point, the voltage will rise to 1.02 pu.

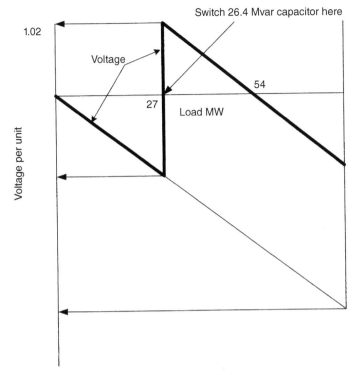

Figure 11.29   Calculation of capacitor bank size and switching control to maintain a certain voltage profile.

Then, as the load develops, the voltage will behave as shown in Fig. 11.29. Thus, by switching a much lower capacitor bank at partial load of 27 MW, the line regulation can be maintained within plus minus 2%. Practically voltage-load profile will not be linear.

The example shows that a proper study is required to determine required shunt capacitor sizes and their switching controls.

Although we have discussed fuse protection and unbalance protection of the capacitor banks, this is vast subject, much involved, and requires careful application and coordination between the system and the capacitor banks, including use of surge arresters. The application of protective relaying is not covered in this book. IEEE Guide for the Protection of Shunt Capacitor Banks [7] is a good starting point and it provides a large bibliography for further reading.

REFERENCES

1. J.J. Grainger and S.H. Lee. "Optimum size and location of shunt capacitors for reduction of losses on distribution feeders," IEEE Transactions on Power and Systems, vol. PAS 100, no. 3, pp. 1105–1118, 1981.

2. S.H. Lee and J.J. Gaines. "Optimum placement of fixed and switched capacitors of primary distribution feeders," IEEE Transactions on Power and Systems, vol. PAS 100, 345–352, 1981.
3. S.K. Chan. "Distribution system automation," Ph.D. dissertation, University of Texas at Arlington, 1982.
4. IEEE Draft P1036/D13a. Draft guide for the application of shunt capacitors, 2006.
5. IEEE Standard 18. Standard for shunt capacitor banks, 2002.
6. UL 810. Capacitors, 2008.
7. IEEE Standard C37.99. Guide for protection of shunt capacitor banks, 2005.
8. ANSI/IEEE Std C37.06-1987 (R 2000). AC high voltage circuit breakers rated on a symmetrical current basis—preferred ratings and related required capabilities, 2000.
9. J.C. Das. Transients in Electrical Systems. McGraw-Hill, New York, 2011.
10. S.R. Mendis, M.T. Bishop, J.C. McCall and W.M. Hurst. "Overcurrent protection of capacitors applied in industrial distribution systems," IEEE Transactions on Industry Applications, vol. 29, no. 3, 1993.
11. R.S. Bayless, J.D. Selmen, D.E. Traux and W.E. Reid. "Capacitor switching and transformer transients," IEEE Transactions on Power Delivery, vol. 3, no. 1, pp. 349–357, 1988.
12. M. McGranagham, W.E. Reid, S. Law and Dgresham. "Overvoltage protection of shunt capacitor banks using MOV arresters," IEEE Transactions on Power and Systems, vol. 104, no. 8, pp. 2326–2336, 1984.
13. IEEE Report by Working Group 3.4.17, "Surge protection of high voltage shunt capacitor banks on ac power systems---Survey results and application considerations," IEEE Transactions on Power Delivery, vol. 6, no. 3, pp. 1065–1072, 1991.
14. R.F. Dudley, C.L. Fellers and J.A. Bonner. "Special design considerations for filter banks in arc furnace installations," IEEE Transactions on Industry Applications, vol. 33, no. 1, pp. 226–233, 1997.
15. S.R. Mendis and D.A. Gonzalez. "Harmonic and transient overvoltage analyses in arc furnace power systems," IEEE Transactions on Industry Applications, vol. 28, no. 2, pp. 336–342, 1992.
16. IEEE Standard 141. IEEE recommended practice for electrical power distribution for industrial plants, 2009.
17. ANSI/IEEE Standard C37.012. Application guide for capacitance current switching for AC high voltage circuit breakers rated on symmetrical current basis, 2005 (Revision of 1979).
18. IEEE Standard C37.04a. IEEE standard rating structure for AC-high voltage circuit breakers rated on symmetrical current basis, Amendment 1: capacitance current switching, 2003.
19. IEEE PC37.06/D-11. Draft standard AC high voltage circuit breakers rated on symmetrical current basis-preferred ratings and related required capabilities for voltages above 1000 volts, 2008.
20. J.C. Das and D.C. Mohla. "Harmonization with the IEC, ANSI/IEEE standards for HV breakers," IEEE Industry Application Magazine, vol. 19, pp. 16–26, 2013.
21. J.C. Das. "Analysis and control of large shunt capacitor bank switching transients," IEEE Transactions on Industry Applications, vol. 41, no. 6, pp. 1444–1451, 2005.
22. J.C. Das. "Effects of medium voltage capacitor bank switching surges in an industrial distribution system," in Conference Record, IEEE IC&PS Conference, Pittsburgh, pp. 57–64, 1992.
23. R.W. Alexander. "Synchronous closing control for shunt capacitors," IEEE Transactions on Power and Systems, vol. PAS-104, no. 9, pp. 2619–2626, 1985.
24. IEEE Power System Relaying Committee. "Static VAR compensator protection," IEEE Transactions on Power Delivery, vol. 10, no. 3, pp. 1224–1233, 1995.
25. H.M. Pflanz and G.N. Lester. "Control of overvoltages on energizing capacitor banks," IEEE Transactions on Power and Systems, vol. PAS-92, no. 3, pp. 907–915, 1973.
26. R.P.O. Leary and R.H. Harner. "Evaluation of methods for controlling overvoltages reduced by energization of shunt power capacitor," CIGRE 13-05, 1988 Paris Meeting.

CHAPTER 12

MODELING OF SYSTEM COMPONENTS FOR HARMONIC ANALYSIS

Harmonic power flow analysis requires that the system components are adequately modeled with respect to frequency. Commercially available softwares have varying capabilities. This chapter provides the fundamental basis of modeling of major system components.

12.1 TRANSMISSION LINES

12.1.1 *ABCD* Constants

A transmission line of any length can be represented by a four-terminal network (Fig. 12.1(a)). In terms of A, B, C, and D constants, the relation between sending and receiving end voltages and currents can be expressed as

$$\begin{vmatrix} V_s \\ I_s \end{vmatrix} = \begin{vmatrix} A & B \\ C & D \end{vmatrix} \begin{vmatrix} V_r \\ I_r \end{vmatrix} \tag{12.1}$$

In the case where sending end voltages and currents are known, the receiving end voltage and current can be found by

$$\begin{vmatrix} V_r \\ I_r \end{vmatrix} = \begin{vmatrix} D & -B \\ -C & A \end{vmatrix} \begin{vmatrix} V_s \\ I_s \end{vmatrix} \tag{12.2}$$

The significance of these constants can be stated as follows:

$A = V_s/V_r$, when $I_r = 0$, that is, the receiving end is open circuited. It is the ratio of two voltages and is dimensionless.

$B = V_r/I_r$, when $V_r = 0$, that is, the receiving end is short circuited. It has the dimensions of an impedance and specified in ohms.

$C = I_r/V_r$, when the receiving end is open circuited and I_r is zero. It has the dimensions of an admittance.

Power System Harmonics and Passive Filter Designs, First Edition. J.C. Das.
© 2015 The Institute of Electrical and Electronics Engineers, Inc. Published 2015 by John Wiley & Sons, Inc.

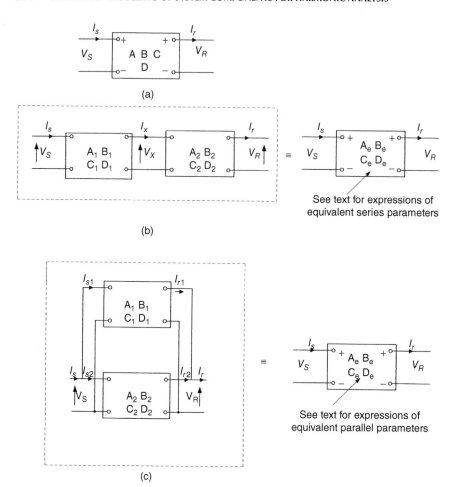

Figure 12.1 (a) Schematic representation of two-terminal network using *ABCD* constants, (b) two-terminal networks in series, and (c) two-terminal networks in parallel.

$D = I_s/I_r$, when $V_r = 0$, that is, the receiving end is short circuited. It is the ratio of two currents and dimensionless.

Two *ABCD* networks in series, Fig. 12.1(b) can be reduced to a single equivalent network:

$$\begin{vmatrix} V_s \\ I_s \end{vmatrix} = \begin{vmatrix} A_1 & B_1 \\ C_1 & D_1 \end{vmatrix} \begin{vmatrix} A_2 & B_2 \\ C_2 & D_2 \end{vmatrix} \begin{vmatrix} V_r \\ I_r \end{vmatrix} = \begin{vmatrix} A_1A_2 + B_1C_2 & A_1B_2 + B_1D_2 \\ C_1A_2 + D_1C_2 & C_1B_2 + D_1D_2 \end{vmatrix} \begin{vmatrix} V_r \\ I_r \end{vmatrix} \quad (12.3)$$

For parallel *ABCD* networks, Fig. 12.1(c), the combined *ABCD* constants are

$$A = \frac{(A_1B_2 + A_2B_1)}{(B_1 + B_2)}$$

$$B = \frac{B_1 B_2}{(B_1 + B_2)}$$

$$C = \frac{(C_1 + C_2) + (A_1 - A_2)(D_2 - D_1)}{(B_1 + B_2)}$$

$$D = \frac{(B_2 D_1 + B_1 D_2)}{(B_1 + B_2)} \tag{12.4}$$

12.1.2 Models with Respect to Line Length

Transmission lines of length less than approximately 50 miles can be considered short lines. The shunt capacitance and shunt conductance of the lines can be neglected. For transmission lines in the range 50–150 miles (80–240 km), the shunt admittance cannot be neglected. There are two models in use, the nominal Π- and nominal T-circuit models. In the T-circuit model, the shunt admittance is connected at the midpoint of the line, while in the Π model, it is equally divided at the sending end and the

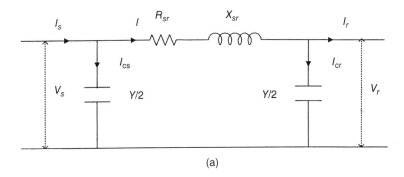

(a)

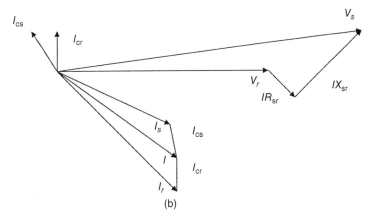

(b)

Figure 12.2   (a) Π model and (b) phasor diagram.

TABLE 12.1 *ABCD* **Constants of Transmission Lines**

Line Length	Equivalent Circuit	A	B	C	D
Short	Series impedance only	1	Z	0	1
Medium	Nominal Π	$1 + \frac{1}{2}\,YZ$	Z	$Y\left[1 + \frac{1}{4}(YZ)\right]$	$1 + \frac{1}{2}\,YZ$
Medium	Nominal T	$1 + \frac{1}{2}\,YZ$	$Z\left[1 + \frac{1}{4}(YZ)\right]$	Y	$1 + \frac{1}{2}\,YZ$
Long	Distributed parameters	$\cosh \gamma l$	$Z_0 \sinh \gamma l$	$\dfrac{\sinh \gamma l}{Z_0}$	$\cosh \gamma l$

receiving end. The Π equivalent circuit and phasor diagram are shown in Fig. 12.2(a) and (b). Similarly, the phasor diagram of the T model can be drawn. *ABCD* constants are shown in Table 12.1.

Example 12.1: Prove the A, B, C, D constants in Table 12.1 for a Π model.

The sending end current is equal to the receiving end current, and the current through the shunt elements $Y/2$ at the receiving and sending ends is

$$I_s = I_r + \frac{1}{2}V_r Y + \frac{1}{2}V_s Y$$

The sending end voltage is the vector sum of the receiving end voltage, and the drop through the series impedance Z is

$$V_s = V_r + \left(I_r + \frac{1}{2}V_r Y\right)Z = V_r\left(1 + \frac{1}{2}YZ\right) + I_r Z$$

The sending end current can, therefore, be written as

$$I_s = I_r + \frac{1}{2}V_r Y + \frac{1}{2}Y\left[V_r\left(1 + \frac{1}{2}YZ\right) + I_r Z\right]$$

$$= V_r Y\left(1 + \frac{1}{4}YZ\right) + I_r\left(1 + \frac{1}{2}YZ\right)$$

or in matrix form

$$\left|\begin{matrix} V_s \\ I_s \end{matrix}\right| = \left|\begin{matrix} \left(1 + \frac{1}{2}YZ\right) & Z \\ Y\left(1 + \frac{1}{4}YZ\right) & \left(1 + \frac{1}{2}YZ\right) \end{matrix}\right| \left|\begin{matrix} V_r \\ I_r \end{matrix}\right|$$

12.1.3 Long-Line Model

Lumping the shunt admittance of the lines is an approximation; for line lengths over 150 miles (240 km), the distributed parameter representation of a line is used. Each elemental section of line has a series impedance and shunt admittance associated with it. The operation of a long line can be examined by considering an elemental section of impedance z per unit length and admittance y per unit length (Fig. 12.3).

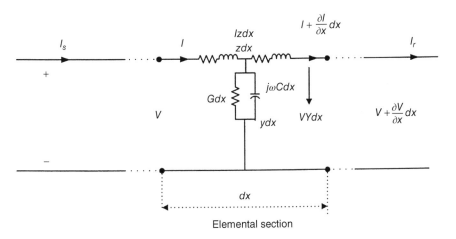

Elemental section

Figure 12.3   Elemental model of a long transmission line.

It can be shown that the voltage at any distance x from sending end can be written as

$$V_x = \left|\frac{V_r + Z_0 I_r}{2}\right| e^{\alpha x + j\beta x} + \left|\frac{V_r - Z_0 I_r}{2}\right| e^{-\alpha x - j\beta x} \qquad (12.5)$$

These equations represent traveling waves. The solution consists of two terms, each of which is a function of two variables, time and distance. At any instant of time, the first term, the incident wave, is distributed sinusoidally along the line, with amplitude increasing exponentially from the receiving end. After a time interval Δt, the distribution advances in phase by $\omega \Delta t / \beta$, and the wave is traveling toward the receiving end. The second term is the reflected wave, and after a time interval Δt, the distribution retards in phase by $\omega \Delta t / \beta$, and the wave is traveling from the receiving end to the sending end. Figure 12.4 depicts the incident and reflected *voltage* waves with load end of the line open circuited. The voltage and current reflection coefficients are discussed in Chapter 13. The incident wave progresses from source to load varying as $e^{\gamma x}$ end and reflected wave from load to source, varying in amplitude as $e^{-\gamma x}$, with initial value = incident voltage. The total incident voltage is the sum of two waves. The figure depicts four incidents of time. Current waves follow the same pattern, but with different reflection coefficients.

In the above equations, γ is the complex propagation constant written as

$$\gamma = \alpha + j\beta \qquad (12.6)$$

where α is defined as the *attenuation constant*. Common units are nepers per mile or per km. Equations for α and β, neglecting shunt conductance of the line, are given by

$$\alpha = |\gamma| \, \cos\left[\frac{1}{2} \tan^{-1}\left(\frac{-r_{sc}}{x_{sc}}\right)\right] \qquad (12.7)$$

$$\beta = |\gamma| \, \sin\left[\frac{1}{2} \tan^{-1}\left(-\frac{r_{sc}}{x_{sc}}\right)\right] \qquad (12.8)$$

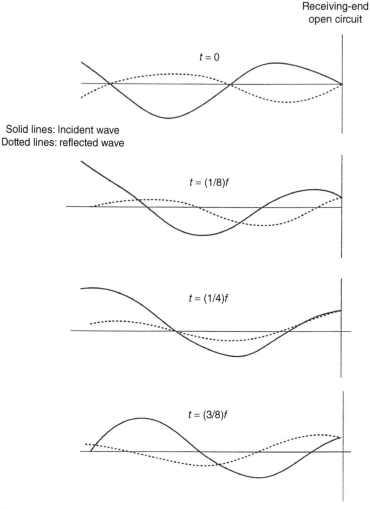

Figure 12.4   Traveling wave phenomena in long transmission lines, incident, and reflected waves.

where β is the *phase constant*, and r_{sc} and x_{sc} are the series resistance and reactance. Common units are radians per mile. The *characteristic impedance* is

$$Z_0 = \sqrt{\frac{z}{y}} \qquad \gamma = \sqrt{zy} \qquad (12.9)$$

A similar explanation holds for the current:

$$I_x = \left| \frac{V_r/Z_0 + I_r}{2} \right| e^{\alpha x + j\beta x} - \left| \frac{V_r/Z_0 - I_r}{2} \right| e^{-\alpha x - j\beta x} \qquad (12.10)$$

These equations can be written as

$$V_x = V_r \left(\frac{e^{\gamma x} + e^{-\gamma x}}{2} \right) + I_r Z_0 \left(\frac{e^{\gamma x} - e^{-\gamma x}}{2} \right)$$

$$I_x = \frac{V_r}{Z_0} \left(\frac{e^{\gamma x} - e^{-\gamma x}}{2} \right) + I_r \left(\frac{e^{\gamma x} + e^{-\gamma x}}{2} \right) \tag{12.11}$$

or in matrix form

$$\begin{vmatrix} V_s \\ I_s \end{vmatrix} = \begin{vmatrix} \cosh \gamma l & Z_0 \sinh \gamma l \\ \frac{1}{Z_0} \sinh \gamma l & \cosh \gamma l \end{vmatrix} \begin{vmatrix} V_r \\ I_r \end{vmatrix} \tag{12.12}$$

This proves the *ABCD* constants for long lines shown in Table 12.1.

12.1.4   Calculations of Line Constants

Practically, the transmission or cable system parameters will be calculated using computer-based subroutine programs. For simple systems, the data are available from various Refs [1–4].

As we have seen, the conductor AC resistance is dependent on frequency and proximity effects, temperature, spiraling, and bundle conductor effects, which increase the length of wound conductor in spiral shape with a certain pitch. The ratio R_{AC}/R_{DC} considering proximity and skin effects is given in Table 8.10. The resistance increases linearly with temperature and is given by the following equation:

$$R_2 = R_1 \left(\frac{T + t_2}{T + t_1} \right) \tag{12.13}$$

where R_2 is the resistance at temperature t_2, R_1 is the resistance at temperature t_1, and T is the temperature coefficient, which depends on the conductor material. It is 234.5 for annealed copper, 241.5 for hard drawn copper, and 228.1 for aluminum.

The *internal* inductance of a solid, smooth, round metallic cylinder of infinite length is due to its internal magnetic field when carrying an alternating current and is given by

$$L_{int} = \frac{\mu_0}{8\pi} \, \text{H/m} \tag{12.14}$$

where μ_0 is the permeability of free air $= 4\pi \times 10^{-7}$ (H/m). Its *external* inductance is due to the flux outside the conductor and is given by

$$L_{ext} = \frac{\mu_0}{2\pi} \ln \left(\frac{D}{r} \right) \text{H/m} \tag{12.15}$$

where D is any point at a distance D from the surface of the conductor, and r is the conductor radius. In most inductance tables in references, D is equal to 1 ft and adjustment factors are tabulated for higher conductor spacings. The total reactance is

$$L = \frac{\mu_0}{2\pi} \left[\frac{1}{4} + \ln \frac{D}{r} \right] = \frac{\mu_0}{2\pi} \left[\ln \frac{D}{e^{-1/4}r} \right] = \frac{\mu_0}{2\pi} \left[\ln \frac{D}{\text{GMR}} \right] \text{H/m} \tag{12.16}$$

where GMR is called the geometric mean radius and is $0.7788r$. It can be defined as the radius of a tubular conductor with an infinitesimally thin wall that has the same external flux out to a radius of 1 ft as the external and internal fluxes of a solid conductor to the same distance.

We can write the inductance matrix of a three-phase line in terms of flux linkages λ_a, λ_b, and λ_c:

$$\begin{vmatrix} \lambda_a \\ \lambda_b \\ \lambda_c \end{vmatrix} = \begin{vmatrix} L_{aa} & L_{ab} & L_{ac} \\ L_{ba} & L_{bb} & L_{bc} \\ L_{ca} & L_{cb} & L_{cc} \end{vmatrix} \begin{bmatrix} I_a \\ I_b \\ I_c \end{bmatrix} \tag{12.17}$$

The flux linkages λ_a, λ_b, and λ_c are given by

$$\lambda_a = \frac{\mu_0}{2\pi} \left[I_a \ln\left(\frac{1}{GMR_a}\right) + I_b \ln\left(\frac{1}{D_{ab}}\right) + I_c \ln\left(\frac{1}{D_{ac}}\right) \right]$$

$$\lambda_b = \frac{\mu_0}{2\pi} \left[I_a \ln\left(\frac{1}{D_{ba}}\right) + I_b \ln\left(\frac{1}{GMR_b}\right) + I_c \ln\left(\frac{1}{D_{bc}}\right) \right]$$

$$\lambda_c = \frac{\mu_0}{2\pi} \left[I_a \ln\left(\frac{1}{D_{ca}}\right) + I_b \ln\left(\frac{1}{D_{cb}}\right) + I_c \ln\left(\frac{1}{GMR_c}\right) \right] \tag{12.18}$$

where $D_{ab}, D_{ac}, \ldots$, are the distances between conductor of a phase with respect to conductors of b and c phases; L_{aa}, L_{bb}, and L_{cc} are the self-inductances of the conductors, and L_{ab}, L_{ba}, and L_{ca} are the mutual inductances. If we assume a symmetrical line, that is, the GMR of all three conductors is equal and also the spacing between the conductors is equal. The equivalent inductance per phase is

$$L = \frac{\mu_0}{2\pi} \ln\left(\frac{D}{GMR}\right) H/m \tag{12.19}$$

The phase-to-neutral inductance of a three-phase symmetrical line is the same as the inductance per conductor of a two-phase line.

A transposed line is shown in Fig. 12.5(a) and (b). Each phase conductor occupies the position of two other phase conductors for one-third of the length. The purpose is to equalize the phase inductances and reduce unbalance. For Fig. 12.5(a), the inductance derived for symmetrical line is still valid, and the equivalence of bundle conductors is in Section 12.1.6. The distance D in Eq. (12.16) is substituted by GMD (geometric mean distance) (Fig. 12.5(a)). It is given by

$$GMD = (D_{ab} D_{bc} D_{ca})^{1/3} \tag{12.20}$$

A detailed treatment of transposed lines with rotation matrices is given in Refs [5,6].

A transmission line with composite conductors is shown in Fig. 12.6. Consider that group X is composed of n conductors in parallel, and each conductor carries $1/n$

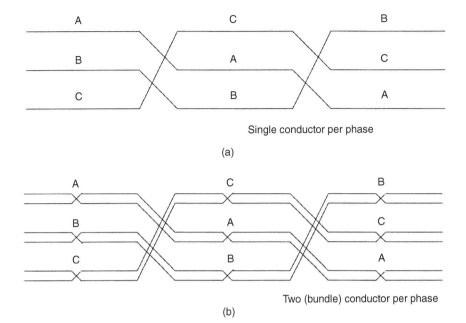

A   C   B

B   A   C

C   B   A

Single conductor per phase

(a)

A   C   B

B   A   C

C   B   A

Two (bundle) conductor per phase

(b)

Figure 12.5   Transposed transmission line representation.

of the line current. The group Y is composed of m parallel conductors, each of which carries $-1/m$ of the return current. Then, L_x, the inductance of conductor group X, is

$$L_x = 2 \times 10^{-7} \ \ln \ \frac{\sqrt[nm]{(D_{aa'}D_{ab'}D_{ac'} \ \cdots \ D_{am}) \ \cdots \ (D_{na'}D_{nb'}D_{nc'} \ \cdots \ D_{nm})}}{\sqrt[n^2]{(D_{aa}D_{ab}D_{ac} \ \cdots \ D_{an}) \ \cdots \ (D_{na}D_{nb}D_{nc}....D_{nn})}} \ \mathrm{H/m}$$

(12.21)

We write Eq. (12.21) as

$$L_x = 2 \times 10^{-7} \ \ln \left(\frac{D_m}{D_{sx}} \right) \ \mathrm{H/m}$$

(12.22)

Similarly,

$$L_y = 2 \times 10^{-7} \ \ln \left(\frac{D_m}{D_{sy}} \right) \ \mathrm{H/m}$$

(12.23)

The total inductance is

$$L = (L_x + L_y) \mathrm{H/m}$$

(12.24)

In three-phase, high-voltage transmission lines which are transposed, the sequence impedance matrix is of the form

$$\overline{Z}_{012} = \frac{1}{3} \begin{vmatrix} Z_s + 2Z_m & 0 & 0 \\ 0 & Z_s - Z_m & 0 \\ 0 & 0 & Z_s - Z_m \end{vmatrix}$$

(12.25)

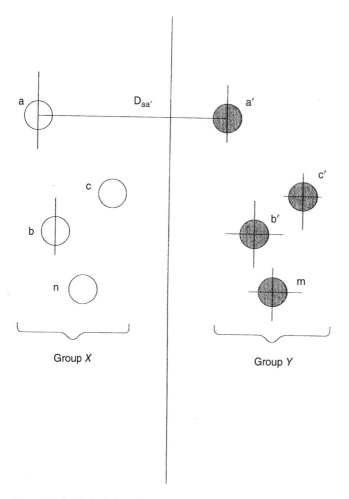

Figure 12.6   Calculation of inductance of composite conductors in a transmission line.

where Z_s is the self-impedance of the phase conductors, considered equal, and Z_m is the mutual impedance between the phase conductors. This is generally true, and the mutual couplings between phases are almost equal. However, the same cannot be said of distribution lines, and these may have unequal off-diagonal terms. In many cases, the off-diagonal terms are smaller than the diagonal terms, and the errors introduced in ignoring these will be small. Sometimes, an equivalence can be drawn by the equations:

$$Z_s = \frac{Z_{aa} + Z_{bb} + Z_{cc}}{3}$$

$$Z_m = \frac{Z_{ab} + Z_{bc} + Z_{ca}}{3} \tag{12.26}$$

12.1.5 Three-Phase Line with Ground Conductors

Consider three-phase input terminals a, b, c and two ground wires w and v and output terminals denoted with single prime.

The following matrix holds for the voltage differentials between terminals w, v, a, b, c and w', v', a', b', c':

$$\begin{vmatrix} \Delta V_a \\ \Delta V_b \\ \Delta V_c \\ \Delta V_w \\ \Delta V_w \end{vmatrix} = \begin{vmatrix} Z_{aa-g} & Z_{ab-g} & Z_{ac-g} & Z_{aw-g} & Z_{av-g} \\ Z_{ba-g} & Z_{bb-g} & Z_{bc-g} & Z_{bw-g} & Z_{bv-g} \\ Z_{ca-g} & Z_{cb-g} & Z_{cc-g} & Z_{cw-g} & Z_{cv-g} \\ Z_{wa-g} & Z_{wb-g} & Z_{wc-g} & Z_{ww-g} & Z_{wv-g} \\ Z_{va-g} & Z_{vb-g} & Z_{vc-g} & Z_{vw-g} & Z_{vv-g} \end{vmatrix} \begin{vmatrix} I_a \\ I_b \\ I_c \\ I_w \\ I_v \end{vmatrix} \tag{12.27}$$

where primed symbols correspond to the output terminals.

The 5×5 matrix of transmission lines with two ground wires can be reduced to a 3×3 matrix-by-matrix manipulation. In the partitioned form, this matrix can be written as

$$\begin{vmatrix} \Delta \overline{V}_{abc} \\ \Delta \overline{V}_{wv} \end{vmatrix} = \begin{vmatrix} \overline{Z}_A & \overline{Z}_B \\ \overline{Z}_C & \overline{Z}_D \end{vmatrix} \begin{vmatrix} \overline{I}_{abc} \\ \overline{I}_{wv} \end{vmatrix} \tag{12.28}$$

Considering that the ground wire voltages are zero

$$\Delta \overline{V}_{abc} = \overline{Z}_A \overline{I}_{abc} + \overline{Z}_B \overline{I}_{wv}$$

$$0 = \overline{Z}_c \overline{I}_{abc} + \overline{Z}_D \overline{I}_{wv} \tag{12.29}$$

Thus,

$$\overline{I}_{wv} = -\overline{Z}_D^{-1} \overline{Z}_C \overline{I}_{abc} \tag{12.30}$$

$$\Delta \overline{V}_{abc} = (\overline{Z}_A - \overline{Z}_B \overline{Z}_D^{-1} \overline{Z}_C) \overline{I}_{abc} \tag{12.31}$$

This can be written as

$$\Delta \overline{V}_{abc} = \overline{Z}_{abc} \overline{I}_{abc} \tag{12.32}$$

$$\overline{Z}_{abc} = \overline{Z}_A - \overline{Z}_B \overline{Z}_D^{-1} \overline{Z}_C = \begin{vmatrix} Z_{aa'-g} & Z_{ab'-g} & Z_{ac'-g} \\ Z_{ba'-g} & Z_{bb'-g} & Z_{bc'-g} \\ Z_{ca'-g} & Z_{cb'-g} & Z_{cc'-g} \end{vmatrix} \tag{12.33}$$

12.1.6 Bundle Conductors

Consider bundle conductors consisting of two conductors per phase (Fig. 12.7). The original circuit of conductors a, b, c and a', b', c' can be transformed into an equivalent conductor system of a'', b'', and c''.

Transform to

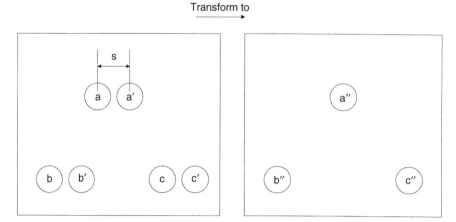

Figure 12.7   Bundle conductors and their one-conductor equivalent.

Each conductor in the bundle carries a different current and has a different self-impedance and mutual impedance because of its specific location. Let the currents in the conductors be I_a, I_b, I_c and I'_a, I'_b, I'_c, respectively. The following primitive matrix equation can be written as

$$\begin{vmatrix} V_a \\ V_b \\ V_c \\ V'_a \\ V'_b \\ V'_c \end{vmatrix} \begin{vmatrix} Z_{aa} & Z_{ab} & Z_{ac} & Z_{aa'} & Z_{ab'} & Z_{ac'} \\ Z_{ba} & Z_{bb} & Z_{bc} & Z_{ba'} & Z_{bb'} & Z_{bc'} \\ Z_{ca} & Z_{cb} & Z_{cc} & Z_{ca'} & Z_{cb'} & Z_{cc'} \\ Z_{a'a} & Z_{a'b} & Z_{a'c} & Z_{a'a'} & Z_{a'b'} & Z_{a'c'} \\ Z_{b'a} & Z_{b'b} & Z_{b'c} & Z_{b'a'} & Z_{b'b'} & Z_{b'c'} \\ Z_{c'a} & Z_{c'b} & Z_{c'c} & Z_{c'a'} & Z_{c'b'} & Z_{c'c'} \end{vmatrix} \begin{vmatrix} I_a \\ I_b \\ I_c \\ I_{a'} \\ I_{b'} \\ I_{c'} \end{vmatrix}$$
(12.34)

This can be partitioned so that

$$\begin{vmatrix} \overline{V}_{abc} \\ \overline{V}_{a'b'c'} \end{vmatrix} = \begin{vmatrix} \overline{Z}_1 & \overline{Z}_2 \\ \overline{Z}_3 & \overline{Z}_4 \end{vmatrix} \begin{vmatrix} \overline{I}_{abc} \\ \overline{I}_{a'b'c'} \end{vmatrix}$$
(12.35)

for symmetrical arrangement of bundle conductors $\overline{Z}_1 = \overline{Z}_4$.

Modify so that the lower portion of the vector goes to zero. Assume that

$$V_a = V'_a = V''_a$$

$$V_b = V'_b = V''_b$$

$$V_c = V'_c = V''_c$$
(12.36)

The upper part of the matrix can then be subtracted from the lower part:

$$
\begin{vmatrix} V_a \\ V_b \\ V_c \\ 0 \\ 0 \\ 0 \end{vmatrix} = \begin{vmatrix} Z_{aa} & Z_{ab} & Z_{ac} & Z_{aa'} & Z_{ab'} & Z_{ac'} \\ Z_{ba} & Z_{bb} & Z_{bc} & Z_{ba'} & Z_{bb'} & Z_{bc'} \\ Z_{ca} & Z_{cb} & Z_{cc} & Z_{ca'} & Z_{cb'} & Z_{cc'} \\ Z_{a'a} - Z_{aa} & Z_{a'b} - Z_{ab} & Z_{a'c} - Z_{ac} & Z_{a'a'} - Z_{aa'} & Z_{a'b'} - Z_{ab'} & Z_{a'c'} - Z_{ac'} \\ Z_{b'a} - Z_{ba} & Z_{b'b} - Z_{bb} & Z_{b'c} - Z_{bc} & Z_{b'a'} - Z_{ba'} & Z_{b'b'} - Z_{bb'} & Z_{b'c'} - Z_{bc'} \\ Z_{c'a} - Z_{ca} & Z_{c'b} - Z_{cb} & Z_{c'c} - Z_{cc} & Z_{c'a'} - Z_{ca'} & Z_{c'b'} - Z_{cb'} & Z_{c'c'} - Z_{cc'} \end{vmatrix} \begin{vmatrix} I_a \\ I_b \\ I_c \\ I_{a'} \\ I_{b'} \\ I_{c'} \end{vmatrix}
$$
(12.37)

We can write it in the partitioned form as

$$
\begin{vmatrix} \overline{V}_{abc} \\ 0 \end{vmatrix} = \begin{vmatrix} \overline{Z}_1 & \overline{Z}_2 \\ \overline{Z}_2^t - \overline{Z}_1 & \overline{Z}_4 - \overline{Z}_2 \end{vmatrix} \begin{vmatrix} \overline{I}_{abc} \\ \overline{I}_{a'b'c'} \end{vmatrix}
$$
(12.38)

$$ I_a'' = I_a + I_a' $$

$$ I_b'' = I_b + I_b' $$

$$ I_c'' = I_c + I_c' $$

The matrix is modified as

$$
\begin{vmatrix} Z_{aa} & Z_{ab} & Z_{ac} & Z_{aa'} - Z_{aa} & Z_{ab'} - Z_{ab} & Z_{ac'} - Z_{ac} \\ Z_{ba} & Z_{bb} & Z_{bc} & Z_{ba'} + Z_{ba} & Z_{bb'} + Z_{bb} & Z_{bc'} - Z_{bc} \\ Z_{ca} & Z_{cb} & Z_{cc} & Z_{ca'} - Z_{ca} & Z_{cb'} - Z_{cb} & Z_{cc'} - Z_{cc} \\ Z_{a'a} - Z_{aa} & Z_{a'b} - Z_{ab} & Z_{a'c} - Z_{ac} & Z_{a'a'} - Z_{aa'} - Z_{a'a} + Z_{aa} & Z_{a'b'} - Z_{ab'} - Z_{a'b} + Z_{ab} & Z_{a'c'} - Z_{ac'} - Z_{a'c} + Z_{ac} \\ Z_{b'a} - Z_{ba} & Z_{b'b} - Z_{bb} & Z_{b'c} - Z_{bc} & Z_{b'a'} - Z_{ba'} - Z_{b'a} + Z_{ba} & Z_{b'b'} - Z_{bb'} - Z_{b'b} + Z_{bb} & Z_{b'c'} - Z_{bc'} - Z_{b'c} + Z_{bc} \\ Z_{c'a} - Z_{ca} & Z_{c'b} - Z_{cb} & Z_{c'c} - Z_{cc} & Z_{c'a'} - Z_{ca'} - Z_{c'a} + Z_{ca} & Z_{c'b'} - Z_{cb'} - Z_{c'b} + Z_{cb} & Z_{c'c'} - Z_{cc'} - Z_{c'c} + Z_{cc} \end{vmatrix} \begin{vmatrix} I_a + I_a' \\ I_b + I_b' \\ I_c + I_c' \\ I_a' \\ I_b' \\ I_c' \end{vmatrix}
$$
(12.39)

or in partitioned form:

$$
\begin{vmatrix} \overline{V}_{abc} \\ 0 \end{vmatrix} = \begin{vmatrix} \overline{Z}_1 & \overline{Z}_2 - \overline{Z}_1 \\ \overline{Z}_2^t - \overline{Z}_1 & (\overline{Z}_4 - \overline{Z}_2) - (\overline{Z}_2^t - \overline{Z}_1) \end{vmatrix} \begin{vmatrix} \overline{I}_{abc}'' \\ \overline{I}_{a'b'c'} \end{vmatrix}
$$
(12.40)

This can now be reduced to the following 3 × 3 matrix as before:

$$
\begin{vmatrix} V_a'' \\ V_b'' \\ V_c'' \end{vmatrix} = \begin{vmatrix} Z_{aa}'' & Z_{ab}'' & Z_{ac}'' \\ Z_{ba}'' & Z_{bb}'' & Z_{bc}'' \\ Z_{ca}'' & Z_{cb}'' & Z_{cc}'' \end{vmatrix} \begin{vmatrix} I_a'' \\ I_b'' \\ I_c'' \end{vmatrix}
$$
(12.41)

12.1.7   Carson's Formula

The theoretical value of $Z_{abc\text{-}g}$ can be calculated by Carson's formula (ca. 1926). This is of importance even today in calculations of line constants. For an n-conductor configuration, the earth is assumed as an infinite uniform solid with a constant resistivity. Figure 12.8 shows image conductors in the ground at a distance equal to the height

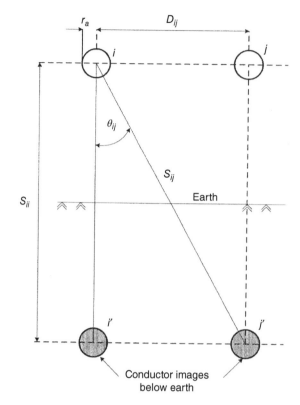

Figure 12.8 OH conductors and their images in ground, Carson's formula.

of the conductors above ground and exactly in the same formation, with the same spacing between the conductors. A flat conductor formation is shown in Fig. 12.8:

$$Z_{ii} = R_i + 4\omega P_{ii}G + j \left[X_i + 2\omega G \ln \frac{S_{ii}}{r_i} + 4\omega Q_{ii}G \right] \Omega/\text{mile} \qquad (12.42)$$

$$Z_{ij} = 4\omega P_{ij}G + j \left[2\omega G \ln \frac{S_{ij}}{D_{ij}} + 4\omega Q_{ij}G \right] \Omega/\text{mile} \qquad (12.43)$$

where

Z_{ii}	=	the self-impedance of conductor i with earth return (ohms/mile)
Z_{ij}	=	mutual impedance between conductors i and j (ohms/mile)
R_i	=	resistance of conductor in ohms/mile
S_{ii}	=	conductor-to-image distance of the ith conductor to its own image
S_{ij}	=	conductor-to-image distance of the ith conductor to the image of the jth conductor
D_{ij}	=	distance between conductors i and j

r_i =   radius of conductor (ft)
ω =   angular frequency
G = 0.1609347×10^{-7} ohm $\cdot$ cm
GMR_i =   geometric mean radius of conductor i
ρ =   soil resistivity
θ_{ij} =   angle as shown in Fig. 12.8

Expressions for P and Q are

$$P = \frac{\pi}{8} - \frac{1}{3\sqrt{2}} k \, \cos\theta + \frac{k^2}{16} \, \cos 2\theta \left(0.6728 + \ln\frac{2}{k} \right) + \frac{k^2}{16}\theta \, \sin\theta$$

$$+ \frac{k^3 \, \cos 3\theta}{45\sqrt{2}} - \frac{\pi k^4 \, \cos 4\theta}{1536} \tag{12.44}$$

$$Q = -0.0386 + \frac{1}{2} \, \ln\frac{2}{k} + \frac{1}{3\sqrt{2}} \, \cos\theta - \frac{k^2 \, \cos 2\theta}{64} + \frac{K^3 \, \cos 3\theta}{45\sqrt{2}}$$

$$- \frac{k^4 \, \sin 4\theta}{384} - \frac{k^4 \, \cos 4\theta}{384} \left(\ln\frac{2}{k} + 1.0895 \right) \tag{12.45}$$

where

$$k = 8.565 \times 10^4 S_{ij} \sqrt{\frac{f}{\rho}} \tag{12.46}$$

S_{ij} is in feet, ρ is the soil resistivity in ohms-meter, and f is the system frequency. This shows the dependence on frequency as well as on soil resistivity.

12.1.8 Approximations to Carson's Equations

These approximations involve P and Q and the expressions are given by

$$P_{ij} = \frac{\pi}{8} \tag{12.47}$$

$$Q_{ij} = -0.03860 + \frac{1}{2} \, \ln\frac{2}{k_{ij}} \tag{12.48}$$

Using these assumptions, $f = 60\,\text{Hz}$ and soil resistivity $= 100\Omega \cdot \text{m}$, the equations reduce to the form

$$Z_{ii} = R_i + 0.0953 + j0.12134 \left(\ln\frac{1}{GMR_i} + 7.93402 \right) \Omega/\text{mile} \tag{12.49}$$

$$Z_{ij} = 0.0953 + j0.12134 \left(\ln\frac{1}{D_{ij}} + 7.93402 \right) \Omega/\text{mile} \tag{12.50}$$

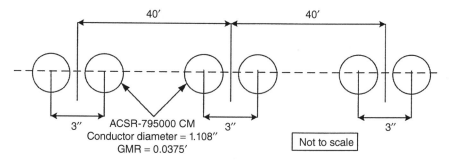

Figure 12.9   Bundle conductors spacing, (Example. 12.1).

Equations (12.49) and (12.50) are of practical significance for calculations of line impedances.

Example 12.2:   This exemplifies the reduction matrix of bundle conductors.

Conductor: ACSR, 795 KCMIL, 26 strands, two layers, strand diameter = 0.1749 in., steel core, 7 strands, strand diameter = 0.1360 in., overall outer conductor diameter = 1.108 in., GMR = 0.0375 ft, approximate ampacity = 900 A, and resistance R_i = 0.117 ohms/mile.

Configuration: Two conductors per phase, spaced 3 in. center line, spacing between phases = 40 in., flat formation, Fig. 12.9.

Then, using Eqs (12.49) and (12.50), matrix Z_1

$$\overline{Z}_1 = \begin{vmatrix} 0.2123 + j1.3611 & 0.0953 + j0.8166 & 0.0953 + j0.7325 \\ 0.0953 + j0.8166 & 0.2123 + j1.3611 & 0.0953 + j0.8166 \\ 0.0953 + j0.7325 & 0.0953 + j0.8166 & 0.2123 + j1.3611 \end{vmatrix}$$

This is also matrix Z_4, because of symmetry.
The matrix Z_2 is

$$\overline{Z}_2 = \begin{vmatrix} 0.0953 + j1.3092 & 0.0953 + j0.8078 & 0.0953 + j0.7280 \\ 0.0953 + j0.8260 & 0.0953 + j1.3092 & 0.0953 + j0.8078 \\ 0.0953 + j0.7372 & 0.0953 + j0.8260 & 0.0953 + j1.3092 \end{vmatrix}$$

Then

$$\overline{Z}_1 - \overline{Z}_2 = \begin{vmatrix} 0.1170 + j0.0519 & j0.0088 & j0.0045 \\ -j0.0094 & 0.1170 + j0.0519 & j0.0088 \\ -j0.0047 & -j0.0094 & 0.1170 + j1.3092 \end{vmatrix}$$

$$\overline{Z}_2^t = \begin{vmatrix} 0.0953 + j1.3092 & 0.0953 + j0.8260 & 0.0953 + j0.7372 \\ 0.0953 + j0.8078 & 0.0953 + j1.3092 & 0.0953 + j0.8260 \\ 0.0953 + j0.7280 & 0.0953 + j0.8078 & 0.0953 + j1.3092 \end{vmatrix}$$

$$\overline{Z}_2' - \overline{Z}_1 = \begin{vmatrix} -0.117 - j0.0519 & j0.0094 & j0.0047 \\ -j0.0088 & -0.117 - j0.0519 & j0.0094 \\ -j0.0045 & -j0.0088 & -0.117 - j0.0519 \end{vmatrix}$$

$$\overline{Z}_k = (\overline{Z}_1 - \overline{Z}_2) - (\overline{Z}_2' - \overline{Z}_1) = \begin{vmatrix} 0.234 + j0.1038 & -j0.0006 & -j0.0002 \\ -j0.0006 & 0.234 + j0.1038 & -j0.0006 \\ -j0.0002 & -j0.0006 & 0.234 + j0.1038 \end{vmatrix}$$

$$\overline{Z}_k^{-1} = \begin{vmatrix} 3.571 - j1.584 & 6.785 \times 10^{-3} + j6.152 \times 10^{-3} & 2.256 \times 10^{-3} + j2.069 \times 10^{-3} \\ 6.785 \times 10^{-3} + j6.152 \times 10^{-3} & 3.571 - j1.584 & 6.785 \times 10^{-3} + j6.152 \times 10^{-3} \\ 2.256 \times 10^{-3} + j2.069 \times 10^{-3} & 6.785 \times 10^{-3} + j6.152 \times 10^{-3} & 3.571 - j1.584 \end{vmatrix}$$

Then

$$(\overline{Z}_2 - \overline{Z}_1)\overline{Z}_k^{-1}(\overline{Z}_2' - \overline{Z}_1) =$$

$$\begin{vmatrix} 0.058 + j0.026 & -1.494 \times 10^{-4} - j8.361 \times 10^{-5} & 2.963 \times 10^{-4} - j1.806 \times 10^{-4} \\ -1.494 \times 10^{-4} - j8.361 \times 10^{-5} & 0.058 + j0.026 & -1.494 \times 10^{-4} - j8.361 \times 10^{-5} \\ 2.963 \times 10^{-4} - j1.806 \times 10^{-4} & -1.494 \times 10^{-4} - j8.361 \times 10^{-5} & 0.058 + j0.026 \end{vmatrix}$$

Then the transformed matrix is matrix Z_1 minus the above matrix:

$$\overline{Z}_{\text{transformed}} = \begin{vmatrix} 0.1543 + j1.3351 & 0.0969 + j0.8167 & 0.0950 + j0.7327 \\ 0.0969 + j0.8167 & 0.1543 + j1.3351 & 0.0969 + j0.8167 \\ 0.0950 + j0.7327 & 0.0969 + j0.8167 & 0.1543 + j1.3351 \end{vmatrix}$$

This can be transformed into sequence impedance matrix:

$$\overline{Z}_{012} = \overline{T}_s \overline{Z}_{abc} \overline{T}_s^{-1}$$

where T_s is the symmetrical component transformation matrix.

$$\overline{Z}_{012} = \begin{vmatrix} 0.347 + j2.932 & 0.024 - j0.015 & -0.025 - j0.013 \\ -0.025 - j0.013 & 0.058 + j0.566 & -0.048 + j0.029 \\ 0.024 - j0.015 & 0.049 + j0.027 & 0.058 + j0.566 \end{vmatrix}$$

12.1.9 Capacitance of OH lines

For harmonic analysis, it is important to model the shunt capacitances of the lines. The shunt capacitance per unit length of a two-wire, single-phase transmission line is

$$C = \frac{\pi \varepsilon_0}{\ln(D/r)} \text{ F/m} \qquad (12.51)$$

where ε_0 is the permittivity of free space $= 8.854 \times 10^{-12}$ F/m, and other symbols are as defined earlier. For a three-phase line with equilaterally spaced conductors, the line-to-neutral capacitance is

$$C = \frac{2\pi \varepsilon_0}{\ln(D/r)} \text{ F/m} \qquad (12.52)$$

For unequal spacing, D is replaced with GMD from Eq. (12.20). The capacitance is affected by the ground, and the effect is simulated by a mirror image of the conductors exactly at the same depth as the height above the ground. These mirror image conductors carry charges that are of opposite polarity to conductors above the ground (Fig. 12.10). From this figure, the capacitance to ground is

$$C_n = \frac{2\pi\varepsilon_0}{\ln(\text{GMD}/r) - \ln\left(\sqrt[3]{S_{ab'}S_{bc'}S_{ca'}}\,/\,\sqrt[3]{S_{aa'}S_{bb'}S_{cc'}}\right)} \tag{12.53}$$

Using the notations in Eq. (12.53), this can be written as

$$C_n = \frac{2\pi\varepsilon_0}{\ln(D_m/D_s)} = \frac{10^{-9}}{18\,\ln(D_m/D_s)}\,\text{F/m} \tag{12.54}$$

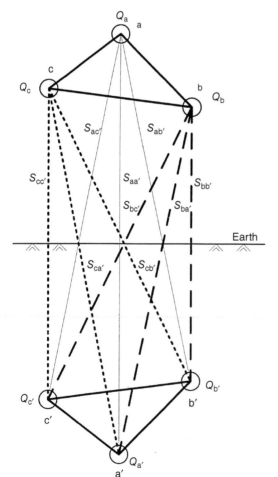

Figure 12.10 Calculations of capacitance of OH line conductors; conductors, their images, spacing, and charges.

The capacitance matrix of a three-phase line is

$$\overline{C}_{abc} = \begin{vmatrix} C_{aa} & -C_{ab} & -C_{ac} \\ -C_{ba} & C_{bb} & -C_{bc} \\ -C_{ca} & -C_{cb} & C_{cc} \end{vmatrix} \tag{12.55}$$

This is diagrammatically shown in Fig. 12.11(a). The capacitance between the phase conductor a and b is C_{ab}, and the capacitance between conductor a and ground is $C_{aa} - C_{ab} - C_{ac}$. If the line is perfectly symmetrical, all the diagonal elements are the same and all off-diagonal elements of the capacitance matrix are identical:

$$\overline{C}_{abc} = \begin{vmatrix} C & -C' & -C' \\ -C' & C & -C' \\ -C' & -C' & C \end{vmatrix} \tag{12.56}$$

Symmetrical component transformation is used to diagonalize the matrix:

$$\overline{C}_{012} = \overline{T}_s^{-1} \overline{C}_{abc} \overline{T}_s = \begin{vmatrix} C - 2C' & 0 & 0 \\ 0 & C + C' & 0 \\ 0 & 0 & C + C' \end{vmatrix} \tag{12.57}$$

The zero, positive, and negative sequence networks of capacitance of a symmetrical transmission line are shown in Fig. 12.11(b). The eigenvalues are $C - 2C'$, $C + C'$, and $C + C'$. The capacitance $C + C'$ can be written as $3C' + (C - 2C')$, that is, it is equivalent to the line capacitance of a three-conductor system plus the line-to-ground capacitance of a three-conductor system.

In a capacitor, $V = Q/C$. The capacitance matrix can be written as

$$\overline{V}_{abc} = \overline{P}_{abc} \overline{Q}_{abc} = \overline{C}_{abc}^{-1} \overline{Q}_{abc} \tag{12.58}$$

where $\overline{P}$ is called the potential coefficient matrix, that is,

$$\begin{vmatrix} V_a \\ V_b \\ V_c \end{vmatrix} = \begin{vmatrix} P_{aa} & P_{ab} & P_{ac} \\ P_{ba} & P_{bb} & P_{bc} \\ P_{ca} & P_{cb} & P_{cc} \end{vmatrix} \begin{vmatrix} Q_a \\ Q_b \\ Q_c \end{vmatrix} \tag{12.59}$$

where

$$P_{ii} = \frac{1}{2\pi\varepsilon_0} \ln \frac{S_{ii}}{r_i} = 11.17689 \ln \frac{S_{ii}}{r_i}$$

$$P_{ij} = \frac{1}{2\pi\varepsilon_0} \ln \frac{S_{ij}}{D_{ij}} = 11.17689 \ln \frac{S_{ij}}{D_{ij}} \tag{12.60}$$

where

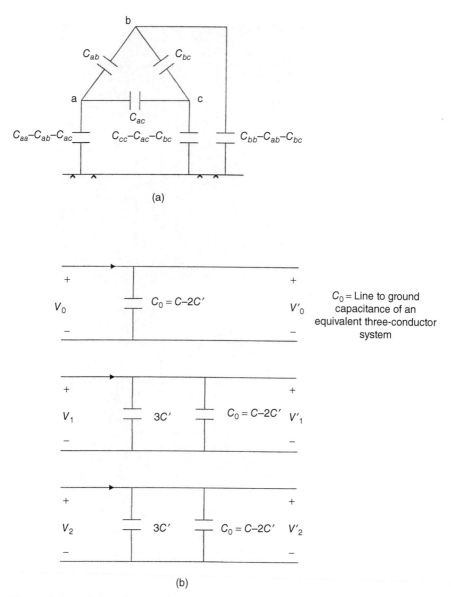

Figure 12.11   (a) Capacitances of a three-phase line and (b) equivalent sequence capacitances.

S_{ij} = conductor-to-image distance below ground (ft)
D_{ij} = conductor-to-conductor distance below ground (ft)
r_i = radius of the conductor (ft)
ε_0 = permittivity of the medium surrounding the conductor = 1.424×10^{-8}

For sine-wave voltage and charge, the equation can be expressed as

$$\begin{vmatrix} I_a \\ I_b \\ I_c \end{vmatrix} = j\omega \begin{vmatrix} C_{aa} & -C_{ab} & -C_{ac} \\ -C_{ba} & -C_{bb} & -C_{bc} \\ -C_{ca} & -C_{cb} & C_{cc} \end{vmatrix} \begin{vmatrix} V_a \\ V_b \\ V_c \end{vmatrix} \tag{12.61}$$

The capacitance of three-phase lines with ground wires and with bundle conductors can be addressed as in the calculations of inductances. The primitive P matrix can be partitioned and reduces to a 3×3 matrix.

Example 12.3: Consider an unsymmetrical overhead line configuration, as shown in Fig. 12.12. The phase conductors consist of 556.5 KCMIL (556,500 circular mils) of ACSR conductor consisting of 26 strands of aluminum, two layers and seven strands of steel. From the properties of ACSR conductor tables, the conductor has a resistance of 0.1807 ohms at 60 Hz and its GMR is 0.0313 ft at 60 Hz; conductor diameter = 0.927 in. The neutral consists of 336.4 KCMIL, ACSR conductor, resistance 0.259 ohms per mile at 60 Hz and 50 °C and GMR 0.0278 ft, and conductor diameter 0.806 in. Calculate the Y capacitance matrix.

The mirror images of the conductors are drawn in Fig. 12.12. This facilitates calculation of the spacings required in Eqs (12.59) and (12.60) for the P matrix. On the basis of the geometric distances and conductor diameter, the primitive P matrix is

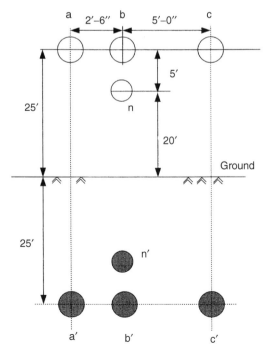

Figure 12.12 Conductor configuration for the calculation of capacitance Y matrix, (Example 12.3).

$$\overline{P} = \begin{vmatrix} P_{aa} & P_{ab} & P_{ac} & P_{an} \\ P_{ba} & P_{bb} & P_{bc} & P_{bn} \\ P_{ca} & P_{cb} & P_{cc} & P_{cn} \\ P_{na} & P_{nb} & P_{nc} & P_{nn} \end{vmatrix}$$

$$= \begin{vmatrix} 80.0922 & 33.5387 & 21.4230 & 23.3288 \\ 33.5387 & 80.0922 & 25.7913 & 24.5581 \\ 21.4230 & 25.7913 & 80.0922 & 20.7547 \\ 23.3288 & 24.5581 & 20.7547 & 79.1615 \end{vmatrix}$$

This is reduced to a 3×3 matrix

$$P = \begin{vmatrix} 73.2172 & 26.3015 & 15.3066 \\ 26.3015 & 72.4736 & 19.3526 \\ 15.3066 & 19.3526 & 74.6507 \end{vmatrix}$$

Therefore, the required $\overline{C}$ matrix is inverse of $\overline{P}$, and $\overline{Y}_{abc}$ is

$$\overline{Y}_{abc} = j\omega \overline{P}^{-1} = \begin{vmatrix} j6.0141 & -j1.9911 & -j0.7170 \\ -j1.9911 & j6.2479 & -j1.2114 \\ -j0.7170 & -j1.2114 & j5.5111 \end{vmatrix} \mu S/mile$$

12.1.10 EMTP Models of OH Lines

EMTP permits a number of transmission line models. The constant parameter (CP) model is a frequency-independent transmission line model for the wave equation of the distributed parameter line. It can be used for harmonic analysis. A frequency-dependent (FD) model is also available. The model CP is good for the frequencies involved in harmonic analysis.

EMTP uses the following transformation matrices. The Clarke's $\alpha\beta_0$ transformation [7,8] matrix for m-phase balanced line is

$$\overline{T}_i = \begin{vmatrix} \dfrac{1}{\sqrt{m}} & \dfrac{1}{\sqrt{2}} & \dfrac{1}{\sqrt{6}} & \cdot & \dfrac{1}{\sqrt{j(j-1)}} & \cdot & \dfrac{1}{\sqrt{m(m-1)}} \\ \dfrac{1}{\sqrt{m}} & -\dfrac{1}{\sqrt{2}} & \dfrac{1}{\sqrt{6}} & \cdot & \dfrac{1}{\sqrt{j(j-1)}} & \cdot & \dfrac{1}{\sqrt{m(m-1)}} \\ \dfrac{1}{\sqrt{m}} & 0 & -\dfrac{2}{\sqrt{6}} & \cdot & \cdot & \cdot & \cdot \\ \cdot & \cdot & 0 & \cdot & \dfrac{-(j-1)}{\sqrt{j(j-1)}} & \cdot & \cdot \\ \cdot & \cdot & \cdot & \cdot & 0 & \cdot & \cdot \\ \cdot & \cdot & \cdot & \cdot & \cdot & \cdot & \cdot \\ \dfrac{1}{\sqrt{m}} & 0 & 0 & \cdot & 0 & \cdot & \dfrac{-(m-1)}{\sqrt{m(m-1)}} \end{vmatrix} \qquad (12.62)$$

Applying this m-phase transformation to matrices of m-phase balanced lines will produce a diagonal matrix of the form:

$$\begin{vmatrix} Z_{\text{g-m}} & & \\ & Z_{\text{L-m}} & \\ & & - \\ & & Z_{\text{L-m}} \end{vmatrix} \tag{12.63}$$

$Z_{\text{g-m}}$ is the ground-mode matrix and $Z_{\text{L-m}}$ is the line-mode matrix. The solution becomes simpler if m-phase transmission line equations (M-coupled equations) can be transformed into M-decoupled equations. Many transposed and even untransposed lines can be diagonalized with transformations to modal parameters based on eigenvalue/eigenvector theory.

Consider a matrix of modal voltages V_m and a transformation matrix T, transforming conductor voltage matrix V:

$$\overline{V} = \overline{T}\,\overline{V}_m \tag{12.64}$$

Then, the wave equation can be decoupled and can be written as

$$\frac{\partial^2 \overline{V}_m}{\partial x^2} = \overline{T}^{-1}[\overline{L}\,\overline{C}]\overline{T}\frac{\partial^2 \overline{V}_m}{\partial t^2}$$

$$= \overline{T}^{-1}\overline{M}^{-1}\overline{T}\frac{\partial^2 \overline{V}_m}{\partial t^2}$$

$$= \overline{\lambda}\frac{\partial^2 \overline{V}_m}{\partial t^2} \tag{12.65}$$

For decoupling matrix, $\overline{\lambda} = \overline{T}^{-1}\overline{M}^{-1}\overline{T}$ must be diagonal. This is done by finding the eigenvalues of $\overline{M}$ from the solution of its characteristic equation. The significance of this analysis is that for n conductor system, the matrices are of the order n, and n number of modal voltages are generated which are independent of each other. Each wave travels with a velocity:

$$v_n = \frac{1}{\lambda_k}, \quad k = 1, 2, \dots, n \tag{12.66}$$

And the actual voltage on the conductors is given by Eq. (12.66).

For a three-phase line:

$$\overline{T} = \begin{vmatrix} 1 & 1 & 1 \\ -1 & 0 & 1 \\ 0 & -1 & 1 \end{vmatrix} \tag{12.67}$$

For a two-conductor line, there are two modes. In the line mode, the voltage and current travel over one conductor returning through the other, none flowing through the ground. In the ground mode, model quantities travel over both the conductors and return through the ground. For a three-conductor line, there are two line modes and one ground mode.

The line-to-line modes of propagation are close to the speed of light and encounter less attenuation and distortion compared to ground modes. The ground mode has more resistance and hence more attenuation and distortion. The resistance of the conductors and earth resistivity play an important role.

Example 12.4: Consider the following parameters of a 400-kV transmission line:

Phase Conductors: ACSR, 30 strands, 500 KCMIL, resistance at $25\,^\circ$C = 0.187 ohms per mile, resistance at $50\,^\circ$C = 0.206 ohms per mile, X_a = 0.421 ohms per mile, outside diameter = 0.904″, GMR = 0.0311 ft.

Ground Wires: 7#8, 115.6 KCMIL, seven strands, resistance at $25\,^\circ$C = 2.44 ohms per mile, resistance at $50\,^\circ$C = 3.06 ohms per mile, X_a = 0.749 ohms per mile, outside diameter = 0.385″, GMR = 0.00209 ft.

Line Configuration: Phase conductors, flat formation, height above ground = 108 ft. Two ground wires, height above ground = 143 ft. Spacing between the ground wires = 50 ft.

Other Data: Soil resistivity = 100 ohm/m, tower footing resistance = 20 ohms, and span = 800 ft.

The calculated modal parameters for a CP, balanced line using EMTP routine, are shown in Fig. 12.13. Note that this figure shows two sets of parameters, one at 3543 Hz and the other at 60 Hz. The CP model will be fairly accurate up to 3543 Hz. Usually, harmonic analysis studies are conducted to 50th harmonic.

For extra high-voltage (EHV) lines, steel saving semiflexible towers, that is, self-supporting towers, are receiving more considerations. Figure 12.14 shows 500-kV tower structures used by various utilities.

High-voltage lines exceeding 138 kV are generally built on self-supporting or rigid steel towers, which are most satisfactory. High-strength aluminum alloy towers have some advantages, that is, resistance to corrosion, but with added problems of greater deflections when conductor's stresses are applied. These have not been extensively used.

12.1.11 Effects of Harmonics

Long-line effects should be represented for lines of length $150/h$ miles, where h is the harmonic number. The effect of higher frequencies is to increase the skin effect and proximity effects. A frequency-dependent model of the resistive component becomes important, though the effect on the reactance is ignored. The resistance can be multiplied by a factor $g(h)$ [9]:

$$R(h) = R_{DC}g(h) \tag{12.68}$$

$$g(h) = 0.035X^2 + 0.938 > 2.4 \tag{12.69}$$

```
        MODAL PARAMETERS FOR BALANCED-LINE TRANSFORMATION MATRICES

   FREQUENCY = 6.0000E+01  HZ                    LENGTH = 3.2187E+02  KM

    R          L          C         ZC       PH(ZC)   ATTENUATION   VELOCITY
 (OHMS/KM)   (MH/KM)  (MICROF/KM)  (OHMS)   (DEGREES) (E**-GAM*L)  (KM/SEC)
 3.1424E-01 2.5878E+00 5.6182E-03 6.9563E+02-8.9244E+00 9.2903E-01 2.5901E+05
 1.1937E-01 1.4559E+00 7.7265E-03 4.3913E+02-6.1331E+00 9.5694E-01 2.9643E+05
 1.1937E-01 1.4559E+00 7.7265E-03 4.3913E+02-6.1331E+00 9.5694E-01 2.9643E+05

 MODAL SHUNT CONDUCTANCE (MHOS/KM):

 2.0000E-10  2.0000E-10  2.0000E-10
```

```
        MODAL PARAMETERS FOR BALANCED-LINE TRANSFORMATION MATRICES

   FREQUENCY = 3.5434E+03  HZ                    LENGTH = 3.2187E+02  KM

    R          L          C         ZC       PH(ZC)   ATTENUATION   VELOCITY
 (OHMS/KM)   (MH/KM)  (MICROF/KM)  (OHMS)   (DEGREES) (E**-GAM*L)  (KM/SEC)
 2.6237E+00 2.2236E+00 5.6182E-03 6.2956E+02-1.5168E+00 5.1123E-01 2.8283E+05
 1.4563E-01 1.4504E+00 7.7265E-03 4.3327E+02-1.2916E-01 9.4733E-01 2.9872E+05
 1.4563E-01 1.4504E+00 7.7265E-03 4.3327E+02-1.2916E-01 9.4733E-01 2.9872E+05

 MODAL SHUNT CONDUCTANCE (MHOS/KM):

 2.0000E-10  2.0000E-10  2.0000E-10
```

Figure 12.13   400- kV line parameters calculated with EMTP.

$$= 0.35X + 0.3 \leq 2.4 \tag{12.70}$$

where

$$X = 0.3884\sqrt{\frac{f_h}{f}}\sqrt{\frac{h}{R_{DC}}} \tag{12.71}$$

where f_h is the frequency and f is the system frequency. Another equation for taking into account the skin effect is

$$R = R_e\left(\frac{j\mu\omega}{2\pi a}\frac{J_z(r)}{\partial J_z(r)/\partial r|_{r=a}}\right) \tag{12.72}$$

where $J_z(r)$ is the current density and a is the outside radius of the conductor.

12.1.12 Transmission Line Equations with Harmonics

In the presence of harmonics:

$$V_{s(h)} = V_{r(h)}\cosh(\gamma l_{(h)}) + I_{r(h)}Z_{0(h)}\sinh(\gamma l_{(h)})$$

$$= V_{r(h)}\cosh(\sqrt{Z_{(h)}Y_{(h)}}) + I_{r(h)}\sqrt{\frac{Z_{(h)}}{Y_{(h)}}}\sinh(\sqrt{Z_{(h)}Y_{(h)}}) \tag{12.73}$$

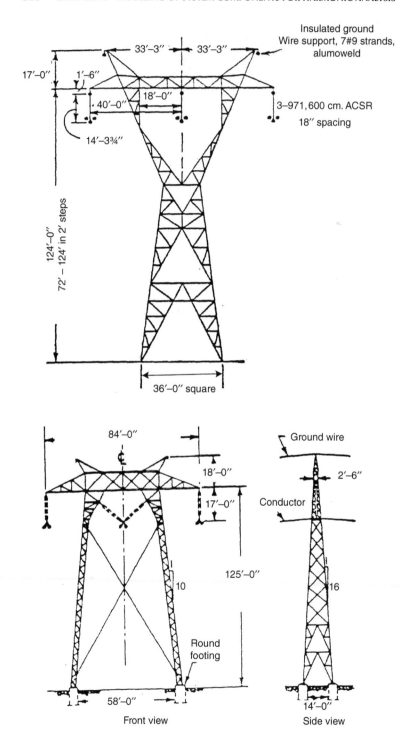

Figure 12.14   Construction of 500-kV OH lines.

$$I_{s(h)} = \frac{V_{r(h)}}{Z_{0(h)}} \sinh(\sqrt{Z_{(h)}Y_{(h)}}) + I_{r(h)} \cosh(\sqrt{Z_{(h)}Y_{(h)}}) \tag{12.74}$$

Similarly, from

$$\begin{vmatrix} V_r \\ I_r \end{vmatrix} = \begin{vmatrix} \cosh(\gamma l) & -Z_0 \sinh(\gamma l) \\ -\frac{\sinh(\gamma l)}{Z_0} & \cosh \gamma l \end{vmatrix} \begin{vmatrix} V_s \\ I_s \end{vmatrix} \tag{12.75}$$

equations of receiving end current and voltages can be written.

The variation of the impedance of the transmission line with respect to frequency is of much interest. Table 12.1 shows the constants for a nominal Π model. An equivalent Π model, which relates the Π model with long-line model, can be derived.

$$Z_s = Z_0 \sinh \gamma l$$

$$Y_p = \frac{Y}{2} \left[\frac{\tanh \gamma l/2}{\gamma l/2} \right] \tag{12.76}$$

Here, we have denoted the series element of the Π-model with Z_s and the shunt element with Y_p. In the presence of harmonics, we can write

$$Z_{s(h)} = Z_{0(h)} \sinh \gamma l_{(h)} = Z_{(h)} \frac{\sinh \gamma l_{(h)}}{\gamma l_{(h)}}$$

$$Y_{p(h)} = \frac{\tanh(\gamma l_{(h)}/2)}{Z_{0(h)}} = \frac{Y_{(h)}}{2} \frac{\tanh(\gamma l_{(h)}/2)}{\gamma l_{(h)}/2} \tag{12.77}$$

And

$$\gamma_{(h)} = \sqrt{z_{(h)}y_{(h)}} \approx \frac{h\omega}{l} \sqrt{LC}$$

$$Z_{0(h)} = \sqrt{z_{(h)}/y_{(h)}} \approx \sqrt{\frac{L}{C}} \tag{12.78}$$

Also

$$\lambda_{(h)} = 2\pi/\beta_{(h)} \approx \frac{l}{hf\sqrt{LC}}$$

$$v_{(h)} = f\lambda_h \approx \frac{l}{\sqrt{LC}}$$

$$f_{osc(h)} = \frac{v_{(h)}}{l} \approx \frac{1}{\sqrt{LC}} \tag{12.79}$$

Thus, the characteristic impedance, velocity of propagation, and frequency of oscillations are all independent of h while wavelength varies inversely with h.

The impedance, resistance, and reactance plots of transmission lines versus frequency depend on the line model used, and there is much variation in these plots and resonant frequencies exhibited depending on the line model (see Section 14.13).

We can also write the following equations:

$$
\begin{vmatrix} I_s \\ I_r \end{vmatrix} = \frac{1}{B} \begin{vmatrix} D & CB - DA \\ 1 & -A \end{vmatrix} \begin{vmatrix} V_s \\ V_r \end{vmatrix}
$$

$$
= \frac{1}{Z_0 \sinh \gamma l} \begin{vmatrix} \cosh \gamma l & -1 \\ 1 & -\cosh \gamma l \end{vmatrix} \begin{vmatrix} V_s \\ V_r \end{vmatrix}
$$

$$
= \begin{vmatrix} \dfrac{1}{Z_s} + Y_p & -\dfrac{1}{Z_s} \\ \dfrac{1}{Z_s} & -\dfrac{1}{Z_s} - Y_p \end{vmatrix} \begin{vmatrix} V_s \\ V_r \end{vmatrix} \tag{12.80}
$$

Note that

$$
CB - DA = \sinh^2(\gamma l) - \cosh^2(\gamma l) = -1
$$

Looking from one end, the impedance of the line is

$$
\frac{1}{Y_{p(h)}} \text{ in parallel with } \left(Z_{s(h)} + \frac{1}{Y_{p(h)}} \right)
$$

$$
= \frac{Z_{s(h)} Y_{p(h)} + 1}{Y_{p(h)}[Z_{s(h)} Y_{p(h)} + 2]} \tag{12.81}
$$

A computer program is required to calculate the impedance along the line using Eq. (12.81) with incremental changes in the frequency to locate the resonance points.

Example 12.5:   Consider a transmission line with the following parameters:

Conductor "FLINT," 740.8 KCMIL, ACSR, 37 strands, spaced horizontally 25 ft apart, one ground wire placed 16 ft above the conductors, height of conductors from ground = 60 ft, soil resistivity = 90 ohms-m.

A computer program is used to calculate the line parameters:

$$
R = 0.14852 \, \Omega/\text{mile}
$$

$$
X = 0.837 \, \Omega/\text{mile}
$$

$$
Y = 5.15 \times 10^{-6} \, \text{S/mile}
$$

Then

$$L = 0.00222\,\mathrm{H/mile}$$

$$C = 0.01366 \times 10^{-6}\,\mathrm{F/mile}$$

$$Z_0 = 403\,\Omega$$

$$\gamma = \sqrt{zy} = [(0.14852 + j0.837)(j5.15 \times 10^{-6}]^{1/2}$$

$$= (0.184 + j2.084) \times 10^{-3}/\mathrm{mile}$$

$$\lambda = \frac{2\pi}{\beta} = 3014\,\mathrm{mile}$$

$$v = f\lambda = 1.808 \times 10^5\,\mathrm{mile/s}$$

$$f_{\mathrm{osc}} = \frac{1}{\sqrt{LC}} = 3016\,\mathrm{Hz}(300\,\mathrm{miles\text{-}line})$$

Also

$$Z_s = Z_c\ \sinh(\gamma l) = \sqrt{\frac{0.14852 + j0.837}{j5.15 \times 10^{-6}}}\ \sinh(0.0552 + j0.6252)$$

$$= (405 - j36)\ \sinh(0.0552 + j0.6252)$$

$$\sinh(0.0552 + j0.6252) = \sinh(0.0552)\ \cos(0.6252) + j\ \cosh(0.0552)\ \sin(0.6252)$$

$$= (0.0552)(0.8108) + j(1.0)(0.5853)$$

$$= 0.045 + j0.5853$$

Then

$$Z_s = (405 - j36)(0.045 + j0.5833)$$

$$= 39.224 + j234.6$$

$$Y_p = \tanh\frac{(\gamma l/2)}{Z_c}$$

$$\tanh\left(\frac{\gamma l}{2}\right) = \frac{\sinh(\gamma l/2)}{\cosh(\gamma l/2)}$$

$$\sinh\left(\frac{\gamma l}{2}\right) = \sinh(0.0275 + j0.326) = (0.0275)(0.947) + j(1)(0.32)$$

$$= 0.026 + j0.32$$

$$\cosh\left(\frac{\gamma l}{2}\right) = \cosh(0.275)(0.947) + j(0.0275)(0.32) = 0.947 + j0.0088$$

$$\tanh\left(\frac{\gamma l}{2}\right) = 0.031 + j0.338$$

$$Y_p = \frac{0.031 + j0.338}{405 - j36} = 2.341 \times 10^{-6} + j8.348 \times 10^{-4} \approx j8.348 \times 10^{-4}$$

Then, the line impedance at fundamental frequency is

$$\frac{Z_{s(h)}Y_{p(h)} + 1}{Y_{p(h)}[Z_{s(h)}Y_{p(h)} + 2]}$$

$$= \frac{(405 - j36)(j8.348 \times 10^{-4}) + 1}{(j8.348 \times 10^{-4})[(405 - j36)(j8.348 \times 10^{-4}) + 2]}$$

$$= \frac{1.03 + j0.338}{-2.822 \times 10^{-4} + 1.695 \times 10^{-3})}$$

$$= 98.46 - j600.36$$

This shows the procedure. A computer program will run these calculations at close increment of frequency to capture the series and shunt resonance frequencies, plots as demonstrated in Fig. 12.15(a) and (b).

Figure 12.15(a) shows the calculated impedance angle and Fig. 12.15(b) shows the impedance modulus versus frequency plots of a 400-kV line, consisting of four bundled conductors of 397.5 KCMIL ACSR per phase in horizontal configuration, GMD = 44.09′, 200 miles long at no-load: looking from the sending end. The line generates 194.8 Mvar of capacitive reactive power. When the line is loaded to 200 MVA at 0.9 PF (no harmonics), the impedance angle and modulus (which are much reduced) are shown in Fig. 12.15(c) and (d). Two ground wires of 7#8 are considered in the calculations, and the earth resistivity is 100 Ωm. Each conductor has a diameter of 0.806 in. (0.317 cm) and bundle conductors are spaced 6 in. (0.15 m) center to center. The plots show a number of resonant frequencies. The impedance angle changes abruptly at each resonant frequency. For harmonic analyses of high-voltage (HV) long transmission lines, a computer program that can model the long-line model with harmonics is required (also see Refs [10–14]).

12.2   CABLES

Cables have higher capacitance than overhead lines; therefore, modeling of cable capacitances is of much significance for harmonic analysis. Much like transmission lines, the equivalent Π model should be used, and not the nominal Π model. A cable length can be considered short if the surge traveling time is lower than approximately 30% of the time constant of the voltage rise time in the system. The cross bonding

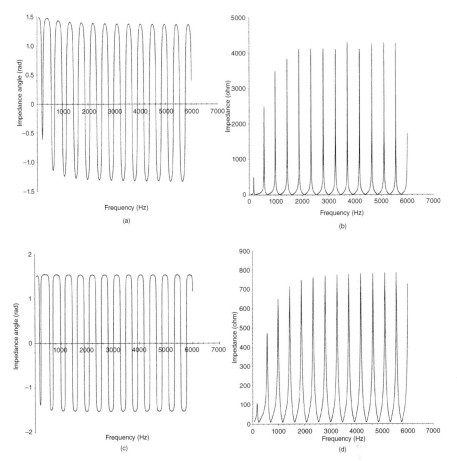

Figure 12.15   (a) Frequency scan of a 400-kV line with bundle conductors and (b) corresponding phase angle of the impedance modulus; (c) and (d) correponding plots with line loaded at 200 MVA, 0.9 lagging power factor.

of the sheaths of the cables presents another consideration, and each section of the cable length has sheath bonding and grounding connections.

A frequency-dependent model for armored power cables can be obtained using FE (finite element method), and the frequency response is obtained by coupling the cable FE domain model and external electrical circuits. The resulting vector-fitting rational function can then be represented by an equivalent network. This frequency response is then fitted with rational function approximation. For harmonic frequencies up to 3000 Hz, the resistance will increase. The slight decrease in inductance and the effect on shunt capacitance can be ignored. Π models are considered adequate for frequency scans (though not for transient analysis).

Cables vary in constructional features, not discussed here (see Refs [1,15]. Extruded dielectric (XD) also called solid dielectric – principally cross-linked polyethylene (XLPE) cables – have been used up to 500 kV. These are replacing pipe-insulted cables and oil-filled paper-insulated cables because of lower costs and less maintenance. Extruded cables include ethylene–propylene rubber (EPR) and low-density polyethylene (LDPE), although XPLE is the most common insulation system used for transmission cables. The XD cables also have a lower capacitance compared to the paper-insulted cables. In the United States, though XD cables are popular for the new installations, approximately 70% of circuit miles in service are pipe-type cables. Triple extrusion (inner shield, insulation, and outer conducting shield) processed in one sealed operation is responsible for developments of XLPE cables for higher voltages. Also a dry vulcanization process with no steam or water is used in the modern manufacturing technology.

Low-pressure oil-filled cables, also called self-contained liquid-filled (SCLF) systems, were developed in 1920 and were extensively used worldwide till 1980, while in the United States, pipe-type cables (described next) were popular. There are many miles in operation at 525 kV, both land and submarine cables. The stranded conductors are formed as a hollow duct in the center of the conductor. These may be wrapped with carbon paper, nonmagnetic steel tape, and conductor screen followed by many layers of oil-filled paper insulation. Insulation screen, lead sheath, bedding, reinforcement tapes, and outer polymeric sheath follow next. When the cable heats up during operation, the oil flows through the hollow duct in an axial direction into oil reservoirs connected with the sealing ends. The oil reservoirs are equipped with gas-filled flexible wall metal cells. The oil flow causes the cells to compress, thus increasing the pressure. This pressure forces the oil back into the cable. High-pressure fluid-filled (HPFF) cables have been a US standard. The system is fairly rugged. A welded catholically protected steel pipe, typically 8.625 or 10.75 in optical density is pressure and vacuum tested and three mass-impregnated cables are pulled into the pipe. The cable consists of copper or aluminum conductors, conductor shield, paper insulation, insulation shield, outer shielding, and skid wire for pulling cables into the pipe. This pipe is pressurized to approximately 200 psig with dielectric liquid to suppress ionization. Expansion and contraction of the liquid require large reservoir tanks and sophisticated pressurizing systems. Nitrogen gas at 200 psi may be used to pressurize the cable at voltages to 138 kV. To state other constructions, mass-impregnated nondraining (MIND) cables are used for high-voltage direct current (HVDC) circuits. Superconductivity using liquid helium at 4 K has been known and in the last couple of years has produced high-temperature superconductors (HTSs). These use liquid nitrogen temperatures (80 K). Special computer subroutines are required to correctly estimate the cable parameters for the harmonic analysis. Furthermore, the cable constants also depend on the geometry of cable installations, underground, overhead, in duct banks, and so on.

In EMTP simulations, the following cable models are available:

- FD cable model
- CP model
- Exact Π model
- Wideband cable model.

FD model provides an accurate representation of distributed nature of cable parameters R, L, G, and C, as well as their frequency dependence in modal quantities.

It is assumed that the characteristic impedance and the propagation function matrices can be diagonalized by a modal transformation matrix. For harmonic frequencies, the frequency-dependent models are not necessary, though the effect on resistance is considerable and should be modeled.

The velocity of propagation on cables is much lower than on OH lines due to higher capacitance and, thus, smaller lengths of cables can be termed "long." Considering the velocity of propagation of about 40%, the 40 miles long cable can be termed "long." Long-line effects should be represented for cables of length $40/h$ miles, where h is the harmonic number. The distributed parameter model akin to transmission lines can be used.

With respect to frequency dependence, the rigorous modeling and mathematical treatment of calculations of cable constants become complex, not discussed here. Much cable data are required for each cable type, its geometry, and method of installation. Much work has been done in LINE CONSTANT routines for EMTP simulations [16–23].

12.2.1 Cable Constants

The inductance per unit length of a single-conductor cable is given by

$$L = \frac{\mu_0}{2\pi} \ln \frac{r_1}{r_2} \, \text{H/m} \tag{12.82}$$

where r_1 is the radius of the conductor and r_2 is the radius of the sheath, that is, the cable outside diameter divided by 2.

When single-conductor cables are installed in magnetic conduits, the reactance may increase by a factor of 1.5. Reactance is also dependent on conductor shape, that is, circular or sector, and on the magnetic binders in three-conductor cables.

In general, the unit inductance of a cable to neutral is given by

$$L = \left(0.1404 \, \log_{10} \frac{2S}{d} + 0.0153 \right) \times 10^{-6} \, \text{H/ft} \tag{12.83}$$

where s is the center-to-center conductor spacing in inches and d is the diameter over conductor in inches:

$$S = \sqrt[3]{ABC} \tag{12.84}$$

where A, B, and C are the spacing in inches between conductors.

12.2.2 Capacitance of Cables

In a single-conductor cable, the capacitance per unit length is given by

$$C = \frac{2\pi\varepsilon\varepsilon_0}{\ln(r_1/r_2)} \, \text{F/m} \tag{12.85}$$

The dielectric constant ε is a specific property of the insulating material and is defined as the capacitance having a specific electrode/dielectric geometry to the capacitance with air as the dielectric. As an example for XPLE, the ε may vary from 2.3 to 6.

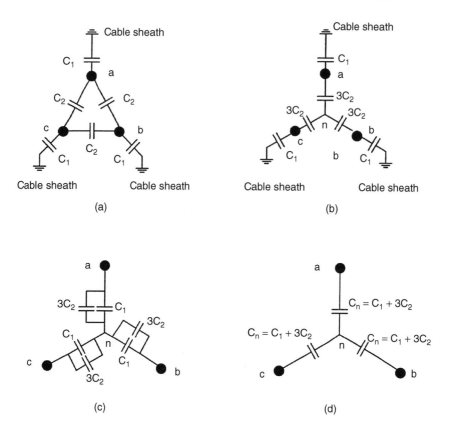

Figure 12.16 (a) Capacitances of a three-conductor cable, (b) and (c) equivalent circuits, and (d) final values.

Note that ε is the permittivity of the dielectric medium relative to air. (The parameter ε is frequency dependent, generally ignored for harmonics) The capacitances in a three-conductor cable are shown in Fig. 12.16. This assumes a symmetrical construction, and the capacitances between conductors and from conductors to the sheath are equal. The circuit of Fig. 12.16(a) is successively transformed through Fig. 12.16(b) and (c), and Fig. 12.16(d) shows that the net capacitance per phase $= C_1 + 3C_2$.

By change of units, Eq. (12.85) can be expressed as

$$C = \frac{7.35\varepsilon}{\log(r_1/r_2)} \, \text{pF/ft} \tag{12.86}$$

This gives the capacitance of a single-conductor shielded cable. Table 12.2 gives the values of ε for various cable insulation types.

Reference [6] provides an EMTP model and parameters of a single-conductor 400 kV XPLE cable.

TABLE 12.2   Typical Values for Dielectric Constants of Cable Insulation

Type of Insulation	Permittivity (ϵ)
Polyvinyl chloride (PVC)	3.5–8.0
Ethylene–propylene (EP) insulation	2.8–3.5
Polyethylene insulation	2.3
Cross-linked polyethylene	2.3–6.0
Impregnated paper	3.3–3.7

Some equations for medium-voltage *tape-shielded* cables are as follows:

$$Z_1 = Z_{aa} - Z_{ab} - \frac{(Z_{as} - Z_{ab})^2}{Z_{ss} - Z_{ab}} \tag{12.87}$$

$$Z_0 = Z_{aa} + 2Z_{ab} - \frac{(Z_{as} + 2Z_{ab})^2}{Z_{ss} + 2Z_{ab}} \tag{12.88}$$

In ohms/1000 ft

where

Z_1 = positive or negative sequence impedance

Z_0 = zero sequence impedance

Z_{aa} = self-impedance of each conductor with earth return in ohms per 1000 ft

Z_{ab} = mutual impedance of each conductor with earth return in ohms per 1000 ft

Z_{as} = mutual impedance of the each tape shield with earth return in ohms per 1000 ft

Z_{ss} = self-impedance of the each tape shield with earth return in ohms per 1000 ft

$$Z_{aa} = R_\varphi + 0.0181 + 0.0529j \, \log_{10}\left(\frac{D_e}{\mathrm{GMR}_\varphi}\right)$$

$$Z_{ss} = R_s + 0.0181 + 0.0529j \, \log_{10}\left(\frac{D_e}{\mathrm{GMR}_\varphi}\right)$$

$$Z_{ab} = 0.0181 + 0.0529j \, \log_{10}\left(\frac{D_e}{\mathrm{GMD}_\varphi}\right)$$

$$Z_{as} = 0.0181 + 0.0529j \, \log_{10}\left(\frac{D_e}{\mathrm{GMR}_s}\right) \tag{12.89}$$

In ohms/1000 ft.

where

R_Φ	=	AC resistance of a phase conductor at operating temperature in ohms per 1000 ft
R_s	=	effective shield resistance in ohms per 1000 ft
D_e	=	mean depth of earth return current in inches
GMR_ϕ	=	GMR of phase conductor in inches
GMR_s	=	radius from center of phase conductor to center of shield in inches
GMD_ϕ	=	distance between the centers of two-phase conductors in inches

Note that these expressions will vary with wire shield concentric neutral cables and single-phase laterals.

12.3   ZERO SEQUENCE IMPEDANCE OF OH LINES AND CABLES

The zero sequence impedance of the lines and cables is dependent on the current flow through a conductor and return through the ground, or sheaths, and encounters the impedance of these paths. The zero sequence current flowing in one phase also encounters the currents arising out of that conductor self-inductance from mutual inductance to other two-phase conductors, from the mutual inductance to the ground and sheath return paths, and from the self-inductance of the return paths. Tables and analytical expressions are provided in Ref. [12]. As an example, the zero sequence impedance of a three-conductor cable with a solidly bonded and grounded sheath is given by

$$z_0 = r_c + r_e + j0.8382 \frac{f}{60} \ \log_{10} \frac{D_e}{GMR_{3c}} \qquad (12.90)$$

where

r_c	=	AC resistance of one conductor (ohms/mile)
r_e	=	AC resistance of earth return (depending on the equivalent depth of earth return, soil resistivity, taken as 0.286 ohms/mile)
D_e	=	distance to equivalent earth path [12]
GMR_{3c}	=	geometric mean radius of conducting path made up of three actual conductors taken as a group (inch)

$$GMR_{3c} = \sqrt[3]{GMR_{1c}S^2} \qquad (12.91)$$

where GMR_{1c} is the geometric mean radius of individual conductor and $S = (d + 2t)$, where d is the diameter of the conductor and t is the thickness of the insulation. Many handbooks provide ready reference tables for cable constants, and also computer data bases have extensive libraries for various cable types and insulations.

12.3.1   Grounding of Cable Shields

Treatment of shields is an important consideration in HV cable installations. For short length of cables, it is customary to ground the shields at the sending and receiving ends. This will erate the cable ampacity, which can be calculated by cable modeling as the shields will carry some induced currents. For single-point grounding, it is recommended in IEEE standards that the shield potential should be limited to 25 V. The induced voltages are dependent on the cable spacing and geometry. The shield resistance and mutual reactance can be calculated, and mathematical equations are available. Then based on the conductor spacing, the losses in the shield can be calculated Refs. [24,25]. Some special bonding techniques are the following:

- Single-point bonding
- Multiple-point bonding
- Impedance bonding
- Sectionalized cross bonding
- Continuous cross bonding.

In all situations, the bonding techniques must address the following:

- Provide grounding for the cable
- Maintain a continuous current return path, either through the shield/sheath or through ground continuity conductor
- Limit normal steady-state shield voltages
- Specifically reduce shield losses
- Limit transient overvoltages to an acceptable level.

Figure 12.17 is reproduced from Ref. [25].

12.4   FILTER REACTORS

The frequency-dependent Q of the filter reactors is especially important, as it affects the sharpness of tuning. Resistance at high frequencies can be calculated by the following expressions:

$$R_h = \left[\frac{0.115h^2 + 1}{1.15}\right] R_f \text{ for aluminum reactors} \qquad (12.92)$$

$$R_h = \left[\frac{0.055h^2 + 1}{1.055}\right] R_f \text{ for copper reactors} \qquad (12.93)$$

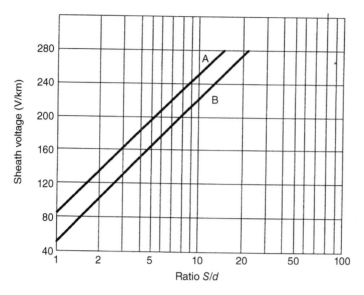

A: Outer cable of group in flat formation

B: Cable in trefoil and center cable of group in flat formation

Figure 12.17   Sheath voltage, single-conductor cables. Source: Ref. [25].

where R_f is the resistance at fundamental frequency.

12.5   TRANSFORMERS

For harmonic analysis, generally, we need not consider high-frequency models of transformers, which can be fairly complex [7,26–28], A two-winding transformer can simply be represented by its % impedance:

$$Z_t = kR_T + jhX_T \qquad (12.94)$$

where R_T and X_T pertain to the fundamental frequency. The factor k takes account of the skin and eddy current effects at higher frequency. We assume that the transformer reactance at a harmonic is linearly proportioned with the frequency, that is, saturation due to higher frequencies is neglected, that is

$$X_{T(f1)} = kX_T \left(\frac{f_1}{f} \right) \qquad (12.95)$$

where X_T is the reactance at fundamental frequency f; and at frequency f_1, the reactance is proportionate, $k = 1$.

The zero sequence impedance varies depending on the transformer construction, shell or core type, three-limb or five-limb, and winding connections.

12.5.1 Frequency-Dependent Models

Fundamental frequency values of resistance and reactance can be found by no-load and short-circuit tests on the transformer. The resistance of the transformer can be modified with an increase in frequency according to Fig. 12.18. While the resistance increases with frequency, the leakage inductance reduces. The magnetizing branch in the transformer model is often omitted if the transformer is not considered a source of the harmonics. This simplified model may not be accurate, as it does not model the nonlinearity in the transformer on account of the following:

- Core and copper losses due to eddy current and hysteresis affects. The core loss is the summation of eddy current and hysteresis loss; both are frequency dependent:

$$P_c = P_e + P_h = K_e B^2 f^2 + K_h B^s f \qquad (12.96)$$

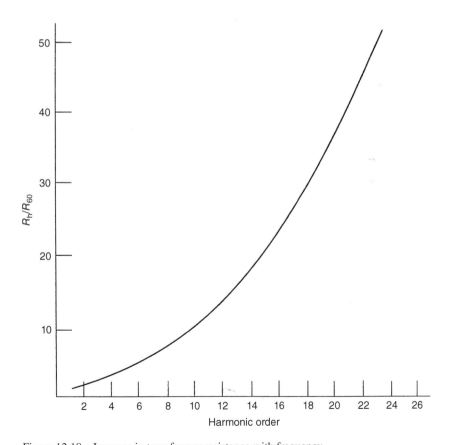

Figure 12.18   Increase in transformer resistance with frequency.

TABLE 12.3 Capacitance of Core-Type Transformers in Nanofarads

MVA Rating of Transformer	C_{hg}	C_{hl}	C_{lg}
1	1.2–14	1.2–17	3.1–16
2	1.4–16	1–18	3–16
5	1.2–14	1.1–20	5.5–17
10	4–7	4–11	8–18
25	2.8–4.2	2.5–18	5.2–20
50	4–6.8	3.4–11	3–24
75	3.5–7	5.5–13	2.8–30
100	3.3–7	5–13	4–40

Note: Bashing capacitance not shown.

where B is the peak flux density, s is the Steinmetz constant (typically 1.5–2.5, depending on the core material), f is the frequency, and K_e and K_h are constants.

- Leakage fluxes about the windings, cores, and surrounding medium
- Core magnetization characteristics, which are the primary source of transformer nonlinearity

A number of approaches can be taken to model the nonlinearities. EMTP transformer models *Satura* and *Hysdat* are examples [8]. Even these models consider only core magnetization characteristics and neglect nonlinearities or frequency dependence of core losses and winding effects. More elaborate models are available.

Capacitances of the transformer windings are generally omitted for harmonic analysis of distribution systems; however, for transmission systems, capacitances are included. Table 12.3 provides approximate values of the capacitances shown in Fig. 12.19. These do not have a significant impact except at high frequencies. Also, hysteresis modeling though important for transient studies such as switching or faults is often neglected for harmonic analyses.

Converter loads may draw DC and low-frequency currents through the transformers, that is, a cycloconverter load, and saturate the transformers.

12.5.2 Three-Winding Transformers

For phase multiplication, three-winding transformers are used, Chapter 6.

The wye equivalent circuit of a three-winding transformer is shown in Fig. 12.20.

We can write the following equations:

$$Z_H = \frac{1}{2}(Z_{HM} + Z_{HL} - Z_{ML})$$

$$Z_M = \frac{1}{2}(Z_{ML} + Z_{HM} - Z_{HL})$$

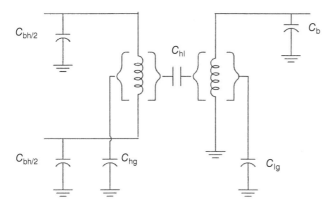

Figure 12.19   Simplified capacitance model of a two-winding transformer.

$$Z_L = \frac{1}{2}(Z_{HL} + Z_{ML} - Z_{HM}) \tag{12.97}$$

where Z_{HM} is the leakage impedance between the H and X windings, as measured on the H winding with M winding short circuited and L winding open circuited; Z_{HL} is the leakage impedance between the H and L windings, as measured on the H winding with L winding short circuited and M winding open circuited; Z_{ML} is the leakage impedance between the M and L windings, as measured on the M winding with L winding short circuited and H winding open circuited.

Equation (12.97) can be written as

$$\begin{vmatrix} Z_H \\ Z_M \\ Z_L \end{vmatrix} = 1/2 \begin{vmatrix} 1 & 1 & -1 \\ 1 & -1 & 1 \\ -1 & 1 & 1 \end{vmatrix} \begin{vmatrix} Z_{HM} \\ Z_{HL} \\ Z_{ML} \end{vmatrix} \tag{12.98}$$

We also see that

$$Z_{HL} = Z_H + Z_L$$

$$Z_{HM} = Z_H + Z_M$$

$$Z_{ML} = Z_M + Z_L \tag{12.99}$$

For the transformers for converter application, the secondary windings are equally rated:

$$Z_{HL} = Z_{HM}$$

$$\therefore$$

$$Z_H = Z_L = 0.5Z_{HL} \tag{12.100}$$

Example 12.6:   A three-phase, three-winding transformer nameplate reads as

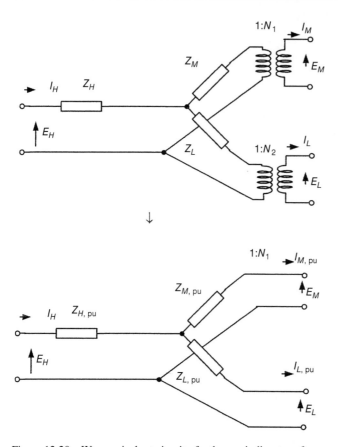

Figure 12.20   Wye equivalent circuit of a three-winding transformer.

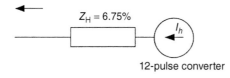

Figure 12.21   Conversion of a three-winding transformer to two-winding transformer for harmonic analysis.

High-voltage winding: 13.8 kV, wye connected rated at 30-MVA, medium-voltage winding 2.4 kV, wye connected rated at 15 MVA, 2.4 kV winding (tertiary winding), delta connected and rated 15 MVA. $Z_{HM} = 9\%$, $Z_{HL} = 9\%$. On the basis of Eq. (12.97), $Z_{ML} = 4.5\%$. Then,

$$Z_H = 0.5(9 + 9 - 4.5) = 6.75\%$$

Thus, the model is simplified and is used for the harmonic analysis, 12-pulse converter, representation as shown in Fig. 12.21.

12.5.3 Four-Winding Transformers

By proper choice of winding impedances, four-winding transformers are used for inductive filtering, which prevents harmonics impacting the source, harmonics limited within the transformer. Figure 12.22(a) shows schematically the circuit of a four-winding transformer, one primary winding, and three secondary windings. We have impedances between windings as follows:

Z_{12}, Z_{13}, Z_{14} – between windings 1 and 2, 1 and 3, and 1 and 4.

Z_{24}, Z_{34}, Z_{23} – between windings 2 and 4, 3 and 4, and 2 and 3, respectively.

Then, winding impedances Z_1, Z_2, Z_3, and Z_4 can be calculated as follows:

$$Z_1 = \frac{1}{2}(Z_{12} + Z_{14} - Z_{24} - K)$$

$$Z_2 = \frac{1}{2}(Z_{12} + Z_{23} - Z_{13} - K)$$

$$Z_3 = \frac{1}{2}(Z_{23} + Z_{34} - Z_{24} - K)$$

$$Z_4 = \frac{1}{2}(Z_{34} + Z_{14} - Z_{13} - K) \tag{12.101}$$

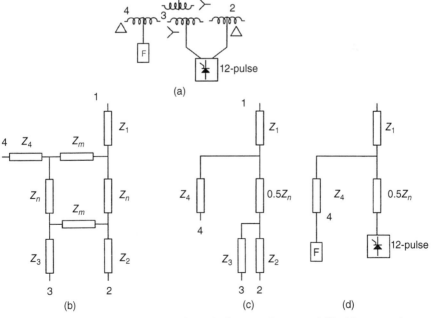

Figure 12.22   (a) Representation of a four-winding transformer and (b)–(d) progressive reduction of equivalent circuit.

where

$$K = \sqrt{K_1 K_2} = \frac{Z_m Z_n}{Z_m + Z_n} \tag{12.102}$$

and

$$K_1 = Z_{13} + Z_{24} - Z_{12} - Z_{34} = \frac{Z_m^2}{Z_m + Z_n}$$

$$K_2 = Z_{13} + Z_{24} - Z_{14} - Z_{23} = \frac{Z_n^2}{Z_m + Z_n} \tag{12.103}$$

Consider that the secondary windings 2 and 3 and tertiary windings are equally rated:

$$Z_{12} = Z_{13}$$

$$Z_{24} = Z_{34} \tag{12.104}$$

Then,

$$K_1 = 0, \quad K = 0$$

$$Z_n = K_2$$

$$Z_2 = Z_3 = 0.5 Z_{23}$$

$$Z_m = 0 \tag{12.105}$$

Then the circuit of Fig. 12.22(a) progressively reduces to that of Fig. 12.22(d).

Example 12.7:   A four-winding transformer has the following vendor's data:

$Z_{12} = Z_{13} = 25\%$, $Z_{14} = 10\%$, $Z_{24} = Z_{34} = 5.5\%$, and $Z_{23} = 10\%$ impedances on their respective MVA base

MVA base = Z_{12}, Z_{13}, and $Z_{14} = 69$ MVA, $Z_{24} = Z_{34} = Z_{23} = 23$ MVA
Voltage of windings 1 = 138 kV, 2, 3, and 4 = 13.8 kV

The equivalent circuit can be developed as follows.
Convert to a common 100-MVA base in pu

$$Z_{12} = Z_{13} = 0.36, \quad Z_{14} = 0.145, \quad Z_{24} = Z_{34} = 0.239, \quad Z_{23} = 0.435$$

Then $K_1 = 0$, $K_2 = 0.019$, $Z_m = 0$, $Z_n = K_2 = 0.019$.
The transformer impedances are

$$Z_1 = 0.133$$

$$Z_2 = Z_3 = 0.2175$$

$$Z_4 = 0.024$$

Generally, $Z_{13} > Z_{14} + Z_{34}$ and Z_4 is negative.
A matrix model of N-winding transformer can be written as

$$\begin{vmatrix} v_1 \\ v_2 \\ v_3 \\ \cdot \\ v_n \end{vmatrix} = \begin{vmatrix} r_{11} & r_{12} & r_{13} & \cdot & r_{1n} \\ r_{21} & r_{22} & r_{23} & \cdot & r_{2n} \\ r_{31} & r_{32} & r_{33} & \cdot & r_{3n} \\ \cdot & \cdot & \cdot & \cdot & \cdot \\ r_{n1} & r_{n2} & r_{n3} & \cdot & r_{nn} \end{vmatrix} \begin{vmatrix} i_1 \\ i_2 \\ i_3 \\ \cdot \\ i_n \end{vmatrix} + \begin{vmatrix} l_{11} & l_{12} & l_{13} & \cdot & l_{1n} \\ l_{21} & l_{22} & l_{23} & \cdot & l_{2n} \\ l_{31} & l_{32} & r_{33} & \cdot & l_{3n} \\ \cdot & \cdot & \cdot & \cdot & \cdot \\ l_{n1} & l_{n2} & l_{n3} & \cdot & l_{nn} \end{vmatrix} \frac{d}{dt} \begin{vmatrix} i_1 \\ i_2 \\ i_3 \\ \cdot \\ i_n \end{vmatrix} \quad (12.106)$$

12.5.4   Sequence Networks of Transformers

The sequence networks of some two-winding and three-winding transformer connections are shown in Tables 12.4 and 12.5. The zero sequence impedance varies from a low value to an open circuit.

12.5.5   Matrix Equations

The three-phase transformer transformation matrices and their developments are discussed in Ref. [29]. The node admittance matrix of a transformer can be divided into submatrices as follows:

$$\overline{Y}_{node} = \begin{vmatrix} \overline{Y}_1 & \overline{Y}_{11} \\ \overline{Y}_{11}^t & \overline{Y}_{111} \end{vmatrix} \quad (12.107)$$

where each 3×3 submatrix depends on the winding connections, as shown in Table 12.6. The submatrices in this table are defined as follows:

$$\overline{Y}_I = \begin{vmatrix} y_t & 0 & 0 \\ 0 & y_t & 0 \\ 0 & 0 & y_t \end{vmatrix} \quad \overline{Y}_{II} = \frac{1}{3}\begin{vmatrix} 2y_t & -y_t & -y_t \\ -y_t & 2y_t & -y_t \\ -y_t & -y_t & 2Y_t \end{vmatrix} \quad \overline{Y}_{III} = \frac{1}{\sqrt{3}}\begin{vmatrix} -y_t & y_t & 0 \\ 0 & -y_t & y_t \\ y_t & 0 & -y_t \end{vmatrix}$$
$$(12.108)$$

Here, y_t is the leakage admittance per phase in per unit. Table 12.6 can be used as a simplified approach to the modeling of common core-type, three-phase transformer in an unbalanced system.

If the off-nominal tap ratio between primary and secondary windings is $\alpha:\beta$, where α and β are the taps on the primary and secondary sides, respectively, in per unit, then the submatrices are modified as follows:

- Divide self-admittance of primary matrix by α^2
- Divide self-admittance of secondary matrix by β^2
- Divide mutual admittance matrixes by $\alpha\beta$.

TABLE 12.4 Equivalent Positive, Negative, and Zero Sequence Circuits of Two-Winding Transformers

No.	Winding connections	Zero-sequence circuit	Positive- or negative-sequence circuit
1	Z_{nH}	H — Z_H — Z_L — L; $3Z_{nH}$; N_0	H — Z_H — Z_L — L; N_1 or N_2
2		H — Z_H — Z_L — L; N_0	H — Z_H — Z_L — L; N_1 or N_2
3		H — Z_H — Z_L — L; N_0	H — Z_H — Z_L — L; N_1 or N_2
4		H — Z_H — Z_L — L; N_0	H — Z_H — Z_L — L; N_1 or N_2
5		H — Z_H — Z_L — L; N_0	H — Z_H — Z_L — L; N_1 or N_2
6		H — Z_H — Z_L — L; N_0	H — Z_H — Z_L — L; N_1 or N_2
7		H — Z_H — Z_L — L; N_0	H — Z_H — Z_L — L; N_1 or N_2
8		H — Z_H — Z_L — L; N_0; Z_M	H — Z_H — Z_L — L; N_1 or N_2
9	Z_{nH} Z_{nL}	H — Z_H — Z_L — L; $3Z_{nH}$ $3Z_{nL}$; Z_M; N_0	H — Z_H — Z_L — L; N_1 or N_2
10	Z_{nH} Z_{nL}	Z_H — Z_L — L; $3Z_{nH}$ $3Z_{nL}$; N_0	H — Z_H — Z_L — L; N_1 or N_2

TABLE 12.5   Equivalent Positive, Negative, and Zero Sequence Circuits of Three-Winding Transformers

No.	Winding connections	Zero-sequence circuit	Positive- or negative sequence circuit
1	H, M, L	Z_H, Z_M, Z_L, M, L, N_0	Z_H, Z_M, Z_L, M, L, N_1 or N_2
2	H, Z_{nH}, M, L	$3Z_{nH}$, Z_H, Z_M, Z_L, M, L, N_0	Z_H, Z_M, Z_L, M, L, N_1 or N_2
3	H, M, L	Z_H, Z_M, Z_L, M, L, N_0	Z_H, Z_M, Z_L, M, L, N_1 or N_2
4	H, Z_{nH}, M, Z_{nM}, L	$3Z_{nH}$, Z_H, $3Z_{nM}$, Z_M, Z_L, M, L, N_0	Z_H, Z_M, Z_L, M, L, N_1 or N_2
5	H, Z_{nH}, M, L	$3Z_{nH}$, Z_H, Z_M, Z_L, M, L, N_0	Z_H, Z_M, Z_L, M, L, N_1 or N_2
6	H, M, L	Z_H, Z_M, Z_L, M, L, N_0	Z_H, Z_M, Z_L, M, L, N_1 or N_2

TABLE 12.6   Submatrices of Three-Phase Transformer Connections

Winding Connections		Self-Admittance		Mutual Admittance	
Primary	Secondary	Primary	Secondary	Primary	Secondary
Wye-G	Wye-G	$\overline{Y}_{\mathrm{I}}$	$\overline{Y}_{\mathrm{I}}$	$-\overline{Y}_{\mathrm{I}}$	$-\overline{Y}_{\mathrm{I}}$
Wye-G Wye Wye	Wye Wye-G Wye	$\overline{Y}_{\mathrm{II}}$	$\overline{Y}_{\mathrm{II}}$	$-\overline{Y}_{\mathrm{II}}$	$-\overline{Y}_{\mathrm{II}}$
Wye-G	Delta	$\overline{Y}_{\mathrm{I}}$	$\overline{Y}_{\mathrm{II}}$	$\overline{Y}_{\mathrm{III}}$	$\overline{Y}'_{\mathrm{III}}$
Wye	Delta	$\overline{Y}_{\mathrm{II}}$	$\overline{Y}_{\mathrm{II}}$	$\overline{Y}_{\mathrm{III}}$	$\overline{Y}'_{\mathrm{III}}$
Delta	Wye	$\overline{Y}_{\mathrm{II}}$	$\overline{Y}_{\mathrm{II}}$	$\overline{Y}'_{\mathrm{III}}$	$\overline{Y}_{\mathrm{III}}$
Delta	Wye-G	$\overline{Y}_{\mathrm{II}}$	$\overline{Y}_{\mathrm{I}}$	$\overline{Y}'_{\mathrm{III}}$	$\overline{Y}_{\mathrm{III}}$
Delta	Delta	$\overline{Y}_{\mathrm{II}}$	$\overline{Y}_{\mathrm{II}}$	$-\overline{Y}_{\mathrm{II}}$	$-\overline{Y}_{\mathrm{II}}$

Y^t_{III} is the transpose of Y_{III}.

Consider a wye-grounded transformer. Then, from Table 12.6:

$$\overline{Y}^{abc} = \begin{vmatrix} \overline{Y}_I & -\overline{Y}_I \\ -\overline{Y}_I & \overline{Y}_I \end{vmatrix} = \begin{vmatrix} y_t & 0 & 0 & -y_t & 0 & 0 \\ 0 & y_t & 0 & 0 & -y_t & 0 \\ 0 & 0 & y_t & 0 & 0 & -y_t \\ -y_t & 0 & 0 & y_t & 0 & 0 \\ 0 & -y_t & 0 & 0 & y_t & 0 \\ 0 & 0 & -y_t & 0 & 0 & y_t \end{vmatrix} \qquad (12.109)$$

For off-nominal taps, the matrix is modified as

$$\overline{Y}^{abc} = \begin{vmatrix} \dfrac{y_t}{\alpha^2} & 0 & 0 & \dfrac{y_t}{\alpha\beta} & 0 & 0 \\ 0 & \dfrac{y_t}{\alpha^2} & 0 & 0 & \dfrac{y_t}{\alpha\beta} & 0 \\ 0 & 0 & \dfrac{y_t}{\alpha^2} & 0 & 0 & \dfrac{y_t}{\alpha\beta} \\ \dfrac{y_t}{\alpha\beta} & 0 & 0 & \dfrac{y_t}{\beta^2} & 0 & 0 \\ 0 & \dfrac{y_t}{\alpha\beta} & 0 & 0 & \dfrac{y_t}{\beta^2} & 0 \\ 0 & 0 & \dfrac{y_t}{\alpha\beta} & 0 & 0 & \dfrac{y_t}{\beta^2} \end{vmatrix} \qquad (12.110)$$

A three-phase transformer winding connection of delta primary and grounded-wye secondary is commonly used. From Table 12.6, its matrix equation is

$$
\overline{Y}^{abc} =
\begin{vmatrix}
\dfrac{2}{3}y_t & -\dfrac{1}{3}y_t & -\dfrac{1}{3}y_t & -\dfrac{y_t}{\sqrt{3}} & \dfrac{y_t}{\sqrt{3}} & 0 \\[2ex]
-\dfrac{1}{3}y_t & \dfrac{2}{3}y_t & -\dfrac{1}{3}y_t & 0 & -\dfrac{y_t}{\sqrt{3}} & \dfrac{y_t}{\sqrt{3}} \\[2ex]
-\dfrac{1}{3}y_t & -\dfrac{1}{3}y_t & \dfrac{2}{3}y_t & \dfrac{y_t}{\sqrt{3}} & 0 & -\dfrac{y_t}{\sqrt{3}} \\[2ex]
-\dfrac{y_t}{\sqrt{3}} & 0 & \dfrac{y_t}{\sqrt{3}} & y_t & 0 & 0 \\[2ex]
\dfrac{y_t}{\sqrt{3}} & -\dfrac{y_t}{\sqrt{3}} & 0 & 0 & y_t & 0 \\[2ex]
0 & \dfrac{y_t}{\sqrt{3}} & -\dfrac{y_t}{\sqrt{3}} & 0 & 0 & y_t
\end{vmatrix}
\qquad (12.111)
$$

where y_t is the leakage reactance of the transformer.

For an off-nominal transformer, the Y matrix is modified as shown in the following equation:

$$
\overline{Y}^{abc} =
\begin{vmatrix}
\dfrac{2}{3}\dfrac{y_t}{\alpha^2} & -\dfrac{1}{3}\dfrac{y_t}{\alpha^2} & -\dfrac{1}{3}\dfrac{y_t}{\alpha^2} & -\dfrac{y_t}{\sqrt{3}\alpha\beta} & \dfrac{y_t}{\sqrt{3}\alpha\beta} & 0 \\[2ex]
-\dfrac{1}{3}\dfrac{y_t}{\alpha^2} & \dfrac{2}{3}\dfrac{y_t}{\alpha^2} & -\dfrac{1}{3}\dfrac{y_t}{\alpha^2} & 0 & -\dfrac{y_t}{\sqrt{3}\alpha\beta} & \dfrac{y_t}{\sqrt{3}\alpha\beta} \\[2ex]
-\dfrac{1}{3}\dfrac{y_t}{\alpha^2} & -\dfrac{1}{3}\dfrac{y_t}{\alpha^2} & \dfrac{2}{3}\dfrac{y_t}{\alpha^2} & \dfrac{y_t}{\sqrt{3}\alpha\beta} & 0 & -\dfrac{y_t}{\sqrt{3}\alpha\beta} \\[2ex]
-\dfrac{y_t}{\sqrt{3}\alpha\beta} & 0 & \dfrac{Y_t}{\sqrt{3}\alpha\beta} & \dfrac{y_t}{\beta^2} & 0 & 0 \\[2ex]
\dfrac{y_t}{\sqrt{3}\alpha\beta} & -\dfrac{y_t}{\sqrt{3}\alpha\beta} & 0 & 0 & \dfrac{y_t}{\beta^2} & 0 \\[2ex]
0 & \dfrac{y_t}{\sqrt{3}\alpha\beta} & -\dfrac{y_t}{\sqrt{3}\alpha\beta} & 0 & 0 & \dfrac{y_t}{\beta^2}
\end{vmatrix}
$$

$$(12.112)$$

where α and β are the taps on the primary and secondary sides in per unit. The equation of a wye-grounded delta transformer and $\alpha = \beta = 1$ can be written as

$$
\begin{vmatrix} I_A \\ I_B \\ I_C \\ I_a \\ I_b \\ I_c \end{vmatrix} =
\begin{vmatrix}
y_t & 0 & 0 & -\dfrac{1}{\sqrt{3}}y_t & \dfrac{1}{\sqrt{3}}y_t & 0 \\[2mm]
0 & y_t & 0 & 0 & -\dfrac{1}{\sqrt{3}}y_t & \dfrac{1}{\sqrt{3}}y_t \\[2mm]
0 & 0 & y_t & \dfrac{1}{\sqrt{3}}y_t & 0 & -\dfrac{1}{\sqrt{3}}y_t \\[2mm]
-\dfrac{1}{\sqrt{3}}y_t & 0 & \dfrac{1}{\sqrt{3}}y_t & \dfrac{2}{3}y_t & -\dfrac{1}{3}y_t & -\dfrac{1}{3}y_t \\[2mm]
\dfrac{1}{\sqrt{3}}y_t & -\dfrac{1}{\sqrt{3}}y_t & 0 & -\dfrac{1}{3}y_t & \dfrac{2}{3}y_t & -\dfrac{1}{3}y_t \\[2mm]
0 & \dfrac{1}{\sqrt{3}}y_t & -\dfrac{1}{\sqrt{3}}y_t & -\dfrac{1}{3}y_t & -\dfrac{1}{3}y_t & \dfrac{2}{3}y_t
\end{vmatrix}
\begin{vmatrix} V_A \\ V_B \\ V_C \\ V_a \\ V_b \\ V_c \end{vmatrix}
$$

$$(12.113)$$

Here, the currents and voltages with capital subscripts relate to primary and those with lowercase subscripts relate to secondary. In the condensed form, we will write it as

$$\bar{I}_{ps} = \bar{Y}_{Y-\Delta}\bar{V}_{ps} \qquad (12.114)$$

Using symmetrical component transformation:

$$
\begin{vmatrix} \bar{I}_p^{012} \\ \bar{I}_s^{012} \end{vmatrix} =
\begin{vmatrix} \bar{T}_s & 0 \\ 0 & \bar{T}_s \end{vmatrix}^{-1}
\bar{Y}_{y-\Delta}
\begin{vmatrix} \bar{T}_s & 0 \\ 0 & \bar{T}_s \end{vmatrix}
\begin{vmatrix} \bar{V}_p^{012} \\ \bar{V}_s^{012} \end{vmatrix}
\qquad (12.115)
$$

Expanding

$$
\begin{vmatrix} \bar{I}_p^{012} \\ \bar{I}_s^{012} \end{vmatrix} =
\begin{vmatrix}
y_t & 0 & 0 & 0 & 0 & 0 \\
0 & y_t & 0 & 0 & y_t < -30^\circ & 0 \\
0 & 0 & y_t & 0 & 0 & y_t < 30^\circ \\
0 & 0 & 0 & 0 & 0 & 0 \\
0 & y_t < 30^\circ & 0 & 0 & y_t & 0 \\
0 & 0 & y_t < -30^\circ & 0 & 0 & y_t
\end{vmatrix}
\begin{vmatrix} \bar{V}_p^{012} \\ \bar{V}_s^{012} \end{vmatrix}
\qquad (12.116)
$$

The positive sequence equations are

$$I_{p1} = y_t V_{p1} - y_t < -30^\circ V_{s1}$$

$$I_{s1} = y_t V_{s1} - y_t < 30^\circ V_{p1} \qquad (12.117)$$

The negative sequence equations are

$$I_{p2} = y_t V_{p2} - y_t < 30^\circ V_{s2}$$

$$I_{s2} = y_t V_{s2} - y_t < -30^\circ V_2 \qquad (12.118)$$

The zero sequence equation is

$$I_{p0} = y_t V_{p0}$$

$$I_{sO} = 0 \tag{12.119}$$

For a balanced system, only the positive sequence component needs to be considered. The power flow on the primary side:

$$S_{ij} = V_i I_{ij}^* = V_i(y_t^* V_i^* - y_t^* < 30° V_j^*)$$

$$= [y_t V_i^2 \cos \theta_{yt} - y_t |V_i V_j| \cos(\theta_i - \theta_{yt} - (\theta_j + 30°))]$$

$$+ j[-y_t V_i^2 \sin \theta_{yt} - y_t |V_i V_j| \sin(\theta_i - \theta_{yt} - (\theta_j + 30°))] \tag{12.120}$$

and on the secondary side:

$$S_{ji} = V_i I_{ji}^* = V_j(y_t^* V_j^* - y_t^* < -30° V_i^*)$$

$$= [y_t V_j^2 \cos \theta_{yt} - y_t |V_j V_i| \cos(\theta_j - \theta_{yt} - (\theta_i - 30°))]$$

$$+ j[-y_t V_j^2 \sin \theta_{yt} - y_t |V_j V_i| \sin(\theta_j - \theta_{yt} - (\theta_i - 30°))] \tag{12.121}$$

Table 12.7 shows modeling guide lines according to CIGRE [30]. The models and applications of the following devices are not discussed:

TABLE 12.7   Modeling Guidelines for Transformers Based on CIGRE Recommendations

Transformer	Group 1 0.1 Hz to ≈ 3 kHz	Group 2 60 Hz to ≈ 30 kHz	Group 3 10 kHz to ≈ 3 MHz	Group 4 100 kHz to ≈ 50 MHz
Short-circuit impedance	Very important	Very important	Important only for surge transfer	Negligible
Saturation	Very important	See note below	Negligible	Negligible
Frequency-dependent series losses	Very important	Important	Negligible	Negligible
Hysteresis and iron losses	Important only for resonance phenomena	Important only for transformer energizing	Negligible	Negligible
Capacitance coupling	Negligible	Important for surge transfer	Very important for surge transfer	Very important for surge transfer

Very important for transformer energizing and load rejection with high-voltage increases, otherwise negligible.

Source: Ref. [1].

- Single-phase and three-phase autotransformers and their sequence impedances
- Phase-shifting transformers
- Step-voltage regulators
- Line drop compensators
- Scot connection of transformers
- Corner-grounded delta–delta or midpoint-grounded delta–delta transformers
- Open delta transformers.

These equipments are used in power systems, see Refs [7,31] for their models and applications.

12.6 INDUCTION MOTORS

Figure 12.23(a) and (b) illustrates the equivalent circuits of an induction motor for the positive and negative sequences. The shunt elements g_m and b_m are relatively large compared to R_1, r_2, X_1, and x_2. Generally, the locked rotor current of the motor is known and is given by the following equation:

$$I_{lr} = \frac{V_1}{(R_1 + r_2) + j(X_1 + x_2)} \tag{12.122}$$

At fundamental frequency, neglecting magnetizing and loss components, the motor reactance is

$$X_f = X_1 + x_2 \tag{12.123}$$

and the resistance is

$$R_f = R_1 + \frac{r_2}{s} \tag{12.124}$$

This resistance is not the same as used in short-circuit calculations. At harmonic frequencies, the reactance can be directly related to the frequency:

$$X_h = hX_f \tag{12.125}$$

This relation is only approximately correct. The reactance at higher frequencies is reduced due to saturation. The stator resistance can be assumed to vary as the square root of the frequency:

$$R_{1h} = \sqrt{h} \cdot (R_1) \tag{12.126}$$

The harmonic slip is given by

$$s_h = \frac{h - 1}{h} \text{ for positive sequence harmonics} \tag{12.127}$$

$$s_h = \frac{h + 1}{h} \text{ for negative sequence harmonics} \tag{12.128}$$

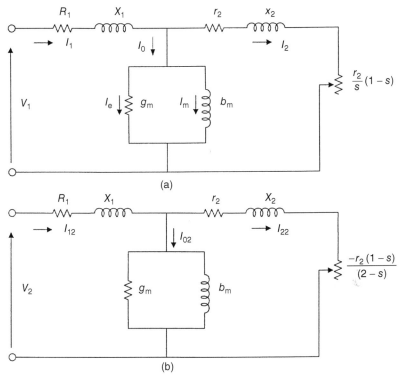

Figure 12.23 (a) Equivalent circuit of an induction motor for balanced positive sequence voltage and (b) negative sequence voltage.

The rotor resistance at harmonic frequencies is

$$r_{2h} = \frac{\sqrt{(1 \pm h)}}{s_h} \qquad (12.129)$$

The motor impedance neglecting magnetizing resistance is infinite for triplen harmonics, as the motor windings are not grounded.

From Fig. 12.23(a), the motor impedance at harmonic h is

$$R_1 + jhX_1 + \frac{jhx_m \left(\frac{r_2}{s_h} + jhx_2 \right)}{(r_2/s_h) + jh(x_m + x_2)} \qquad (12.130)$$

as seen from the input terminals of the motor, where

$$s_h = 1 - \frac{n}{hn_s} \qquad h = 3n + 1$$

$$s_h = 1 + \frac{n}{hn_s} \qquad h = 3n - 1 \qquad (12.131)$$

Example 12.8: An induction motor is rated 2.3-kV, four-pole, 2500-hp, full-load power factor = 0.92, full load efficiency = 0.93%, locked rotor current = six times the full-load current at 20% power factor; calculate its impedance at fundamental frequency and at fifth harmonic.

From the given data, motor full-load current = 547.18A

Therefore, locked rotor current = 3283.1 A at a power factor of 20%.

$$(R_1 + r_2) + j(X_1 + x_2) = \frac{2.3 \times 10^3}{\sqrt{3}(3283.1)(0.2 - j0.9798)} = 0.081 + j0.396\,\Omega$$

Assume that $\begin{array}{l} R_1 = r_2 = 0.0405\,\Omega \\ X_1 = x_2 = 0.1980\,\Omega \end{array}$

For the induction motor, the x_m can be assumed to be 12 ohms approximately. Also assume full-load slip = 3%.

Then impedance at fundamental frequency is

$$0.0405 + j0.1980 + \frac{j12.0\left(\frac{0.0405}{0.03} + j0.1980\right)}{\frac{0.0405}{0.03} + j(12 + 0.1980)}$$

$$= 1.3365 + j0.4340$$

Fifth harmonic is negative. Then, $S_h = 1.20$ (consider $n_s = n$). Again using the same equation, impedance at fifth harmonic is

$$0.0405 + j4.90 + \frac{j60.0\left(\frac{0.0405}{1.2} + j4.90\right)}{\frac{0.0405}{1.2} + j(60 + 4.90)}$$

$$= 0.0695 + j9.43\,\Omega$$

12.7 SYNCHRONOUS GENERATORS

The synchronous generators do not produce harmonic voltages and can be modeled with shunt impedance at the generator terminals. An empirical linear model is suggested, which consists of subtransient reactance at a power factor of 0.2. The average inductance experienced by harmonic currents, which involve both the direct axis and quadrature axis reactances, is approximated by

$$\text{Average inductance} = \frac{L''_d + L''_q}{2} \qquad (12.132)$$

At harmonic frequencies, the fundamental frequency reactance can be directly proportioned. The resistance at harmonic frequencies is given by

$$R_h = R_{DC} \left[1 + 0.1 \left(\frac{h_f}{f} \right)^{1.5} \right] \tag{12.133}$$

This expression can also be used for the calculation of harmonic resistance of transformers and cables having copper conductors. Model generators for harmonics with

$$Z_{0(h)} = R_h + jhX_0 \quad h = 3, 6, 9 \ldots$$

$$Z_{1(h)} = R_h + jhX_d'' \quad h = 1, 4, 7 \ldots$$

$$Z_{2(h)} = R_h + jhX_2 \quad h = 2, 5, 8 \ldots \tag{12.134}$$

X_0, X_d'', and X_2 are the generator zero sequence, subtransient, and negative sequence reactances at fundamental frequency.

EMTP models use Park transformations in $dq0$ axis, not discussed here. The steady-state and transient models of the synchronous machines and Park transformations are described in Ref. [32].

Reference [33] describes harmonic interaction between the generator and the system. For transmission lines that may be strongly unbalanced at harmonic frequencies; resonances may appear for single modes so that harmonic unbalance is created and amplified. These references develop an admittance matrix for the synchronous machine for harmonic analysis.

12.8   LOAD MODELS

Figure 12.24(a) shows a parallel RL load model. It represents bulk power load as an RL circuit connected to ground. The resistance and reactance components are calculated from fundamental frequency voltage, reactive volt-ampere, and power factor:

$$R = \frac{V^2}{S \cos \phi} \quad L = \frac{V^2}{2\pi f S \sin \phi} \tag{12.135}$$

The reactance is frequency dependent, and resistance may be constant or it can also be frequency dependent. Alternatively, the resistance and reactance may remain constant at all frequencies. Figure 12.24(b) is for leading loads, S is load in kVA.

Figure 12.24(c) shows a CIGRE (Conference Internationale des Grands Reseaux Electriques à Haute Tension) type C load model [34], which represents bulk power, valid between 5th and 30th harmonics. Here, the following relations are applicable:

$$R_s = \frac{V^2}{P} \quad X_s = 0.073 h R_s \quad X_p = \frac{h R_s}{6.7 \frac{Q}{P} - 0.74} \tag{12.136}$$

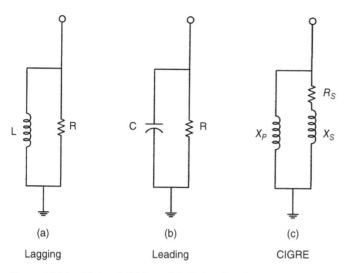

Figure 12.24 (a) Parallel RL model, (b) Leading load model and (c) CIGRE type C model.

This load model was derived by experimentation.

A delta-connected three-phase load (balanced) can be modeled as

$$\begin{vmatrix} I_a \\ I_b \\ I_c \end{vmatrix} = \frac{1}{Z_{\text{delta}}} \begin{vmatrix} 2 & -1 & -1 \\ -1 & 2 & -1 \\ -1 & -1 & 2 \end{vmatrix} \begin{vmatrix} V_a \\ V_b \\ V_c \end{vmatrix} \tag{12.137}$$

where Z_{delta} is the impedance/phase of the delta load.

A wye-connected, neutral isolated three-phase load has sequence impedances:

$$Z_{012} = \begin{vmatrix} \infty & 0 & 0 \\ 0 & Z_{\text{wye}} & 0 \\ 0 & 0 & Z_{\text{wye}} \end{vmatrix} \tag{12.138}$$

In general, for a three-phase load:

$$Z_{abc} = \begin{vmatrix} Z_{aa} & Z_{ab} & Z_{ac} \\ Z_{ba} & Z_{bb} & Z_{bc} \\ Z_{ca} & Z_{cb} & Z_{cc} \end{vmatrix} \tag{12.139}$$

Z_{012} can be written using symmetrical component transformation.

12.8.1 Study Results with PQ and CIGRE Load Models

Figure 12.25(a) and (b) shows the calculated impedance modulus versus frequencies on two buses with PQ load model and CIGRE load model. The difference is noteworthy.

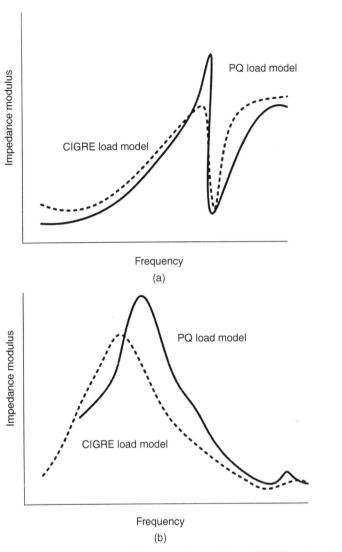

Figure 12.25   (a) and (b) Differences in the PQ and CIGRE load models on two buses, a harmonic analysis study.

12.9   SYSTEM IMPEDANCE

The system impedance to harmonics is not a constant number. Figure 12.26(a) shows the $R - X$ plot of system impedance. The fundamental frequency impedance is inductive, its value representing the *stiffness* of the system. The resonances in the system make the $R - X$ plots a spiral shape, and the impedance may become capacitive. Such spiral-shaped impedances have been measured for high-voltage systems, and resonances at many frequencies are common. These frequencies at resonance points and also at a number of other points on the spiral-shaped curves are indicated as shown

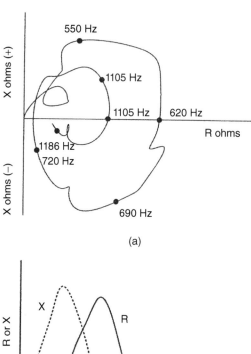

Figure 12.26 (a) Spiral R–X plot of supply system source impedance, (b) separate R and X plot.

in Fig. 12.26(a). (Approximate plots of R and X with respect to frequency are in Fig. 12.26(b).) At the resonance, the impedance reduces to a resistance. The system impedance can be ascertained by the following means:

- A computer solution can be used to calculate the harmonic impedances.
- In noninvasive measurements, the harmonic impedance can be calculated directly from the ratio of harmonic voltage and current reading.
- In another measurement method, shunt impedance is switched in the circuit and the harmonic impedance is calculated by comparing the harmonic voltages and currents before and after switching.

The spiral-shaped impedance plots can be bounded in the Z-plane by a circle on the right side of the Y-axis and tangents to it at the origin with an angle of 75°. This configuration can also be translated in the Y-plane [35] (Fig. 12.27).

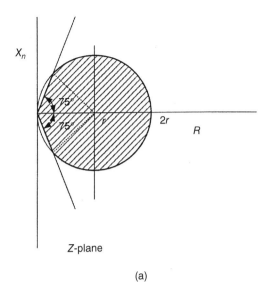

Z-plane

(a)

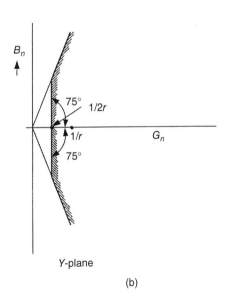

Y-plane

(b)

Figure 12.27   Generalized impedance plot, (a) $R-X$ plane and (b) Y-plane.

Figure 12.28 is self-impedance and mutual impedance loci matrix of a test system from Ref. [36]. The frequency response must include network asymmetry and mutual coupling. These loci are converted into equivalent circuits.

12.10   THREE-PHASE MODELS

The power system elements are not perfectly symmetrical. Asymmetry is involved in the circuit loading and mutual couplings and unbalanced self-impedance and

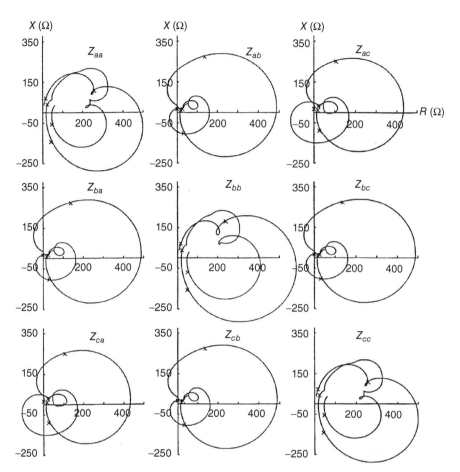

Figure 12.28   Impedance matrix plots of self-impedance and mutual impedance. Source: Ref. [27].

mutual impedance result. The problem is similar to the three-phase fundamental frequency load flow, and is compounded by the nonlinearities of the harmonic loads. Single-phase models are not adequate in the following cases:

- Telephone interference is of concern. The influence of zero sequence harmonics is important, which gives rise to most of the interference with the communication circuits.

- There are single-phase capacitor banks in the system.

- There are single-phase or unbalanced harmonic sources.

- Triplen harmonics are to be considered, ground currents are important, and significant unbalanced loading is present.

For unbalanced distribution circuits and transmission line unbalance loads, three-phase models are required. When supplied with unbalanced voltages, power converters can generate uncharacteristic harmonics. The harmonic orders do not follow the sequence order of balanced cases.

When the harmonic distortion of the buses is greater than 5%, this can impact the harmonic producing equipment. An iterative approach can overcome this problem. Suppose on the first run, the voltage distortion exceeds 5%. Then, in an iterative harmonic analysis study, harmonic mitigation devices are applied and the distortion is reduced to 1%. The results of this iteration together with harmonic compensating equipment are, generally, acceptable.

The network asymmetries and mutual couplings should be included, which require a 3×3 impedance matrix at each harmonic. Three-phase models of transformers with neutral grounding impedances, mutually coupled lines, distributed parameter transposed, and untransposed transmission lines should be possible in the harmonic analysis program.

The networks can be divided into subsystems. A subsystem may be defined as any part of the network, divided so that no subsystem has any mutual coupling between its constituent branches and those of the rest of the network. Following the example of a two-port network,

$$
\begin{vmatrix} \overline{V}_s \\ \overline{V}_r \end{vmatrix} = \begin{vmatrix} \overline{A} & \overline{B} \\ \overline{C} & \overline{D} \end{vmatrix} \begin{vmatrix} \overline{V}_r \\ \overline{I}_r \end{vmatrix}
\tag{12.140}
$$

the current and voltages are now matrix quantities, the dimensions depending on the section being considered, $3, 6, 9 \cdots$. All sections contain the same number of mutually coupled elements, and all matrices are of the same order. Then, the equivalent model Y matrix can be written as

$$
\overline{Y} = \begin{vmatrix} [\overline{D}][\overline{B}]^{-1} & \vdots & [\overline{C}] - [\overline{D}][\overline{B}]^{-1}[\overline{A}] \\ \cdots & \vdots & \cdots \\ [\overline{B}]^{-1} & \vdots & -[\overline{B}]^{-1}[\overline{A}] \end{vmatrix}
\tag{12.141}
$$

12.11   UNCHARACTERISTIC HARMONICS

Uncharacteristic harmonics will be present. These may originate from variations in ignition delay angle. As a result, harmonics of the order $3q$ are produced in the direct voltage and $3q \pm 1$ are produced in the AC line current, where q is an odd integer. Third harmonic and its odd multiples are produced in the DC voltage and even harmonics in the AC line currents. The uncharacteristic harmonics may be amplified, as these may shift the times of voltage zeros and result in more unbalance of firing angles. A firing angle delay of $1°$ causes approximately 1% third harmonic. The modeling of any user-defined harmonic spectrum and their angles should be possible in a harmonic analysis program.

On a simplistic basis for a 12-pulse converter model noncharacteristic harmonics such as $5, 7, 17, 19, \ldots$ at 15% of the level of the harmonics computed for a 6-pulse converter. Also model triplen harmonics $3, 9, 15, \ldots$ as 1% of the fundamental. However, this does not account for unbalance. If we write the unbalance voltages as

$$V_a = V \sin \omega t$$

$$V_b = V(1+d) \sin(\omega t - 120)°$$

$$V_c = V \sin(\omega t + 120°) \tag{12.142}$$

where d is the unbalance, then conduction intervals for the current are

$$t_a = t_c = 120° - e$$

$$t_b = 120° + 2e \tag{12.143}$$

where

$$e = \tan^{-1} \frac{\sqrt{3}(1+d)}{3+d} - 30° \tag{12.144}$$

The instantaneous currents ignoring overlap are

$$i_a = \frac{4}{\pi} I_d \frac{(-1)^h}{h} \sin\left(\frac{ht_a}{2}\right) \sin(h\omega t)$$

$$i_b = \frac{4}{\pi} I_d \frac{(-1)^h}{h} \sin\left(\frac{ht_b}{2}\right) \left(h\omega t - 120° - \frac{e}{2}\right)$$

$$i_c = \frac{4}{\pi} I_d \frac{(-1)^h}{h} \sin\left(\frac{ht_c}{2}\right) \sin(h\omega t + 120° - e) \tag{12.145}$$

If there is 1% unbalance $d = 0.1$, then third harmonic currents are

$$i_{a3} = 0.01586k < 180°$$

$$i_{b3} = 0.03169k < -2.36°$$

$$i_{c3} = 0.01586k < 175.27° \tag{12.146}$$

where

$$k = \frac{2\sqrt{3}I_d}{\pi} \tag{12.147}$$

The positive and negative sequence components of these currents will be injected into the network even if a delta-connected transformer is used [37].

12.12 CONVERTERS

Chapter 1 discusses the harmonic generation from the converters. There is an interaction between the AC voltages and the DC current in weak AC/DC interconnections. Reference [38] describes voltage instabilities, control system instabilities, transient dynamic overvoltages, temporary and harmonic overvoltages, and low-order harmonic resonance. Reference [39] describes the inadequacy of commercial harmonic analysis programs to model transient system harmonics and development of converter models and nonlinear resistor models to represent mercury arc and sodium vapor lamps. The computational algorithm to determine the harmonic components of the converter current and reactive power drawn considers overlap angle, calculated V_d and I_d, reiterates with overlap angle till $V_d I_d = P$.

The harmonic currents produced by nonlinear loads are derived on the assumption of a perfectly sinusoidal source that is a strong sinusoidal voltage system. These harmonic currents are then injected into the AC system to determine levels of voltage distortion. However, when the injected harmonic frequency is close to the parallel resonant frequency, the calculation algorithm often diverges and a transient converter simulation (TCS) is required. This requires AC system equivalents responding accurately to power and harmonic frequencies. An equivalent circuit consisting of number of tuned RLC branches has been proposed [40], as a solution to HVDC studies.

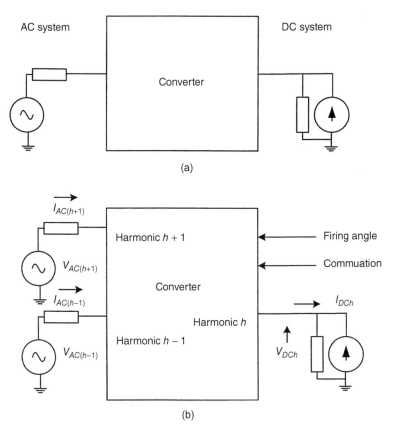

Figure 12.29   (a) Converter interconnecting AC and DC systems and (b) noncharacteristic frequency interactions.

This describes frequency-dependent equivalent circuits suitable for integration in the time-domain solutions.

The converter and its AC system and the converter and the DC system can be reduced to Thevénin or Norton equivalent. Referring to Fig. 12.29(a), the harmonic relationships can be written as

$$\overline{V}_{DC} = \overline{A}\,\overline{V}_{AC} + \overline{B}\,\overline{\beta} + \overline{C}\,\overline{I}_{DC}$$

$$\overline{I}_{AC} = \overline{D}\,\overline{I}_{DC} + \overline{E}\,\overline{\beta} + \overline{F}\,\overline{V}_{AC}$$

$$\overline{V}_{AC} = \overline{Z}_{AC}\overline{I}_{AC} + \overline{V}_{AC0}$$

$$\overline{I}_{DC} = \overline{Y}_{DC}\overline{V}_{DC} + \overline{I}_{DC0} \qquad (12.148)$$

where $\overline{A}$, $\overline{B}$, $\overline{C}$, … are matrices representing converter transfer functions from AC voltage to DC voltage, firing angle to DC voltage, DC current to AC current, and so on. $\overline{Z}_{AC}$ and $\overline{Y}_{DC}$ are the diagonal matrices of AC-side harmonic impedance and DC-side harmonic admittances. $\overline{I}_{DC}$ and $\overline{V}_{DC}$ are the vectors of harmonic current sources on DC side and harmonic voltage sources on DC side. The elements of these matrices are derived by numerical techniques [41,42]. The frequency-dependent behavior of the converter may be defined as returned distortion (current or voltage) as a result of applied distortion at the same frequency. Manipulations of Eq. (12.148) lead to the interaction, as shown in Fig. 12.29(b).

REFERENCES

1. H.W. Beaty and D.G. Fink (Eds). Standard Handbook for Electrical Engineers, 15th Edition, McGraw-Hill, New York, 2007.
2. C. Croft. American Electrician's Handbook, 9th Edition, McGraw-Hill, New York, 1970.
3. Central Station Engineers. Electrical Transmission and Distribution Reference Book, 4th Edition, Westinghouse Corporation, East Pittsburgh, PA, 1964.
4. The Aluminum Association. Aluminum Conductor Handbook, 2nd Edition, The Aluminum Association, Washington, DC, 1982.
5. J.G. Anderson. Transmission Reference Book, Edison Electric Company, New York, 1968.
6. J.C. Das. Transients in Electrical Systems, McGraw-Hill, New York, 2010.
7. J.C. Das. Power System Analysis-Short-Circuit Load Flow and Harmonics. 2nd Edition, Florida, CRC Press, 2012.
8. Canadian/American EMTP User Group. ATP Rule Book, Oregon, Portland, 1987–1992.
9. EPRI. HVDC-AC system interaction for AC harmonics. EPRI Report 1: 7.2-7.3, 1983.
10. EPRI. Transmission Line Reference Book—345 kV and Above, EPRI, Palo Alto, CA, 1975.
11. EHV Transmission. IEEE Trans 1966 (special issue), No. 6, PAS-85, pp. 555–700, 1966.
12. P.M. Anderson. Analysis of Faulted Systems, Iowa State University Press, Ames, IA, 1973.
13. L.V. Beweley. Traveling Waves on Transmission Systems, 2nd Edition, John Wiley and Sons, New York, 1951.
14. T. Gonen. Electrical Power Transmission System Engineering, Analysis and Design, 2nd Edition, Florida, CRC Press, 2009.
15. L. M. Wedepohl and D.J. Wilcox. "Transient analysis of underground power transmission systems: system model and wave propagation characteristics," Proceedings of IEE, vol. 120, pp. 252–259, 1973.

16. G. Bianchi and G. Luoni. "Induced currents and losses in single-core submarine cables," IEEE Transactions on Power and Systems, vol. PAS-95, pp. 49–58, 1976.

17. L. Marti. "Simulation of electromagnetic transients in underground cables with frequency-dependent model transformation matrices," Ph.D. thesis, University of British Columbia, Vancouver, 1986.

18. G.W. Brown and R.G. Rocamora, "Surge propagation in three-phase pipe-type cables, Part I-Unsaturated pipe," IEEE Transactions on Power and Systems, vol. PAS-90, pp. 1287–1294, 1971.

19. G.W. Brown and R.G. Rocamora. "Surge propagation in three-phase pipe-type cables, Part II-Duplication of field tests including the effect of neutral wires and pipe saturation," IEEE Transactions on Power and Systems, vol. PAS-90, pp. 826–833, 1977.

20. A. Ammentani. "A general formation of impedance and admittance of cables," IEEE Transactions on Power and Systems, vol. PAS-99, pp. 902–910, 1980.

21. A. Semlyen. "Overhead line parameters form handbook formulas and computer programs," IEEE Transactions on Power and Systems, vol. PAS-104, p. 371, 1985.

22. D.R. Smith and J.V. Bager. "Impedance and circulating current calculations for UD multi-wire concentric neutral circuits," IEEE Transactions on Power and Systems, vol. PAS-91, pp. 992–1006, 1972.

23. P. de Arizon and H.W. Dommel, "Computation of cable impedances based on subdivision of conductors," IEEE Transactions on Power Delivery, vol. PWRD-2, pp. 21–27, 1987.

24. IEEE Standard 525. IEEE guide for the design and installation of cable systems in substations, 1987.

25. IEEE P575/D12. Draft guide for bonding shields and sheaths of single-conductor power cables rated 5 kV through 500 kV, 2013.

26. C.E. Lin, J.B. Wei, C.L. Huang and C.J. Huang. "A new method for representation of hysteresis loops," IEEE Transactions on Power Delivery, vol. 4, pp. 413–419, 1989.

27. J.D. Green and C.A. Gross. "Non-linear modeling of transformers," IEEE Transactions on Industry Applications, vol. 24, pp. 434–438, 1988.

28. W.J. McNutt, T.J. Blalock and R.A. Hinton. "Response of transformer windings to system transient voltages," IEEE Transactions on Power and Systems, vol. 9, pp. 457–467, 1974.

29. M. Chen and W.E. Dillon, "Power system modeling," Proceedings of IEEE Power and Systems, vol. 62, pp. 901–915, 1974.

30. CIGRE Working Group 33.02, "Guidelines for representation of network elements when calculating transients," CIGRE brochure 39, 1990.

31. M. Heathcote. J&P Transformer Handbook, 13th Edition, A reprint of Elsevier Ltd, Burlington, MA, 2007.

32. P.M. Anderson and A.A. Fouad. Power System Control and Stability, IEEE Press, New Jersey, 1994.

33. A. Semlyen, J.F. Eggleston and J. Arrillaga. "Admittance matrix model of a synchronous machine for harmonic analysis," IEEE Transactions on Power Systems, vol. PWRS-2, no. 4, pp. 833–839, 1987.

34. CIGRE Working Group 36–05. "Harmonic characteristic parameters, methods of study, estimating of existing values in the network," Electra, pp. 35–54, 1977.

35. E.W. Kimbark. Direct Current Transmission, John Wiley and Sons, New York, 1971.

36. N.R. Watson and J. Arrillaga. "Frequency-dependent AC system equivalents for harmonic studies and transient converter simulation," IEEE Transactions on Power Delivery, vol. 3, no. 3, pp. 1196–1203, 1988.

37. M. Valcarcel and J.G. Mayordomo. "Harmonic power flow for unbalance systems," IEEE Transactions on Power Delivery, vol. 8, pp. 2052–2059, 1993.

38. C. Hatziadoniu and G.D. Galanos. "Interaction between the AC voltages and DC current in weak AC/DC interconnections," IEEE Transactions on Power Delivery, vol. 3, pp. 1297–1304, 1988.

39. J.P. Tamby and V.I. John. "Q'Harm-A harmonic power flow program for small power systems," IEEE Transactions on Power and Systems vol. 3, no. 3, pp. 945–955, 1988.

40. J. Arrillaga, D.A. Bradley and P.S. Bodger. Power System Harmonics, John Wiley, New York, 1985.

41. J. Arrillaga and N.R. Watson. Power System Harmonics, 2nd Edition, New York, John Wiley, 2004.

42. E.V. Larsen, D.H. Baker and J.C. Mclver. "Low order harmonic interaction on AC/DC systems," IEEE Transactions on Power Delivery, vol. 4, no. 1, pp. 493–501, 1989.

CHAPTER 13

HARMONIC MODELING OF SYSTEMS

13.1 ELECTRICAL POWER SYSTEMS

Electrical power systems can be broadly classified in to generation, transmission, sub-transmission, and distribution systems. Individual power systems are organized in the form of electrically connected areas of *regional* grids, which are interconnected to form *national* grids and also international grids. Each area is interconnected to another area in terms of some contracted parameters like generation and scheduling, tie line power flow, and contingency operations. The business environment of power industry has entirely changed due to deregulation and decentralization, more emphasis on economics and cost centers.

The irreplaceable sources of power generation are petroleum, natural gas, oil, and nuclear fuels. The fission of heavy atomic weight elements like uranium and thorium and fusion of lightweight elements, that is, deuterium offer almost limitless reserves. Replaceable sources are elevated water, pumped storage systems, solar, geothermal, wind and fuel cells, which in recent times have received much attention driven by strategic planning for independence from foreign oil imports. Nuclear generation in USA has much public concern due to a possible meltdown and disposal of nuclear waste, though France relies heavily on this source. U238 has a half life of 4.5×10^{10} years. Disposal of nuclear waste products is one public deterrent against nuclear generation. The coal based plants produce SO_2, nitrogen oxides, CO, CO_2, hydrocarbons, and particulates. Single shaft steam units of 1500 MW are in operation, and superconducting single units of 5000 MW or more are a possibility. On the other hand dispersed generating units, integrated with grid may produce only a few kilowatt of power.

The generation voltage is low 13.8–25 kV because of problems of inter-turn and winding insulation at higher voltages in the generator stator slots, though a 110 kV Russian generator is in operation. The transmission voltages have risen to 765 kV or higher and many HVDC links around the world are in operation.

Power System Harmonics and Passive Filter Designs, First Edition. J.C. Das.
© 2015 The Institute of Electrical and Electronics Engineers, Inc. Published 2015 by John Wiley & Sons, Inc.

Maintaining acceptable voltage profile and load frequency control are major issues. Synchronous condensers, shunt capacitors, static var compensators and FACTS devices are employed to improve power system stability and enhance power handling capability of transmission lines. The sub-transmission voltage levels are 23 kV to approximately 69 kV, though for large industrial consumers voltages of 230 kV and 138 kV are in use. Sub-transmission systems connect high voltage substations to local distribution substations. The voltage is further reduced to 12.47 kV and several distribution lines and cables emanate from distribution stations. At the consumer level the variety of load types, their modeling and different characteristics present a myriad of complexities. A pulp mill may use single synchronous motors of 30,000 hp or more and ship propulsion can use even higher ratings.

In the years to come power industry will undergo profound changes, need based – environmental compatibility, reliability, improved operational efficiencies, integration of renewable energy sources and dispersed generation. The dynamic state of the system will be known at all times and under any disturbance. The technologies driving smart grids are:

- RAS (remedial action schemes).
- SIPS (system integrated protection systems).
- WAMS (wide area measurement systems).
- FACTS (flexible AC transmission systems).
- EMS (energy management systems).
- PMU's (phasor measurement units).
- Overlay of 750 kV lines and HVDC links.

After November 1965 great blackout, National Reliability Council was created in 1968 and later, its name changed to North American Reliability Council (NERC). It lays down guidelines and rules for utility companies to follow and adhere to. Yet it seems that adequate dynamic system studies are not carried out during the planning stage. This is attributed to lack of resources, data or expertise (lack of models for dispersed generation, verified data and load models, models for wind generation). In a modern complex power system the potential problem area may lay hidden, and may not be possible to identify intuitively or some power system studies. It is said that cascading type blackouts can be minimized and time frame prolonged but these cannot be entirely eliminated. There does not seem to be a thrust for harmonic evaluations on large scale basis, the most important standard being IEEE 519.

The electrical power systems are highly nonlinear and dynamic – perhaps the most nonlinear man-made systems of the highest order on earth. The systems are constantly subjected to internal switching transients (faults, closing and opening of breakers, charging of long transmission lines, varying generation in response to load demand, etc.) and of external origin, that is, lightning surges. Electromagnetic and electromechanical energy is constantly being redistributed among power system components. Figure 13.1 shows a general concept of the electrical power systems.

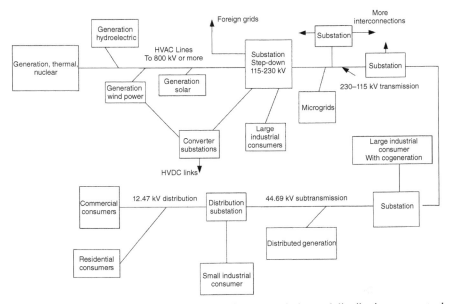

Figure 13.1 Electrical power systems-generation, transmission and distribution, conceptual configuration.

If an aerial map is taken of the electrical power lines and interconnections in a dense area, it will be denser than all the routes of highways, sub highways lanes and by-lanes combined together.

With respect to harmonics, these exist at all levels of the power systems – in transmission, sub transmission and distribution, industrial and commercial installations. Though standards of harmonic emission have been established, the permissible pollution that a consumer may inject at his PCC, add together and the effect of harmonics can be seen at a distance. Another consideration is that as yet the harmonic emission limits may not be strictly enforced. This is especially true of the power systems in developing countries.

13.1.1 Harmonic Considerations

It is a stupendous task to chase and control the harmonics in interconnected systems, consisting of thousand of buses, substations and generating facilities. It requires computing resources which may not be available. Practically, a system has to be truncated and boundary conditions established.

Under balanced conditions it is sufficient to model characteristic harmonics; presence of SVC requires modeling of odd harmonics except triplen, and negative sequence networks. If transformer saturation is modeled triplen harmonics will be present. The current-source model may not be effective. A three-phase modeling is warranted [1]. Thus, the modeling is dependent on the network being studied, and the nonlinear loads. Networks vary in complexity and size and generally it is not possible to include the detailed model of every component in the study.

13.1.2 Effective Designs of Power Systems

It is imperative that a power system, large or small, must be effectively designed before a harmonic analysis is conducted. It must meet the requirements of national ANSI/IEEE, NEC. OSHA and other national or international standards like IEC. In general, all electrical power systems should be:

- Secure.
- Safe.
- Expandable.
- Maintainable.
- Reliable.
- And economical.

The integrity of the electrical equipment should be maintained with emphasis on type of enclosures, insulation coordination, operating mechanisms, grounding and protective relaying. Yet, the power system designs may fall short from personnel safety considerations. Even the functionality for which these systems are designed for adequate performance may be compromised. It is not unusual to see inadequately designed systems, lacking in some respect or the other. Competition and economical constraints can make even expert designers and planners to cut corners, which may ultimately result in spending more funds for the short-term fixes and long-term upgrades.

Designing with respect to harmonics opens another consideration. Even when the harmonics are mitigated to meet the standard requirements, say with the application of passive filters, these must flow to the location of the filters in the power system. That means a number of transformers and cables will carry the full blast of harmonic spectrums and these must be derated to carry nonlinear loads. The harmonic mitigation will be not 100% and the effect of harmonic flows throughout the distribution system must be considered. As discussed in previous chapters all motors, loads, generators absorb a certain percentage of harmonics. Cases are on record for failure of power system components due to harmonic loading. This implies, that a system design must be pre-and post evaluated with harmonic analysis studies.

13.2 EXTENT OF NETWORK MODELING

Two examples of extent of modeling are discussed in Ref. [2]. It appears common to represent the external network at least 5 buses and two transformations back from the point of interest. The buses with significant level of capacitive compensation should be modeled. A logic place to truncate or equivalence is in the vicinity of generating plants which serve as sinks due to their low subtransient reactance. The balance of the system may be represented at the boundary buses by short-circuit equivalents.

TABLE 13.1   Positive Sequence Driving Point Impedances

Harmonic Order	Z and f	Three Bus External System			Five Bus External System		
		Phase *a*	Phase *b*	Phase *c*	Phase *a*	Phase *b*	Phase *c*
4	Z Ohms	804	864	1018	649	550	475
	f (Hz)	235	244	235	227	226	227
5	Z Ohms	665	688	729	405	885	536
	f (Hz)	299	318	299	281	280	280

Reference [2] describes modeling of network external to NYPP system with respect to a particular bus, "Fraser." Two models are used:

- 56 transmission lines, 66 single phase transformers, 9-three phase capacitor banks, 25-three phase source equivalents, and three current sources.
- The second model is extended, and includes NYPP system 5 buses away from the Fraser 345 kV bus. It consists of 68 transmission lines, 81 single phase transformers, 9 three-phase capacitor banks, 44 three-phase source equivalents and three current sources.

The comparative results are shown in Table 13.1, Ref. [2]. An extended system model is needed for studying detailed harmonic impedances and voltages around resonance points.

Another example of the effect of the extent of system modeling is shown in Fig. 13.2 for a 200-MW HVDC tie in a 230-kV system, Ref. [2]. A 20-bus model shows resonance at the 5th and 12th harmonics, rather than at the 6th and 13th when 110 buses at a very large generation site, with multiple machines, are represented in the system. Additional details beyond this point do not significantly alter the driving point impedance. This illustrates the risk of inadequate modeling.

In EMTP simulations, frequency scan option will provide the frequency dependent driving point impedance of a network as seen from a bus. It can also provide voltage (current) transfer function between any two points in the system.

13.3   IMPACT OF LOADS AND GENERATION

The loads connected in a power system absorb harmonics and similarly the generators, and motors. Also the harmonic distortion at PCC depends upon the ratio of the nonlinear load with respect to total load.

Figure 13.3 shows a simple system configuration to illustrate the effect of loads and generation. A 2.5 MVA, 6-pulse ASD load is connected at 480-V bus, the harmonic injection current waveform as shown in Fig. 13.4. The harmonics absorbed by the static load, motor, and generator; all of the same 1500-KVA rating is shown in Table 13.2. Though the ratings of these equipments are identical, the harmonics

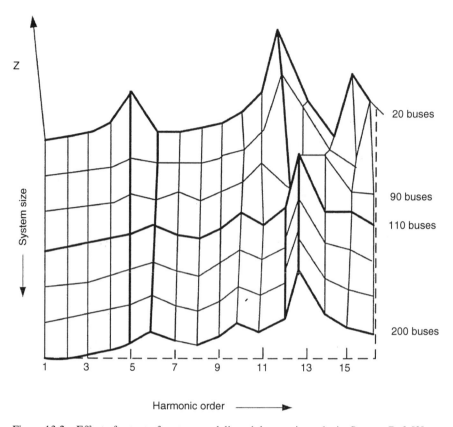

Figure 13.2 Effect of extent of system modeling oh harmonic analysis. Source: Ref. [2].

absorbed by them vary, because the harmonic impedances are different. Figure 13.5 shows the spectrum of currents Fig. 13.6 depicts the waveforms. TDD at PCC, secondary side of the 15 MVA transformer is 2.23%, while at 480-V injection bus, it is 12.55%.

13.4 SHORT-CIRCUIT AND FUNDAMENTAL FREQUENCY LOAD FLOW CALCULATIONS

Fundamental frequency short-circuit and load flow calculations are required to establish the ratio I_s/I_r, based on which the IEEE 519 distortion limits are calculated, Chapter 10. In fact, these studies are a must before harmonic analyses studies are conducted. Much computer input data are common between these three study types.

Normal operating voltages must be established throughout the power system. The operating system voltages should be preferably 1–2% higher than the system voltage. The voltages must be maintained within a certain plus–minus band around

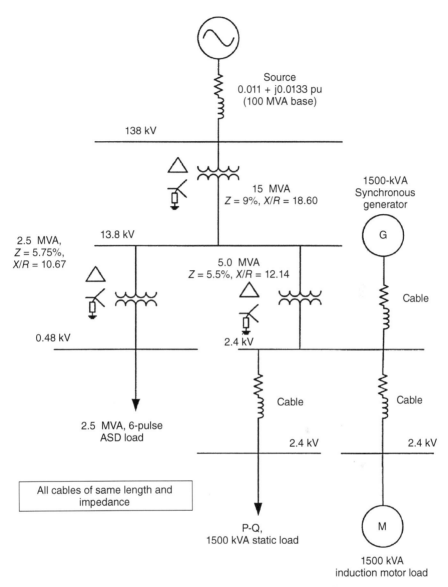

Figure 13.3   A distribution system for study of harmonic flow in different system components (Example 13.1).

the rated equipment voltage, ideally from no load to full load, and under varying loading conditions. Sudden load impacts (starting of a large motor) or load demands under contingency operating conditions, when one or more tie-line circuits may be out of service, result in short-time or prolonged voltage dips. ANSI C84.1 [3] specifies the preferred nominal voltages and operating voltage ranges *A* and *B* for utilization

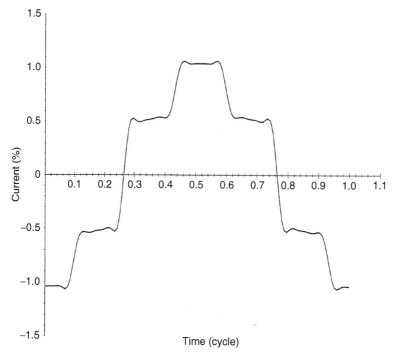

Figure 13.4 The waveform of the 6-pulse ASD current injected in Fig. 13.3.

TABLE 13.2

Harmonics in % of Fundamental	5	7	11	13	17	19	23
Injected	17.46	12.32	7.19	5.13	2.88	1.54	0.51
Generator	1.49	1.05	0.60	0.42	0.23	0.12	0.04
Static load	0.81	0.79	0.71	0.59	0.42	0.25	0.10
Motor	1.33	0.94	0.54	0.38	0.21	0.11	0.04
PCC	9.71	6.84	3.97	2.82	1.57	0.84	0.28

and distribution equipment operating from 120–34,500 V. For transmission voltages over 34,500 V only nominal and maximum system voltages are specified. Range B allows limited excursions outside range A limits. As an example, for a 13.8 kV nominal voltage, range $A = 14.49–12.46$ kV and range $B = 14.5–13.11$ kV. Flicker from cyclic loads, for example, arc furnaces, must be controlled to an acceptable level (see Table 13.3). The electrical apparatuses have a certain maximum and minimum operating voltage, range in which normal operation is maintained, that is, induction motors are designed to operate successfully under the following conditions [4]:

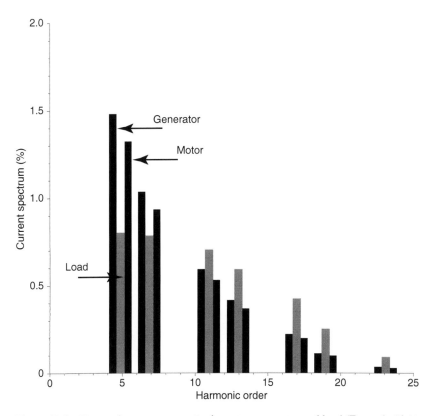

Figure 13.5   Harmonic current spectrum in motor, generator and load (Example 13.1).

1. Plus or minus 10% of rated voltage, with rated frequency.
2. A combined variation in voltage and frequency of 10% (sum of absolute values) provided that the frequency variations do not exceed ±5% of rated frequency.
3. Plus or minus 5% of frequency with rated voltage.

A certain balance between the reactive power consuming and generating apparatuses is required. This must consider losses which may be a considerable percentage of the reactive load demand.

The voltage regulation is achieved by:

- On-load and off-load tap changers.
- Redistribution of loads.
- Adding additional installed capacity, transformers and generators.
- Parallel feeders and redundant systems.
- Voltage regulators.
- SVCs, TCRs, and STATCOMS.
- Shunt connected capacitor banks.

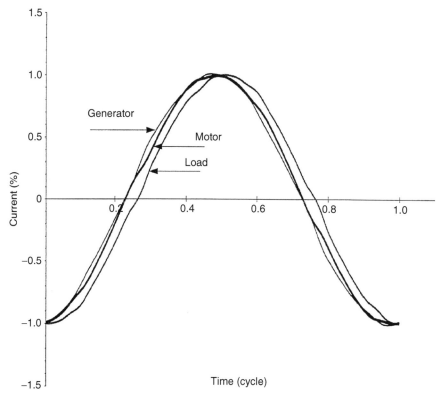

Figure 13.6 Distorted waveforms of the currents in generator, motor and load (Example 13.1).

Consider that off-load taps on main transformer of 138–13.8 kV is set to provide a 5% voltage boost on the secondary side. Downstream in the distribution, 13.8–4.16 kV and 4.16–0.48 kV transformers are again set to provide 5% voltage boost on secondary side. This may give acceptable load voltages on 4.16 kV and 480-V systems under operating loads. But under no-load conditions, the 480-V system voltage will be 15% higher and at 4.16 kV system it will be 10% higher. This is not acceptable.

The estimation of requirements of shunt capacitor banks for reactive power compensation and voltage support at fundamental frequency is required for harmonic load flow; these capacitor banks are, then, converted into harmonic filters.

13.5 INDUSTRIAL SYSTEMS

Industrial systems vary in size and complexity. While a saw mill with a load of 1000 kVA served from 480-V pole mounted transformer may be called an industrial system, the largest utility substation with an installed capacity of 260 MW, in the state of Georgia, USA serves a newsprint mill at 230 kV. Some industrial plants may generate

TABLE 13.3 Standard Nominal System Voltages and Voltage Ranges

Voltage Class	Nominal System Voltage			Nominal Utilization Voltage			Voltage Range A			Voltage Range B		
	2-wire	3-wire	4-wire	2-wire	3-wire	4-wire	Maximum Utilization and Service Voltage	Minimum Service Voltage	Minimum Utilization Voltage	Maximum Utilization and Service Voltage	Minimum Service Voltage	Minimum Utilization Voltage
Single-Phase Systems												
Low voltage	120			115			126	114	108	127	110	104
		120/240			115/230		126/252	114/228	108/216	127/254	110/220	104/208
Three-Phase Systems												
			208Y/120			200	218Y/126	197Y/114	187Y/108	220Y/127	191Y/110	180Y/104
		240/120			230/115		252/126	228/114	216/108	254/127	220/110	208/104
	240			230			252	228	216	254	220	208
			480Y/277			460Y/266	504Y/291	456Y/263	432Y/249	508Y/293	440Y/254	416Y/240
	480			460			504	456	432	508	440	416
	600			575			630	570	540	635	550	520
Medium voltage	2400						2520	2340	2160	2540	2280	2080
			4160Y/2400				4370Y/2520	4050Y/2340	3740Y/2160	4400Y/2540	3950Y/2280	3600Y/2080
	4160						4370	4050	3740	4400	3950	3600
	4800						5040	4680	4320	5060	4560	4160
	6900						7240	6730	6210	7260	6560	5940
			8320Y/4800				8730Y/5040	8110Y/4680		8800Y/5080	7900Y/4560	
			12000Y/6930				12600Y/7270	11700Y/6760		12700Y/7330	11400Y/6580	
			12470Y/7200				13090Y/7560	12160Y/7020		13200Y/7620	11850Y/6840	
			13200Y/7620				13860Y/8000	12870Y/7430		13970Y/8070	12504Y/7240	
			13800Y/7970				14490Y/8370	13460Y/7770		14520Y/8380	13110Y/7570	

(continued)

TABLE 13.3 *(Continued)*

Voltage Class	Nominal System Voltage 2-wire	Nominal System Voltage 3-wire	Nominal System Voltage 4-wire	Nominal Utilization Voltage 2-wire	Nominal Utilization Voltage 3-wire	Nominal Utilization Voltage 4-wire	Voltage Range A — Maximum Utilization and Service Voltage	Voltage Range A — Minimum Service Voltage	Voltage Range A — Minimum Utilization Voltage	Voltage Range B — Maximum Utilization and Service Voltage	Voltage Range B — Minimum Service Voltage	Voltage Range B — Minimum Utilization Voltage
Medium voltage		13800					14490	13460	12420	14520	13110	11880
			20780Y/12000				21820Y/12600	20260Y/11700		22000Y/12700	19740Y/11400	
			22860Y/13200				24000Y/13860	22290Y/12870		24200Y/13970	21720Y/12540	
		23000					24150	22430		24340	21850	
			24940Y/14400				26190Y/15120	24320Y/14040		26400Y/15240	23690Y/13680	
			34500Y/19920				36230Y/20920	33640Y/19420		36510Y/21080	32780Y/18930	
		34500					36230	33640		36510	32780	
							Maximum Voltage	See Ref. [3] for further details and explanations.				
		46000					48300					
		69000					72500					
High voltage		115000					121000					
		138000					145000					
		161000					169000					
		230000					242000					
Extra-high voltage		345000					362000					
		400000					420000					
		500000					550000					
		765000					800000					
Ultra-high voltage		1100000					1200000					

Source: Ref. [3].

their own power and may operate in cogeneration mode with the utility systems at high voltages. The utility service may be at voltages of 115, 138, and 230 kV and the load demand at these voltages can be upward of 100–200 MW. It is usual to represent the utility source by it's short-circuit impedance, however nearby harmonic loads should be considered. The large rolling mill or arc furnace loads can impact an adjacent industrial system which does not have any harmonic producing load. Consider an example of a large distribution system:

- Two (2) utility interconnecting transformer of 50 MVA at 230 kV.
- Four (4) generators totaling 120 MW.
- Load demand of 100 MW, at certain times the excess generated power can be supplied into the utility system.
- Seventy six (76) secondary unit substations (0.5–2.5 MVA) at low-voltage, some of them double-ended.
- Two hundred and eighty (280) Low-voltage MCCs (motor control centers).
- Two-thousand one hundred (2100) low-voltage motors, ranging from 5-hp to 250 hp.
- A number of emergency tie connections between buses for alternate supply of power.
- Six (6) redundant storage battery systems, with duplicate charges. Uninterruptible power supply systems for critical loads.
- Auto-switching and transfer of power for critical process loads and generation auxiliary loads.
- Twenty six (26) primary unit substations at 2.4 kV or 4.16 kV (2.5–7.5 MVA) serving medium voltage motor loads, induction and synchronous, 5000-hp. (Some pulping mill operations use synchronous refiner motors of 45,000-hp).
- 10 miles of cable interconnections at medium and low voltage.
- 5 miles of 13.8 kV overhead lines.
- Six current limiting reactors to control short-circuit levels to acceptable limits within ratings of the circuit breakers.
- 25% of operating load consisting of drive systems, 6-pulse to 18-pulse.
- Three harmonic filters totaling 30 Mvar for voltage support and harmonic mitigation.

The location of an industrial plant impacts the modeling strategy. Many a times the utility high voltage substations, 230–66 kV (sometimes operated by the industrial plant personnel) are located within the industrial plant facilities. An industrial plant may be located close to a large generating station or there may be nearby harmonic sources or capacitors which will impact the extent of external system modeling. Generators and large rotating loads may be modeled individually, while an equivalent AC motor model can be derived connected through fictitious transformer impedance representing a number of transformers to which the motors are connected.

This aggregating of loads is fairly accurate, provided that harmonic source buses and the buses having capacitor compensations are modeled in detail.

The major sources of harmonics in industrial systems are ASDs and the harmonic emission can vary over wide limits depending upon topology. The sources of nonlinear loads vary depending upon processes. Electrolyzing and metallurgical plants will have large chunks of rectifier loads. Cement plants have gearless LCI fed large synchronous motors. Steel rolling mills may have large electronically fed DC motors. Some auxiliary nonlinear loads like UPS systems, battery charging and administrative office loads form a minor percentage of the total loads, rarely exceeding 1–2% for large industrial facilities. However, in any harmonic analysis study these cannot be ignored.

13.6 DISTRIBUTION SYSTEMS

The primary distribution system voltage levels are 4–44 kV, while the secondary distribution systems are of low voltage (<600 V). Distribution embraces all electrical utility system between the bulk power supply source and the consumer disconnect switch/ meter. Figure 13.7 shows the fundamental configuration of distribution systems. A distribution system consists of:

- Sub-transmission circuit operating from 12.47 to 345 kV, which deliver energy to distribution substations.
- Distribution substations, which convert energy to lower voltages for local distribution and regulate the voltage delivered to load centers, that is, use of induction voltage regulators, shunt power capacitors.
- Primary circuits for feeders, which may operate between 2.4 and 34.5 kV and supply loads to certain areas.
- Pole or pad-mounted distribution transformers or transformers, rated 10–2500 KVA located in underground vaults, close to consumer loads which convert energy to the voltage of utilization.
- Secondary circuits at utilization voltage which deliver the energy along the street or alleys; within short distance of users.
- Finally the service drops or underground circuits which bring the energy from the secondary mains to the consumer service switches.

The sub-transmission circuits can be:
- Simple radial circuits.
- Parallel or loop circuits.
- Interconnected circuits forming a grid or network.

13.6.1 The Radial System

The radial form is the simplest, costs less, used when power is transmitted over one radial circuit, Fig. 13.8. The fault on the single circuit feeder will result in interrupting

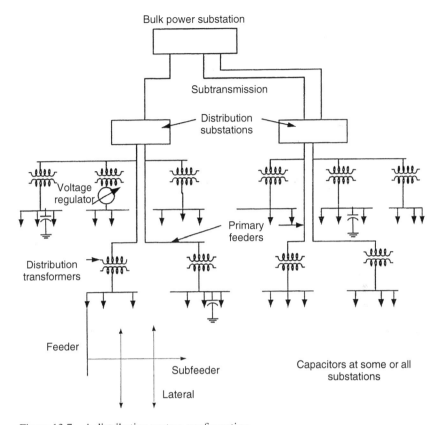

Figure 13.7   A distribution system configuration.

a large block of load and the electrical service to many customers will be interrupted. This is not acceptable.

A modified form of radial system with duplicate automatic and manual throw over switches/circuit breakers is shown in Fig. 13.9. In the event of failure of one circuit, the faulty circuit is quickly relayed open and the alternate normally open breaker can be manually or automatically closed. This arrangement does not preclude large scale short-time interruption, but the power supply can be quickly restored.

13.6.2   The Parallel or Loop System

A parallel or loop system is shown in Fig. 13.10. There are duplicate or parallel feeders to each of the distribution substations. All the circuit breakers shown in this figure are normally closed and therefore the load has two routes of power flow. Note that the circuit to substations 1, 2, and 3 forms a loop. Now consider a fault, say at location F1 between substations 1 and 2. By appropriate discriminative relaying, breakers B2 and C2 will be opened, while all other breakers remain closed. The power supply to any of the substations is not interrupted, though the fault clearance may cause a voltage

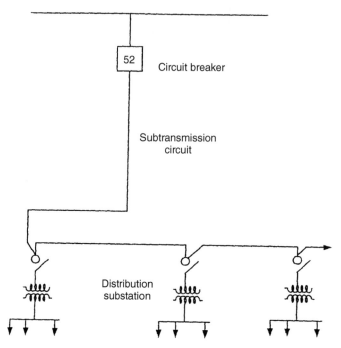

Figure 13.8   A radial distribution system.

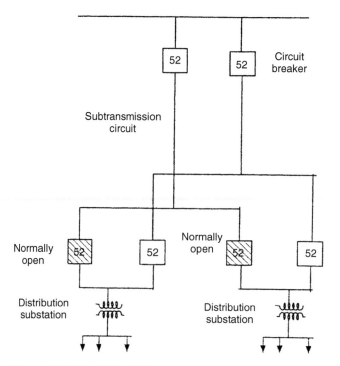

Figure 13.9   Modified radial system with automatic/manual switches or circuit breakers.

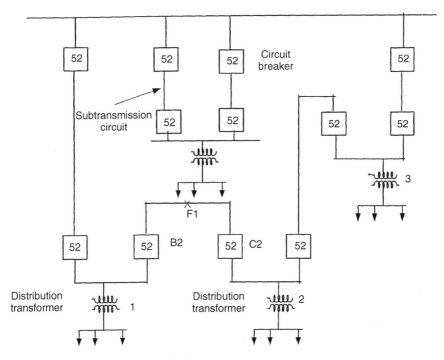

Figure 13.10   A parallel or loop distribution system.

dip, which will be experienced by all the consumers, depending upon their location and impedance to fault point.

13.6.3   Network or Grid System

A grid or network system is shown in Fig. 13.11. This system provides the best reliability of the power supply. Generally more than one bulk power supply source buses will be tied together. In this system, the power can flow from any source to any substation. Network construction permits adding new substations without much capital expenditure or construction. However, it has the disadvantage that there are many circuit breakers, and the selective protective relaying is complex. The reactive power compensation through capacitors may be spread throughout the system, locations decided by power system studies.

13.6.4   Primary Distribution System

The primary distribution system takes energy from the low-voltage buses of the distribution substations (Figs. 13.6–13.8) and delivers it to the primaries of the distribution transformers.

Most primary distribution circuits are radial. These consist of primary feeders which take the energy to a load area, sub feeders that distribute the energy to laterals and finally the laterals that connect to the individual transformers.

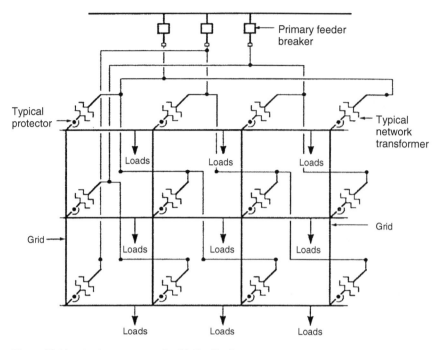

Figure 13.11   An interconnected grid distribution system.

Figure 13.12 shows a tree type of radial circuit. The current tapers off from the substation to the distal radials and the conductor size can be reduced as one proceeds from the trunk of the tree to the branches. It is an economical and simple circuit, but the voltages drops may increase toward the remote laterals from the substation. The maximum voltage drop is limited to 5% or even less.

Figure 13.13 shows another typical distribution system. Note the reclosure and switched power capacitors. Many faults are transient in nature, for example, an insulator on OH line may temporarily flash over to ground, creating a temporary line-to-ground fault. A reclosure will open on such a fault and reclose after a short-delay. If the fault was of a temporary nature, the service will be restored quickly; however, if the fault is of permanent nature, it will again trip. As many as 3–4 attempts may be made to restore the service, before the reclosure will lockout. The power capacitors are used to support the voltage on heavy loads and automatic voltage dependent switching is resorted to. Alternate circuit connections can restore power in case of a fault in one section. This figure also shows fuses at single phase tap-points and transformers with internal or external fusing. The voltage profile is maintained for about 50 km from the nearest to the farthest customer by application of line regulators or shunt capacitors, and in some instances both these devices are used.

When a failure occurs on the radial system, action must be taken by utility employees to restore power, which may not occur automatically, prolonging the period of shutdown. In some densely populated areas and large cities, utilities

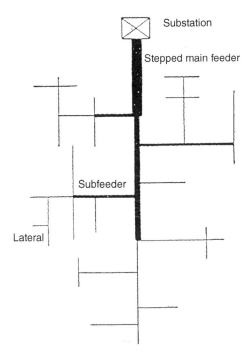

Figure 13.12   A radial type of tree distribution circuits with laterals to consumer loads.

have installed secondary networks, consisting of multiple distribution feeders; each serving an underground transformer installed in vaults. The low-voltage secondary of these transformers are interconnected and the service from these secondaries provide high reliability. *Network protectors* are installed due to high short-circuit current and current limiters on the secondaries.

Investments in distribution systems constitute 50% of the capital investment of a typical utility system.

13.6.5   Distribution System Harmonic Analysis

A distribution system harmonic study is undertaken to investigate harmonic problems, resonances, and distortion in the existing system or to investigate the effect of adding another harmonic source. The distribution systems are impacted with harmonics from commercial buildings, residential loads, industrial systems, presence of power factor improvement capacitors and the like. Single- or three-phase models can be used as demanded by the study. At low-voltage distribution levels the unbalance loads and harmonic sources are of particular importance, demanding a three-phase representation. The distribution systems are tied into the interconnected power network, and the system may be too large to be modeled. In most cases it will be sufficiently accurate to represent a transmission network by it's short-circuit impedance. This impedance may change, depending upon modifications to the system or system operation and is not a fixed entity. When capacitors or harmonic sources are present, a more detailed model will be necessary. The transmission system itself can be a

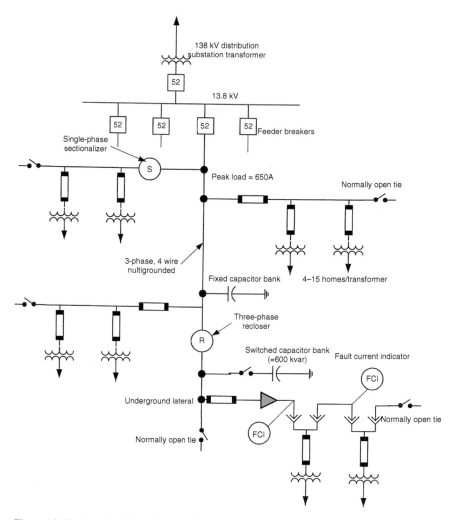

Figure 13.13   A typical distribution with reclosures, capacitors and switching arrangements.

significant source of harmonic impacts in the distribution system and measurement at points of interconnection may be necessary to ascertain it. The load models are not straightforward, if these are lumped with harmonic producing loads. The motor loads can be segregated from the nonrotating loads which can be modeled as equivalent parallel $R-L$ impedance. Again it is not feasible to model each load individually and feeder loads can be aggregated into large groups without loss of much accuracy.

The studies generally involve finite harmonic sources and the background harmonic levels are often ignored: this may lead to errors in the results. These can be measured and analysis, generally, combines modeling and measurements for accuracy. A study shows that the harmonic currents at higher frequencies have widely varying phase angles, which result in their cancellation. At lower frequencies up to

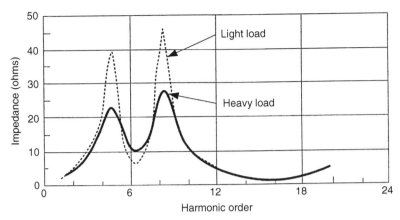

Figure 13.14   Frequency response characteristics of distribution loads toward the end of a feeder. Source: Ref. [7].

the 13th the cancellation is not complete. Unbalance loads on the feeder result in high harmonic currents in the neutral and ground paths. At fifth and seventh harmonics, the loads can be modeled as: (1) harmonic sources, (2) a harmonic source with a parallel *RL* circuit and (3) a harmonic source with a series *RL* circuit. Radial distribution systems will generally exhibit a resonance or cluster of resonances between fifth and seventh harmonics. See also Refs. [5,6]. Figure 13.14 from Ref. [7] shows the frequency response characteristics of the distribution system loads near the end of the feeder.

For the distribution systems modeling:

- Cables are represented as equivalent π model.

- For short lines model capacitance at the bus.

- Transformers represented by their equivalent circuits ignoring saturation.

- Power factor correction capacitors modeled at their locations.

- Harmonic filters and generators represented as discussed in Chapter 12.

- All impedances should be uncoupled.

Loads are modeled as per their composition and characteristics.

13.7   TRANSMISSION SYSTEMS

A high-voltage grid system may incorporate hundreds of generators, transmission lines, and transformers. Thus, the extent to which a system should be modeled has to be decided. The system to be retained and deriving an equivalent of the rest of the system can be addressed properly only through a sensitivity study, Chapter 14. Transmission systems have higher *X/R* ratios and lower impedances and the harmonics can be propagated over much longer distances. The capacitances of transformers and lines need to be included. The operating configuration range of a transmission system is

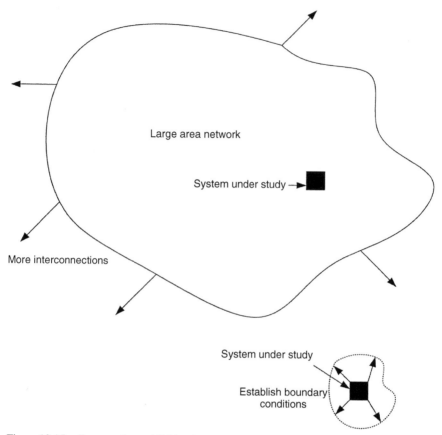

Figure 13.15   Concepts in establishing boundary limits of harmonic studies for a part of large area network.

much wider than that of a distribution system. A study may begin by identifying a local area which must be modeled in detail. The distant portions of the system are represented as lumped equivalents, Fig. 13.15. Equivalent impedance based on short-circuit impedance is one approach, the second approach uses a frequency versus impedance curve of the system, and there is a third intermediate area, whose boundaries must be carefully selected for accuracy. These can be based on geographical distance from the source bus. Series line impedance and the number of buses distant from the source are some other criteria. Sensitivity methods provide a better analytical tool. The considerations are:

- Experience based on previous studies [5,6].
- Distance from source is considered as modeling criteria. Geographical distance, line impedance and number of buses distant from source has been used, Ref. [7]. Problem arises as the distant buses may be itself sources of harmonics and then go further away from them, encompassing a greater and greater system for study.

- The sensitivity analysis can be applied to the remote source short-circuit or open circuit modeling.
- An approach which uses a frequency response curve representing changes in the impedance of the remote system with respect to frequency variations can be used, Ref. [8]. Yet, it is difficult to perform switching studies involving components in the equivalent network
- The transmission systems use SVC/s TCR's, STATCOMs, TCSC, and other FACTs controllers, which need to be properly modeled.

13.7.1 Ferranti Effect

As the transmission line length increases, the receiving end voltage rises above the sending end voltage, due to line capacitance. This is called the Ferranti effect. In a long line model, at no load ($I_R = 0$), the sending end voltage is

$$V_s = \frac{V_r}{2}e^{\alpha l}e^{j\beta l} + \frac{V_r}{2}e^{-\alpha l}e^{-j\beta l} \tag{13.1}$$

At $l = 0$, both incident and reflected waves are equal to $V_r/2$. As l increases, the incident wave increases exponentially, while the reflected wave decreases. Thus, the receiving end voltage rises. Another explanation of the voltage rise can be provided by considering that the line capacitance is lumped at the receiving end. Let this capacitance be Cl; where l is the line length then, on open circuit the sending end current is:

$$I_s = \frac{V_s}{\left(j\omega Ll - \frac{1}{j\omega Cl}\right)} \tag{13.2}$$

C is small in comparison with L. Thus, ωLl can be neglected. The receiving end voltage can then be written as

$$V_r = V_s - I_s(j\omega Ll)$$

$$= V_s + V_s\omega^2 CLl^2$$

$$= V_s(1 + \omega^2 CLl^2) \tag{13.3}$$

This gives a voltage rise at the receiving end:

$$|V_s|\omega^2 CLl^2 = |V_s|\omega^2 l^2/v^2 \tag{13.4}$$

where v is the velocity of propagation. Considering that v is constant, the voltage rises with the increase in line length.

Also, from Eq. (13.1) the voltage at any distance x terms of the sending end voltage, with the line open circuited and resistance neglected, is

$$V_x = V_s\frac{\cos \beta(l - x)}{\cos \beta l} \tag{13.5}$$

and the current is

$$I_x = j\frac{V_s}{Z_0}\frac{\sin\beta(l-x)}{\cos\beta l} \tag{13.6}$$

Example 13.1: A 230-kV three-phase transmission line has 795 KCMIL, ACSR conductors, one per phase. Neglecting resistance, $z = j0.8\,\Omega$ per mile and $y = j5.4 \times 10^{-6}$ Siemens (same as mhos) per mile. Calculate the voltage rise at the receiving end for a 400 mile long line.

Using the expressions developed above:

$$Z_0 = \sqrt{\frac{z}{y}} = \sqrt{\frac{j0.8}{j5.4 \times 10^{-6}}} = 385\,\Omega$$

$\beta = \sqrt{zy} = 2.078 \times 10^{-3}\,\text{rad/mile} = 0.119°/\text{mile}$; $\beta_l = 0.119 \times 400 = 47.6°$. The receiving end voltage rise from Eq. (13.5):

$$V_r = V_s\frac{\cos(l-l)}{\cos 47.6°} = \frac{V_s}{0.674} = 1.483V_s$$

The voltage rise is 48.3% and at 756 miles, one-quarter wavelength, it will be infinite. As the load is thrown off, the sending end voltage will rise before the generator voltage regulators and excitation systems act to reduce the voltage, further increasing the voltages on the line. This points to the necessity of compensating the transmission lines.

The sending end charging current from Eq. (13.2) is 1.18 per unit and falls to zero at the receiving end. This means that the charging current flowing in the line is 118% of the line natural load, (see Section 13.7.2).

The cables have even higher shunt capacitances, though these may not be used in very long lengths. Regardless of voltage, conductor size, or spacing of a line, the series reactance is approximately 0.8Ω/mile and the shunt-capacitive reactance is 0.2MΩ/mile. This gives a β of 1.998×10^{-3}/mile or 0.1145°/mile. Table 13.4 shows comparative line and cable constants for 230 kV transmission line and solid dielectric cable.

Shunt reactors have been used to control the voltages at light load. As the load increases the voltage drops. SVC's and STATCOMS have been employed. Transmission line voltage regulation is of much importance.

13.7.2 Surge Impedance Loading

The surge impedance loading (SIL) of the line is defined as the power delivered to a purely resistive load equal in value to the surge impedance of the line:

$$\text{SIL} = V_r^2/Z_0 \tag{13.7}$$

For 400 ohms surge impedance, SIL in kW is 2.5 multiplied by the square of the receiving end voltage in kV. The surge impedance is a real number and therefore

TABLE 13.4   Typical Electrical Characteristics, 230 kV OH lines Versus Underground Cables

Parameter	Overhead Line	Underground XLPE	Underground HPFF
Shunt capacitance μF/mile	0.015	0.30	0.61
Series inductance mH/mile	2.0	0.95	0.59
Series reactance ohms/mile	0.77	0.36	0.22
Charging current, A/mile	1.4	15.2	30.3
Dielectric loss, kW/mile	0+	0.2	2.9
Reactive charging power, MVA/mile	0.3	6.1	12.1
Capacitive energy kJ/mile	0.26	2.3	7.6
Surge impedance, ohms	375	26.8	14.6
Surge impedance loading limit, MW	141	1975	3623

the power factor along the line is unity, that is, no reactive power compensation is required. The SIL loading is also called the natural loading of the transmission line. Table 13.4 shows that the SIL limits for cables are much higher.

13.7.3   Transmission Line Voltages

There are many transmission lines operating at 800 kV around the globe. UHV (ultra-high voltages) to 1200 kV are in testing or development stage. See Fig. 13.16. A double circuit 1000 kV transmission line tower rises to approximately 108 m from the ground, the lowest conductor being at a clearance of 50 m from the ground, with conductor to conductor spacing of approximately 21 m.

With increasing transmission voltages, it should not be assumed that at the high voltage transmission lines there will be no harmonic problems. The harmonics generated at the loads, distort the distribution systems, substations, sub-transmission and finally transmission systems. Note the stricter limits of harmonic distortions specified in IEEE 519, as the voltage increases, Chapter 10.

13.8   COMPENSATION OF TRANSMISSION LINES

Transmission lines are compensated to maintain an acceptable voltage profile and concurrently the stability is improved, the maximum power handling capability increases and the length of line over which the power can be transmitted increases. A 750 kV transmission line require approximately 230 Mvar of capacitive reactive power compensation per 100 km.

13.8.1 Z_0 Compensation

As stated above a flat voltage profile can be achieved with a SIL loading of V^2/Z_0. The surge impedance can be modified with addition of capacitors and reactors so that

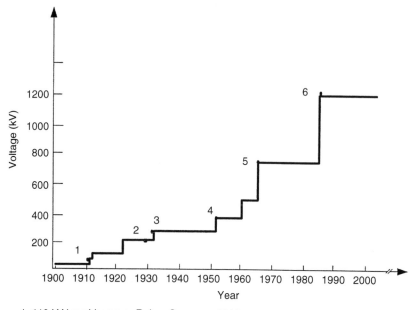

1. 110 kV Lauchhammer-Reisa, Germany, 1911
2. 220 kV Brauweiler-Hoheneck, Germany, 1929
3. 287 kV Boulder Dam, Los Angeles, USA, 1932
4. 380 kV Harspranget- Halsberg, Sweden, 1952
5. 735 kV Montreal-Manicouagan, Canada 1965
6. 1200 kV Ekibastuz-Kokchetave, USSR 1985

Figure 13.16   Developments in transmission voltages.

the desired power transmitted is given by the ratio of the square of the voltage and modified surge impedance:

$$P_{\text{new}} = V^2/Z_{\text{modified}} \qquad (13.8)$$

However, the load can suddenly change, and ideally, the compensation should also adjust instantaneously according to Eq. (13.8). This is not a practical operating situation and stability becomes a consideration. Passive and active compensators are used to enhance stability. The passive compensators are shunt reactors, shunt capacitors, and series capacitors. We discussed the series compensation in Chapter 5 in connection with subsynchronous resonance. The active compensators generate or absorb reactive power at the terminals where these are connected; like STATCOM or SVC's.

13.8.2   Line Length Compensation

It can be shown that the line has a flat voltage profile if

$$\beta = \frac{2\pi f}{v} = \frac{2\pi}{\lambda} \qquad (13.9)$$

As $1/\sqrt{LC} \approx$ velocity of light, the line length λ is 3100, 6200, ... miles or $\beta = 0.116°$ per mile. The quantity β_l is called the *electrical length* of the line. The length calculated above is too long to avail this ideal property. Even under ideal conditions, the natural load (SIL) cannot be transmitted $> \lambda/4$ (=775 miles). Practical limits are much lower. As $\beta = \omega\sqrt{LC}$, the inductance can be reduced by series capacitors, thereby reducing β. The phase-shift angle between the sending end and receiving end voltages is also reduced and the stability limit is, therefore, improved. This compensation may be done by sectionalizing the line. For power lines, series and shunt capacitors for heavy load conditions and shunt reactors under light load conditions are used to improve power transfer and line regulation.

13.8.3 Compensation by Sectionalizing the Line

The line can be sectionalized, so that each section is independent of one other, that is, meets its own requirements of flat voltage profile and load demand. This is compensation by sectionalizing, achieved by connecting constant-voltage compensators along the line. These are active compensators, that is, thyristor switched capacitors (TSCs), thyristor controlled reactors (TCRs), and synchronous condensers. All three types of compensating strategies may be used in a single line.

Consider that a *distributed* shunt inductance L_{shcomp} is introduced. This changes the effective value of the shunt capacitance as

$$j\omega C_{comp} = j\omega C + \frac{1}{j\omega L_{shcomp}}$$

$$= j\omega C(1 - K_{sh}) \qquad (13.10)$$

where

$$K_{sh} = \frac{1}{\omega^2 L_{shcomp} C} = \frac{X_{sh}}{X_{shcomp}} = \frac{b_{shcomp}}{b_{sh}} \qquad (13.11)$$

where K_{sh} is the degree of shunt compensation. It is negative for a shunt capacitance addition.

Similarly, let a distributed *series* capacitance C_{srcomp} be added. The degree of series compensation is given by K_{sc}:

$$K_{sc} = \frac{X_{srcomp}}{X_{sr}} = \frac{b_{sr}}{b_{srcomp}} \qquad (13.12)$$

The series or shunt elements added are distinguished by subscript "comp" in the above equations. Combining the effects of series and shunt compensations:

$$Z_{0\ comp} = Z_o \sqrt{\frac{1 - K_{sc}}{1 - K_{sh}}}$$

$$P_{0\ comp} = P_o \sqrt{\frac{1 - K_{sh}}{1 - K_{sc}}} \qquad (13.13)$$

Also,

$$\beta_{comp} = \beta\sqrt{(1 - K_{sh})(1 - K_{sc})} \qquad (13.14)$$

The effects are summarized as follows:

- Capacitive shunt compensation increases β and power transmitted and reduces surge impedance. Inductive shunt compensation has the opposite effect, reduces β and power transmitted and increases surge impedance. A 100% inductive shunt capacitance will theoretically increase the surge impedance to infinity. Thus, at no load, shunt reactors can be used to cancel the Ferranti effect.

- Series capacitive compensation decreases surge impedance and β and increases power transfer capacity. Series compensation is applied more from the steady state and transient stability consideration rather than from power factor improvement. It provides better load division between parallel circuits, reduced voltage drops, and better adjustment of line loadings. It has practically no effect on the load power factor. Shunt compensation, on the other hand, directly improves the load power factor. Both types of compensations improve the voltages and, thus, affect the power transfer capability. The series compensation reduces the large shift in voltage that occurs between the sending and receiving ends of a system and improves the stability limit.

The performance of a symmetrical line is discussed in detail in Ref. [9]. At no load, the midpoint voltage of a symmetrical compensated line is given by:

$$V_m = \frac{V_s}{\cos(\beta l/2)} \qquad (13.15)$$

Therefore, series capacitive and shunt inductive compensation reduce the Ferranti effect, while shunt capacitive compensation increases it.

The reactive power at the sending end and receiving end of a symmetrical line is:

$$Q_s = -Q_r = \frac{\sin\theta}{2}\left[Z_0 I_m^2 - \frac{V_m^2}{Z_0}\right] \qquad (13.16)$$

This equation can be manipulated to give the following equation in terms of natural load of the line and $P_m = V_m I_m$ and $P_0 = V_0^2/Z_0$:

$$Q_s = -Q_r = P_0\frac{\sin\theta}{2}\left[\left(\frac{PV_0}{P_0 V_m}\right)^2 - \left(\frac{V_m}{V_0}\right)^2\right] \qquad (13.17)$$

For $P = P_0$ (natural loading) and $V_m = 1.0$ per unit, $Q_s = Q_r = 0$.
If the terminal voltages are adjusted so that $V_m = V_0 = 1$ per unit:

$$Q_s = -Q_r = P_0\frac{\sin\theta}{2}\left[\left(\frac{P}{P_0}\right)^2 - 1\right] \qquad (13.18)$$

At no load:

$$Q_s = -Q_r = -P_0 \tan \frac{\theta}{2} \approx -P_0 \frac{\theta}{2} \qquad (13.19)$$

If the terminal voltages are adjusted so that for a certain power transfer, $V_m = 1$ per unit, then the sending end voltage is

$$V_s = V_m \left(1 - \sin^2 \frac{\theta}{2} \left[1 - \left(\frac{P}{P_0} \right)^2 \right] \right)^{1/2} = -V_r \qquad (13.20)$$

When series and shunt compensation are used, the reactive power requirement at *no load* is approximately given by

$$Q_s = -P \frac{\beta l}{2} (1 - K_{sh}) = -Q_r \qquad (13.21)$$

If K_{sh} is zero, the reactive power requirement of a series compensated line is approximately the same as that of an uncompensated line, and the reactive power handling capability of terminal synchronous machines becomes a limitation. Series compensation schemes, thus, require SVCs or synchronous condensers/shunt reactors. The derivation of all the above equations is not strictly presented, see Ref. [9], which provides a detailed description of the transmission line compensation and voltage profiles.

The modeling for harmonics must consider all the line compensating devices.

13.8.4 Reflection Coefficients

The relative values of sending end and receiving end voltages, V_1 and V_2 depend on the conditions at the terminals of the line. The reflection coefficient at the load-end is defined as the ratio of the amplitudes of the backward and forward traveling waves. For a line terminated in a load impedance Z_L:

$$V_2 = \left(\frac{Z_L - Z_0}{Z_L + Z_0} \right) V_1 \qquad (13.22)$$

Therefore, the voltage reflection coefficient at the load end is

$$\rho_L = \frac{Z_L - Z_0}{Z_L + Z_0} \qquad (13.23)$$

The current reflection coefficient is negative of the voltage reflection coefficient. For a short-circuited line, the current doubles and for an open circuit line the voltage doubles. The reflected wave at an impedance discontinuity is a mirror image of the incident wave moving in the opposite direction. Every point in the reflected wave is the corresponding point on the incident wave multiplied by the reflection coefficient, but a mirror image. At any time the total voltage is the sum of the incident and reflected waves. The reflected wave moves toward the source and is again reflected, see Fig. 12.4. The source reflection coefficient, akin to the load reflection coefficient, can be defined as

$$\rho_s = \frac{Z_s - Z_0}{Z_s + Z_0} \tag{13.24}$$

A forward traveling wave originates at the source, as the backward traveling wave originates at the load. At any time the voltage or current at any point on the line is the sum of all voltage or current waves existing at the line at that point and time.

For harmonic analysis there is a certain load which will give the maximum harmonic distortion. Define the reflection coefficient for current at receiving end as:

$$\rho_r = \frac{Z_R - Z_0}{Z_R + Z_0} \tag{13.25}$$

Then the total current I_R is:

$$I_R = I_R^+ - \rho_r I_R^+ \tag{13.26}$$

where I_R^+ is the incident current at receiving end and $I_R^- = \rho_r I_R^+$ is the reflected current at receiving end. When the angles of incident and reflected currents are equal, then I_R will be maximum.

$$\theta_R^+ + \beta x = \theta_R^- - \beta x \tag{13.27}$$

where x is the distance from the receiving end.

$$x = \frac{\theta_R^- + \theta_R^+}{2\beta} \tag{13.28}$$

Current will also be a local maximum at intervals of every one half wavelength for the length of the line:

$$x = \frac{\theta_R^- + \theta_R^+}{2\beta} + n\frac{\lambda}{2} = \frac{\theta_R^- + \theta_R^+}{2\beta} + \frac{n}{2}\frac{2\pi}{\beta} \tag{13.29}$$

Reference [10] shows plots of 19th and 29th harmonic currents versus the line length for a 345 kV transmission line.

13.9   COMMERCIAL BUILDINGS

The effect of harmonics can be represented with a relatively simple system model as shown in Fig. 13.17, where the building loads are broken into various types. Example 6.1 illustrated that though the harmonic emission levels from individual load types may show considerable distorted waveforms, but harmonic cancellation occurs and overall distortion at PCC is reduced; Fig. 6.5. Neutral overloading can be a concern.

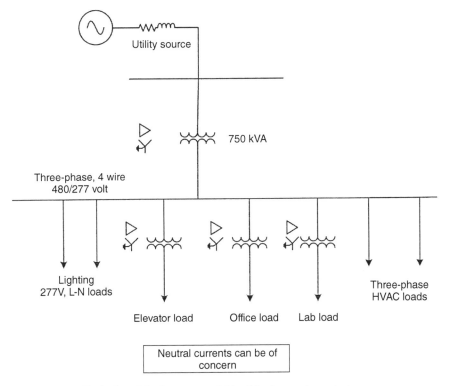

Figure 13.17   Typical model of a commercial load for harmonic analysis.

13.10   RESIDENTIAL LOADS

Residential load types and their spectrums are discussed in Chapter 4. Figure 13.18 illustrates that several residences are supplied from the same distribution transformer. Thus, PCC is at the secondary of the distribution transformer. Harmonic distortion levels from each household can be metered at the service entrance metering point. The short-circuit levels will vary depending upon the size of the distribution transformer, a range of 30–300 MVA is shown. Considering an average household load of 6 kVA, high I_{sr}/I_L will result. TDD for the consumer current should be less than 15%. IEC 61000-3-2 provides harmonic limits that are applicable to appliances up to 16A, Chapter 10. A growing percent of household load is electronic and uses SMPs for PCs, copy machines, TV sets, which impact the distortion levels and must be properly modeled.

13.11   HVDC TRANSMISSION

The cumulative megawatts of HVDC systems around the world approach 100 GW. HVDC has long history, but a transition point occurred when thyristor valves took

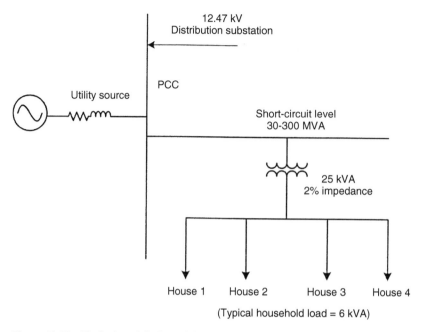

Figure 13.18   Typical model of a residential load for harmonic analysis. (Ref. [7].)

over mercury arc rectifiers in late 1970. The major technology leap was in Brazil with the 3150 MW± 600 kV Itaipu project commissioned from 1984 to 1987. The overhead line is 800 km long and each 12-pulse converter is rated 790 MW, 300 kV. HVDC is finding major applications in countries like India and China, and a large number of thyristor valve based systems are planned – the power levels and distances are such that ±800 kV may be needed. HVDC project list worldwide can be seen on web site [11]. Another web site of interest is CIGRE Study Committee B4, HVDC and Power Electronic Equipment [12]. Other references are [13–16].

13.11.1   HVDC Light

The IGBTs for motor drives have begun to find applications in HVDC systems at the lower end of power usage. These operate using PWM techniques; there is practically no or little need for reactive power compensation as the converters can generate active and reactive power. The systems have found applications in off-shore wind farms and short-distance XLPE type cable systems. The largest system to date is the 330-MW Cross-Sound DC link between Connecticut and Long Island.

13.11.2   HVDC Configurations and Operating Modes

Figure 13.19 shows common DC transmission system configurations, which are only partially explained as follows; referring to configurations marked (a)–(f) in this figure:

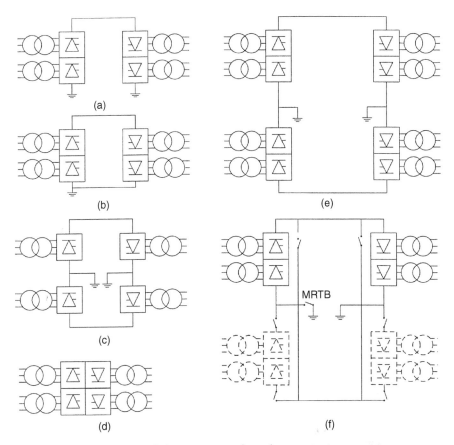

Figure 13.19   HVDC transmission, system configurations, see text.

(a) Monopolar systems are the simplest and most economical for moderate power transfer. Only two converters and one high voltage connection is required. These have been used with low-voltage electrode lines and sea electrodes to carry return currents.

(b) In congested areas, or soils of high resistivity conditions may not be conductive to monopolar systems. In such cases a low-voltage cable is used for the return path and DC circuit uses local ground connection for potential reference.

(c) An alternative of monopolar systems with metallic return is that the midpoint of a 12-pulse converter can be connected to earth and two-half voltage cables or line conductors can be used. The converter is operated only in 12-pulse mode, so that there are no stray currents.

(d) Back-to-back systems are used for interconnection of asynchronous networks and use AC lines to connect on either side. The power transfer characteristics are limited by the relative capabilities of adjacent AC systems. There are no DC lines. The purpose is to provide bi-directional exchange of power, easily and quickly. An AC link will have limitations in control over direction and amount of power flow. 12-pulse bridges are used. It is preferable to connect two back-to-back systems in parallel between the same AC buses.

(e) The most common configuration is 12-pulse bipolar converter for each pole at the terminal. This gives two independent circuits each of 50% capacity. For normal balanced operation there is no earth current. Monopolar earth return operation can be used during outage of the opposite pole.

(f) The earth return option can be minimized during monopolar operation by using opposite pole line for metallic return through pole/converter bypass switches at each end. This requires a metallic return transfer breaker in the ground electrode line at one of the DC terminals to commutate the current from relatively low resistance of earth into that of DC line conductor. This metallic return facility is provided for most DC transmission systems.

For voltages above ± 500 kV series connected converters are used to reduce energy unavailability for individual converter outage or partial line insulation failure. By using two-series connected converters per pole in a bipolar system, only 25% of the line capability is lost for a converter outage or if the line insulation is degraded and it can support only 50% of the rated line voltage.

13.11.3 DC Filters

Figure 13.20 shows a practical layout of a terminal. While reactive power compensation and harmonic filters have been previously discussed, DC filters are required to limit interference with communication circuits, which may be inductively coupled

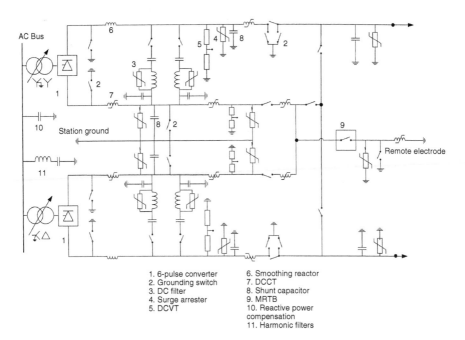

1. 6-pulse converter
2. Grounding switch
3. DC filter
4. Surge arrester
5. DCVT
6. Smoothing reactor
7. DCCT
8. Shunt capacitor
9. MRTB
10. Reactive power compensation
11. Harmonic filters

Figure 13.20   Typical terminal layout of a 12-pulse HVDC terminal.

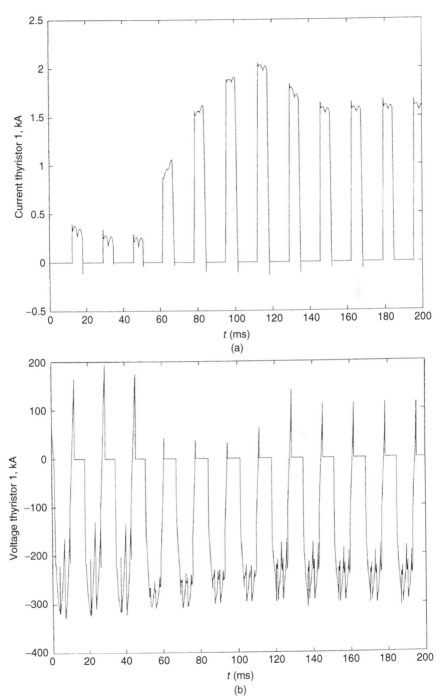

Figure 13.21   (a) and (b). Thyristor current and voltages, respectively, for a 6-pulse CSI bridge.

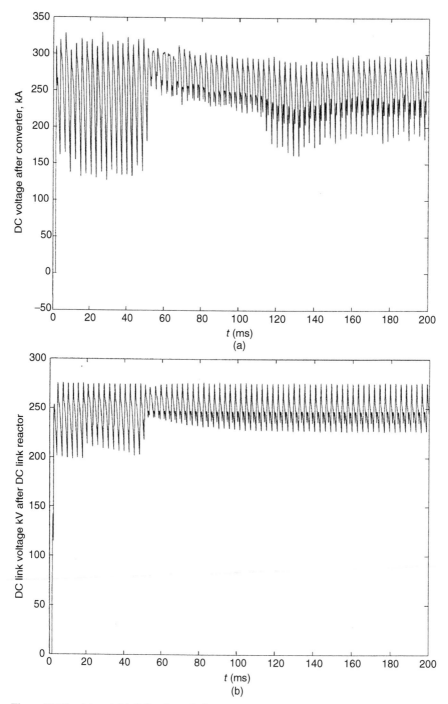

Figure 13.22 (a) and (b). DC voltage before and after the DC reactor, respectively for a 6-pulse CSI bridge.

to the DC line. The parameters to be considered are separation between the DC and communication lines, their shielding, the presence of ground wires and the soil resistivity. This criterion is expressed as equivalent disturbing current. Disturbance effects are lower in bipolar designs. The filter design must account for all operating modes and harmonic sources. In chapter 4 we noted that while the converter throws odd harmonics on to the AC supply system, the even harmonics of input frequency are transmitted to the load.

Example 13.2:   Figure 13.21(a) and (b) depict the thyristor current and voltages for a 6-pulse CSI bridge, while Figs. 13.22 (a) and (b) depict the DC link voltage before and after the DC link reactor, EMTP simulations. This shows the necessity of providing AC and DC filters, both. In Chapter 15, it is demonstrated how the DC and AC voltages will be smoothed out by provision of suitable filters.

Generally 12th harmonic band pass filter with active filtering of higher order harmonics are provided.

We discussed highly polluting loads of arc furnaces, welding and induction heating and their modeling in Chapter 6.

REFERENCES

1. W. Xu, J. Marti, and H.W. Dommel. "A multi-phase harmonic load flow solution technique," IEEE Transactions on Plasma Science, vol. 6, pp. 174–182, 1991.
2. IEEE. Task Force on Harmonic Modeling and Simulation. "Modeling and simulation of propagation of harmonics in electrical power systems-Part II: Sample systems and examples," IEEE Transactions on Power Delivery, vol. 11, pp. 466–474, 1996.
3. ANSI. Voltage rating for electrical power systems and equipments (60 Hz), 1988. Standard C84.1.
4. NEMA MG1 – Part 20. Large machines-induction motors, 1993.
5. T Hiyama, M.S.A.A. Hammam, T.H. Ortmeyer, "Distribution system modeling with distributed harmonic sources," IEEE Transactions on Power Delivery, vol. 4, pp. 1297–1304, 1989.
6. M.F. McGranaghan, R.C. Dugan, W.L. Sponsler. "Digital simulation of distribution system frequency-response characteristics," IEEE Transactions on PAS, vol. 100, pp. 1362–1369, 1981.
7. IEEE P519.1/D9a. IEEE guide for applying harmonic limits on power systems, Unapproved Draft, 2004.
8. M.F Akram, T.H Ortmeyer, J.A Svoboda. "An improved harmonic modeling technique for transmission network," IEEE Transactions on Power Delivery, vol. 9, pp. 1510–1516, 1994.
9. T.J.E. Miller. Reactive Power Control in Electrical Systems, John Wiley, New York, 1982.
10. R.D. Shultz, R.A. Smith, and G.L Hickey. "Calculations of maximum harmonic currents and voltages on transmission lines," IEEE Transactions on PAS, vol. 102, no. 4, pp. 817–821, 1983.
11. www.ece.uidaho.edu/HVDCfacts: IEEE/PES transmission and distribution committee, Web site.
12. www.cigre-b4.org: CIGRE study committee B4, HVDC and power electronic equipment,Web site.
13. EHV Transmission. IEEE Transactions, vol. 85, no. 6, pp. 555–700, 1966 (special issue).
14. M.S. Naidu and V. Kamaraju. High Voltage Engineering. 2nd Edition, McGraw Hill, New York, 1999.
15. S. Rao. EHV-AC, HVDC Transmission and Distribution Engineering, Khanna Publishers, New Delhi, 2004.
16. K.R. Padiyar. HVDC Power Transmission Systems, 2nd Edition, New Academic Science, Waltham, MA, 2011.

CHAPTER 14

HARMONIC PROPAGATION

The terms harmonic load flow, power flow, harmonic penetration, and propagation have been used interchangeably. A very popular algorithm is based on harmonic power flow (Section 14.5.3). Harmonic propagation or penetration seems to be better terminology, as the harmonics from the source propagate (penetrate) into the power system.

The purpose is to ascertain the distribution of harmonic currents, voltages, harmonic distortion indices, and harmonic resonance in a power system. This analysis is then applied to the harmonic filter designs and also to study other effects of harmonics on the power system, that is, notching and ringing, neutral currents, derating of transformers, and overloading of system components. In addition to models of loads, transformers, generators, motors, and so on, the models for harmonic injection sources, arc furnaces, converters, SVCs, and so on are included. These are not limited to characteristic harmonics and a full spectrum of load harmonics can be modeled. These harmonic current injections will be at different locations in a power system. As a first step, a frequency scan is obtained that plots the variation of impedance modulus and phase angle at a selected bus with a variation of frequency or generates $R - X$ plots of the impedance. This enables the resonant frequencies to be ascertained. The application of frequency scans has been demonstrated in previous chapters. The harmonic current flows in the lines are calculated, and the network, assumed to be linear at each step of the calculations with added constraints, is solved to obtain the harmonic voltages. The calculations may include the following:

- Calculation of harmonic distortion indices.
- Calculation of TIF, KVT, and IT.
- Induced voltages on communication lines.
- Sensitivity analysis, that is, the effect of variation of a system component.

This is rather a simplistic approach. The rigorous harmonic analysis gets involved because of the interaction between harmonic producing equipment and the power system, the practical limitations of modeling each component in a large power system, the extent to which the system should be modeled for accuracy, and the types of component and nonlinear source models. Furnace arc impedance varies erratically and is asymmetrical. Large power high-voltage DC (HVDC) converters and flexible alternating current transmission system (FACTS) devices have large

Power System Harmonics and Passive Filter Designs, First Edition. J.C. Das.
© 2015 The Institute of Electrical and Electronics Engineers, Inc. Published 2015 by John Wiley & Sons, Inc.

nonlinear loads and superimposition is not valid. Depending on the nature of the study, *simplistic methods may give erroneous results.*

14.1 HARMONIC ANALYSIS METHODS

There are a number of methodologies for calculation of harmonics and effects of nonlinear loads. Direct measurements can be carried out using suitable instrumentation. In a noninvasive test, the existing waveforms are measured. An EPRI project describes two methods, and Refs [1,2] provide a summary of the research project sponsored by EPRI and BPA (Bonneville Power Administration) on HVDC system interaction from AC harmonics. One method uses harmonic sources on the AC network, and the other injects harmonic currents into the AC system using HVDC converter. In the later case, the system harmonic impedance is a ratio of the harmonic voltage to current. In the first case, a switchable shunt impedance such as a capacitor or filter bank is required at the point where the measurement is taken. By comparing the harmonic voltages and currents before and after switching the shunt device, the network impedance can be calculated. This research project implements (1) development of data acquisition system to measure current and voltage signals, (2) development of data processing package, (3) calculation of harmonic impedances, and (4) development of a computer model for impedance calculations. The analytical analysis can be carried out in the frequency and time domains. Another method is to model the system to use a state-space approach. The differential equations relating current and voltages to system parameters are found through basic circuit analysis. Thus, the techniques for determining harmonic impedances can be divided into three categories:

- Frequency and time domain methods [3,4].
- Direct injection, switching of a capacitor, which assumes that Norton's equivalents are not affected by the switching operations; however, the phase angle variations does impact the results [5].
- Analysis of transient inrush current and voltage waveforms produced by normal switching operations [6].

14.2 FREQUENCY DOMAIN ANALYSIS

For calculations in the frequency domain, the harmonic spectrum of the load is ascertained and the current injection is represented by a Norton's equivalent circuit. Harmonic current flow is calculated throughout the system for each of the harmonics. The system impedance data are modified to account for higher frequency and are reduced to their Thevenin equivalent. The principle of superposition is applied. If all nonlinear loads can be represented by current injections, the following matrix equations are applicable:

$$\overline{V}_h = \overline{Z}_h \overline{I}_h \tag{14.1}$$

$$\overline{I}_h = \overline{Y}_h \overline{V}_h \tag{14.2}$$

The distribution of harmonic voltages and currents is no different for networks containing one or more sources of harmonic currents. During the steady state, the harmonic currents entering the network are considered as being produced by ideal sources that operate without repercussion. The entire system can then be modeled as an assemblage of passive elements. Corrections will be applied to the impedance elements for dynamic loads, for example, generators and motors; frequency-dependent characteristics at each incremental frequency chosen during the study can be modeled. The system harmonic voltages are calculated by direct solution of the linear matrix equations (14.1) and (14.2).

In a power system, the harmonic injection will occur only on a few buses. These buses can be ordered last in the Y matrix, and a reduced matrix can be formed. For n nodes and $n - j + 1$ injections, the reduced Y matrix is

$$\begin{vmatrix} I_j \\ \cdot \\ \cdot \\ I_n \end{vmatrix} = \begin{vmatrix} Y_{jj} & \cdot & Y_{jn} \\ \cdot & \cdot & \cdot \\ Y_{nj} & \cdot & Y_{nn} \end{vmatrix} \begin{vmatrix} V_i \\ \cdot \\ \cdot \\ V_n \end{vmatrix} \tag{14.3}$$

where diagonal elements are the self-admittances and the off-diagonal elements are transfer impedances as in the case of load flow calculations.

Linear transformation techniques are used. The admittance matrix is formed from a primitive admittance matrix by transformation:

$$\overline{Y}_{abc} = \overline{A}' \overline{Y}_{prim} \overline{A} \tag{14.4}$$

$\overline{A}'$ denotes transpose of matrix $\overline{A}$ and the symmetrical component transformation is given by

$$\overline{Y}_{012} = \overline{T}_s^{-1} \overline{Y}_{abc} \overline{T}_s \tag{14.5}$$

The vector of nodal voltages is given by

$$\overline{V}_h = \overline{Y}_h^{-1} \overline{I}_h = \overline{Z}_h \overline{I}_h \tag{14.6}$$

For the injection of a unit current at bus k:

$$\overline{Z}_{kk} = \overline{V}_k \tag{14.7}$$

where Z_{kk} is the impedance of the network seen from bus k. The current flowing in branch jk is given by

$$\overline{I}_{jk} = \overline{V}_{jk}(\overline{V}_j - \overline{V}_k) \tag{14.8}$$

where Y_{jk} is the nodal admittance matrix of the branch connected between j and k.

Variation of the bus admittance matrix, which is produced by a set of modifications in the change of impedance of a component, can be accommodated by modifications to the Y-bus matrix as discussed earlier. For harmonic analysis, the admittance matrix must be built at each frequency of interest for component-level RLC parameters for circuit models of lines, transformers, cables, and other equipment. Thus, the harmonic voltages can be calculated. A new estimate of the harmonic injection currents is then obtained from the computed harmonic voltages, and the process is iterative until the convergence on each bus is obtained. Under resonant conditions, large distortions may occur and the validity of assumption of linear system components is questionable. From Eq. (14.1), we see that the harmonic impedance is important in the response of the system to harmonics. There can be interaction between harmonic sources throughout the system, and if these are ignored, the single-source model and the superposition can be used to calculate the harmonic distortion factors and filter designs. The assumption of constant system impedance is not valid as the system impedance always changes, say due to switching conditions, operation, or future additions. These impedance changes in the system may have a more profound effect on the ideal current source modeling than the interaction between harmonic sources. A weak AC/DC interconnection defined with a short-circuit ratio (short-circuit capacity of the AC system divided by the DC power injected by the converter into converter bus) of < 3 may have voltage and power instabilities, transient and dynamic overvoltages, and harmonic overvoltages [7].

14.3 FREQUENCY SCAN

In a multiport (N-port) network, we can write

$$
\begin{vmatrix} I_1 \\ I_i \\ I_j \\ \cdot \\ I_n \end{vmatrix} = \begin{vmatrix} y_{11} & -y_{1i} & -y_{1j} & \cdot & -y_{1n} \\ -y_{i1} & y_{ii} & -y_{ij} & \cdot & -y_{in} \\ -y_{j1} & -y_{ji} & y_{jj} & \cdot & -y_{jn} \\ \cdot & \cdot & \cdot & \cdot & \cdot \\ -y_{n1} & -y_{ni} & -y_{nj} & \cdot & y_{nn} \end{vmatrix} \begin{vmatrix} v_1 \\ v_i \\ v_j \\ \cdot \\ v_n \end{vmatrix}
\tag{14.9}
$$

where I_i, v_i I_1, v_1 are the inputs and I_n, I_j, v_j, v_n are the outputs.

The complex admittances at a known frequency can be determined:

$$
y_{ij} = \left. \frac{I_i}{v_j} \right|_{I_k = 0} \qquad k = 1....n, \ \ k \neq i
\tag{14.10}
$$

where the off-diagonal elements, that is, y_{ij} is the transfer admittance and y_{ii} is the self-admittance.

For harmonic analysis, the admittance matrix is formed at each frequency every time, as the matrix built at one frequency cannot be applied to other frequencies. The matrix is built from component models of transformers, transmission lines, and other power system components. The frequency scan is conducted through repeated applications of Eq. (14.9) with admittance formed at each frequency:

$$
I_h = Y_h V_h
\tag{14.11}
$$

For a single harmonic current injection, say at node 1, Y matrix in Eq. (14.9) can be solved to find the voltages at each node. If the injection current is assigned with

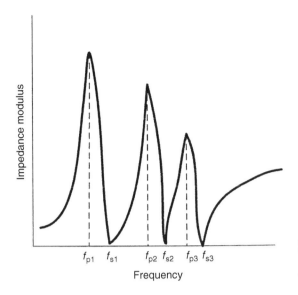

Figure 14.1 Frequency scan showing parallel and series resonance frequencies.

a value of $1 < 0°$, the values of the voltages thus determined represent the driving point and transfer impedances. As the admittance matrix is assumed to have passive elements, linearity can be applied to scale the results to any desired value.

A frequency scan is merely a repeated application in certain incremental steps of some initial value of frequency to the final value, these two values spanning the range of harmonics are to be considered. The procedure is equally valid whether there are single or multiple harmonic sources in the system, so long as the principle of superimposition is held valid. Varying the frequency gives a series of impedances that can be plotted to provide an indication of the resonant conditions. Figure 14.1 shows a frequency scan of impedance modulus versus frequency. The parallel resonances occur at peaks, which give the maximum impedances, and the series resonances occur at the lowest points of the impedance plots. Figure 14.1 shows parallel resonance at two frequencies f_{p1} and f_{p2} and series resonance at f_{s1} and f_{s2}. We will see in Chapter 15 that such a frequency scan is obtained with two single-tuned shunt filters. Multiphase frequency scans can identify the harmonic resonance caused by single-phase capacitor banks.

14.4 VOLTAGE SCAN

A voltage scan may similarly be carried out by applying unit voltage $1 < 0°$ to a node and calculating the voltages versus frequency in the rest of the system. The resulting voltages represent the voltage transfer function to all other nodes in the system. This analysis is commonly called a *voltage transfer function* study. The peaks in the scan identify the frequencies at which the voltages will be magnified and the lowest points indicate frequencies where these will be attenuated. Figure 14.2 shows a voltage transfer function.

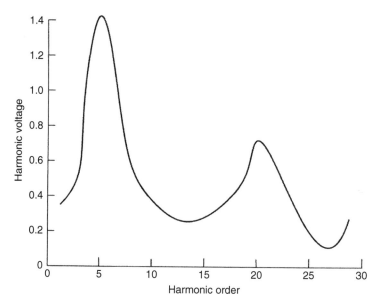

Figure 14.2   Voltage scan, voltage transfer function.

The admittance matrix can be formed based on sequence networks or phase-variable networks; the phase angles of the voltage or current injections are important. In phase variables, a three-phase scan will be conducted by injecting transpose of a vector $[1 < 0°, 1 < -120°,$ and $1 < 120°]$ into a three-phase bus. For a zero sequence scan, all angles will be 0.

14.5   HARMONIC ANALYSIS METHODS

14.5.1   Current Injection Method

Generally, the harmonic analysis programs use linear models, where nonlinear loads are represented by ideal constant current harmonic sources independent of voltage distortions [8].

- Form system admittance model including contributions of all sources and linear loads.
- Construct a current injection vector – it is a representation of nonlinear loads, where each entry is a term of known frequency in Fourier series representation with its angle.
- Solve Eq. (14.6) for harmonic voltages at all network buses.
- Start at the lowest frequency and repeat for each incremental frequency to cover the entire spectrum of interest.

The concern is that of accuracy, as generated harmonic currents can be dependent on voltage distortion. If the voltage distortion exceeds 5%, the results will be inaccurate. For a 6-pulse converter, the maximum harmonic generation will occur when the firing angle is zero and output power P is the maximum. The DC voltage will be

$$V_d = \frac{3\sqrt{3}\,V_m}{\pi} = \frac{P}{I_d} \tag{14.12}$$

For the ideal case, with AC-side inductance $= 0$ and $\cos \mu = 1$, the harmonic current is

$$I_h = \frac{P}{\sqrt{3}V_h} \tag{14.13}$$

Many methods of harmonic load flow have been proposed in frequency and time domain. The problem is how to account for nonlinearity and interaction between converters. The simplest way is to assume no harmonic interaction between network and nonlinear devices [8]. Iterative harmonic analysis was the first modification when considering harmonic voltage influence on nonlinear device behavior [9]. The harmonic voltages impact the fundamental frequency voltages [10]. The nonlinear device treatment and harmonic voltage calculations are included in Refs [11,12]. A three phase harmonic AC-DC load flow is discussed in Ref. [13], which writes 26 equations for interaction between AC and DC converters (also, see Refs [14–19]). A direct approach is described in Ref. [20]. Furthermore, hybrid and probabilistic methods and possibility theory approach (Fuzzy logic) have been proposed.

14.5.2 Forward and Backward Sweep

Forward and backward harmonic flow method that can be applied to three-phase distribution systems is an off-shoot of solving ladder networks on fundamental frequency load flow.

The basic principle is illustrated with reference to Fig. 14.3.

A distribution system forms a ladder network, with loads teed-off in a radial manner (Fig. 14.3(a)). It is a nonlinear system, as most loads are of constant kW and kvar. However, linearization can be applied.

For a linear network, assume that the line and load impedances are known and the source voltage is known. Starting from the last node, and assuming a node voltage of V_n, the load current is given by

$$I_n = \frac{V_n}{Z_{1n}} \tag{14.14}$$

With respect to harmonic currents:

$$\overline{I}_h = \overline{Y}_h \overline{V}_h \tag{14.15}$$

Note that I_n is also the line current, as this is the last node. Therefore, the voltage at node $n-1$ can be obtained simply by subtracting the voltage drop:

$$V_{n-1} = V_n - I_{n-1,n}Z_{n-1,n} \tag{14.16}$$

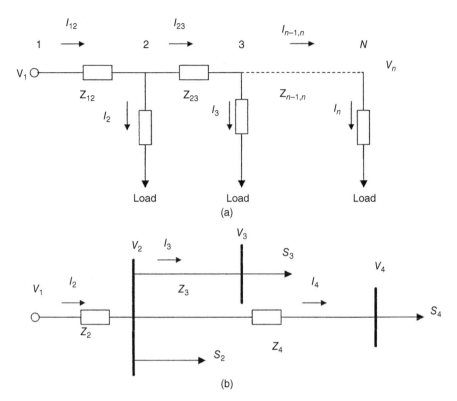

Figure 14.3 (a) Tapped loads in a radial feeder, (b) equivalent ladder network.

Or for harmonic voltages, we can write

$$V_{h(n-1)} = V_{h(n)} - I_{h(n-1,n)}Z_{h(n-1,n)} \qquad (14.17)$$

This process can be carried out until the sending end node is reached. The calculated value of the sending end voltage will be different from the actual applied voltage. Since the network is linear, all the line and load currents and node voltages can be multiplied by the ratio:

$$V_{actual}/V_{calculated} = V_s/V_{calculated} \qquad (14.18)$$

Actually, the node current must be calculated on the basis of complex power at the load:

$$I_{node} = (S_{node}/V_{node})^* \qquad (14.19)$$

Starting from the last node, the sending end voltage is calculated in the *forward sweep*, as in the linear case. This voltage will be different from the sending end voltage. Using this voltage, found in the first iteration, a *backward sweep* is performed, that is, the voltages are recalculated starting from the first node to the nth node. This new voltage is used to recalculate the currents and voltages at the nodes in the second forward sweep. This process can be repeated until the required tolerance is achieved.

A lateral circuit, Fig. 14.3(b), can be handled as follows [21,22]:

- Calculate the voltage at node 2, starting from node 4, ignoring the lateral to node 3. Let this voltage be V_{2h} $(h=1 \text{ to max})$.
- Consider that the lateral is isolated and is an independent ladder. Now the voltage at node 3 can be calculated and therefore current I_3 is known.
- The voltage at node 2 is calculated back, that is, voltage drop $I_{3h}Z_{3h}$ is added to voltage V_{3h}. Let this voltage be V_2'. The difference between V_{2h} and V_2' must be reduced to an acceptable tolerance. The new node 3 voltage is $V_{3(\text{new})} = V_3 - (V_2 - V_2')$. The current I_3 is recalculated and the calculations are iterated until the desired tolerance is achieved.

A flow chart is shown in Fig. 14.4.

14.5.3 Iterative Newton–Raphson Method

An application of iterative Newton–Raphson method applied to harmonic load flow is described in Refs [11,12]. This is based on the balance of active power and reactive volt-amperes, whether at fundamental frequency or at harmonics. The active and reactive power balance is forced to zero by the bus voltage iterations. This method is essential for balanced systems, where linear and nonlinear loads are treated in terms of power. It is the reformation of conventional Newton–Raphson method for fundamental frequency load flow to include nonlinear loads.

Consider a system with $n + 1$ buses, bus 1 is a slack bus, buses 2 through $m - 1$ are conventional load buses, and buses m to n have nonsinusoidal loads. It is assumed that the active power and the reactive volt-ampere balance are known at each bus and that the nonlinearity is known. The power balance equations are constructed so that ΔP and ΔQ at all nonslack buses is zero for all harmonics. The form of ΔP and ΔQ as a function of bus voltage and phase angle is the same as in conventional load flow, except that Y_{bus} is modified for harmonics. The specified active and reactive powers (P^s and Q^s, $s = 1, 5, 7, \ldots$) are known at buses 2 through $m - 1$, but only active power is known at buses m through n. Two additional parameters are required: the current balance and volt-ampere balance. The current balance for fundamental frequency is written and the equation is modified for buses with harmonic injections. The harmonic response of buses 1 through $m - 1$ is modeled in admittance bus matrix.

The third equation is the apparent volt-ampere balance at each bus:

$$S_L^2 = \sum_s P_L^2 + \sum_s Q_L^2 + \sum D_L^2 \qquad (14.20)$$

where the third term of the equation denotes distortion power at bus L, which is not considered as an independent variable, as it can be calculated from real and imaginary components of currents.

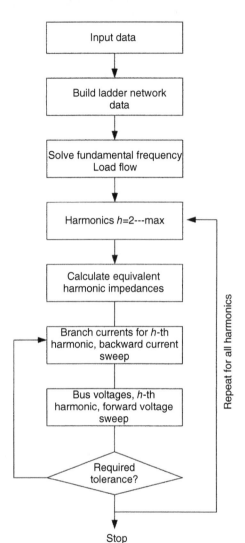

Figure 14.4   Flow chart for harmonic load flow, ladder network, forward, and backward sweeps.

The final equations for the harmonic power flow become

$$
\begin{vmatrix} \Delta W \\ \Delta I^1 \\ \Delta I^5 \\ \Delta I^7 \\ \cdots \end{vmatrix} = \begin{vmatrix} J^1 & J^5 & J^7 & \cdots & 0 \\ YG^{1,1} & YG^{1,5} & YG^{1,7} & \cdots & H^1 \\ YG^{5,1} & YG^{5,5} & YG^{5,7} & \cdots & H^5 \\ YG^{7,1} & YG^{7,5} & YG^{7,7} & \cdots & H^7 \\ \cdots & \cdots & \cdots & \cdots & \cdots \end{vmatrix} \begin{vmatrix} \Delta V^1 \\ \Delta V^5 \\ \Delta V^7 \\ \\ \Delta \alpha \end{vmatrix} \tag{14.21}
$$

where all elements in Eq. (14.21) are subvectors and submatrices partitioned from ΔM (apparent mismatches), J, and ΔU, that is, $\Delta M = J \Delta U$.

$$\Delta V^k = (V_1^k \Delta \theta_1^K, \Delta V_1^k, ..., \Delta V_n^k)^t \quad k = 1, 5, 7....$$ (14.22)

where k is the order of harmonic and θ is the phase angle.

$$\Delta \alpha = (\Delta \alpha_m, \Delta \beta_m ..... \Delta \beta_n)^t$$ (14.23)

α = the firing angle
β = the commutation parameter
ΔW = mismatch active and reactive volt–amperes

$$\Delta I^1 = (I_{r,m}^1 + g_{r,m}^1, I_{i,m}^1 + g_{i,m}^1, ..., I_{i,n}^1 + g_{i,n}^1)^t$$ (14.24)

= mismatch fundamental current, where $I_{r,m}^1$, $I_{i,m}^1$ are the real and imaginary parts of the current injection at bus m and $g_{r,m}^1$, $g_{i,m}^1$ are the real and imaginary parts of current balance equation

ΔI^k = mismatch harmonic current at kth harmonic
J^1 = conventional power flow Jacobian
J^k = Jacobian at harmonic k

$$= \left[\frac{0_{2(m-1),2n}}{\text{Partial derivatives of } P \text{ and } Q \text{ with respect to } V^k \text{and } \theta^k} \right]$$

$0_{2(m-1), 2n}$ denotes a $2(m-1) \times 2n$ array of zeros.

$$(YG)^{k,j} = \begin{array}{l} Y^{k,k} + G^{k,k} \quad (k = j) \\ G^{k,j} (k \neq j) \end{array}$$ (14.25)

where $Y^{k,k}$ is an array of partial derivatives of injection currents at the kth harmonic with respect to the kth harmonic voltage; $G^{k,j}$ are the partial derivatives of the kth harmonic load current with respect to the jth harmonic supply voltage; and H^k are the partial derivatives of nonsinusoidal loads for real and imaginary currents with respect to α and β.

$G^{k,j}$ is given by

$$\left[\begin{array}{c|c|c|c} 0_{2(m-1),2(m-1)} & 0_{2(m-1),2N} & & \\ \hline & \dfrac{\partial g_{r,m}^k}{V_m^j \partial \theta_m^j} \quad \dfrac{\partial g_{r,m}^k}{\partial V_m^j} & & \\ 0_{2N,2(m-1)} & \dfrac{\partial g_{i,m}^k}{V_m^j \partial \theta_m^j} \quad \dfrac{\partial g_{i,m}^k}{\partial V_m^j} & 0 & 0 \\ \hline & 0 & \cdots & 0 \\ \hline & & & \dfrac{\partial g_{r,n}^k}{V_n^j \partial \theta_n^j} \quad \dfrac{\partial g_{r,n}^k}{\partial V_n^j} \\ & 0 & 0 & \dfrac{\partial g_{i,n}^k}{V_n^j \partial \theta_n^j} \quad \dfrac{\partial g_{i,n}^k}{\partial V_n^j} \end{array} \right]$$ (14.26)

And H^k is given by

$$H^k = \text{diag} \begin{vmatrix} \dfrac{\partial g^k_{r,t}}{\partial \alpha_t} & \dfrac{\partial g^k_{r,t}}{\partial \beta_t} \\ \dfrac{\partial g^k_{i,t}}{\partial \alpha_t} & \dfrac{\partial g^k_{i,t}}{\partial \beta_t} \end{vmatrix} \quad \begin{array}{l} t = m, ..., n \\ k = 1, 5, 7, ... \end{array} \tag{14.27}$$

A flow chart is shown in Fig. 14.5.

Reference [8] discusses the impedance matrix method of harmonic analysis. Equations from (14.1) through (14.7) describe the concepts of impedance method.

14.5.4   A Three-Phase Harmonic Load Flow

A multiphase harmonic load flow technique is described in Ref. [14] related to the *unbalance* operation of the power system. It considers network, generator, and load equations, expressing these in general form:

$$F(|x|) = 0$$

where

$$|x| = |V \ I_V \ I_{L2} \ I_M \ I_{L3} \ E_p \ y|^t$$

$$|F| = |f_1 \ f_2 \ f_3 \ f_4 \ f_5 \ f_6 \ f_7|^t$$

$$|I_u| = |I_V \ I_{L2} \ I_M \ I_{L3} \ |^t \tag{14.28}$$

and

$|I_V|$ = the current vector from voltage sources
$|I_{L2}|$ = the vector of single-phase PQ load currents
$|I_M|$ = the vector of machine currents
$|I_{L3}|$ = the vector of static load currents
$|E_p|$ = the vector of machine internal voltages
$|y|$ = the vector of static load parameter y.

The network equation is formed as

$$|Y||V| + |I_s| + |I_u| = 0 \tag{14.29}$$

where
$|Y|$ = the node admittance matrix
$|V|$ = the vector of node voltages
$|I_s|$ = the vector of current sources leaving the node
$|I_u|$ = the vector of unknown currents leaving each node.

These equations are a set of nonlinear algebraic equations solved iteratively by rectangular coordinates using Newton–Raphson method. The load flow algorithms are not discussed in this book.

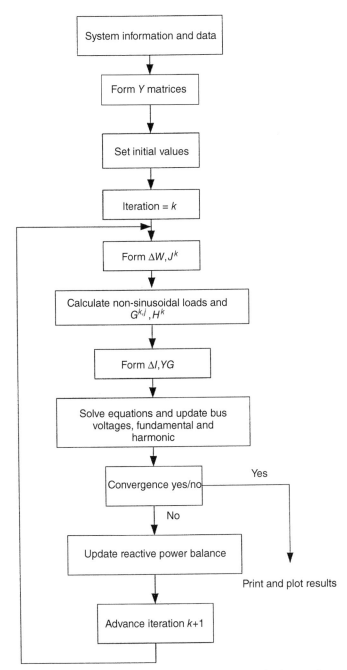

Figure 14.5   Flow chart for Newton–Raphson iterative harmonic power flow algorithm based on Refs [11,12].

14.6 TIME DOMAIN ANALYSIS

The simplest harmonic model is a rigid harmonic source and linear system impedance. A rigid harmonic source produces harmonics of a certain order with constant magnitude and phase, and the linear impedances do not change with frequencies. Multiple harmonic sources are assumed to act in isolation and the principle of superimposition applies. These models can be solved by iterative techniques. For arc furnaces and electronic converters under resonant conditions, an ideal current injection may cause significant errors. The nonlinear and time-varying elements in the power system can significantly change the interaction of the harmonics with the power system. Consider the following:

1. Most harmonic devices that produce uncharacteristic harmonics as terminal conditions are in practice not ideal, for example, converters operating with unbalanced voltages.

2. There is interaction between AC and DC quantities, and there are interactions between harmonics of different order given by switching functions (defined later).

3. Gate control of converters can interact with harmonics through synchronizing loops.

Time domain analyses have been used for transient stability studies, transmission lines, and switching transients. It is possible to solve a wide range of differential equations for the power system using computer simulation and to build up a model for harmonic calculations, which could avoid many approximations inherent in the frequency domain approach. Harmonic distortions can be directly calculated, and making use of FFT, these can be converted into frequency domain. The graphical results are waveforms of zero crossing, ringing, high dv/dt, and commutation notches. The transient effects can be calculated, for example, the part-winding resonance of a transformer can be simulated. The synchronous machines can be simulated with accurate models to represent saliency, and the effects of frequency can be dynamically simulated. EMTP is one very widely used program for simulation in the time domain.

For analysis in the time domain, a part of the system of interest may be modeled in detail. This detailed model consists of three-phase models of system components, transformers, harmonic sources, and transmission lines, and it may be coupled with a network model of lumped RLC branches at interconnection buses to represent the driving point and transfer impedances of the selected buses. The overall system to be studied is considerably reduced in size and time domain simulation is simplified.

14.7 SENSITIVITY METHODS

The purpose is to ascertain the sensitivity of the system response when a component parameter varies. Adjoint network analysis can be used. The network N consisting of linear passive elements is excited at the bus of primary interest by a unit current at the harmonic source bus, and branch currents I_1 I_2, ... , I_n are obtained. The transfer impedance T is defined as the voltage output across the bus of primary interest divided by the harmonic current of the input bus. The adjoint network N^*, which has the same topology as the original network, is excited by a unit current source from the output to obtain adjoint network branch currents $I_1^*, I_2^*, \ldots I_n^*$. The sensitivity of a transfer impedance T is defined as

$$S_x^T = \frac{\partial T}{\partial x} \left(\frac{x}{T} \right) \qquad (14.30)$$

where x is any parameter R, L, or C at frequency denoted by S_x^T. The sensitivities can be calculated using

$$\frac{\partial T}{\partial x} = I_x \cdot I_x^* \qquad (14.31)$$

where I_x and I_x^* are x element branch currents from two-network analysis of N and N^*, respectively (Fig. 14.6(a)). The efficacy of the method is limited to small variations in the parameter. When large variations occur in external system equivalents, these can cause serious changes in the transfer function and the bilinear theorem can be applied. These large changes in the transfer function of a two-port network to changes in an internal parameter are analyzed by pulling out Z of the network, effectively forming a three-port network (Fig. 14.6(b)). For the transfer impedance, the following equation is obtained:

$$T = \frac{V_2}{I_1} = \frac{Z_{xin} T(0) + ZT(\infty)}{Z + Z_{xin}} \qquad (14.32)$$

where $T(0)$ is the transfer impedance when $Z = 0$, $T(\infty)$ is the transfer impedance when $Z = \infty$ (open circuited), and $Z_{x\,in}$ is the input impedance looking into the network from the nodes of Z [18].

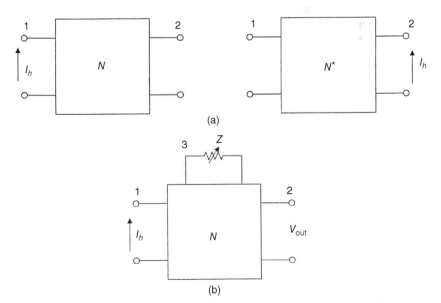

Figure 14.6   (a) Two-port N networks with a source and its adjoint network, (b) three-port network.

14.8 UNBALANCED AC SYSTEM AND HVDC LINK

In HVDC systems, problems arise because of harmonics fed into the AC system as a result of converter operation. Noncharacteristic harmonics can lead to undesirable results because filter circuits are installed for the characteristic harmonics of the HVDC converter. The HVDC systems can excite subsynchronous oscillations of large turbogenerators as a result of harmonic transfer through the DC interconnection. Referring to Fig. 14.7, the steady-state operation of HVDC interconnection can be written as

$$F(\chi, V, \theta) = 0 \tag{14.33}$$

where the vector χ contains 26 unknowns and can be written as

$$\chi = \begin{bmatrix} E_s^1, E_s^2, E_s^3, \varphi_s^1, \varphi_s^2, \varphi_s^3, I_s^1, I_s^2, I_s^3, \omega_s^1, \omega_s^2, \omega_s^3 \\ U_{12}, U_{23}, U_{31}, C_1, C_2, C_3, \alpha_1, \alpha_2, \alpha_3, a_1, a_2, a_3, V_d, I_d \end{bmatrix} \tag{14.34}$$

where

$E_s < \phi_s$ = fundamental frequency voltage at the secondary side of the rectifier transformer

$U_{12} < C_1$ = phase-to-phase voltage of the *converter referred to the secondary side*, and C_1 is zero crossing for the timing of firing pulses

a_1 = off-nominal tap ratio of the primary side of the transformer

α_1 = firing delay angle measured from the respective zero crossing

$I_s < \omega_s$ = fundamental frequency line current at the transformer secondary

V_d = average DC voltage

I_d = average DC current.

$V < \theta$ = vector of three converter voltage arguments.

Similar set of equations apply to inverter also.

For three-phase load flow, the system admittance matrix should include the effect of each separate phase as well as the coupling between phases. The active and

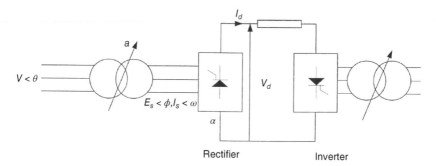

Figure 14.7 Interaction of AC and DC harmonics in a converter, equivalent circuit.

reactive powers are specified separately for each phase:

$$(\Delta P_{\text{gen}})_j = (P_{\text{gen}})_j^{\text{SP}} - (P_{\text{gen}})_j$$

$$(\Delta V_{\text{reg}})_j = f(V_j^1, V_j^2, V_j^3) = 0 \qquad (14.35)$$

The superscript "SP" denotes specified values.
For $j = 1, \ldots, n_g - 1$, where n_g is the number of generators at slack generator internal buses,

$$(\Delta V_{\text{reg}})_{\text{SL}} = f(V_{\text{SL}}^1, V_{\text{SL}}^2, V_{\text{SL}}^3) = 0 \qquad (14.36)$$

where superscript "SL" specifies slack bus.
At every generator terminal bus and load bus:

$$\Delta P_i^P = (P_i^P)^{\text{SP}} - (P_i^P)^{\text{AC}} = 0$$

$$\Delta Q_i^P = (Q_i^P)^{\text{SP}} - (Q_i^P)^{\text{AC}} = 0 \qquad (14.37)$$

where superscript "P" is number of phases, and $i = 1, \ldots n_b$, where n_b is the number of buses.

These equations are solved with fast decoupled Newton–Raphson method, commonly used for fundamental frequency load flow.

For each converter, a set of equations is written that includes 26 variables, Eq. (14.34).

Eq. (14.37) is modified at the converter terminals:

$$\Delta P_i^P = (P_i^P)^{\text{SP}} - (P_i^P)_{\text{AC}} - (P_i^P)_{\text{DC}} = 0$$

$$\Delta Q_i^P = (Q_i^P)^{\text{SP}} - (Q_i^P)_{\text{AC}} - (Q_i^P)_{\text{DC}} = 0 \qquad (14.38)$$

where the DC terms are functions of AC terminal conditions. A five-step flow chart is shown in Fig. 14.8.

14.9   HYBRID FREQUENCY AND TIME DOMAIN CONCEPT

A hybrid three-phase harmonic load flow is proposed in Ref. [23].

- Electronic devices, such as converters, are represented in time domain in a flexible way. Input data of these devices is given in input PSpice-like format and state equations are formed.
- Time domain steady-state solution is obtained using an accelerating technique that takes advantage of piecewise linear nature of devices.
- The working point is found by taking into account the power and voltage specifications.

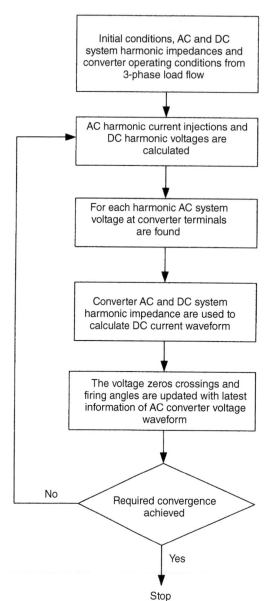

Figure 14.8   Flow chart for three-phase harmonic load flow considering AC and DC interaction in a converter.

- Linear elements are modeled in the frequency domain. Equivalents of the grid for nonlinear elements are used.
- Harmonic interactions, that is, influence of voltage harmonics on the current harmonic production of nonlinear devices is solved using a full Newton method.

A fast three-phase load flow method is described in Ref. [24]. It is based on sequence decoupling load flow method: sequence decoupling compensation Newton–Raphson (SDCNR) and sequence decoupling compensation fast-decoupled (SDCFD) methods. Only SDCNR is described. The mathematical model used is

$$J_1 \Delta V_1 = \Delta S_1$$

$$Y_2 V_2 = I_2$$

$$Y_0 V_0 = I_0 \tag{14.39}$$

where J_1 is the positive sequence Jacobian, such as fundamental frequency load flow, and ΔS_1 is the mismatch of vector of positive sequence powers. Positive sequence currents injected into asymmetrical line terminals buses or shunt elements are transformed into active and reactive powers that are included in ΔS_1. ΔV_1 is the error vector of positive sequence voltage and phase angles. Vectors I_2 and I_0 include compensation currents injected into PQ buses, with linear loads and asymmetrical lines or unbalanced shunt elements, using their sequence models (Chapter 12). In an iterative Newton–Raphson algorithm, the sequence voltages are used to calculate the phase voltages, which allow updating of positive sequence active and reactive powers and current injections of negative and zero sequences for the loads, and updating of compensation currents due to unbalanced elements. Negative and zero sequence bus impedance matrices are invariant and need not be updated after each iteration.

The above equations can be replaced with

$$Y_2 \Delta V_2 = \Delta I_2$$

$$Y_0 \Delta V_0 = \Delta I_0$$

$$\Delta I_2(k + 1) = I_2(k + 1) - I_2(k)$$

$$\Delta I_0(k + 1) = I_0(k + 1) - I_0(k) \tag{14.40}$$

Similarly, voltages can be updated

$$\Delta V_2(k + 1) = V_2(k + 1) - V_2(k)$$

$$\Delta V_0(k + 1) = V_0(k + 1) - V_0(k) \tag{14.41}$$

In the hybrid interaction analysis algorithm, the following nonlinear equation is solved:

$$I_{eq} = Y_{eq} V_{NL} = I_{NL} \tag{14.42}$$

where
V_{NL} = vector of harmonic phase voltages at nonlinear buses
Y_{eq} = three-phase harmonic admittance matrix of Norton equivalent
I_{eq} = vector of harmonic current sources
I_{NL} = current demanded by nonlinear device dependent on harmonic voltage.

The iterative equation to be solved is

$$|Y_{eq} + W^{(i)}|\Delta V_{NL}^{(i+1)} = I_{eq} - Y_{eq}V_{nL}^{(i)} - F(V_{NL}^{(i)}) \tag{14.43}$$

where W is a sensitivity matrix between different harmonic variables:

$$W^{(i)} = \frac{dF(V_{NL})}{dV_{nL}}\Bigg|_{1 \text{ to } i} \tag{14.44}$$

This matrix is composed of submatrices $W_{km}^{(i)}$, which represent sensitivity sub-matrix of harmonic k of the current with respect to harmonic m of voltage at iteration i. A formulation of the sensitivity matrix for periodically excited piecewise linear circuits is

$$W_{km}^{(i)} = \frac{\partial I_{NL}(k)}{\partial V_{NL}(m)}\Bigg|_{i} = \frac{1}{N}\sum_{n=0}^{N-1} Y_n(m)\exp\{-j(2\pi/N)n(k-m)\} = P_{k-m}(m) \tag{14.45}$$

where N is the number of samples of currents and voltages equally spaced along a period.

$P_{k-m}(m)$ can be defined as the harmonic term of order $(k - m)$ of the evaluation with the time of the harmonic admittance matrix of order m of the nonlinear system in the steady state. These can be found by FFT. All harmonics in Eq. (14.45) are coupled. A flow chart of the calculations is shown in Fig. 14.9.

14.10 PROBABILISTIC CONCEPTS

The harmonic phasors can be represented in polar of Cartesian coordinates, $x + jy$. The x and y are dependent on each other (current and voltage harmonics), and their distribution is described by a joint probability distribution function (jpdf), $p_{XY}(x, y)$, which is related to joint cumulative distribution $P_{XY}(x, y)$ by

$$p_{XY}(x, y) = \frac{\partial^2 P_{XY}(x, y)}{\partial x \partial y} \tag{14.46}$$

The characteristic measures include five parameters: the mean value and standard deviation of x and y, $(\mu_x, \mu_y, \sigma_x, \sigma_y)$, and joint second moment about the mean value (Chapter 7) or covariance defined by

$$\sigma_{xy}^2 = \int_{-\infty}^{\infty}\int_{-\infty}^{\infty} xy p_{XY}(x, y)dxdy - \mu_x\mu_y \tag{14.47}$$

The correlation coefficient ρ that measures the dependence between x and y is defined as

$$\rho = \frac{\sigma_{xy}^2}{\sigma_x\sigma_y} \tag{14.48}$$

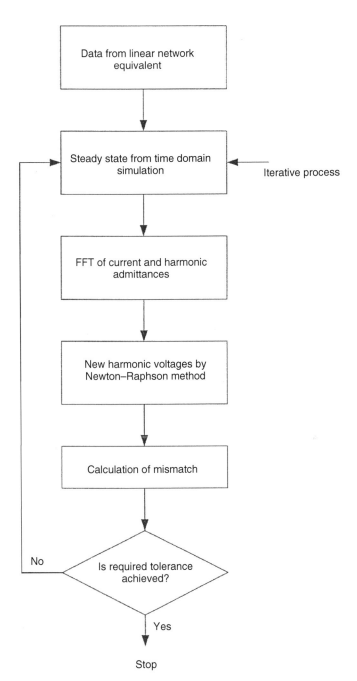

Figure 14.9   Hybrid frequency and time domain three-phase harmonic load flow chart.

The marginal pdf of one of the projections of the phasor is calculated by integrating its jpdf over the other variable. Consider a phasor with fixed magnitude a, while its angle varies with a uniform distribution between 0 and 2π. The marginal pdf of x or y is given by

$$pv(u) = \frac{1}{\pi\sqrt{a^2 - u^2}}, \quad u = x, y \ for \ |u| < a \tag{14.49}$$

and $pv(u) = 0$ elsewhere. If the magnitude also varies randomly with a uniform distribution between 0 and maximum value, then pdf of $x - y$ components become

$$pv(u) = \frac{1}{\pi a}\ln\left(\frac{a + \sqrt{a^2 - u^2}}{|u|}\right), \quad u = x, y \ a > u > -a \tag{14.50}$$

If we consider a sum of n independent phasors of same frequency:

$$a = \sum_i^n a_i = \left(\sum_i^n x_i\right) + j\left(\sum_i^n y_i\right) \tag{14.51}$$

The pdf of x and y are obtained by convolution:

$$pv(u) = pv_1(u_1) * pv_2(u_2) * \ldots pv_n(u_n) \tag{14.52}$$

Another analytical solution method is to find Fourier transform $F_i(s)$ of each pdf and take inverse of these functions, that is,

$$pv(u) = F^{-1}\left(\prod_1^n F_i(s)\right) \tag{14.53}$$

Both the convolution and Fourier transform methods are known to be difficult to solve analytically and numerically. An alternate method is Monte Carlo simulation. It is a repeated process to generate a deterministic solution corresponding to a set of deterministic values of random variables. Thousands of such simulations are required, which are not a problem with modern fast computers.

If the phasors to be added are large and no single phasor is dominant, an accurate solution can be found by *central limiting theorem* of probability; which states that the sum approaches a normal distribution regardless of the distributions of individual variables, so long as the number of variables is sufficiently large.

Practically, analytical expressions describing joint distributions of harmonics are not so simple and easily available as actual distributions are spread over complex surfaces and in nonuniform manner. The data available generally include scatter plots, from which one can extract simple distributions that fit standard functions such as elliptical shapes. When correlations exist between the real and imaginary parts of an elliptical distribution, a multivariable expression for the ellipse needed as a model is

$$\frac{1}{1-\rho^2}\left[\left(\frac{x-\mu_x}{\sigma_x}\right)^2 - 2\rho\left(\frac{x-\mu_x}{\sigma_x}\right)\left(\frac{y-\mu_y}{\sigma_y}\right) + \left(\frac{y-\mu_y}{\sigma_y}\right)^2\right] = 1 \qquad (14.54)$$

There are several publications that have analyzed a cluster of specific loads.

As discussed previously, the estimation of harmonic voltages in linear systems, where the bus voltages are not distorted more than 5%, is relatively easily. This is the most common case. In cases where harmonic current depends on harmonic voltage, an analytical expression for the two signals is needed, for example, nonlinear harmonic algorithms as discussed in previous sections. Such studies have been extended to probabilistic ones. Reference [25] may be seen for further discussions. This reference also describes 5th harmonic voltage calculations at a particular bus in a 14-bus system, which are shown in Fig. 14.10. The solid curve is Monte Carlo simulation, while the other two curves illustrate effect of transfer impedance on pdf of harmonic voltage. Although the curves are approximately Gaussian, both the mean and standard deviation are sensitive to the relative magnitude of transfer function.

A flow chart for the probabilistic Monte Carlo simulation is shown in Fig. 14.11 [26,27]. For each random input datum, a value is generated according to its pdf.

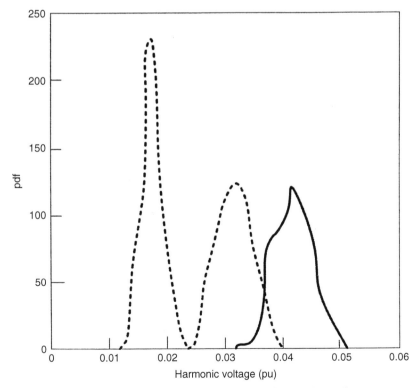

Figure 14.10   Pdf of magnitude of sum of phasors and independent polar components. Source: Ref. [26].

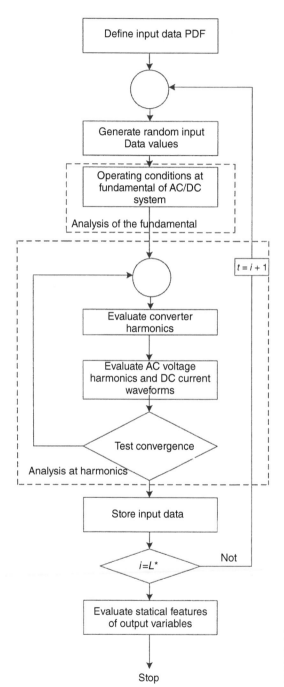

Figure 14.11 Harmonic load flow using probabilistic concepts, Monte Carlo simulation. Source: Ref. [28].

According to these values, the operating conditions at fundamental frequency of AC/DC power systems are first evaluated. Then, the analysis at harmonic frequencies is performed evaluating iteratively converter harmonics and the AC system voltages together with the DC current waveforms. Once the convergence is reached, the data are stored.

The procedure is repeated a certain number of times, minimum L^* to obtain a good estimate of the probability of the output.

The error in estimating probability Π of a given event by the relative frequency f_L of the occurrences of that event in L independent trails can be written as

$$\text{prob}\{|f_L - \Pi|\} > \varepsilon^* < \lambda^* \tag{14.55}$$

where the error ε^* is bound to be lower than a given value λ^*.

With the central limit theorem:

$$L > [\Pi(1 - \Pi)][\varphi^{-1}(\lambda^*/2)]^2/(\varepsilon^*)^2 \tag{14.56}$$

where

$$\varphi(z) = \frac{1}{\sqrt{2\pi}} \int_{-\infty}^{z} e^{-(u^2/2)} du \tag{14.57}$$

φ^{-1} is inverse of the above expression.

The right-hand side of Eq. (14.56) depends on Π, which is unknown. If a value of ½ is given to Π, which maximizes the product $\Pi(1 - \Pi)$, then the limit L^* is

$$L^* = [\varphi^{-1}(\lambda^*/2)]^2/(2\varepsilon^*)^2 \tag{14.58}$$

14.11 COMPUTER-BASED PROGRAMS

The computer-based programs vary in their capabilities. The desirable features are as follows:

- Three-phase modeling
- Modeling of firing angles and overlap angles for converters that may differ from phase to phase
- Frequency-dependent modeling of power system components
- Capability of modeling of SVCs, TCRs, FACTS controllers, and HVDC
- Systems with multiple swing buses, capability of handling looped, radial, and isolated subsystems. System where some buses may remain de-energized or may be de-energized to study a changed configuration
- Multiple loading and generation capabilities and diversity factors
- Fundamental frequency load flow calculations, automatic adjustments of load tap changers on transformers
- Effects of machine and transformer winding connections and their grounding systems

- Short- and long-line models of transmission lines
- Modeling of positive, negative, and zero sequence harmonics, up to at least 71st.
- Generation of harmonic sources, user selectable
- Calculations of various harmonic indices as per IEEE 519, TDD, voltage distortion, total arithmetic sum value, TIF, IT product, and so on for the buses and branch currents
- Frequency, voltage scans, and powerful harmonic flow algorithms such as Newton–Raphson
- Capability to model various filter types, some built in library of filters with their economic analysis. Check and verify filter loading and loading of filter components, current spectra through filters
- Capability to plot waveforms, frequency scans, voltage scans, phase angles, harmonic voltages, and currents anywhere in the system
- Detailed text reports
- Ascertaining transformer K factor ratings based on the harmonic loading
- Online display of capacitor and filter currents
- Detailed printouts of the results. The violations could be flagged for immediate attention and recalculations.

14.12   HARMONIC ANALYSES OF A LARGE INDUSTRIAL SYSTEM

Figure 14.12 shows an industrial system with 6-pulse ASD load and low-voltage and medium-voltage AC motor loads. The harmonic producing load is a large percentage of the total load. The loads are connected to transformers of equivalent ratings. A 50-MW generator operates in synchronism with a utility tie transformer, which is provided with ULTC (under load tap changing equipment). The output of the generator is constrained to no more than 40 MW due to steam utilization in the process plant. It is required that the system should operate at a power factor higher than 0.85 as measured on the 138 kV side and the harmonic emission should meet IEEE 519 [28] requirements at 138-kV bus 1 designated as PCC.

14.12.1   Objectives of Study

It is very common that more than single objective is required to be met, for example, in this study – the control of both power factor and harmonic emission is required.

The impedance data of the components are not shown in Fig. 14.12. The short-circuit currents at the PCC should be calculated to apply IEEE criteria of TDD at PCC. The calculation shows that

- Short circuit at 138 kV = 31.1 kA rms symmetrical
- Short circuit at 13.8 kV bus = 39.2 kA with generator in operation.

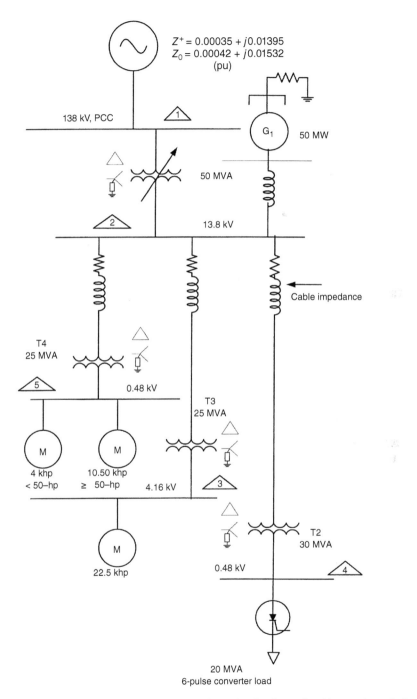

Figure 14.12   An industrial system configuration for the study of harmonic analysis.

The fundamental frequency load flow situation is depicted in Fig. 14.13, with G1 in service. The source (138 kV) power factor is only 61.9%. The total load demand is 47.52 MW (61.54 MVA) at a power factor of 77.43 lagging. The drive system non-linear load of 6-pulse is 16.28 MW at 80% lagging power factor. Approximately, 6.10 Mvar losses occur in the system. It is imperative that the correct operating load power factors are modeled in the load flow study. Figure 14.13 shows that the generator operates at 5% higher than rated voltage. If the generator is operated at its rated voltage, its reactive power output will be reduced and the power factor at PCC will be only 22%. According to ANSI/IEEE standards [29], a generator can be operated at ±5% variations with respect to its rated voltage subject to some qualifications. By suitable tap settings on the transformers, the voltages throughout the distribution can be maintained close to the rated voltages.

We should also consider the outage of the generator; this load flow is shown in Fig. 14.14.

Operating voltages on the buses are adjusted by ULTC on the utility tie transformer. The source power factor at the 138-kV bus is now 75.14%. To improve the power factor to the desired level of more than 85%, a capacitor bank of 12 MVA is required at the 13.8-kV bus. The load flow with the capacitor bank connected in service and without and with generator in service is shown in Figs 14.15 and 14.16, respectively.

The power factor at 138 kV is

- 86% lagging when generator is out of service
- Almost unity when generator is in service.

The utilities may not accept the leading power generation from the consumer. Note that there are two voltage regulating elements, ULTC on the transformer and voltage regulator of the generator, see Ref. [30], for reactive power control.

These can hunt with respect to each other if the voltage in the system varies due to load change or change in the utility source voltage. The ULTC is put as a secondary voltage control, allowing the generator voltage regulator to take precedence; sometimes, ULTC is put under manual control and the generator voltage regulator is solely allowed to regulate the voltage. ULTC is used manually in cases of sudden loss of generator.

These studies are required before harmonic analysis study is conducted.

14.12.2 Harmonic Emission Model

Figure 14.17 shows the waveform, and Fig. 14.18 shows the spectrum of the 6-pulse load harmonics modeled on secondary of the transformer T2. Table 14.1 shows the spectrum in tabulated form.

14.12.3 Harmonic Propagation – Case 1

The harmonic load flow is carried out *with no* 12-Mvar capacitor on 13.8-kV bus and generator G1 in operation. First, a frequency scan is conducted, and the results of the impedance modulus and angle up to 50th harmonic are plotted in Figs 14.19 and 14.20, respectively. These are straight lines and show no resonance.

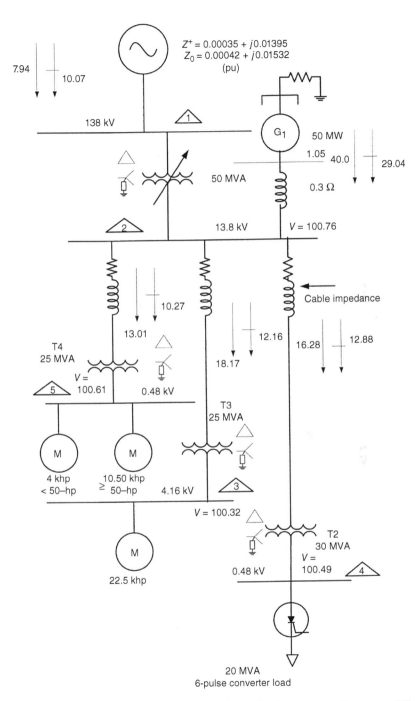

Figure 14.13   Load flow with generator G1 in service, in system configuration of Fig. 14.12.

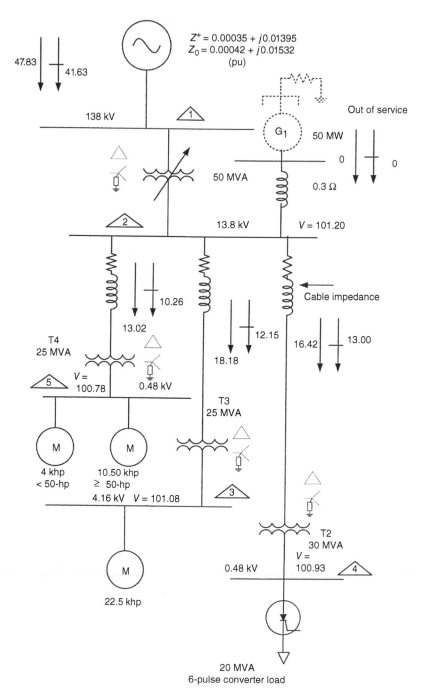

Figure 14.14   Load flow with generator G1 out of service, in system configuration of Fig. 14.12.

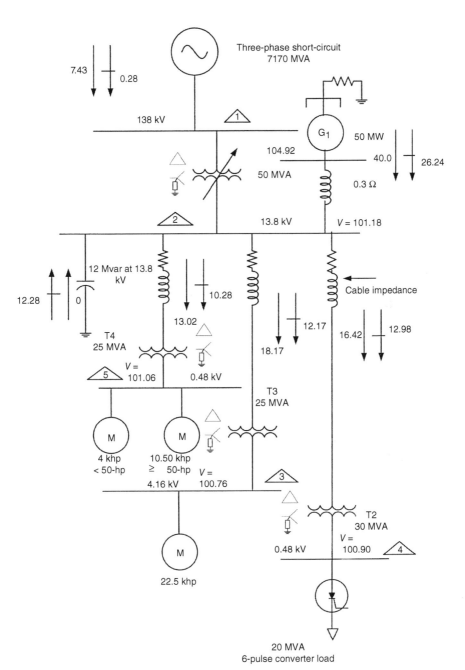

Figure 14.15   Load flow with 12.0-Mvar capacitor bank and generator G1 in service, in system configuration of Fig. 14.12.

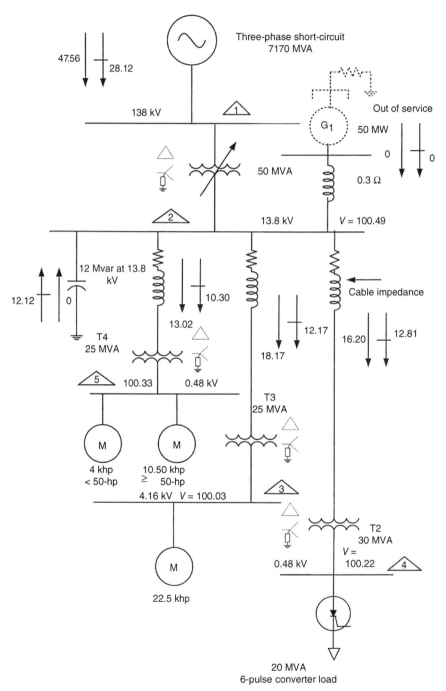

Figure 14.16 Load flow with 12-Mvar capacitor bank and generator G1 out of service, in system configuration of Fig. 14.12.

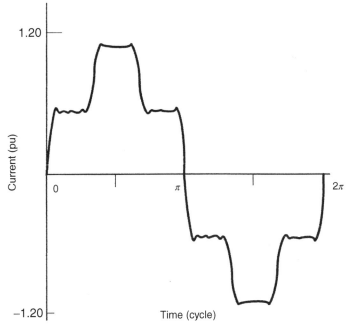

Figure 14.17   Line current waveform of the six-pulse nonlinear load, industrial system for harmonic analysis.

Generally, no resonance with load-generated harmonics is expected even in large distribution systems. Yet, stray system capacitances are modeled because when capacitors are provided, these impact the resonance frequency and harmonic filter tuning. Long transmission lines are an exception.

The spectrum of the harmonics at PCC is shown in Fig. 14.21 and its waveform in Fig. 14.22.

Table 14.2 shows harmonic current flows throughout the system, in percentage of the fundamental frequency current. The TDD at PCC is 26.22%. For the 138-kV system with short-circuit level of 31 kA and load demand = 46.34A, $I_s/I_r = 669$. Referring to Table 10.2, the permissible TDD = 7.5%. Also distortion at a number of harmonics is beyond the permissible levels in IEEE 519.

Table 14.3 compiles harmonic voltages at PCC and at 13.8-kV bus 2, again in percentage of the nominal voltages. It is interesting to note that the voltage distortion (THD_V) at PCC (138 kV) is only 0.31%, though the TDD is high, 26.22%. This is dependent on the stiffness of the utility system. See Chapter 16, case study 3, where high-voltage distortion occurs in a weak utility system, though the TDD is relatively small. THD_V at 13.8-kV bus is 4.41%. Figure 14.23 depicts the voltage waveforms at PCC and also at 13.8-kV bus.

14.12.4   Harmonic Propagation, Case 2

This case conducts harmonic analysis with 12-Mvar capacitor bank in service on 13.8-kV bus 2 and the generator in operation. First, a frequency scan is conducted

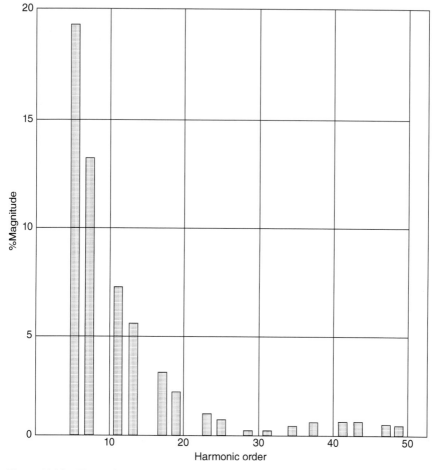

Figure 14.18 Harmonic spectrum of the six-pulse nonlinear load, industrial system for harmonic analysis.

and Fig. 14.24 shows resonance close to 7th harmonic, around 518 Hz. Table 14.4 shows the impedance modulus and parallel resonance frequencies, and Table 14.5 is the printout of a typical frequency scan.

The results of the harmonic load flow are in Tables 14.6 and 14.7. Figures 14.25 and 14.26 are the current spectrum and its waveform at the PCC. Note that due

TABLE 14.1 Harmonic Spectrum of the 6-Pulse Nonlinear Load (% of Fundamental)

Harmonic	5	7	11	13	17	19	23	25
Magnitude (%)	19.10	13.10	7.20	5.60	3.30	2.40	1.20	0.80
Harmonic	29	31	35	37	41	43	47	49
Magnitude (%)	0.20	0.20	0.40	0.50	0.50	0.40	0.40	0.40

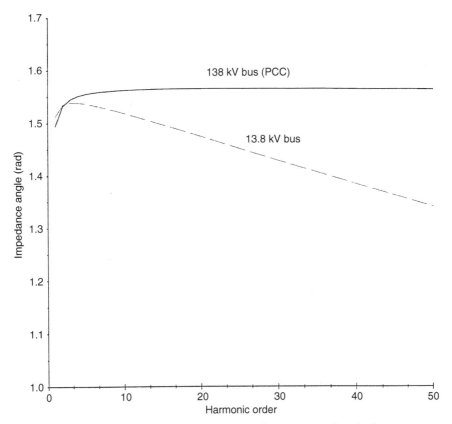

Figure 14.19   Frequency scan, impedance angle (radians), no capacitors in the system, industrial system for harmonic analysis.

to resonance the 7th harmonic current has risen to 53.8% of the fundamental. The TDD at PCC has gone up from 26.22% to 69.77%. The voltage distortion at PCC goes up from 0.31% to 0.53% and at bus 2 from 4.41% to 7.38%. The application of capacitors has seriously increased the distortion throughout the distribution system.

14.12.5   Harmonic Propagation, Case 3

Case 3a   The 12-Mvar capacitor bank at 13.8-kV bus is turned into an ST harmonic filter tuned to 4.85th harmonic and *generator G1 is out of service*. (This may be too large for switching transient considerations, and practically, it may have to be split into two filter banks of 6 Mvar each). Consider 8.20-kV, 400-kvar capacitor units, 11 units in parallel per phase. This gives 4.034 Mvar per phase and a total three-phase wye-ungrounded connected bank size of 12.10 Mvar. The TDD at PCC is reduced to 2.88%. Furthermore, the harmonic distortions at each of the harmonics meet the

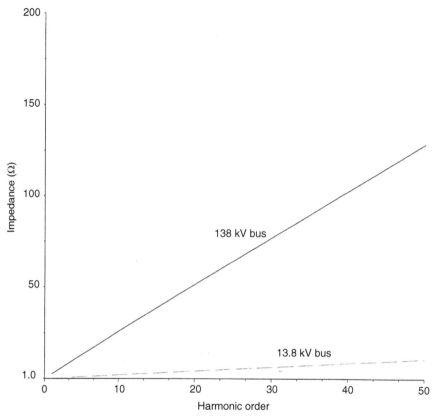

Figure 14.20 Frequency scan impedance modulus, no capacitors in the system; industrial system for harmonic analysis.

IEEE 519 limits, calculated as follows:

$I_s/I_r = 125.7$ at the PCC. As per IEEE 519, the permissible TDD is 7.5%. The distortion at the individual frequencies is

- $< 11 : 6\%$
- $11 \leq h \leq 17 \; 2.75\%$
- $17 \leq h \leq 23 \; 2.5\%$
- $23 \leq h \leq 35 \; 1.0\%$
- $35 \leq h \; 0.05\%$

The harmonic spectrum in Fig. 14.27 confirms that these limits are met. Also Fig. 14.28 shows the current waveform at the PCC. The shifted parallel resonance frequency is 271.9 Hz.

Table 14.8 shows the harmonic currents at PCC, in percentage and also in amperes at PCC and through the filter, amply illustrating the efficacy of the filter to shunt a major portion of the 5th harmonic. Table 14.9 plots the harmonic voltages at PCC and at 13.8-kV bus. Harmonic voltage distortion at PCC is 0.24%.

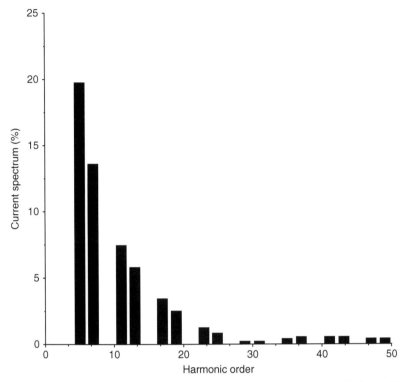

Figure 14.21   Spectrum of the harmonics injected into the secondary windings of the utility tie transformer of 50-MVA industrial system for harmonic analysis.

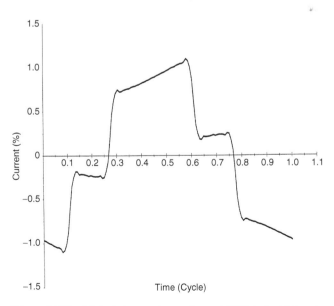

Figure 14.22   Distorted current waveform at PCC, industrial system for harmonic analysis.

TABLE 14.2 Harmonic Current Flows (% of Fundamental), with 12 Mvar Capacitor Bank, and Generator G1 in Service

ID	FundA	5th	7th	11th	13th	17th	19th	23rd	25th	29th	31st	35th	37th	41st	43rd	47th	49th
PCC	46.34	19.80	13.58	7.46	5.80	3.42	2.49	1.24	0.83	0.21	0.21	0.41	0.52	0.52	0.52	0.41	0.41
T4	931.60	1.88	1.29	0.71	0.55	0.32	0.24	0.12	0.08	0.02	0.02	0.04	0.05	0.05	0.05	0.04	0.04
T3	706.17	1.72	1.18	0.65	0.51	0.30	0.22	0.11	0.07	0.02	0.02	0.04	0.05	0.05	0.05	0.04	0.04
T2	842.06	18.98	13.02	7.15	5.56	3.28	2.38	1.19	0.79	0.20	0.20	0.40	0.50	0.50	0.40	0.40	0.40
Gen React	2033.79	1.89	1.30	0.71	0.56	0.33	0.24	0.12	0.08	0.02	0.02	0.04	0.05	0.05	0.05	0.04	0.04

TDD at PCC = 26.22%.

TABLE 14.3 Harmonic Voltages as % of Nominal Bus Voltage

ID	FundKV	5th	7th	11th	13th	17th	19th	23rd	25th	29th	31st	35th	37th	41st	43rd	47th	49th
Bus 1	138	0.15	0.15	0.13	0.12	0.09	0.07	0.04	0.03	0.01	0.01	0.02	0.03	0.03	0.03	0.03	0.03
Bus 2	13.8	2.13	2.04	1.77	1.63	1.26	1.02	0.62	0.45	0.13	0.14	0.32	0.42	0.47	0.49	0.43	0.45

THD (voltage) bus 1 = 0.31%, bus 2 = 4.41%.

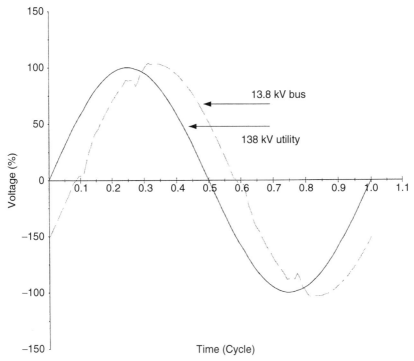

Figure 14.23   Voltage waveforms at 138-kV utility and 13.8-kV buses, industrial system for harmonic analysis.

Case 3b The harmonic load flow is repeated with generator in service and 5th harmonic filter as in case 3a in service. *This is the normal operating condition.*

The harmonic spectrum in Fig. 14.29 shows that the TDD at PCC has increased manifolds from 2.88% in case 3a to 17.18%. Also the waveform of current at PCC, Fig. 14.30, looks much distorted.

Note that when the generator is in service, the load demand from the PCC is reduced from 225.90 to 32.10 A. Here, $I_s / I_r = 971$, but the same limits of distortion as in case 3a are applicable. Thus, *relatively* the harmonics as a percentage of load demand increase, and there is no other change in the distribution system or the ST filter at 13.8-kV bus.

Table 14.10 shows that when the results are compared *in amperes* with that of Table 14.8 for current injections at PCC, these are *even lower* when the generator is not in service. Also the voltage distortions at PCC and 13.8-kV bus are lower as compared to case 3a (Table 14.11).

When the loads are mainly served from the plant generators and the utility source acts as a standby for transfer of loads in case of failure of plant generators, it is hard to control the TDD to IEEE 519 limits: specially so if the utility is a stiff source. There can be zero load flow from the utility source, or in cogeneration mode some power can be supplied into the utility system.

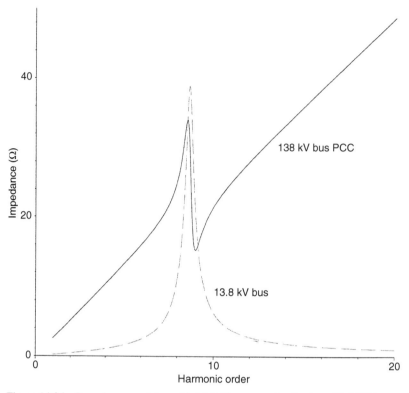

Figure 14.24 Impedance modulus with 12.0-Mvar capacitor bank on 13.8-kV bus and G1 in service; industrial system for harmonic analysis.

TABLE 14.4 Impedance Modulus and Shifted Resonance Frequencies

Bus	Z, mag (Ω)	Harmonic	Hz
1	34.10	8.53	512
2	38.92	8.63	518
3	2.29	8.63	518
4	0.05	8.63	518
5	0.04	8.63	518

The harmonic spectrum in Fig. 14.29 and its waveform in Fig. 14.30 suggest that additional 7th and 11th ST filters are required.

The harmonic analysis is an iterative process. All these iterations are not documented. The required objective is obtained by

- Providing a 0.25 Ω, 2500 A reactor on the 13.8-kV side of the utility tie transformer. This increases the impedance to the harmonic power flow to PCC.

TABLE 14.5 Frequency Scan, a Specimen Computer Printout

Bus 2

Freq.	Mag.	Angle	Freq.	Mag.	Angle	Freq.	Mag.	Angle	Freq.	Mag.	Angle
492.00	15.96	1.14	494.00	17.06	1.11	496.00	18.28	1.08	498.00	19.66	1.03
500.00	21.22	0.99	502.00	22.97	0.93	504.00	24.93	0.87	506.00	27.10	0.79
508.00	29.46	0.71	510.00	31.94	0.60	512.00	34.40	0.48	514.00	36.58	0.34
516.00	38.19	0.19	518.00	38.92	0.03	520.00	38.63	−0.13	522.00	37.40	−0.29
524.00	35.47	−0.43	526.00	33.17	−0.56	528.00	30.76	−0.67	530.00	28.42	−0.76
532.00	26.23	−0.84	534.00	24.25	−0.90	536.00	22.47	−0.96	538.00	20.88	−1.01
540.00	19.47	−1.05	542.00	18.22	−1.09	544.00	17.10	−1.12	546.00	16.10	−1.15
548.00	15.21	−1.18	550.00	14.40	−1.20	552.00	13.67	−1.22	554.00	13.01	−1.24
556.00	12.41	−1.25	558.00	11.86	−1.27	560.00	11.36	−1.28	562.00	10.89	−1.29
564.00	10.47	−1.30	566.00	10.07	−1.32	568.00	9.71	−1.32	570.00	9.37	−1.33
572.00	9.05	−1.34	574.00	8.76	−1.35	576.00	8.48	−1.36	578.00	8.22	−1.36
580.00	7.98	−1.37	582.00	7.75	−1.38	584.00	7.53	−1.38	586.00	7.33	−1.39
588.00	7.14	−1.39	590.00	6.95	−1.40	592.00	6.78	−1.40	594.00	6.62	−1.41
596.00	6.46	−1.41	598.00	6.31	−1.41	600.00	6.17	−1.42	602.00	6.04	−1.42
604.00	5.91	−1.42	606.00	5.78	−1.43	608.00	5.67	−1.43	610.00	5.55	−1.43
612.00	5.44	−1.44	614.00	5.34	−1.44	616.00	5.24	−1.44	618.00	5.14	−1.44
620.00	5.05	−1.45	622.00	4.96	−1.45	624.00	4.88	−1.45	626.00	4.80	−1.45
628.00	4.72	−1.45	630.00	4.64	−1.46	632.00	4.57	−1.46	634.00	4.49	−1.46
636.00	4.42	−1.46	638.00	4.36	−1.46	640.00	4.29	−1.47	642.00	4.23	−1.47
644.00	4.17	−1.47	646.00	4.11	−1.47	648.00	4.05	−1.47	650.00	4.00	−1.47
652.00	3.94	−1.47	654.00	3.89	−1.48	656.00	3.84	−1.48	658.00	3.79	−1.48
660.00	3.74	−1.48	662.00	3.70	−1.48	664.00	3.65	−1.48	666.00	3.61	−1.48
668.00	3.56	−1.48	670.00	3.52	−1.49	672.00	3.48	−1.49	674.00	3.44	−1.49
676.00	3.40	−1.49	678.00	3.36	−1.49	680.00	3.33	−1.49	682.00	3.29	−1.49
684.00	3.25	−1.49	686.00	3.22	−1.49	688.00	3.19	−1.49	690.00	3.15	−1.49
692.00	3.12	−1.50	694.00	3.09	−1.50	696.00	3.06	−1.50	698.00	3.03	−1.50
700.00	3.00	−1.50	702.00	2.97	−1.50	704.00	2.94	−1.50	706.00	2.91	−1.50
708.00	2.89	−1.50	710.00	2.86	−1.50	712.00	2.83	−1.50	714.00	2.81	−1.50
716.00	2.78	−1.50	718.00	2.76	−1.50	720.00	2.73	−1.51	722.00	2.71	−1.51
724.00	2.69	−1.51	726.00	2.66	−1.51	728.00	2.64	−1.51	730.00	2.62	−1.51
732.00	2.60	−1.51	734.00	2.58	−1.51	736.00	2.56	−1.51	738.00	2.54	−1.51
740.00	2.52	−1.51	742.00	2.50	−1.51	744.00	2.48	−1.51	746.00	2.46	−1.51
748.00	2.44	−1.51	750.00	2.42	−1.51	752.00	2.40	−1.51	754.00	2.39	−1.51
756.00	2.37	−1.51	758.00	2.35	−1.51	760.00	2.33	−1.52	762.00	2.32	−1.52
764.00	2.30	−1.52	766.00	2.29	−1.52	768.00	2.27	−1.52	770.00	2.25	−1.52
772.00	2.24	−1.52	774.00	2.22	−1.52	776.00	2.21	−1.52	778.00	2.19	−1.52
780.00	2.18	−1.52	782.00	2.16	−1.52	784.00	2.15	−1.52	786.00	2.14	−1.52
788.00	2.12	−1.52	790.00	2.11	−1.52	792.00	2.10	−1.52	794.00	2.08	−1.52
796.00	2.07	−1.52	798.00	2.06	−1.52	800.00	2.05	−1.52	802.00	2.03	−1.52
804.00	2.02	−1.52	806.00	2.01	−1.52	808.00	2.00	−1.52	810.00	1.98	−1.52
812.00	1.97	−1.52	814.00	1.96	−1.52	816.00	1.95	−1.52	818.00	1.94	−1.53
820.00	1.93	−1.53	822.00	1.92	−1.53	824.00	1.91	−1.53	826.00	1.90	−1.53

TABLE 14.6 Harmonic Current Flows (% of fundamental), with 12 Mvar Capacitor Bank, and Generator G1 in Service

ID	FundA	5th	7th	11th	13th	17th	19th	23rd	25th	29th	31st	35th	37th	41st	43rd	47th	49th
PCC	33.80	40.77	53.80	16.38	6.27	1.62	0.88	0.28	0.15	0.03	0.02	0.04	0.04	0.03	0.03	0.02	0.02

TDD at PCC = 69.77%.

TABLE 14.7 Harmonic Voltages as % of Nominal Bus Voltage

ID	FundkV	5th	7th	11th	13th	17th	19th	23rd	25th	29th	31st	35th	37th	41st	43rd	47th	49th
Bus 1	138	0.23	0.42	0.20	0.09	0.03	0.02	0.01	0	0	0	0	0	0	0	0	0
Bus 2	13.8	3.20	5.91	2.83	1.27	0.44	0.26	0.10	0.06	0.01	0.01	0.02	0.02	0.02	0.02	0.01	0.01

THD (voltage) bus 1 = 0.53%, bus 2 = 7.38%.

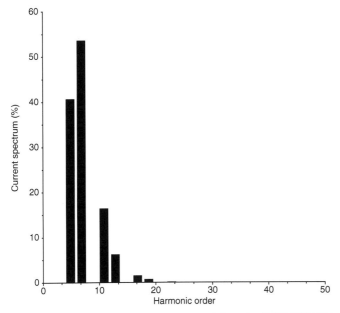

Figure 14.25   Harmonic current spectrum injected at PCC (138-kV bus), industrial system for harmonic analysis.

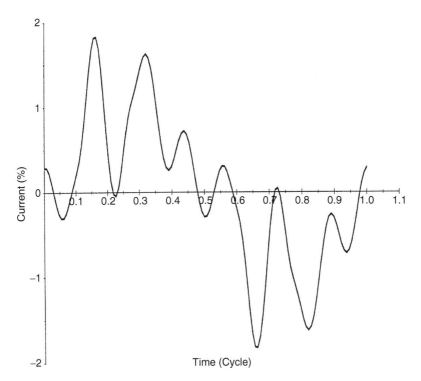

Figure 14.26   Waveform of the current at PCC; industrial system for harmonic analysis.

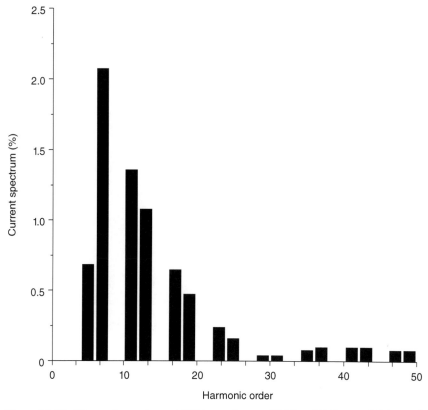

Figure 14.27   Harmonic spectrum at PCC, case 3a, one 12-Mvar ST filter, and G1 out of service; industrial system for harmonic analysis.

- Providing the ST filters of 5th, 7th, and 11th harmonics at 13.8-kV bus 2, as follows:

  ○ 5th harmonic filter: 5, 400-kvar, 8.32-kV capacitor units in parallel per phase, give a three-phase wye-connected bank of 5.405 Mvar, notch frequency 4.85th harmonic.

  ○ 7th harmonic filter: 4, 400-kvar, 8.32-kV capacitor units in parallel per phase, give a three-phase wye-connected bank of 4.404 Mvar, notch frequency 6.85th harmonic.

  ○ 11th harmonic filter: 3, 400-kvar, 8.32-kV capacitor units in parallel per phase, give a three-phase wye-connected bank of 3.303 Mvar, notch frequency 10.9th harmonic.

  ○ This gives a total of 13.112 Mvar of filters. Some reactive power will be consumed in the 0.25 Ω series reactor.

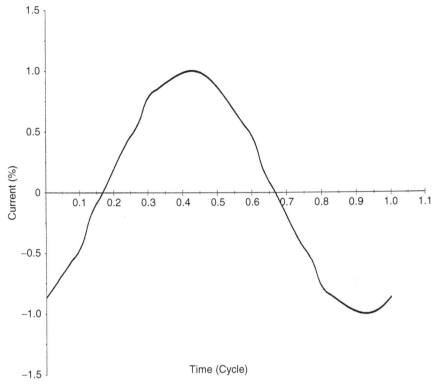

Figure 14.28   Voltage waveforms at PCC and 13.8-kV bus; industrial system for harmonic analysis.

This configuration is shown in Fig. 14.31. The harmonic spectrum at PCC is shown in Fig.14.32, and its waveform is shown in Fig. 14.33. The current waveform through the 5th, 7th and 11th harmonic filters is plotted in Fig. 14.34.

It is documented that

- The total TDD at PCC is now 6.53%. Also the distortions at each of the harmonics meet the IEEE 519 requirements.
- The voltage distortion at PCC is reduced by 0.06%, and at 13.8-kV bus 2, it is 1.57%.
- When the system is operated with generator out of service, the TDD at PCC is 1.05%.

Thus, the objectives of the study have been met. The distortion limits meet IEEE 519 requirements under all operating conditions, and load power factor at PCC is always more than 0.85 as planned.

TABLE 14.8 Harmonic Current Flows (% of fundamental and also in A), 12-Mvar ST Filter, and Generator G1 Out of Service

ID	FundA	5th	7th	11th	13th	17th	19th	23rd	25th	29th	31st	35th	37th	41st	43rd	47th	49th
PCC	225.90	0.68%	2.07%	1.35%	1.08%	0.65%	0.47%	0.24%	0.16%	0.04%	0.04%	0.08%	0.10%	0.10%	0.10%	0.08%	0.08%
	225.90	5.12A	6.64A	3.92A	3.07A	1.82A	1.34A	0.66A	0.44A	0.12A	0.12A	0.22A	0.27A	0.27A	0.27A	0.22A	0.22A
ST filter	569.73	27.42%	9.53%	4.02%	2.99%	1.69%	1.22%	0.60%	0.40%	0.10%	0.10%	0.20%	0.25%	0.25%	0.25%	0.20%	0.20%
	569.73	92.1A	21.60A	8.45A	6.22A	3.49A	2.51A	1.23A	0.82A	0.20A	0.20A	0.41A	0.51A	0.51A	0.51A	0.41A	0.41A

TDD at PCC = 2.88%.

TABLE 14.9 Harmonic Voltages as % of Nominal Bus Voltage

ID	FundkV	5th	7th	11th	13th	17th	19th	23rd	25th	29th	31st	35th	37th	41st	43rd	47th	49th
Bus 1	138	0.03	0.15	0.11	0.11	0.10	0.08	0.07	0.04	0.03	0.01	0.04	0.02	0.03	0.03	0.03	0.03
Bus 2	13.8	0.36	1.54	1.58	1.45	1.17	0.96	0.59	0.43	0.12	0.13	0.30	0.40	0.45	0.47	0.41	0.43

THD (voltage) bus 1 = 0.24%, bus 2 = 3.33%.

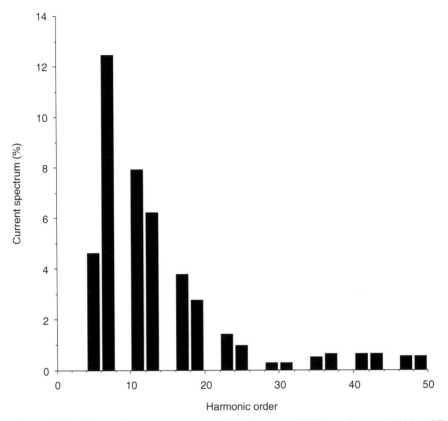

Figure 14.29   Harmonic current spectrum at PCC, generator G1 in service, one 12-Mvar ST filter at 13.8-kV bus; industrial system for harmonic analysis.

14.13   LONG TRANSMISSION LINE

Consider a 230-kV transmission line of 200 miles long, with the following parameters:

ACSR conductors, 500 kcmil, horizontal spacing of 25 ft, height above ground = 108 ft, two ground wires, spaced symmetrically 35 ft above the phase conductors at a height of 158 ft (Fig. 14.35). The ground wires are 115.6 kcmil (7#8), earth resistivity = 100 Ω m. Resistance 25°C = 0.187 Ω/mile, Outside diameter = 0.904″, GMR = 0.0311 ft.

The following line parameters are calculated:

$$R_{abc} = \begin{vmatrix} 68.118 & 31.040 & 30.646 \\ 31.040 & 68.118 & 31.040 \\ 390.646 & 31.040 & 68.118 \end{vmatrix} \ \Omega/\text{mile} \qquad (14.59)$$

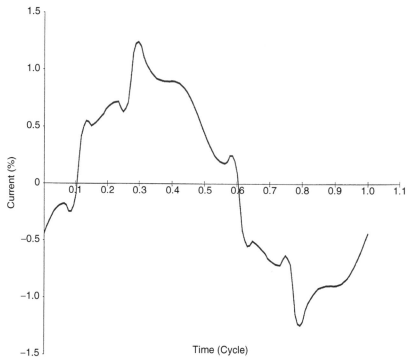

Figure 14.30   Waveform of current injected at PCC, G1 out of service, case 3a; industrial system for harmonic analysis.

$$X_{abc} = \begin{vmatrix} 254.239 & 91.477 & 75.134 \\ 91.477 & 254.239 & 91.477 \\ 75.134 & 91.477 & 254.239 \end{vmatrix} \ \Omega/\text{mile} \tag{14.60}$$

$$Y_{abc} = \begin{vmatrix} 862.794 & -166.883 & -79.808 \\ -166.883 & 862.794 & -166.883 \\ -79.808 & -166.883 & 862.794 \end{vmatrix} \ \mu S/\text{mile} \tag{14.61}$$

$$R_{012} = \begin{vmatrix} 130.166 & 4.283 & -4.644 \\ -4.644 & 37.439 & -9.613 \\ 4.283 & 31.040 & 37.439 \end{vmatrix} \tag{14.62}$$

$$X_{012} = \begin{vmatrix} 426.005 & -2.890 & -2.264 \\ -2.264 & 167.916 & 5.623 \\ -2.890 & 5.566 & 167.916 \end{vmatrix} \tag{14.63}$$

$$Y_{012} = \begin{vmatrix} 596.673 & 9.715 & 9.715 \\ 9.715 & 1010.247 & -33.822 \\ 9.715 & -33.822 & 1010.247 \end{vmatrix} \tag{14.64}$$

These matrices are not symmetrical due to mutual couplings.

TABLE 14.10 Harmonic Current Flows (% of Fundamental and also in A), 12-Mvar ST Filter, and Generator G1 in Service

ID	FundA	5th	7th	11th	13th	17th	19th	23rd	25th	29th	31st	35th	37th	41st	43rd	47th	49th
PCC	32.10A	4.55%	12.27%	7.78%	6.16%	3.69%	2.70%	1.35%	0.90%	0.23%	0.23%	0.45%	0.57%	0.57%	0.57%	0.45%	0.45%
	32.10A	4.52A	5.30A	3.08A	2.41A	1.43A	1.04A	0.52A	0.35A	0.09A	0.09A	0.17A	0.22A	0.22A	0.22A	0.17A	0.17A
ST filter	536.38	25.12%	7.73%	3.17%	2.35%	1.32%	0.96%	0.47%	0.31%	0.08%	0.08%	0.15%	0.19%	0.19%	0.19%	0.16%	0.16%
	536.38	92.1A	21.60A	8.45A	6.22A	3.49A	2.51A	1.23A	0.82A	0.20A	0.20A	0.41A	0.51A	0.51A	0.51A	0.41A	0.41A

TDD at PCC = 17.18%.

TABLE 14.11 Harmonic Voltages as % of Nominal Bus Voltage

ID	FundKV	5th	7th	11th	13th	17th	19th	23rd	25th	29th	31st	35th	37th	41st	43rd	47th	49th
Bus 1	138	0.02	0.09	0.09	0.07	0.05	0.03	0.03	0.02	0.01	0.01	0.02	0.02	0.02	0.03	0.02	0.02
Bus 2	13.8	0.63	1.26	1.26	1.18	0.93	0.76	0.46	0.34	0.10	0.11	0.24	0.32	0.35	0.37	0.33	0.34

THD (voltage) bus 1 = 0.19%, bus 2 = 2.64%

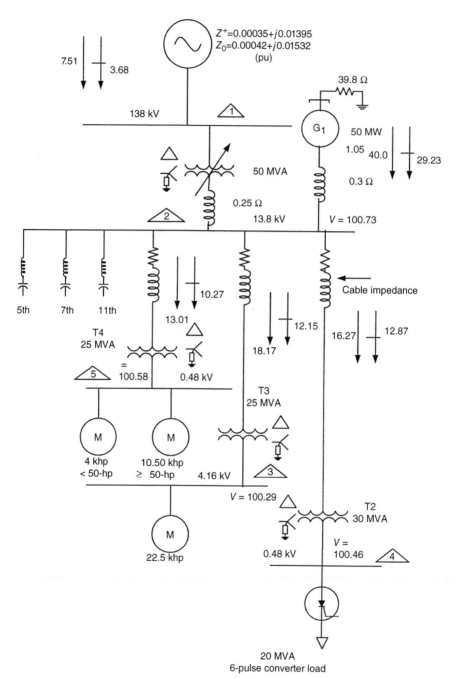

Figure 14.31   Modified system configuration with a reactor on 50-MVA transformer secondary and three harmonic filters; industrial system for harmonic analysis.

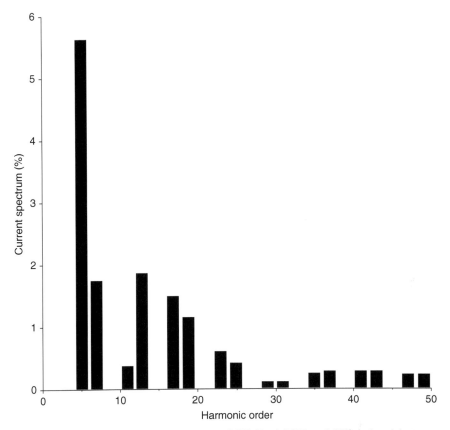

Figure 14.32 Harmonic current spectrum at PCC. Total TDD = 6.53%; industrial system for harmonic analysis.

The total current at any point is the sum of incident and reflected current, see Section 13.8.4 for a maxima.

The line is terminated in a 100-MVA 230–115 kV transformer, with a 6-pulse 100-MVA converter load, operating at $\alpha = 30°$ and $\mu = 6°$.

Practically; there will be multiple harmonic loads with different spectra, and these will be some percentage of the total load. This example, therefore, does not portray a practical situation, and is illustrative of the frequencies that can be generated in a long line model. Figure 14.36 shows the system configuration for harmonic analysis.

A load flow analysis shows that under no-load, the voltage at 115-kV bus 3 rises to 114.98% of the nominal voltage due to Ferranti effect. A 30-Mvar reactor is required to bring down the voltage to approximately rated voltage. When the line is loaded, the voltage dips by 23%. To raise it to a rated voltage, 45 Mvar of capacitive power must be injected. Thus, a variable reactive power compensation device range +30 to −45 Mvar at bus 3 is required for load voltage regulation. An SVC

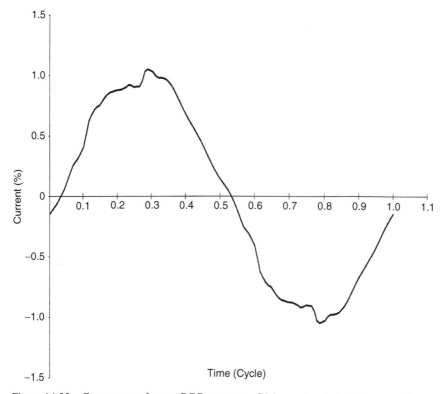

Figure 14.33 Current waveform at PCC, generator G1 in service; industrial system for harmonic analysis.

or STATCOM can be used. For the purpose of this simulation, a 43-Mvar capacitive compensation is considered, with SVC or STATCOM not producing any further harmonic pollution, which may not be true.

Three models for the transmission line are used:

- Distributed parameter, untransposed
- Distributed parameter, frequency dependent and transposed
- Nominal Π model.

A three-phase harmonic load flow is conducted to illustrate the differences in the results with these three line models.

The results of the study are shown in the following figures and tables.

Figures 14.37–14.40 show the impedance, reactance, resistance, and angle versus frequency at bus 1.

Figures 14.41–14.44 show the impedance, reactance, resistance, and angle versus frequency at bus 2.

It is seen that there is a vast difference in the nature of plots, the resonance frequencies, and angles depending on the transmission line model.

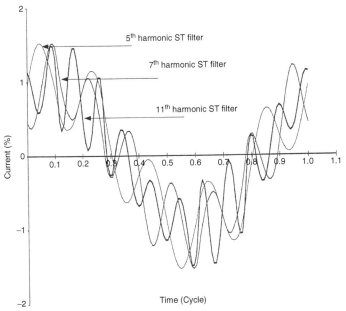

Figure 14.34   Waveforms of the currents through the filters, industrial system for harmonic analysis.

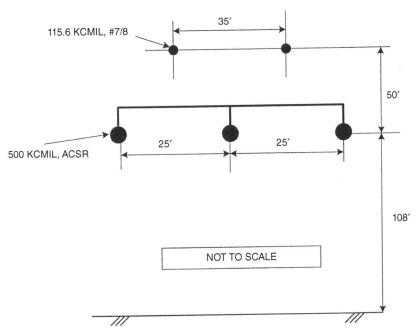

Figure 14.35   Transmission line configuration for the calculation of line parameters.

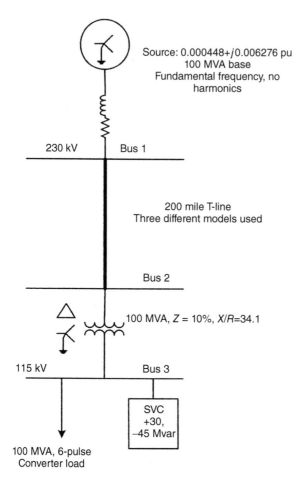

Figure 14.36   Long transmission line configuration with transformer and SVC and harmonic load.

This is confirmed by Fig. 14.45, which illustrates the waveforms of current flow in phases *a*, *b*, and *c* through the transmission line.

For the untransposed line model, Table 14.12 tabulated the currents through the transmission line and their phase angles. Due to untransposed model, the currents in phases *a*, *b*, and *c* differ. There is some unbalance neutral current also, which is not shown. Table 14.13 shows the harmonic voltage distortions at buses 1 and 2 and again these differ in phases *a*, *b*, and *c*.

For the frequency-dependent distributed parameter line model, the currents in phases *a*, *b*, and *c* in the transmission line are balanced (as shown in Table 14.14) and their phase angles differ when compared to untransposed line model. As an example, the 7th harmonic current in untransposed model is 39.74, 41.27, and 39.05 A in phases *a*, *b*, and *c*; while for the frequency-dependent model, it is 47.58 A in all three phases.

Also the voltage distortions between the two models differ. For the frequency-dependent model, the voltage distortion is plotted in Table 14.15. Comparing it with Table 14.13, differences can be seen.

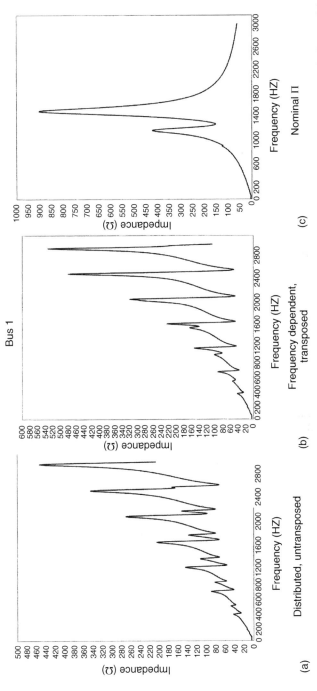

Figure 14.37 Impedance (Ω) versus frequency, bus 1: (a) distributed untransposed; (b) frequency-dependent transposed; and (c) nominal Π models of the transmission line.

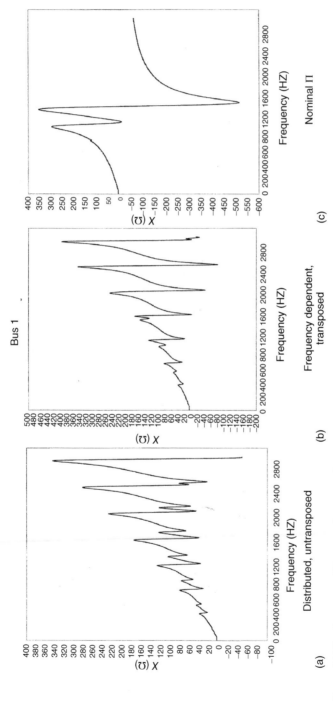

Figure 14.38 Reactance (Ω) versus frequency, bus 1: (a) distributed untransposed; (b) frequency-dependent transposed; and (c) nominal Π models of the transmission line.

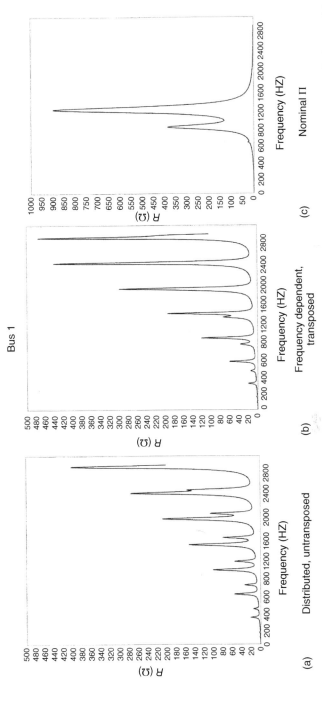

Figure 14.39 Resistance (Ω) versus frequency, bus 1: (a) distributed untransposed; (b) frequency-dependent transposed; and (c) nominal Π models of the transmission line.

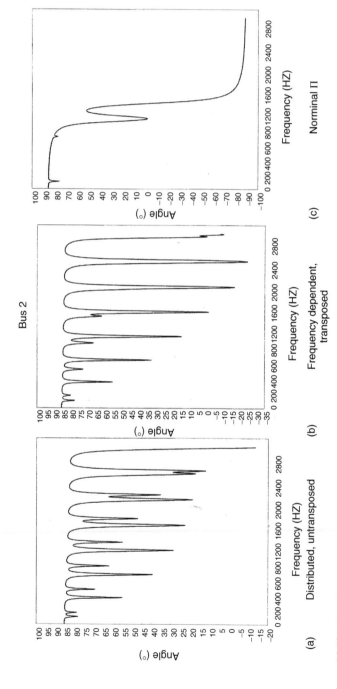

Figure 14.40 Impedance angle versus frequency, bus 1: (a) distributed untransposed; (b) frequency-dependent transposed; and (c) nominal Π models of the transmission line.

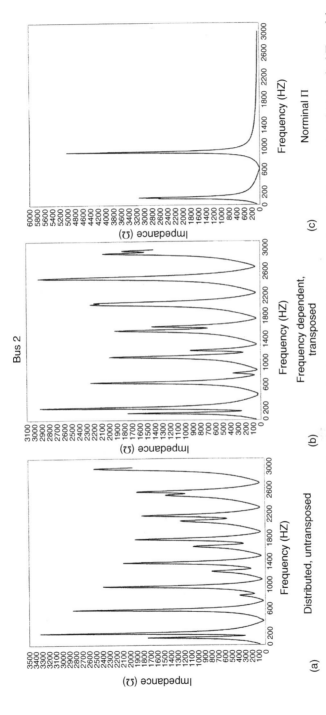

Figure 14.41 Impedance (Ω) versus frequency, bus 2: (a) distributed untransposed; (b) frequency-dependent transposed; and (c) nominal Π models of the transmission line.

665

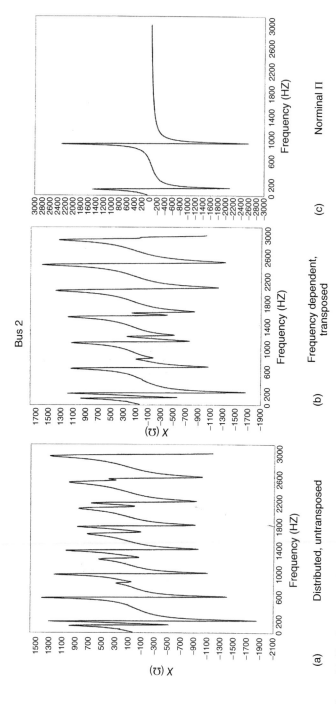

Figure 14.42 Reactance (Ω) versus frequency, bus 2: (a) distributed untransposed; (b) frequency-dependent transposed; and (c) nominal Π models of the transmission line.

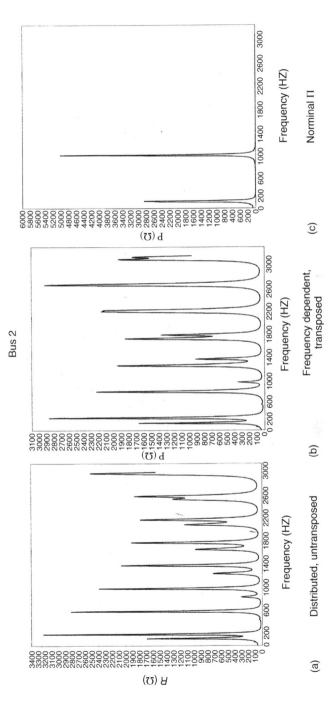

Figure 14.43 Resistance (Ω) versus frequency, bus 2: (a) distributed untransposed; (b) frequency-dependent transposed; and (c) nominal Π models of the transmission line.

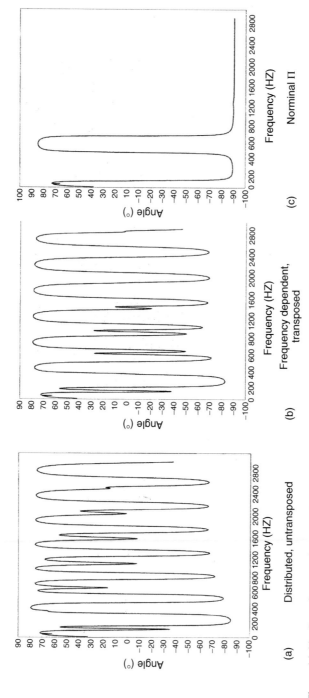

Figure 14.44 Impedance angle versus frequency, bus 2: (a) distributed untransposed; (b) frequency-dependent transposed; and (c) nominal Π models of the transmission line.

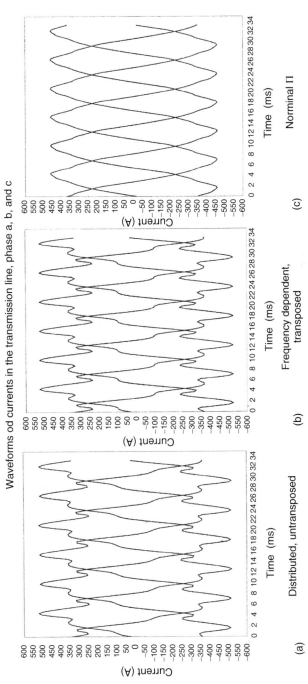

Figure 14.45 Waveforms of the currents in the transmission line: (a) distributed untransposed; (b) frequency-dependent transposed; and (c) nominal Π models of the transmission line.

669

TABLE 14.12 Harmonic Current Flow Transmission Line, Untransposed Distributed Parameter

h	Phase a Current A	Angle	Phase b Current A	Angle	Phase c Current A	Angle
Fund (100%)	309.30	27.46	309.30	−92.54	309.30	147.46
5	23.38	−48.05	25.09	73.67	23.16	−164.00
7	39.74	−42.93	41.27	−165.20	39.05	73.96
11	4.75	−175.17	5.25	−55.11	4.62	67.98
13	6.78	−163.98	7.54	70.42	6.59	−51.36
17	1.62	63.63	1.90	177.98	1.53	−62.18
19	1.60	71.94	1.78	−52.37	1.59	−172.56
23	0.71	−68.96	0.86	63.89	0.93	171.04
25	0.53	−53.84	0.59	−176.57	0.55	65.77
29	0.61	177.15	0.59	−60.19	0.65	55.68
31	0.23	−179.72	0.27	59.26	0.25	−55.31
35	0.79	127.25	0.82	−128.96	1.1	−0.09
37	0.14	54.50	0.16	−62.17	0.15	−172.67
41	0.13	68.76	0.14	−173.71	0.14	−49.18
43	0.15	−58.17	0.16	−162.53	0.13	75.59

Harmonic current flows are shown in percentage of the fundamental frequency current.

TABLE 14.13 Harmonic Voltage Distortion, Untransposed and Distributed Parameter Line

h	Bus 1 Phase a	Phase b	Phase c	Bus 2 Phase a	Phase b	Phase c
5	0.39	0.42	0.39	6.05	6.21	5.93
7	0.93	0.96	0.91	2.14	2.41	2.47
11	0.17	0.19	0.17	1.44	1.54	1.38
13	0.29	0.33	0.29	1.41	1.59	1.45
17	0.09	0.11	0.09	0.39	0.48	0.36
19	0.10	0.11	0.10	0.47	0.52	0.08
23	0.05	0.07	0.07	0.06	0.08	0.15
25	0.04	0.05	0.05	0.15	0.17	0.16
29	0.06	0.06	0.06	0.09	0.07	0.05
31	0.02	0.03	0.03	0.04	0.05	0.04
35	0.09	0.10	0.12	0.22	0.21	0.24
37	0.02	0.02	0.02	0.02	0.01	0.01
41	0.02	0.02	0.02	0.04	0.05	0.04
43	0.02	0.02	0.02	0.03	0.02	0.04

Harmonic voltages are shown in percentage of the nominal system voltage.

TABLE 14.14   Harmonic Current Flow Transmission Line, Transposed Frequency-Dependent Distributed Parameter

h	Phase a		Phase b		Phase c	
	Current A	Angle	Current A	Angle	Current A	Angle
Fund (100%)	309.30	27.46	309.30	−92.54	309.30	147.46
5	24.43	−46.30	24.43	73.70	24.43	−166.30
7	47.58	−48.35	47.58	−168.35	47.58	71.65
11	5.05	−174.62	5.05	−54.62	5.05	65.39
13	8.60	−170.11	8.60	69.89	8.60	−50.11
17	1.70	58.77	1.70	178.77	1.70	−61.23
19	1.92	68.36	1.92	−51.63	1.92	−171.63
23	0.78	−66.60	0.78	53.40	0.78	173.40
25	0.60	−55.65	0.60	−175.65	0.60	64.35
29	0.50	170.70	0.50	−69.31	0.50	50.70
31	0.25	−179.70	0.25	59.69	0.25	−60.31
35	0.64	65.65	0.64	−174.35	0.64	−54.35
37	0.13	55.76	0.13	−64.24	0.13	−175.76
41	0.22	63.88	0.22	−176.12	0.22	−56.12
43	0.10	−64.02	0.10	−175.98	0.10	55.98

Harmonic current flows are shown in percentage of the fundamental frequency current.

TABLE 14.15   Harmonic Voltage Distortion, Transposed, Frequency-Dependent Distributed Parameter Line

h	Phase a Bus 1	Phase a Bus 2
5	0.41	5.96
7	1.11	2.11
11	0.19	1.47
13	0.37	1.63
17	0.10	0.42
19	0.12	0.54
23	0.06	0.10
25	0.05	0.17
29	0.05	0.03
31	0.03	0.05
35	0.07	0.13
37	0.02	0.01
41	0.03	0.07
43	0.01	0.02

Harmonic voltage distortion is shown in percentage of the nominal voltage.

TABLE 14.16 Harmonic Current Flow Transmission Line, Nominal Π Model

	Phase a		Phase b		Phase c	
h	Current A	Angle	Current A	Angle	Current A	Angle
Fund (100%)	309.30	27.46	309.30	−92.54	309.30	147.46
5	11.65	−43.93	11.65	76.06	11.65	−163.94
7	4.43	−28.12	4.43	−148.12	4.43	91.88
11	1.93	4.20	1.93	124.20	1.93	−115.80
13	2.15	19.29	2.15	−100.71	2.15	139.29
17	1.49	−120.90	1.49	−0.90	1.49	119.11
19	0.56	−108.32	0.56	131.68	0.56	11.68
23	0.29	−90.96	0.29	29.04	0.29	149.04
25	0.37	−131.57	0.37	108.43	0.37	−11.57
29	0.03	−178.97	0.03	−58.97	0.03	61.03
31	0.01	−166.56	0.01	73.44	0.01	−46.56
35	0.003	−135.90	0.003	−15.90	0.003	104.10
37	0.002	−119.58	0.002	120.42	0.002	0.42
41	0.001	−86.17	0.001	33.83	0.001	153.83
43	0.00	−69.25	0.00	170.75	0.00	50.75

Harmonic current flows are shown in percentage of the fundamental.

Much difference occurs with nominal Π model of the line. The harmonic content in the transmission line is much reduced (see waveform in Fig. 14.45). This is also tabulated in Table 14.16. For comparison:

- 5th and 7th harmonic currents untransposed distributed parameter model: 25.09 and 41.27 A, respectively, maximum in a phase.
- 5th and 7th harmonic currents frequency dependent, transposed distributed parameter model: 24.43 and 47.58 A, respectively.
- 5th and 7th harmonic currents nominal Π model: 11.65 and 4.43 A, respectively.

Finally, the harmonic distortions at buses 1 and 2 for the nominal Π model are plotted in Table 14.17. Again a specimen comparison is

- 5th and 7th harmonic voltage distortion, bus 1, untransposed distributed parameter model: 0.42% and 0.96%, respectively.
- 5th and 7th harmonic voltage distortion, bus 1, frequency dependent, transposed, distributed parameter model: 0.41% and 1.11%, respectively.
- 5th and 7th harmonic voltage distortion, bus 1, nominal Π model: 0.19% and 0.10%, respectively.

This illustrates the importance of proper transmission line modeling, especially for long lines above 100 miles (Fig. 14.36).

TABLE 14.17 Harmonic Voltage Distortion, Nominal Π Model of the Line

h	Phase a Bus 1	Phase a Bus 2
5	0.19	6.71
7	0.10	3.43
11	0.07	2.06
13	0.09	2.46
17	0.08	1.66
19	0.04	0.55
23	0.02	0.15
25	0.03	0.09
29	0.00	0.04
31	0.00	0.03
35	0.00	0.01
37	0.00	0.01
41	0.00	0.00
43	0.00	0.00

The harmonic voltage distortion is shown as a percentage of the nominal voltage.

14.14 34.5 KV UG CABLE

Figure 14.47 illustrates 10-mile long, 34.5-kV, single-conductor 350-kcmil cable suitable for direct burial in the ground, triplexed, modeled with 10 (ten) Π sections. The cable serves a 12-pulse load. Alternatively, the cable is modeled with a single Π section.

Frequency scans showing impedance phase angle and modulus are in Figs 14.48 and 14.49, respectively. Note that a single Π model seriously attenuated the higher harmonics.

Table 14.18 plots the harmonic voltage distortion in the two cases, and Table 14.19 plots the harmonic currents injected into the 10-MVA transformer. Considerable differences in the harmonic distortions can be seen. A single Π model attenuates the higher harmonics.

See Section 12.2 for number of Π models in series and shield cross bonding.

14.15 5-BUS TRANSMISSION SYSTEM

A harmonic load flow study for a 5-bus hypothetical transmission system at 400 kV, with a 500-MVA generator, 6 transmission lines, 5 buses, and a utility source of 40-kA rms symmetrical short-circuit current, as shown in Fig. 14.49 is conducted. In this

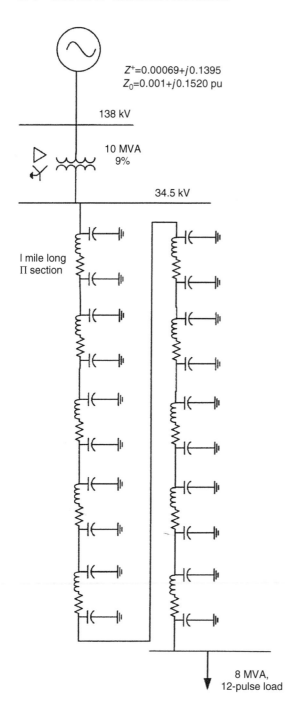

$Z^+ = 0.00069 + j0.1395$
$Z_0 = 0.001 + j0.1520$ pu

138 kV

10 MVA
9%

34.5 kV

I mile long
Π section

8 MVA,
12-pulse load

Figure 14.46   A 10-mile long, 34.5-kV cable represented with 10 (ten) Π sections in series.

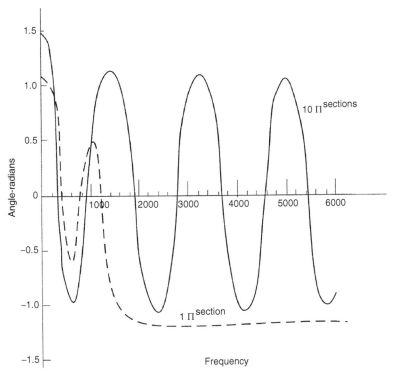

Figure 14.47   Frequency scan, one Π section and 10 (ten) Π sections, impedance angle.

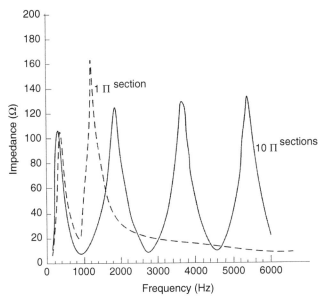

Figure 14.48   Frequency scan, one Π section and 10 (ten) Π sections, impedance modulus.

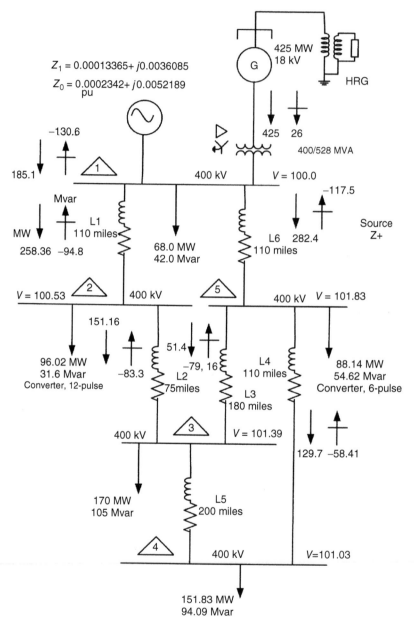

Figure 14.49   A 5-bus 400-kV transmission system with two converter loads and one 500-MVA generator. Load flow results superimposed.

TABLE 14.18 Harmonic Voltage Distortion at Bus 2, Fig. 14.46

Harmonic Order	Single Π Model	10, Π Models in Series
5	1.49	1.76
7	0.81	0.64
11	1.97	1.44
13	1.20	0.83
17	0.07	0.03
19	0.07	0.03
23	0.41	0.25
25	0.19	0.27
29	0.01	0.04
31	0	0.04
35	0.01	0.19
37	0.01	0.13
41	0	0.01
43	0	0.01
47	0	0.05
49	0	0.05

The harmonic voltage distortion is shown as a percentage of the nominal voltage.

TABLE 14.19 Harmonic Current Distortion, Current into 10-MVA Transformer, Fig. 4.45

Harmonic Order	Single Π Model	10, Π Models in Series
Fundamental	27.54 A (138 kV) = 100%	
5	4.95	5.84
7	1.90	1.51
11	2.94	2.13
13	1.51	1.04
17	0.06	0.03
19	0.29	0.02
23	0.12	0.17
25	0	0.17
29	0	0.02
31	0.01	0.02
35	0	0.08
37	0	0.05
41	0	0
43	0	0
47	0	0.01
49	0	0.01

The harmonic current is shown as a percentage of the fundamental frequency current.

figure, bus 2 has a 100-MVA 12-pulse converter load, while bus 5 has a 100-MVA 6-pulse converter load spectrum as in Fig. 14.17.

The line lengths and other pertinent data are shown in Fig. 14.49. All line parameters are the same, ACSR/30 strands, one conductor per phase, 35′ horizontal phase-to-phase spacing, and 7#8 ground wire. The calculated line parameters in ohms per mile are as follows:

$$Z_+ = Z_- = 0.18716 + j0.88038,$$

$$Z_0 = 0.61239 + j2.09945$$

Also, shunt capacitance is $Y_1 = 4.6144$ μS per mile positive and negative sequences and $Y_0 = 3.06053$ μS per mile zero sequence. Long-line distributed parameter model is used for lines > 100 miles.

The total applied constant impedance static load is 578.7 MW and 330.27 Mvar (at 86.2 lagging power factor). The generator operates at 425 MW and 26 Mvar. Yet, a reactive power of 130.6 Mvar is supplied into the 400 kV source at the slack bus. *This means that the distributed shunt capacitance of the lines generates approximately 435 leading Mvar.* Figure 14.49 shows that operating voltages at all the buses can be maintained slightly higher than the rated voltages, without *any reactive power compensation.*

Now, consider a condition where the generator is in operation, and all the loads in the 5-bus transmission system are removed. The calculated voltages on the system are approximately 20–30% higher than the nominal voltages. This means that regulated shunt reactive power is required as the load varies. Due to these higher no-load voltages, the capacitive Mvar output of the lines increases to 728 Mvar.

The frequency scan angle and modulus are shown in Figs 14.50 and 14.51. This shows a number of harmonic resonant frequencies. Table 14.20 tabulates impedance modulus and frequency for three buses, bus 1, bus 3, and bus 5.

The calculated harmonic voltage distortion is shown in Table 14.21. It exceeds the IEEE limits at four buses out of five. The harmonic current through the lines is tabulated in Table 14.22, again IEEE harmonic limits are exceeded. Figure 14.52 shows waveforms of the current through lines 1, 2, and 3. Table 14.23 shows the harmonics absorbed in the 500-MVA generator. These are in percentage of generator load current, which is 13.65 kA.

The study shows that harmonic distortions, current, and voltage are much above permissible limits. The troublesome harmonics are 5th, 7th, and 11th. ST filters for these harmonics with a second-order high-pass filter can be provided at the nonlinear loads – that will mitigate the flow of these harmonics in the lines. The addition of these filters will add more capacitive reactive power and the bus voltages will rise. Thus, controllable inductive reactive power is required through a SVC or STATCOM [31].

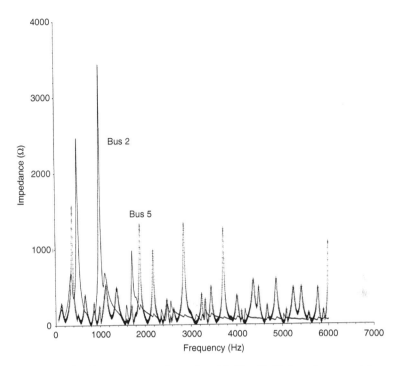

Figure 14.50   Frequency scan, buses 2 and 5, Fig. 14.49; impedance modulus.

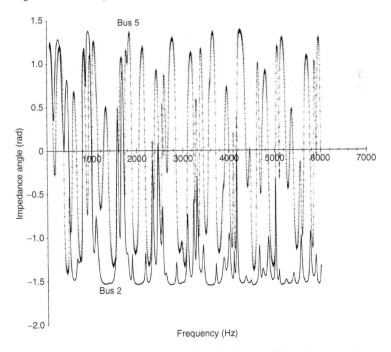

Figure 14.51   Frequency scan, buses 2 and 5, Fig. 14.49 impedance angle.

TABLE 14.20   Frequency Scan Results, 5-Bus Transmission at 400 kV

Bus ID	Z, Modulus (Ω)	Harmonic	Bus ID	Z, Modulus (Ω)	Harmonic
1	37.07	6.107	5	272.89	2.200
	69.1	8.067		1563.42	6.267
	67.59	10.833		136.62	8.330
	139.3	13.3		367.43	11.067
	293.76	16.523		282.83	14.410
	143.87	22.510		238.81	16.512
	483.76	26.357		519.10	11.967
	348.15	28.533		480.01	22.801
	120.67	39.487		64.32	25.467
	445.92	36.312		243.99	28.512
	912.23	39.233		1315.61	31.567
	311.4	41.433		979.86	36.845
	301.44	47.433		58.49	39.257
	1047.23	51.734		322.07	41.401
	432.71	53.987		115.68	42.733
	271.74	55.234		201.21	43.867
	66.69	57.342		1321.33	47.567
	504.97	61.833		393.51	53.977
	562.37	64.623		326.92	56.167
	699.75	66.757		490.63	57.333
	158.98	68.57		1291.0	61.912
	87.50	62.60		82.30	61.342
3	354.54	2.233		175.05	68.512
	372.33	5.901		39.62	69.501
	155.35	12.220			
	699.61	14.508			
	165.97	15.533			
	173.22	18.367			
	577.52	22.912			
	367.92	26.512			
	302.44	28.604			
	180.59	30.204			
	302.86	31.733			
	141.68	35.914			
	266.84	39.432			
	107.13	41.543			
	200.23	62.001			
	302.04	47.633			
	340.8	51.925			
	117.38	54.067			
	271.56	60.767			
	188.34	64.947			
	83.00	67.023			
	116.23	68.467			

TABLE 14.21   Calculations of Voltage Distortion, 5-Bus Transmission at 400 kV

Bus	Voltage	THDV	IEEELimits
1	400	0.72	1.5
2	400	5.50	1.5
3	400	5.39	1.5
4	400	4.19	1.5
5	400	7.67	1.5

TABLE 14.22   Calculations of TDD, 5-Bus Transmission at 400 kV

Line	I-Fund	5	7	11	13	17	19	23	25
1	399.88	7.33	3.06	0.70	1.30	1.79	0.72	0.08	0.18
2	250.37	5.37	0.84	1.16	0.10	1.85	0.76	0.59	0.11
3	136.25	14.39	13.42	8.52	0.20	1.70	1.90	1.05	0.26
4	202.95	13.97	1.31	5.38	0.40	0.42	1.34	0.76	0.16
5	124.19	21.71	13.10	12.55	1.01	0.63	3.32	1.75	0.19
6	446.08	8.82	4.03	5.34	0.93	0.51	0.95	1.08	0.32
IEEE limits	$Is/Ir > 50$	3.0	3.0	1.5	1.5	1.15	1.15	0.45	0.45

All harmonics shown as percentage of fundamental frequency current.

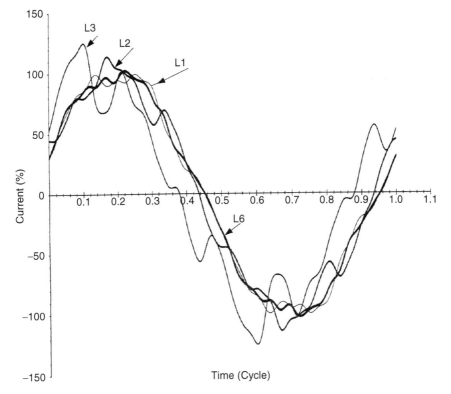

Figure 14.52   Waveforms of currents through some of the lines L1, L2, and L3 in Fig. 14.49.

TABLE 14.23 Harmonic Flow into 500-MVA Generator, 5-Bus Transmission at 400 kV

5	7	11	13	17	19	23	25
0.08	0.05	0.18	0.01	0.04	0.01	0.04	0.02

The harmonics are in terms of percentage of fundamental current.

REFERENCES

1. G.D. Breuer, et al., "HVDC-AC interaction. Part-1 development of harmonic measurement system hardware and software," IEEE Transactions on Power Apparatus and Systems, vol. 101, no. 3, pp. 701–706, March 1982.
2. G.D. Breuer, et al., "HVDC-AC interaction. Part-II AC system harmonic model with comparison of calculated and measured data," IEEE Transactions on Power Apparatus and Systems, vol. 101, no. 3, pp. 709–717, March 1982.
3. IEEE. Task Force on Harmonic Modeling and Simulation, "Modeling and simulation of propagation of harmonics in electrical power Systems-Part I: concepts, models and simulation techniques," IEEE Transactions on Power Delivery, vol. 11, pp. 452–465, 1996.
4. IEEE. Task Force on Harmonic Modeling and Simulation, "Modeling and simulation of propagation of harmonics in electrical power systems-Part II: sample systems and examples," IEEE Transactions on Power Delivery, vol. 11, pp. 466–474, 1996.
5. G.D. Brewer, J.H. Chow, T.J. Gentile, et al. "HVDC –AC Harmonic interaction. Development of a harmonic measurement system hardware and software," IEEE Transactions on Power Apparatus and Systems vol. 101, pp. 701–708, 1982.
6. M. Nagpal, W. Zu, J. Swada, "Harmonic impedance measurement using three-phase transient," IEEE Transactions on Power Delivery, vol. 13, no. 1, pp. 8–13, 1998.
7. C. Hatziadoniu, G.D. Galanos. "Interaction between the AC voltages and DC current in weak AC/DC interconnections," IEEE Transactions on Power Delivery, vol. 3, pp. 1297–1304, 1988.
8. A.A. Mohmoud, R.D. Shultz, "A method of analyzing harmonic distribution in AC power systems," IEEE Transactions on Power Apparatus and Systems 101, 1815–1824, 1982.
9. B.C. Smith, J. Arrilaga, A.R. Wood, and N.R. Watson. "A review of iterative harmonic analysis for AC-DC power systems," IEEE Transactions on Power Delivery, vol. 13, pp. 180–185, Jan. 1998.
10. M. Valcarcel, J.G. Mayordomo, "Harmonic power flow for unbalanced systems," IEEE Transactions on Power Delivery, vol. 8, pp. 2052–2059, Oct. 1993.
11. D. Xia, G.T. Heydt, "Harmonic power flow studies Part-1-formulation and solution," IEEE Transactions on Power Apparatus and Systems vol. 101, pp. 1257–1265, 1982.
12. D. Xia, G.T. Heydt, "Harmonic power flow studies-Part II, implementation and practical applications," IEEE Transactions on Power Apparatus and Systems vol. 101, pp. 1266–1270, 1982.
13. J. Arrillaga, C.D. Callaghan, "Three-phase AC-DC load and harmonic flows," IEEE Transactions on Power Delivery, vol.6, no.1, pp. 38–244, Jan 1991.
14. W. Xu, J.R. Marti, H.W. Dommel, "A multiphase harmonic load flow solution technique," IEEE Transactions on Power Systems, vol. 6, n. 1, pp. 174–182, Feb. 1991.
15. S. Herraiz, L. Sainzand J. Clua. "Review of harmonic load flow formulations," IEEE Transactions on Power Delivery, vol. 18, no. 3, pp. 1079–1087, July 2003.
16. J.P. Tamby, V.I. John. "Q'Harm-A harmonic power flow program for small power systems," IEEE Transactions on Power Systems vol. 3, no. 3, 945–955, Aug. 1988.
17. N.R. Watson, J. Arrillaga, "Frequency dependent AC system equivalents for harmonic studies and transient converter simulation," IEEE Transactions on Power Delivery, vol. 3, no. 3, 1196–1203, July 1988.
18. M.F. Akram,T.H. Ortmeyer, and J.A. Svoboda, "An improved harmonic modeling technique for transmission network," IEEE Transactions on Power Delivery, vol.9, pp.1510–1516, 1994.
19. C. Dzienis, A. Bachry, Z. Stycezynski. "Full harmonic load flow calculation in power systems for sensitivity investigations," in Conf record 17th Zurich Symposium on Electromagnetic Compatibility, pp. 646–649, Feb. 2006.

20. S.M. Mohd. Shokri, Z. Zakaria, "A direct approach used for solving the distribution system harmonic load flow solutions," in Conf. Record, IEEE 7th International Power Engineering and Optimization Conference (PE)C)), Langkawi, Malaysia, pp. 708–713, June 2013.

21. J-H Teng, C-Y Chang, "A fast load flow method for industrial distribution systems," Proceedings, International Conference on Power Systems Technology (PowerCon), pp. 1149–1154,Perth, WA, 2000.

22. W.H. Kersting, D.L. Medive, "An application of ladder network to the solution of three-phase radial load flow problems," in Conf. Record, IEEE PES Winter Meeting, New York, 1976.

23. M.A. Moreno L.de.Saa, and U. Garcia. "Three-phase harmonic load flow in frequency and time domain" IEE Proceedings, Electrical Power Applications, vol. 150, no 3, pp.295–300, May 2003.

24. X-P Zhang, "Fast three-phase load flow methods," IEEE Transactions on Power Systems, vol. 11, no. 3, pp.1547–1554, August 1996.

25. IEEE Task Force On Probabilistic Aspects of Harmonics (Y. Baghzouz-Chair). "Time varying harmonics Part II-harmonic summation and propagation," IEEE Transactions on Power Delivery, Vol. 17, no.1, pp. 279–285, 2002.

26. IEEE Task Force for Modeling and Simulation. "Test system for harmonic modeling and simulation," IEEE Transactions on Power Delivery, vol. 14, no.2, pp. 574–587, 1999.

27. P. Caremia, G. Carpinelli, F. Rossi, P. Verde, "Probabilistic iterative harmonic analysis of power systems," IEE Proceedings, Generation, Transmission and Distribution, vol. 141, no. 4, pp. 329–338, July 1994.

28. IEEE 519. IEEE Recommended Practice and Requirements for Harmonic Control in Electrical Systems, 1992.

29. ANSI C50.13. Requirements for Cylindrical Rotor 50 Hz and 60 Hz Synchronous Generators Rated 10 MVA and Above, 2005.

30. T.J. E. Miller. Reactive Power Control in Electrical Power Systems, John Wiley, New York, 1982.

31. J.G. Mayordomo, M.Izzeddine, and R. Asensi, "Load and voltage balancing in harmonic power flows by means of static var compensators," IEEE Transactions on Power Delivery, vol. 17, no. 3, pp. 761–769, July 2002.

CHAPTER 15

PASSIVE FILTERS

Passive filters use passive components, such as inductors, capacitors, and resistors. These cannot increase the signal energy; the frequency range for harmonic filters is limited to approximately 3000 Hz. It is common to characterize the frequency-selective filters with respect to their passbands.

15.1 FILTER TYPES

A low-pass (LP) filter passes the low-frequency components and suppresses the high-frequency components.

Their loss characteristic is given by

$$A(\omega) = 0, \quad 0 \leq \omega < \omega_c$$

$$= \infty, \quad \omega_c < \omega < \infty \tag{15.1}$$

The frequency from 0 to ω_c is the passband and from ω_c to ∞ is stopband. The boundary between passband and stopband $= \omega_c$ is the cutoff frequency. However, there cannot be a sudden transition from passband to stopband. Practically, passband loss is not zero, and the stopband loss in not infinite. There is a gradual transition between passband and stopband. Then, for the LP filter, the loss characteristic is

$$A(\omega) \leq A_p, \quad 0 \leq \omega \leq \omega_p$$

$$\geq A_a, \quad \omega_a \leq \omega \leq \infty \tag{15.2}$$

A high-pass filter acts in the reverse manner, suppresses the low frequency, and passes the high frequency. For an ideal filter

$$A(\omega) = \infty, \quad 0 \leq \omega < \omega_c$$

$$= 0, \quad \omega_c < \omega < \infty \tag{15.3}$$

Power System Harmonics and Passive Filter Designs, First Edition. J.C. Das.
© 2015 The Institute of Electrical and Electronics Engineers, Inc. Published 2015 by John Wiley & Sons, Inc.

For a practical filter, the loss characteristic is

$$A(\omega) \geq A_a, \quad 0 \leq \omega \leq \omega_a$$

$$\leq A_p, \quad \omega_p \leq \omega \leq \infty \qquad (15.4)$$

The bandpass filter passes frequencies within a certain band and blocks the low and high frequencies. Ideally,

$$A(\omega) = \infty, \quad 0 \leq \omega < \omega_{c1}$$

$$= 0, \quad \omega_{c1} < \omega < \omega_{c2}$$

$$= \infty, \quad \omega_{c2} \leq \omega < \infty \qquad (15.5)$$

For a practical filter, the loss characteristic is

$$A(\omega) \geq A_a, \quad 0 \leq \omega \leq \omega_{a1}$$

$$\leq A_p, \quad \omega_{p1} \leq \omega \leq \omega_{p2}$$

$$\geq A_a, \quad \omega_{a2} \leq \omega \leq \infty \qquad (15.6)$$

A band stop filter has reverse characteristics of a bandpass filter. If the stop band is narrow, it is called a notch filter. Ideally

$$A(\omega) = 0, \quad 0 \leq \omega < \omega_{c1}$$

$$= \infty, \quad \omega_{c1} < \omega < \omega_{c2}$$

$$= 0, \quad \omega_{c2} \leq \omega < \infty \qquad (15.7)$$

For a practical filter, the loss characteristic is

$$A(\omega) \leq A_p, \quad 0 \leq \omega \leq \omega_{p1}$$

$$\geq A_a, \quad \omega_{a1} \leq \omega \leq \omega_{a2}$$

$$\leq A_p, \quad \omega_{p2} \leq \omega \leq \infty \qquad (15.8)$$

The loss function referred earlier can be determined as follows:
A filter represented by voltage transfer function:

$$\frac{V_o(s)}{V_i(s)} = H(s) = \frac{N(s)}{D(s)} \qquad (15.9)$$

where $V_i(s)$ and $V_o(s)$ are Laplace transforms of the input and output voltages and $N(s)$ and $D(s)$ are polynomials in s.

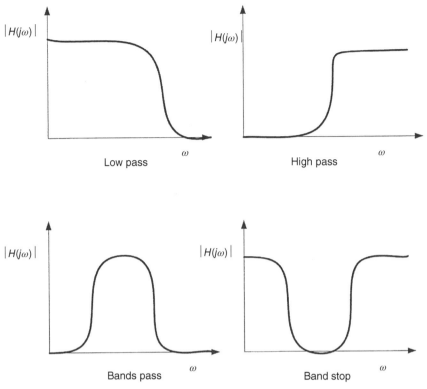

Figure 15.1 Frequency response of low-pass, high-pass, bandpass, and stopband (notch) filters.

The loss or attenuation is in decibels:

$$A(\omega) = 20 \log \left| \frac{V_i\,(j\omega)}{V_o(j\omega)} \right| = 20 \log \frac{1}{|H(j\omega)|} \qquad (15.10)$$

Figure 15.1 shows the magnitude characteristics of these filter types. Figure 15.2(a) shows the LP filter characteristics, and Fig. 15.2(b) shows its attenuation specifications.

We mentioned Butterworth filter in Chapter 6. Its attenuation characteristics are shown in Fig. 15.2(c), which is dependent on the order of the filter. As the order increases, the filter becomes more complex. The magnitude function squared for a Butterworth filter is

$$|H(j\omega)|^2 = \frac{1}{1 + k^2(\omega/\omega_c)^2} \qquad (15.11)$$

where k is a constant that determines variations in the pass band $0 - \omega_c$.

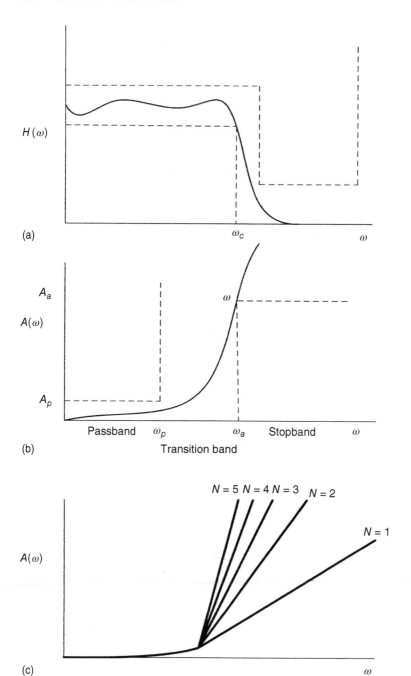

Figure 15.2   (a) and (b) Frequency response and attenuation specifications of a low-pass filter, (c) attenuation response of a Butterworth filter of different orders.

The attenuation is written as

$$A_{\mathrm{dB}} = 10\log\left[1 + \left(\frac{f_{\mathrm{stop}}}{f_c}\right)^{2N}\right]$$
(15.12)

where N is a positive integer defining the filter order.

Generally, the LP filter transfer function contains only the poles:

$$H(s) = \frac{1}{s^n + a_{n-1}s^{n-1} + \ldots + a_1 s + 1}$$
(15.13)

Reference [1] describes a Butterworth function for the common-mode noise mitigation associated with induction motor drives.

15.1.1   Shunt and Series Filters

The filters for harmonic mitigation are generally of shunt type to offer a low-impedance path to a certain harmonic or harmonics so that these are bypassed into the filter and their flow is minimized into the system, as discussed in the following section. These may use resonance in the filter components to offer minimum impedance to a particular harmonic or a band of harmonics. This does not mean that we do not use series filters, that is, filters connected in series with the converter to impede the flow of a certain harmonic (see Section 15.7 for application of bandpass filters).

15.1.2   Location of Harmonic Filters

Passive filters at suitable locations, preferably close to the source of harmonic generation, can be provided so that much of the harmonic currents are trapped at the source and the harmonics propagated to the point of common coupling (PCC) are reduced. Active filters, hybrid combination of active and passive filters, and phase multiplication to reduce harmonic emission at the source are discussed in Chapter 6. By reduction of harmonics at the source, the electrical equipment need not be oversized, losses are minimized, voltage distortions are reduced, the filters can be specifically sized for the loads, and load-dependent switching controls can be provided. Conversely, when filters are located away from the harmonic producing loads, the harmonics must flow to the filter through system impedances with the resultant derating of electrical equipment. Yet, it may not be practical or economical to provide filters at each source of harmonic emission.

The key considerations are the following:

- Harmonic limitations at PCC must meet IEEE 519 requirements, but it is desirable to limit harmonic distortions throughout the power systems.
- Reactive power compensation may be simultaneously required (Section 14.12).
- Normal and contingency conditions of the plant operation, along with ambient harmonics, must be considered.

- Normal and contingency filter conditions must be considered.
- Harmonic emission must be estimated correctly under various operating conditions
- System interaction with harmonic emissions must be considered (Chapter 14).
- A three-phase modeling may be necessary where large unbalances exist.

15.2 SINGLE-TUNED FILTERS

The single-tuned (ST) filters are efficient filters and will bypass a certain harmonic to which these are tuned. These are most widely used filters in all applications of harmonic mitigation. However, care is required in their design, so that the components are not overloaded, and overvoltages due to their applications are controlled. Many times a group of ST filters are applied, each tuned to a specific frequency.

The operation of an ST shunt filter is explained with reference to Fig. 15.3. (Any other type of filter connected in the shunt can be termed a shunt filter.) Figure 15.3(a) shows a system configuration with nonlinear load, and Fig. 15.3(b) shows the equivalent circuit. Harmonic current injected from the source through impedance Z_c divides into filter and system equivalent impedance Z_{eq}. This system impedance can be found by circuit reduction – this is in fact the short-circuit equivalent impedance at bus 1. The current I_s divides into three parallel paths: the current at PCC is the current flowing through utility source, and utility transformer is series:

$$I_h = I_f + I_s \tag{15.14}$$

where I_h is the harmonic current injected into the system, I_f is the current through the filter, and I_s is the current through the system impedance. Also,

$$I_f Z_f = I_s Z_s \tag{15.15}$$

that is, the harmonic voltage across the filter impedance (Z_f) equals the harmonic voltage across the equivalent power system impedance (Z_s).

$$I_f = \left[\frac{Z_s}{Z_f + Z_s}\right] I_h = \rho_f I_h \tag{15.16}$$

$$I_s = \left[\frac{Z_f}{Z_f + Z_s}\right] I_h = \rho_s I_h \tag{15.17}$$

where ρ_f and ρ_s are complex quantities that determine the distribution of harmonic current in the filter and system impedance. These equations can also be written in terms of admittances.

A properly designed filter will have ρ_f close to unity, typically 0.995, and the corresponding ρ_s for the system will be 0.05. The impedance angles of ρ_f and ρ_s may be of the order of $-81°$ and $-2.6°$, respectively.

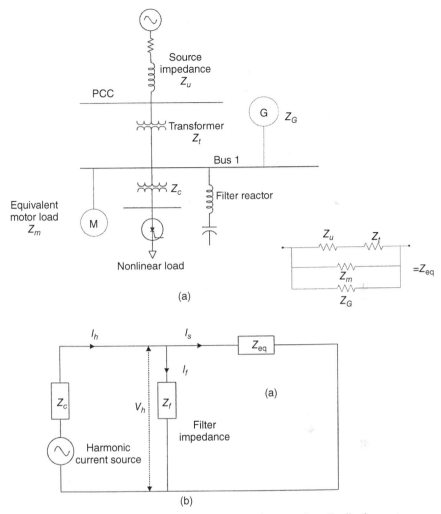

Figure 15.3   (a) Connections of an ST filter, harmonic source in a distribution system;
(b) equivalent circuit looking from harmonic injection as the source.

The harmonic voltages should be as low as possible. The equivalent circuit of
Fig. 15.3(b) shows that system impedance plays an important role in the harmonic
current distribution. For infinite system impedance, that is, a system with very low
short-circuit power, the filtration is perfect, as no harmonic current flows through
the system impedance. Conversely, for a system of zero harmonic impedance, that
is, a source of high short-circuit power, all the harmonic current will flow into the
system and none in the filter. In the case where there is no filtration, all the harmonic current passes on to the system. The lower the system impedance, that is, the
higher the short-circuit current, the smaller is the voltage distortion, provided the filter
impedance is lowered so that it absorbs most of the harmonic current. We alluded to

this concept in Chapter 6 in connection with active filters, the IEEE harmonic limits of TDD [2] are based on this concept. The higher is the short-circuit power of the source, the higher is the permissible TDD.

In an ST filter, as the inductive and capacitive impedances are equal at the resonant frequency, the impedance is given by the resistance R:

$$Z = R + j\omega_n L + \frac{1}{j\omega_n C} \tag{15.18}$$

At resonant frequency ω_n, $Z = R$.
The following parameters can be defined:
ω_n is the tuned angular frequency in radians and is given by

$$\omega_n = \frac{1}{\sqrt{LC}} \tag{15.19}$$

X_0 is the reactance of the inductor or capacitor at the tuned angular frequency. Here, $n = f_n/f$, where f_n is the filter-tuned frequency and f is the power system frequency.

$$X_0 = \omega_n L = \frac{1}{\omega_n C} = \sqrt{\frac{L}{C}} \text{ and } \omega_n = \sqrt{\frac{1}{LC}} \tag{15.20}$$

The quality factor of the tuning reactor is defined as

$$Q = \frac{X_0}{R} = \frac{\sqrt{L/C}}{R} \tag{15.21}$$

It determines the sharpness of tuning, see Chapter 3. The pass band is bounded by frequencies at which

$$|Z_f| = \sqrt{2}R \tag{15.22}$$

$$\delta = \frac{\omega - \omega_n}{\omega_n} \tag{15.23}$$

$$\omega = \omega_n(1 + \delta)$$

At these frequencies, the net reactance equals resistance, capacitive on one side, and inductive on the other side. If it is defined as the deviation per unit from the tuned frequency, then for small frequency deviations, the impedance is approximately given by

$$|Z_f| = R\sqrt{1 + 4\delta^2 Q^2} = X_0\sqrt{Q^{-2} + 4\delta^2} \tag{15.24}$$

To minimize the harmonic voltage, Z_f should be reduced or the filter admittance should be high as compared to the system admittance.

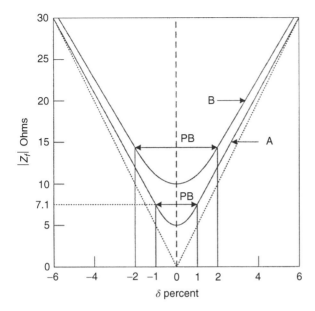

Figure 15.4   Response of an ST shunt filter showing pass band and asymptotes.

The plot of the impedance is shown in Fig. 15.4 [3]. The sharpness of tuning is dependent on R as well as on X_0, and the impedance of the filter at its resonant frequency can be reduced by reducing these. The asymptotes are at

$$|X_f| = \pm 2X_0|\delta| \qquad (15.25)$$

The edges of the pass band are at $\delta = \pm 1/2Q$ and width $= 1/Q$. In Fig. 15.4, curve A is for $R = 5$ ohms, $X_0 = 500$ ohms, and $Q = 100$, with asymptotes and pass band, as shown. Curve B is for $R = 10$ ohms, $X_0 = 500$ ohms, and $Q = 50$. These two curves have the same asymptotes. The resistance, therefore, affects sharpness of tuning.

In terms of admittances

$$Y_f = G_f + jB_f$$

$$= \frac{Q}{X_0(1 + 4\delta^2 Q^2)} - \frac{2\delta Q^2}{X(1 + 4\delta^2 Q^2)} \qquad (15.26)$$

The harmonic voltage at filter bus is

$$V_h = \frac{I_h}{Y_h} \qquad (15.27)$$

For minimum voltage distortion, the overall admittance of filter should be increased. The impedance loci indicate that generally the harmonic impedances can be defined in a region of R, jX, determined by two straight lines and a circle passing through the origin (see Figs 12.27 through 12.29).

15.2.1 Tuning Frequency

ST filter is not tuned exactly to the frequency of the harmonic it is intended to suppress:

- The system frequency may change, causing harmonic frequency to change. The tolerance on filter reactors and capacitors may change due to aging or temperature effects.

- The tolerance on commercial capacitor units is ±20% and on reactors ±5%. For filter applications, it is necessary to specify closer tolerances on capacitors and reactors. When a number of capacitor units are connected in series or parallel, these are carefully formed with tested values of the capacitance so that large phase unbalances do not occur. Any such unbalances between the phases will result in overvoltage stress; in addition, the neutral will not be at ground potential in ungrounded wye-connected banks. A tolerance of ±2.0% on reactors and +5% on capacitors (no negative tolerance) in industrial environment is practical. Closer tolerances may be required for high-voltage direct current (HVDC) applications.

- Tuning to exact harmonic, which is intended to be bypassed, may attract harmonics from the adjacent facilities and overload the filters.

A change in L or C of 2% causes the same detuning as a change of system frequency by 1% [3]:

$$\delta = \frac{\Delta_f}{f_n} + \frac{1}{2}\left(\frac{\Delta L}{L_n} + \frac{\Delta C}{C_n}\right) \tag{15.28}$$

The filter impedance can also be written as

$$Z = R\left(1 + jQ\delta\frac{2+\delta}{1+\delta}\right) \tag{15.29}$$

15.2.2 Minimum Filter

A filter designed to control only the harmonic distortion, without the limitation of meeting a certain reactive power demand, is termed the minimum filter. More often, the filters are also required to meet a certain reactive power demand for power factor (PF) improvement. It may happen that

- The minimum filter may have to be sized larger than the requirements of meeting the reactive power demand only.

- Conversely, it is also true that the requirements for meeting the certain reactive power requirement may increase the size of the minimum filter.

- Both these situations have to be simultaneously considered.

Figure 15.5 shows the $R - X$, $Z - \omega$, and phase angle plots of ST filters in isolation and in parallel.

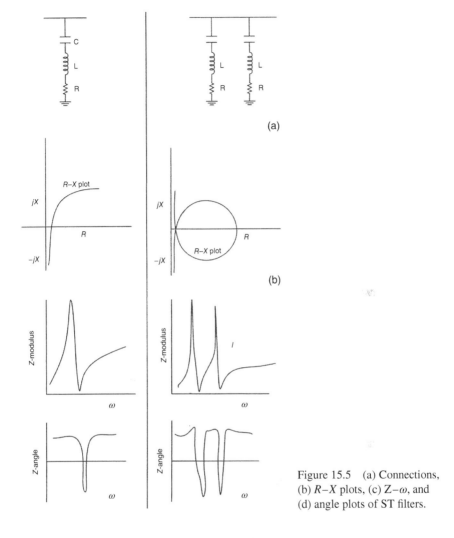

Figure 15.5   (a) Connections, (b) $R-X$ plots, (c) $Z-\omega$, and (d) angle plots of ST filters.

15.2.3   Shifted Resonant Frequencies

- With the application of an ST filter, the resonance is not eliminated. It will always shift to a frequency that is lower than the selected tuned frequency. It is given by

$$f_{11} = \frac{1}{2\pi} \sqrt{\frac{1}{(L_s + L)C}} \qquad (15.30)$$

where L_s is the system reactance.

- The resonance peak has its own value of Q given by

$$Q = \frac{\sqrt{(L_s + L)/C}}{(R_s + R)} \qquad (15.31)$$

where R_s is the system resistance.

- Each parallel ST filter gives rise to a shifted resonant frequency, below its own tuned frequency, also see Fig. 9.25. If the shifted resonance frequency coincides with one of the characteristics, noncharacteristic, or triplen harmonics present in the system, current magnification at these frequencies will occur. The switching inrush current of a transformer is rich in even and third harmonics. As the transformers are switched in and out, harmonic current injections into the system and filters will increase, though this will last for the switching duration of the transformers. It is possible that these currents are sufficiently magnified to give rise to large harmonic voltages.

- High overvoltages can occur if the system is sharply tuned to the harmonic that is being excited by the transformer inrush current (second, third, fourth, and even harmonics). Capacitor banks could also fail prematurely. This places a constraint in the design of ST filters. While designing filters for furnace installations, the furnace transformer may be switched in and out very frequently, and these frequencies must be accounted for.

When capacitors and transformers are switched together, these increase the decay time of the switching transient and harmonic resonance can occur (Fig. 8.13).

The shifted resonance frequencies should have at least *30 cycles* difference between the adjacent and odd or even harmonics. This recommendation is based on practical applications of filter designs. Even then, some amplification of the transformer switching inrush current will occur.

15.2.4 Effect of Tolerances on Filter Components

The tolerances on capacitors and reactors will result in detuning. Consider that components of the following tolerances are selected:

Capacitors: +5%

Reactors: ±2%

Let the capacitance of the fifth and seventh filters increase by 5% and the inductance by 2%. This is quite a conservative assumption for checking the detuning effect and resulting current distribution. The series-tuned frequencies of the fifth and seventh filters will shift to a lower value. An effect of the tolerances on tuning frequency is shown in the following equation [4]:

$$f_{\text{tuned}} = f_{\text{nominal}} \left(\frac{1}{\sqrt{(1 + t_r)\,(1 + t_c)}} \right) \qquad (15.32)$$

where

f_{tuned} = actual tuning frequency
$f_{nominal}$ = specified tuned frequency
t_r = reactor tolerance per unit
t_c = capacitor tolerance per unit

15.2.5 Iterative Design Requirements

The above description demonstrates the necessity of iterating the design with required tolerances, filter performance to mitigate harmonics, placement of shifted harmonic frequencies, and fine-tuning the filters. Closer tolerances on the components are an option, but that may not be practical and economically justifiable. Capacitors with metalized film construction lose capacitance as they age, resulting in a gradual increase in the tuning frequency. Nonmetalized electrode capacitors have a fairly stable capacitance. Tuning a harmonic filter more sharply than that required to attain the desired performance unnecessarily stresses the components and generally makes the filter more prone to overload from other harmonic sources.

Considerations should also be applied to future increase in the loads and consequent increase in harmonics. Transformer energizing and clearing of nearby faults will lead to temporary surge of the filter harmonic current. The faults and energizing of transformer may result in saturation of transformers, which gives additional harmonic loading. If a harmonic filter is not going to be removed automatically during a system outage, it is desirable to do a system study to determine the filter performance. Switching transients are of concern and a rigorous switching analysis is required for the selection of appropriate switching devices and to ensure that these do not precipitate a shutdown.

15.2.6 Outage of One of the Parallel Filters

Consider that we have three-step ST filters for 5th, 7th, and 11th harmonics. Outage of one of the parallel ST filters should be considered. It will have the following effects:

- The current loading of remaining filters in service may increase substantially and the capacitors and reactors may be overloaded.
- The resonant frequencies will shift and may result in harmonic current amplification.
- The harmonic distortion will increase.

It may be necessary that remaining parallel filters are also removed from service. The filter protection and switching scheme are designed so that, with the outage of one of the parallel capacitor filters, the complete system is shut down. This brings another consideration, that is, redundancy in the filter applications so that the harmonic emission at PCC is controlled within IEEE limits. Alternatively, enough spare parts and services should be available to bring the faulty unit in service in a short time (see case study 1, Chapter 16).

15.2.7 Operation with Varying Loads

When load-dependent switching is required for reactive power compensation, multiple capacitor banks are switched in an ascending order, that is, 5th, 7th, and 11th. Generally, this will occur during startup conditions; however, if sustained operation at reduced loads is required, it is necessary to control the harmonic distortion at each of the operating loads and switching steps. The harmonic loads may or may not decrease in proportion to the overall plant load. This adds another step in designing an appropriate passive filtering scheme to meet the TDD requirements.

15.2.8 Division of Reactive kvar Between Parallel Filter Banks

When multiple parallel filters are required and the total kvar requirements are also known, it remains to find out the most useful distribution of kvar among the parallel filters. Suppose we require 5th, 7th, and 11th harmonic filters, sizing them all with equal kvar is too simplistic approach, rarely implemented. As filters should be sized to handle the harmonic loading, one approach would be to divide the required kvar based on the percentage of harmonic current that each filter will carry. This will not be known in advance. The other method is to proportionate the filters with respect to harmonic current generation, that is, the lower order harmonics are higher in magnitude, so more kvar are allocated to a lower order filter. Again some iteration will be required to optimize the sizes initially chosen based on the actual fundamental and harmonic current loadings and the desired reactive power compensation, see the extensive study in Chapter 16.

15.2.9 Losses in the Capacitors

The power capacitors have some active power loss component, though small. Figure 15.6 shows the average losses versus ambient temperature for capacitors based on the published data of one manufacturer for his special dielectric filled capacitors. At an operating temperature of 40 °C, the loss is approximately 0.10 W/kvar

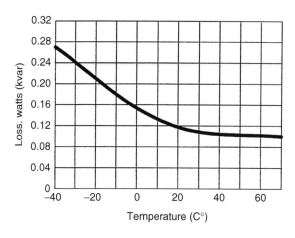

Figure 15.6 Average losses in film-foil capacitors, with variation of temperature, data of a specific capacitor element type, and a specific manufacturer.

and increases to 0.28 W/kvar at $-40\,^\circ$C. This loss should be considered in the filter design by an equivalent series resistance inserted in the circuit.

15.3   HARMONIC FILTER DETUNING AND UNBALANCE

The operation of an internal or external fuse or shorting of elements of a fuseless capacitor banks changes the capacitance of the filter and subjects it to higher overvoltage. Unbalance detection systems, discussed in Chapter 11, are applied and some considerations are as follows:

- The resonant frequency will change. Fuse operation in a parallel-connected capacitor bank will decrease the capacitance and increase the resonant frequency. There is a possibility that shorting of capacitor elements of an externally fused capacitor bank increases the capacitance and decreases the resonant frequency. It is desirable to ascertain the maximum plus/minus capacitance change that can be tolerated. The detuning may be a more stringent condition than the overvoltage on remaining units.

- Ambiguous indications are a possibility. For example, a negligible current will flow through a CT connecting the neutrals of a balanced ungrounded-double-wye bank, and this will not change if equal number of fuse failure or elements short-outs occurs in the same phases of the wye banks. When such a possibility exists, an alarm on first failure of a fuse or first failure of an element in fuseless bank is desirable to be provided.

- An arcing fault external to a capacitor unit can result in large change in filter capacitance and detuning. The unbalance protection may not always operate depending on filter configurations. A phase overcurrent relaying scheme can be designed according to IEEE Standard C37.99 [5]. Furthermore, harmonic currents should not adversely affect trip or alarm relays. The modern relays include filters and algorithms to respond on fundamental currents and voltages.

- To simulate the fuse failure in one capacitor unit in a phase and its effect on detuning, three-phase harmonic load flow modeling and analyses are required.

15.4   RELATIONS IN AN ST FILTER

The reactive power output of a capacitor at fundamental frequency is V^2/X_c. In the presence of a filter reactor, it is given by

$$S_f = \frac{V^2}{X_L - X_c}$$

$$= \frac{V^2}{X_c/n^2 - X_c}$$

$$= \frac{n^2}{n^2 - 1} \times (\text{reactive power without reactor}) \qquad (15.33)$$

The reactive power output with a filter reactor tuned to, say $4.85f$, is approximately 4% higher than without the reactor. This is so because the voltage drop in the reactor is added to the capacitor voltage, and its operating voltage is

$$V_c = V + V_L = \frac{Vj\omega L}{\left(j\omega L - \frac{1}{j\omega C}\right)} = \frac{n^2}{n^2 - 1}V \qquad (15.34)$$

- The capacitors in a fifth harmonic filter tuned to $4.85f$ operate at approximately 4% higher than the system voltage.

The steady-state fundamental frequency voltage can be

$$V_r = V\left(\frac{n^2}{n^2 - 1}\right) + \sum_{h=2}^{\infty} I_h X_{ch} \qquad (15.35)$$

where V_r is the rated voltage, I_h is the harmonic current, V is the maximum system voltage across capacitor excluding voltage rise across the reactor, and X_{ch} is the capacitive reactance at the harmonic order.

The fundamental loading of the capacitors is given by

$$\frac{V_c^2}{X_c} = \frac{V^2}{X_c}\left[\frac{n^2}{n^2 - 1}\right]^2 = s_f\left[\frac{n^2}{n^2 - 1}\right] \qquad (15.36)$$

and the harmonic loading is

$$\frac{I_h^2 X_c}{h} = \frac{I_h^2 V^2}{s_f}\frac{n^2}{n^2 - 1} \qquad (15.37)$$

When harmonic voltages and current flows are known from harmonic simulation, the harmonic loading can be found from

$$\sum_{h=2}^{h=\infty} V_h I_h \qquad (15.38)$$

The fundamental frequency loading of the filter reactor is

$$\frac{V_L^2}{X_L} = \left[\frac{V_c}{n^2}\right]^2\left[\frac{n^2}{X_c}\right] = \frac{V_c^2}{n^2 X_c} = \frac{s_f}{n^2}\left[\frac{n^2}{n^2 - 1}\right] \qquad (15.39)$$

The harmonic loading for the reactor is the same as for the capacitor.

Example 15.1: Consider that a 5-Mvar capacitor bank is required to be formed for an ST filter application at 13.8 kV. The ST filter is wye connected and ungrounded and has the harmonic loading $I_1 = 209$ A, $I_5 = 150$ A, $I_7 = 60$ A, $I_{11} = 20$ A, and $I_{13} = 9$ A, $I_{17} = 4$A, and higher order harmonics neglected. It is turned to 4.7th harmonic. Calculate the voltage at the junction of the series filter reactor and capacitors.

The capacitive reactance at power frequency is given by

$$X_{ch} = \left(\frac{n^2}{n^2 - 1}\right)\left(\frac{kV^2}{Q_{eff}\,(Mvar)}\right) \tag{15.40}$$

where Q_{eff} is the effective reactive power of the filter.

$$X_{ch} = \left(\frac{4.7^2}{4.7^2 - 1}\right)\left(\frac{13.8^2}{5}\right) = 39.89\ \Omega$$

From Eq. (15.35), under steady-state operation, the capacitors will experience a voltage of

$$209 \times 39.89 + 150 \times \frac{39.89}{5} + 60 \times \frac{39.89}{7} + 20 \times \frac{39.89}{11} + 9 \times \frac{39.89}{13}$$

$$+ 4 \times \frac{39.89}{17} = 9.99\ kV$$

Example 11.3 illustrates rated voltage selection criteria for the number of switching transients and maximum calculated transient overvoltage, phase unbalance, and so on. This criterion of calculation of voltage to which the capacitors will be exposed under steady state must be added. Selection of rated voltage is an important criterion – the safety, protection, and integrity of the banks depend on it. Many failures can be attributed to improper selection of capacitor elements' rated voltage.

15.5   SELECTION OF Q FACTOR

Equation (15.21) defines the filter Q based on the inductive or capacitive reactance at the tuned frequency (these are equal). Apart from its impact on the filter performance, the Q factor determines the fundamental frequency losses and this could be an overriding consideration, especially when the reactors at medium-voltage level are required to be located indoors in metal or fiberglass enclosures and space is at a premium. Consider a second harmonic filter that requires a filter reactor = 5.1687 ohms. An X/R of 50 (Q) gives a reactor resistance of 0.1032 ohms. If the fundamental frequency rms current is 1280 A, it gives a loss of approximately 507 kW/h, (= 4441 MW/year), which is very substantial. Mostly the filter reactors are installed in environmentally controlled rooms, and the heat load must be carefully considered in the designing of air-conditioning equipment.

The fundamental frequency losses and heat dissipation are of major consideration, but this does not mean that the effect on filter performance can be ignored.

The higher the value of Q, the more pronounced is the valley at the tuned frequency. For industrial systems, the value of R can be limited to the resistance built in the reactor itself, that is, the reactors are specified to have a certain Q factor – the higher the Q factor, the higher is the cost of the reactor. Yet, there are practical limitations of the limits of Q for the reactors.

The X/R of tuning reactors at 60 Hz is given by 3.07 $K^{0.377}$, where K is the three-phase kVA $= 3I^2X$ (I is the rated current in amperes and X is the reactance in ohms). X/R of a 1500-kVA reactor will be 50 while that of a 10-MVA reactor will be 100. High X/R reactors can be purchased at a cost premium. Thus, selection of X/R of the reactor depends on

- initial capital investment;
- active energy losses;
- effectiveness of the filtering.

For industrial and commercial power systems, the Q of the filter reactor is not so critical. This is not so for high-voltage (HV) transmission.

The optimization of filter admittance and Q for the impedance angle of the network and δ are required for the transmission systems. The optimum value of Q is given by Ref. [3]:

$$Q = \frac{1 + \cos\ \phi_m}{2\delta_m + \sin\ \phi_m} \tag{15.41}$$

where ϕ_m is the network impedance angle. Consider a frequency variation of $\pm 1\%$, a temperature coefficient of 0.02% per degree Celsius, and a temperature variation of $\pm 30\,°C$ on the inductors and capacitors, then from Eq. (15.28), $\delta = 0.006$. For an impedance angle $\phi_m = 80°$, the optimum Q from Eq. (15.41) is 99.31. *The higher the tolerances on components and frequency deviation, the lower are the value of Q.*

15.6   DOUBLE-TUNED FILTER

A double-tuned filter is derived from two ST filters and is shown in Fig. 15.7. Its $R-X$ plot and $Z-\omega$ plots are identical to that of two ST filters in parallel, as shown in Fig. 15.5. The advantage with respect to two ST filters is that the power loss at fundamental frequency is less and one inductor instead of two is subjected to full impulse voltage. In Fig. 15.7, the BIL (basic insulation level) on reactor L_2 is reduced while reactor L_1 sees the full impulse voltage.

This is an advantage in HV applications. The following equations [6] transform two ST filters of different frequencies into a single double-tuned filter:

$$C_1 = C_a + C_b \tag{15.42}$$

$$L_2 = \frac{(L_aC_a - L_bC_b)^2}{(C_a + C_b)^2(L_a + L_b)} \tag{15.43}$$

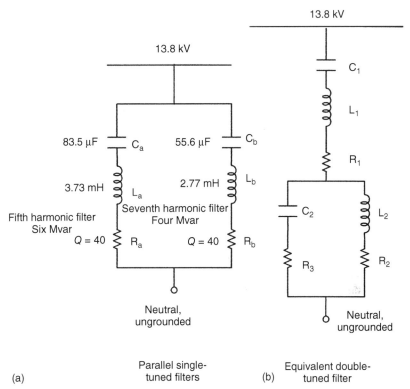

(a) Parallel single-tuned filters   (b) Equivalent double-tuned filter

Figure 15.7   (a) Circuits of two ST parallel filters of 6 Mvar (fifth harmonic) and 4 Mvar (seventh harmonic) and (b) equivalent circuit of one single double-tuned filter.

$$R_2 = R_a \left[\frac{a^2 \left(1 - x^2\right)}{(1 + a)^2 (1 + x^2)} \right] - R_b \left[\frac{1 - x^2}{(1 + a)^2 (1 + x^2)} \right]$$

$$+ R_1 \left[\frac{a (1 - a) (1 - x^2)}{(1 + a)^2 (1 + x^2)} \right] \tag{15.44}$$

$$C_2 = \frac{C_a C_b (C_a + C_b)(L_a + L_b)^2}{(L_a C_a - L_b C_b)^2} \tag{15.45}$$

$$R_3 = -R_a \left[\frac{a^2 x^4 \left(1 - x^2\right)}{(1 + ax^2)^2 (1 + x^2)} \right] + R_b \left[\frac{\left(1 - x^2\right)}{(1 + ax^2)^2 (1 + x^2)} \right]$$

$$+ R_1 \left[\frac{\left(1 - x^2\right)(1 - ax^2)}{(1 - x^2)(1 - ax^2)} \right] \tag{15.46}$$

$$L_1 = \frac{L_a L_b}{L_a + L_b} \tag{15.47}$$

where

$$a = \frac{C_a}{C_b} \quad x = \sqrt{\frac{L_b C_b}{L_a C_a}} \qquad (15.48)$$

Generally, R_1 is omitted and R_2 and R_3 are modified so that the impedance near resonance is practically the same. Note that inductor L_1 will have some resistance, which is considered in the above equations.

These types of filters are applied at all power system voltage levels: transmission, distribution, industrial, and commercial systems. For example, HVDC transmission converters use a number of filter sections tuned to discrete frequencies and are connected in parallel at the AC terminals of each converter to provide reactive power and suppress harmonics. On distribution system, these are applied for a capacitor bank detuning to control the natural resonant frequencies. A second-order high-pass filter may be added to attenuate higher harmonics.

Figure 15.7 shows the parameters of the fifth and seventh harmonic filters in parallel. A reader can convert these to a double-tuned filter using the expressions given earlier.

15.7 BANDPASS FILTERS

Bandpass filter is a new breed of filters for harmonics. Referring to Fig. 15.8(a), a simple LC circuit can act as a bandpass filter, but requires large components. It is free of resonance problems, but at no-load the output voltage can be high and PF is leading at all loads [7].

An improved LLCL filter is shown in Fig. 15.8(b). Filter capacitors C_f are delta connected and damping resistors are connected to C_f. The filter output terminals V_0 are connected to the rectifier terminals and L_0 is relatively small, 3–5%. At dominant rectifier harmonics, the large filter input reactor L_1 provides high impedance with respect to shunt filter impedance over a wide frequency range. Thus, it impedes the flow of harmonic currents generated by the rectifier into the AC lines and also

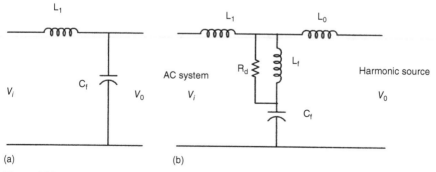

(a) (b)

Figure 15.8 (a) An RC circuit as a bandpass filter and (b) an improved broadband filter for harmonic mitigation.

minimizes the effect of line voltage harmonics on the rectifier. The filter parallel resonance frequency is given by

$$f_p = \frac{1}{2\pi\sqrt{(L_1 + L_f)C_f}} \qquad (15.49)$$

The parallel resonant frequency is selected between the fundamental frequency
and the first dominant frequency of the rectifier circuit, fifth harmonic. The voltages
are maintained in a narrow band on load variations. The components C_f and L_f provide a low-impedance path to the dominant rectifier harmonic, such as an ST filter.
    In the design process, L_0 (Fig. 15.8(b)) is selected as 4% reactor:

$$L_0 = 0.04\frac{V^2}{\omega P} \qquad (15.50)$$

where P is the rated power and V is the line-to-line voltage.
    The other filter components are selected based on total harmonic distortion
(THD), PF, and voltage excursions, no-load to full-load. This is rather a complex
mathematical problem. Reference [8] applies GA (genetic algorithm, see Section
15.14) for the optimization of these parameters using the fitness function:

$$\text{Fitness} = \text{THD} + \Delta V_o + \left(\frac{1}{PF}\right) \qquad (15.51)$$

The parameters thus calculated for a 5.5-kW adjustable speed drive (ASD) are

$$L_1(\text{mH}) = 10.1$$

$$L_f(\text{mH}) = 8.1$$

$$C_f(\mu\text{F}) = 21$$

The THD is limited to 6–6.5%, PF 0.98–0.99, and $\Delta V_0 = 3.0$–3.2% for the
5.5-kW ASD. The parameters must be calculated for each rating of ASD. The filter
has limitation when a number of different nonlinear sources are present.

15.8   DAMPED FILTERS

Figure 15.9 shows four types of damped filters. The first-order filter is not used as
it has excessive loss at fundamental frequency and requires a large capacitor. The
second-order high pass is generally used in composite filters for higher frequencies.
The filter is more commonly described as a second-order high-pass filter.
    If it were to be used for the full spectrum of harmonics, the capacitor size would
become large and fundamental frequency losses in the resistor would be of consideration. The third-order filter has a substantial reduction in fundamental frequency
losses due to the presence of C_2, which increases the filter impedance; C_2 is very

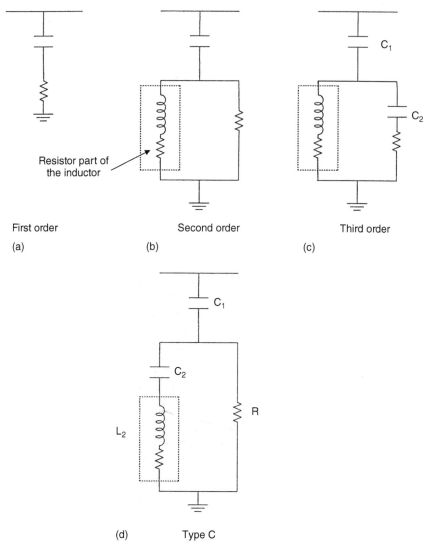

Resistor part of the inductor

First order

(a)

Second order

(b)

Third order

(c)

(d)   Type C

Figure 15.9   Circuits of damped filters. (a) First-order filter, (b) second-order filter, (c) third-order filter, and (d) type C filter.

small compared to C_1. The filtering performance of type C filters lies between that of second-order and third-order filters. C_2 and L_2 are series tuned at fundamental frequency and the fundamental frequency loss is reduced.

The bandpass filters give rise to a shifted resonance frequency, while damped filters do not. This advantage of damped filters can be exploited and possible resonances at shifted frequencies can be avoided. Unlike ST parallel multiple filters, there are no parallel branches, yet the component sizing becomes comparatively large and it may not be possible to exploit this advantage in every system design.

The performance and loading is less sensitive to tolerances. The behavior of damped filters can be described by the following two parameters (see Ref. [6]), which is the fundamental work on these types of filters:

$$m = \frac{L}{R^2 C} \tag{15.52}$$

$$f_0 = \frac{1}{2\pi CR} \tag{15.53}$$

The impedance can be expressed in the parallel equivalent form:

$$Y_f = G_f + jB_f \tag{15.54}$$

where

$$G_f = \frac{m^2 x^4}{R_1[(1 - mx^2)^2 + m^2 x^2]} \tag{15.55}$$

$$B_f = \frac{x}{R_1}\left[\frac{1 - mx^2 + m^2 x^2}{\left(1 - mx^2\right)^2 + m^2 x^2}\right] \tag{15.56}$$

where

$$x = \frac{f}{f_0} \tag{15.57}$$

Considering that the filter is in parallel with an AC system of admittance $Y_a <$ $\pm\phi_a(\max)$, then the minimum total admittance as ϕ_a and Y vary is

$$Y = B_f \cos \phi_a + G_f \sin \phi_a \tag{15.58}$$

provided that the sign of each term is taken as positive and x is less than the value that gives

$$|\cot \phi_f| = \left|\frac{G_f}{B_f}\right| = |\tan \phi_a| \tag{15.59}$$

For a given C, select parameters f_0 and m to obtain a sufficiently high admittance (low impedance) over the required frequency range. Values of m are generally between 0.5 and 2.

15.8.1 Second-Order High-Pass Filter

The characteristics of a second-order high-pass filter are shown in Fig. 15.10, with its $R - X$ and $Z - \omega$ plots. It has low impedance above a corner frequency; thus, it will shunt a large percentage of harmonics at or above the corner frequency. The sharpness of tuning in high-pass filters is the reciprocal of ST filters:

$$Q = \frac{R}{(L/C)^{1/2}} = \frac{R}{X_{LN}} = \frac{R}{X_{CN}} \tag{15.60}$$

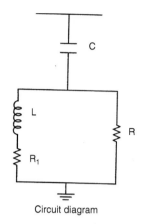

(a)   Circuit diagram

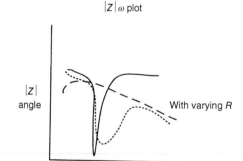

(b) ω

$|z|\,\omega$ plot

(c) ω

Figure 15.10   Second-order high-pass filter. (a) Circuit, (b) R–X plot, and (c) z–ω plot.

$X_{LN} = X_{CN}$ at tuned frequency. Filter impedance is given by

$$Z = \frac{1}{j\omega C} + \left(\frac{1}{R} + \frac{1}{j\omega L} \right)^{-1} \tag{15.61}$$

The higher the resistance, the greater is the sharpness of tuning. The Q value may vary from 0.5 to 2 and there is no optimum Q, unlike with bandpass filters.

The reactive power of the capacitor at fundamental frequency is the same as for an ST filter. The loading at harmonic h is

$$I_h^2 \frac{X_c}{h} = \frac{1}{S_f} \frac{I_h^2}{h} V^2 \left[\frac{n^2}{n^2 - 1} \right] \tag{15.62}$$

Thus, the total harmonic loading is

$$V^2 \frac{n^2}{S_f(n^2 - 1)} \sum_{h=\min}^{h=\max} \frac{I_h^2}{h} \tag{15.63}$$

The reactor loading at fundamental frequency can be calculated by assuming that current through the parallel resistor is zero, that is, current through the inductor is the same as through the capacitor; then, fundamental frequency loading is

$$I_L^2 X_L = I_h^2 \frac{X_c}{n^2} = \frac{S_f}{n^2} \left[\frac{n^2}{2} - 1 \right] \tag{15.64}$$

At harmonic h, the harmonic current I_h divides into the resistance and inductance. The inductive component of the current is

$$I_{hL} = I_h \frac{R}{R + j\omega L} = I_h \frac{Q}{[Q^2 + (h/n)^2]^{1/2}} \tag{15.65}$$

The total harmonic loading is, therefore

$$= Q^2 \frac{V^2}{S_f} \left[\frac{n^2}{n^2 - 1} \right] \sum_{h=\min}^{h=\max} \left[h \frac{I_h^2}{Q^2 n^2 + h^2} \right] \tag{15.66}$$

The loss in the resistor can be calculated as follows:

$$R = QhX_L \tag{15.67}$$

$$|I_R| = \frac{|I_L| X_L}{R} = \frac{I_L}{Qn} \tag{15.68}$$

Thus, the power loss is

$$I_R^2 R = \frac{1}{Qn} I_L^2 X_L \tag{15.69}$$

$$= \frac{1}{Qn} (\text{Mvar loading}) \tag{15.70}$$

$$= \frac{S_f}{Qn^3} \left[\frac{n^2}{n^2 - 1} \right] \tag{15.71}$$

It can be shown that, generally, to mitigate harmonics of the lower order, a much larger second-order high-pass filter is required. Practically, one or more ST filters are used for lower order harmonics, and second-order high-pass filter provides filtering of higher order harmonics and notch reduction. Sometimes, parallel ST filters, tuned to specific low-order harmonics, are provided with a second-order high-pass filter (Fig. 15.11). In this figure, R_1 are the resistors associated with the filter reactors. The effect of variation of resistor R on the $Z - \omega$ plot of a second-order high-pass filter acting alone is shown in Fig. 15.12. The second-order high-pass filter can be effectively designed for the higher order harmonics, while ST filters cater for lower order harmonics. As damped filters do not give rise to a shifted resonant frequency, there application is much desirable when a band of interharmonics are present. Case study 3 in Chapter 16 shows the design and application of this type of filter for mitigation of a wide band of frequencies.

15.9   TYPE C FILTER

C filter was first introduced in France–England HVDC interconnection project [9–11], and then in Intermountain and Quebec–New England HVDC projects. It can replace conventional ST filters effectively and finds its use in arc furnace and ladle furnace installations [12,13]. Figure 15.13(a) shows the equivalent circuit of a type C filter. Neglecting the resistance of the reactor, the impedance of a C-type filter is given by

$$Z(\omega) = \left(\frac{1}{R} + \frac{1}{j\omega L + 1/(j\omega C)} \right)^{-1} + \frac{1}{j\omega C_1}$$

$$= \frac{R(\omega^2 LC - 1)^2 + jR^2 \omega C(\omega^2 LC - 1)}{(R\omega C)^2 + (\omega^2 LC - 1)^2} - j\frac{1}{\omega C_1} \qquad (15.72)$$

The impedance varies with the frequency. To avoid power loss at fundamental frequency f in damping resistor R, the components L and C are tuned to fundamental frequency:

$$\omega_f^2 LC = 1 \qquad (15.73)$$

Therefore, impedance of filter at fundamental frequency is determined by C_1:

$$Z(\omega_f) = \frac{-j}{\omega_f C_1} = -j\frac{V_s}{Q_f} \qquad (15.74)$$

where Q_f is the reactive power requirement at fundamental frequency and V_s is the nominal system voltage. This allows a straightway calculation of C_1.

As the frequency increases, L starts to resonate with $C + C_1$, which makes the filter act as an ST filter with a damping resistor.

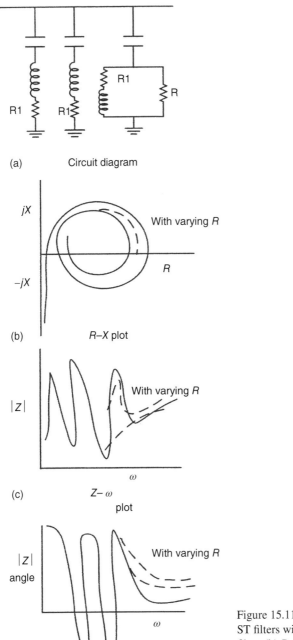

(a)   Circuit diagram

(b) $R-X$ plot

(c) $Z-\omega$ plot

(d)

Figure 15.11   (a) Circuit of two parallel
ST filters with a second-order high-pass
filter, (b) $R-X$ plot, (c) $z-\omega$ plot, and
(d) angle plot.

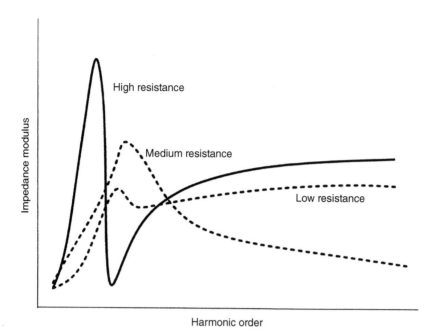

Figure 15.12   Effect of magnitude of resistance R on impedance characteristics in a second-order high-pass filter.

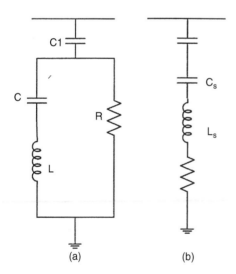

Figure 15.13   (a) Type C filter and (b) equivalence with an ST filter.

The impedance varies with the order of harmonic:

$$\frac{R[jX(h-1/h)]}{R+ +[jX(h-1/h)]} - j\frac{X_1}{h} \tag{15.75}$$

At the tuned frequency, total reactance of the filter is zero:

$$\frac{R^2\omega_0 C(\omega_0^2 LC - 1)}{(R\omega_0 C)^2 + (\omega_0^2 LC - 1)^2} - \frac{1}{\omega_0 C_1} = 0 \tag{15.76}$$

where ω_0 is the tuned frequency in radians.
And the total resistance of the filter is

$$r = \frac{R(\omega_0^2 LC - 1)^2}{(R\omega_0 C)^2 + (\omega_0^2 LC - 1)^2} \tag{15.77}$$

The filter behaves as an equivalent resistance at the tuned frequency. From the above equations:

$$\frac{\omega_0 RC}{\omega_0^2 LC - 1} = \frac{1}{r\omega_0 C_1} \tag{15.78}$$

and

$$r = \frac{R}{\frac{1}{(r\omega_0 C_1)^2} + 1} \tag{15.79}$$

Then, at the tuned frequency,

$$r^2 - Rr + \frac{1}{(\omega_0 C_1)^2} = 0 \tag{15.80}$$

If h_0 is the tuned *harmonic*, then

$$h_0 = \omega_0 \sqrt{LC} \tag{15.81}$$

If we introduce a term

$$R_0 = \frac{2}{\omega_0 C_1} \tag{15.82}$$

and R must be $> R_0$, say $R = mR_0$, for $m \geq 1$, then Eq. (15.80) becomes

$$r^2 - mR_0 r + \frac{R_0^2}{4} = 0 \tag{15.83}$$

One positive root is

$$r = \frac{m - \sqrt{m^2 - 1}}{2} R_0 \tag{15.84}$$

On the basis of the value of r, the filter parameters L and C are as follows:

$$C = \frac{h_0^2 - 1}{m^2 - m\sqrt{m^2 - 1}} \frac{Q_f}{2V_s^2 \omega_f}$$

$$L = \frac{m^2 - m\sqrt{m^2 - 1}}{h_0^2 - 1} \frac{2V_s^2}{\omega_f Q_f} \tag{15.85}$$

L and C can be selected by assuming $m > 1$, but these are not the final or optimal results.

Because L and C are tuned to the fundamental frequency, the fundamental current I_f will entirely flow through components L and C:

$$I_f = \frac{V_s}{\sqrt{3}} \omega_f C_1 \tag{15.86}$$

The reactive power through L and C must be the same; this means the larger the L and the smaller the C or vice versa. If their capacities can be reduced to a minimum, the total filter investment will be reduced.

Examine function

$$g(m) = m^2 - m\sqrt{m^2 - 1} \tag{15.87}$$

Its derivative is always negative and monotonous decreasing function. Its maximum is given by:

$$g(m) = m^2 - m\sqrt{m^2 - 1} \to 0.5, \text{ as } m \to \infty \tag{15.88}$$

This gives parameters L and C as

$$C = \frac{(h_0^2 - 1)Q_f}{\omega_f V_s^2}$$

$$L = \frac{V_s^2}{(h_0^2 - 1)\omega_f Q_f} \tag{15.89}$$

This is based on the assumption that m is infinite. This means that there is no parallel resistance R in the C type filter. Then, the filter returns to an ST filter, with equivalence as shown in Fig. 15.13(b).

L_s is the same as L in C type filter:

$$L_s = \frac{V_s^2}{(h_0^2 - 1)\omega_f Q_f} \tag{15.90}$$

$$\frac{1}{C_s} = \frac{1}{C} + \frac{1}{C_1} = \frac{h_0^2 - 1}{h_0^2 \omega_f} \frac{Q_f}{V_s^2} \tag{15.91}$$

This helps in calculating the C filter parameters.

Finally, to get R, the filter quality factor Q_{filter} can be made use of. It is given by

$$Q_{filter} = \frac{R\omega_0 C}{\omega_0^2 LC - 1} = Rh_0 \frac{Q_f}{V_s^2}$$

$$R = \frac{Q_{filter} V_s^2}{h_0 Q_f} \qquad (15.92)$$

Practically, for many HVDC projects, Q varies from 1 to 2. Case study 2 in Chapter 16 shows the application of a type C filter to an arc furnace installation. Table 15.1 shows some parameters of C type filters applied in HVDC projects. The characteristics with respect to frequency are shown in Fig. 15.14.

TABLE 15.1   Parameters of C-Type Filters in Some HVDC Projects

Projects↓Parameters→	f	h_0	kV	Q	Q_{filter}	R (ohms)	L (mH)	C (μF)	C_1 (μF)
France–England Interconnection	50	3	400	130	1.64	666	424	23.89	2.586
Intermountain Power Project	60	3	345	58	2	1300	658	10.7	1.3
Quebec–New England Radisson	60	3	315	49	2	1349	671	10.48	1.31
Nicolet terminal	60	3	230	38	2	928	462	15.24	1.91
Sandy Pond terminal	60	3	345	88	1	450	450	15.63	1.954
Longqan–Zhenping	60	3	525	118	2.3	1800	929.39	10.929	1.363

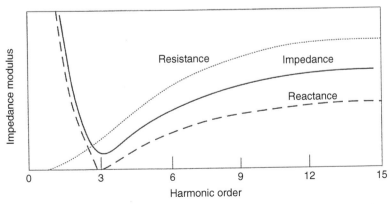

Figure 15.14   Characteristics of type C filter, resistance, reactance, and impedance versus frequency.

15.10   ZERO SEQUENCE TRAPS

Zigzag transformers and delta–wye transformers will act as zero sequence traps when connected in the neutral circuit of a three-phase four-wire system. Figure 15.15 illustrates a delta–wye transformer serving single-phase nonlinear loads of switched-mode power supplies, PCs, printers, and fluorescent lighting. As discussed in Chapter 3, the neutral can carry excessive harmonic currents. A zigzag or delta–wye transformer connected as shown in Figure 15.15 will reduce harmonic currents and voltage.

As we discussed in Chapter 1, the zero sequence impedance of the core-type delta–wye transformer is low as the zero sequence flux seeks a high-reluctance path through air or the transformer tank. The delta winding carries the zero sequence currents to balance the primary ampere turns. In an unbalanced system, the positive and negative sequence components will also be present, and these will not be suppressed. In a zigzag transformer, all windings have the same number of turns, but each pair of windings on a leg is wound in the opposite direction. A zigzag transformer has low zero sequence impedance and works in the same manner as a delta–wye transformer.

The neutral currents have two parallel paths, both of low impedance, through the delta–wye or zigzag transformer and also through the grounded neutral. The neutral voltage rise will be much less, though it will not be completely stable.

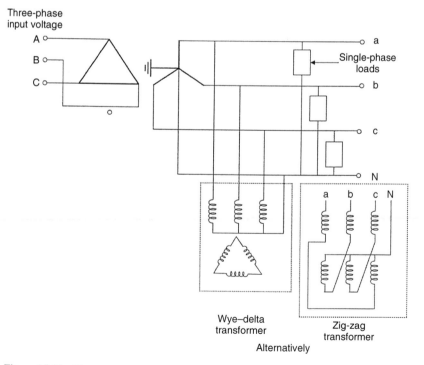

Figure 15.15   Zero sequence traps.

15.11   SERIES-TYPE LOW-PASS FILTER

The basic configuration is shown in Fig. 15.16. This type of filter when properly designed should not result in resonance problems. The capacitor acts as a low-impedance path to the harmonics generated by the load. The high-frequency currents are absorbed by the capacitor. At fundamental frequency, the capacitor supplies reactive power and causes a voltage rise across the reactor. This voltage can be controlled by a regulating transformer to reduce the voltage to the load. If the voltage is not controlled, it can cause loads such as ASDs to trip on overvoltage.

The following equations can be written as

$$X_0 = \sqrt{X_L X_C} = \sqrt{\frac{L}{C}} \tag{15.93}$$

Filter size is

$$Q_{\text{filter}} = \frac{V^2}{X_C - X_L} = \frac{h_n^2}{h_n^2 - 1} Q_c \tag{15.94}$$

The series impedance at harmonic h is

$$Z_h = R_h + j\left(hX_L - \frac{X_C}{h}\right) \tag{15.95}$$

Voltage at fundamental frequency across capacitor is

$$V_{c,f} = V_{\text{bus},f} \frac{h_n^2}{h_n^2 - 1} \tag{15.96}$$

where $V_{\text{bus},f}$ is the fundamental frequency voltage on the bus.

And the harmonic voltage at the capacitor at the tuned frequency is

$$-j\frac{X_n}{R} = -jQ \tag{15.97}$$

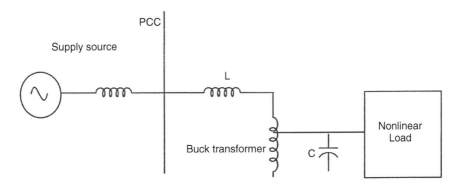

Figure 15.16   Series type low-pass filter.

15.12 TRANSFER FUNCTION APPROACH FOR FILTER DESIGNS

Transfer function approach using Laplace transform is one method of passive filter designs. This book does not discuss Laplace transform. Laplace transform converts a time domain function into complex frequency ($s = \sigma + j\omega$), the Fourier transform converts it into imaginary frequency of $j\omega$ [14].

Laplace transformation transforms the exponential and transcendental functions and their combinations into algebraic equations. Differentiation and integration are transformed into multiplication and division. Effective use of step and impulse responses is made. The boundary values are considered in the formation of the transform. After a solution of these geometric equations is found, the inverse Laplace transform is applied to obtain the solution in the time domain. This three-step process of solving equations may simplify the calculations.

If $f(t)$ is a function defined for all positive values of t, then

$$F(s) = \int_0^\infty e^{-st} f(t) dt \tag{15.98}$$

is called the Laplace transform of $f(t)$, providing the integral exists. We write it in the form:

$$L[f(t)] = F(s) \tag{15.99}$$

If $F(s)$ is the Laplace transform of $f(t)$ and denoted by above equation, then

$$f(t) = L^{-1} F(s) \tag{15.100}$$

is called the inverse Laplace transform of $F(s)$.

Reverting to Fig. 15.3(b), it can be drawn as Fig. 15.17. Then the system/filter transfer function is

$$F(s) = Z(s) = \frac{V(s)}{I(s)} \tag{15.101}$$

where the quantities shown are in s domain. Then

$$F(s) = \frac{V(s)}{I_f(s) + I_s(s)} = \frac{1}{(1/Z_f(s)) + (1/Z_s(s))} \tag{15.102}$$

Define $F_{cds}(s)$ as the ratio of the system current to the injected current and $F_{cdf}(s)$ as the ratio of the filter current to the injected current. Then

$$F_{cds}(s) = \frac{Z_f(s)}{Z_f(s) + Z_s(s)}$$

$$F_{cdf}(s) = \frac{Z_s(s)}{Z_f(s) + Z_s(s)} \tag{15.103}$$

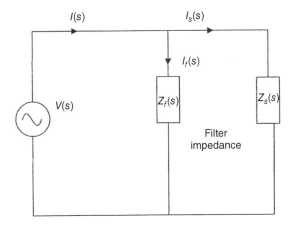

Figure 15.17   Circuit of Fig. 15.3(b) drawn in Laplace transform.

Therefore,

$$F_{cdf}(s) = \frac{Z_s(s)}{Z_f(s)} F_{cds}(s) \tag{15.104}$$

From Fig. 15.17, the filter transformation is

$$F_{fs}(s) = \frac{Z_f(s)Z_s(s)}{Z_f(s) + Z_s(s)} \tag{15.105}$$

Then

$$F_{cds}(s) = \frac{1}{Z_s(s)} H_{fs}(s)$$

$$F_{cdf}(s) = \frac{1}{Z_f(s)} H_{fs}(s) \tag{15.106}$$

These equations can be compared with Eqs (15.16) and (15.17). For a series resonant circuit

$$F_f(s) = R + sL + \frac{1}{sC}$$

$$= \frac{1}{sC}\left[1 + \frac{1}{Q}\left(\frac{s}{\omega_0}\right) + \left(\frac{s}{\omega_0}\right)^2\right] \tag{15.107}$$

The series impedance plot is as shown in Fig. 15.18.

The system and filter current divider transfer function $F_{cds}(s)$ is evaluated at high and low frequencies to determine asymptotes. If the source resistance is ignored and the source impedance L_s is considered $> L$, the situation is shown in Fig. 15.19,

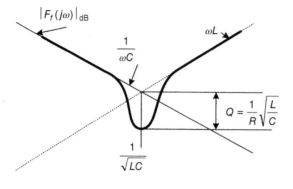

Figure 15.18 Impedance transfer function of a series RLC filter.

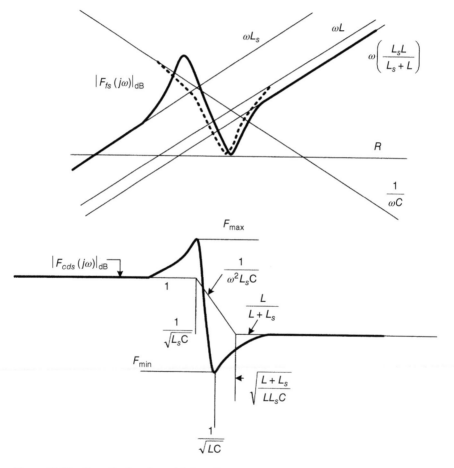

Figure 15.19 Transfer function with $L_s > L$.

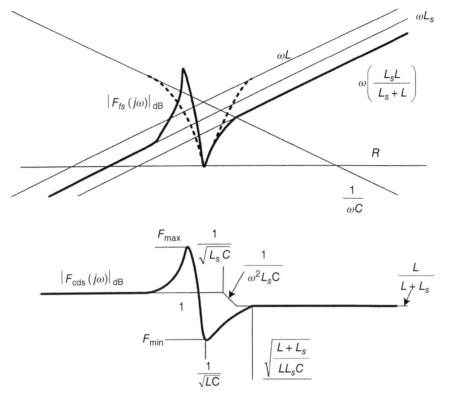

Figure 15.20 Transfer function with $L_s < L$.

and for $L_s < L$ it is shown in Fig. 15.20. F_{max} does not occur at the parallel resonant frequency formed by filter capacitance and system inductance:

$$F_{max} \neq |F_{cds}(j\omega)| \quad \text{at} \quad \omega = \frac{1}{\sqrt{L_s C}}$$

$$F_{min} \approx |F_{cds}(j\omega)| \quad \text{at} \quad \omega = \frac{1}{\sqrt{L_s C}} \qquad (15.108)$$

The transfer function can be plotted graphically to determine the approximate maxima or the derivatives of the functions can be evaluated numerically:

$$\frac{d}{d\omega}|F_{cds}(j\omega)| = 0$$

$$\frac{d}{d\omega}|F_{fs}(j\omega)| = 0 \qquad (15.109)$$

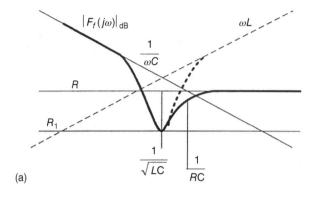

(a)

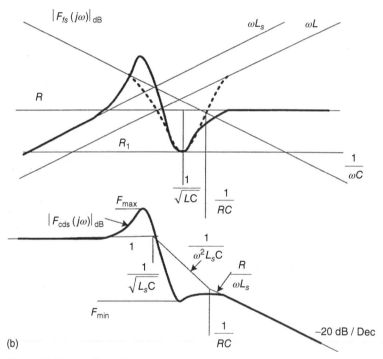

(b)

Figure 15.21 (a) Second-order damped filter impedance transfer function, (b) second-order damped filter/system impedance and current divider transfer function.

For a second-order high-pass filter (Fig. 15.21), the transfer function can be expressed as

$$F_f(s) = \frac{A}{s\left(1 + \frac{s}{\omega_p}\right)}\left[1 + \frac{1}{Q_p}\left(\frac{s}{\omega_0}\right) + \left(\frac{s}{\omega_0}\right)^2\right] \qquad (15.110)$$

where

$$A = \frac{1}{C}$$

$$\omega_0 = \sqrt{\frac{RR_1}{RLC}} \approx \frac{1}{\sqrt{LC}}$$

$$Q_p = \frac{RR_1}{RR_1 CL\omega_0}$$

$$\omega_p = \frac{RR_1}{L} \approx \frac{R}{L}$$

Depending on R, many transfer function characteristics are possible. High Q_p allows more series resonant attenuation and less high pass and lower Q_p: lesser series resonant and higher bypass response. Q_p are commonly in the range $0.5 \le Q_p \le 2.0$. Figure 15.21(a) shows the filter impedance transfer function and Fig. 15.21(b) depicts the filter/system impedance and current divider transfer functions (see Refs [15,16]).

Higher order filters are also used. Any number of series branches can be combined to numerically define $F_f(s)$. We described poles and zeros in Chapter 9, and a factored closed form of equation can be derived for poles and zeros. The transfer function approach becomes attractive, as it required less complex mathematics. The solution can then be plotted graphically, and an iterative design procedure was used for optimization.

$$F_f(s) = \frac{1}{\frac{1}{Z_{fs}(s)} + \frac{1}{Z_{f1}(s)} + \frac{1}{Z_{f11}(s)} + \cdots + \frac{1}{Z_{fn}(s)}} \qquad (15.111)$$

15.13   OPTIMIZATION TECHNIQUES OF FILTER DESIGNS

The application of computer optimization techniques in power systems is a powerful analytical tool, reaching new dimensions with new algorithm for reliability, speed, and applicability. Optimization can be aimed at reducing something undesirable, that is, harmonic distortion or maximizing a certain function, for example, PF. Such maxima and minima are always subject to certain constraints. The problem of optimization is thus translated into the problem of constructing a reliable mathematical model aimed at maximizing or minimizing a certain function within the specified constraints.

Linear programming [17–19] deals with situations where a maximum or minimum of a certain set of linear functions is desired. The equality and inequality constraints define a region in the multidimensional space. Any point in the region or boundary will satisfy all the constraints; thus, it is a region enclosed by the constraints and not a discrete single-value solution. Given a meaningful mathematical function of one or more variables, the problem is to find a maximum or minimum, when the values of the variables vary within some certain allowable limits. The variables may react with each other or a solution may be possible within some acceptable violations, or a solution may not be possible at all.

Mathematically, we can minimize

$$f(x_1, x_2, \ldots, x_n) \tag{15.112}$$

subject to

$$g_1(x_1, x_2, \ldots, x_n) \le b$$

$$g_2(x_1, x_2, \ldots, x_n) \le b_2$$

$$\ldots$$

$$g_m(x_1, x_2, \ldots, x_n) \le b_n \tag{15.113}$$

The linear programming is a special case of an objective function (Eq. (15.112)) when all the constraints (Eq. (15.113)) are linear.

15.13.1 Interior Penalty Function Method

It is a constrained optimization method [20,21]. The unconstrained optimization equation for a constrained function $f(x)$ with $g_j(x)$ on x is

$$\phi(x) = f(x) + r_k \sum_{j=1}^{j=m} G_j(g_j(x)) \tag{15.114}$$

where the second term is called the penalty term, which is used to represent constraints on x.

The Cauchy unconstrained method of modified steepest descent is used to optimize the unconstrained equation formed by penalty function method:

$$S = -\nabla\phi \tag{15.115}$$

From an initial starting point, a step of optimal length is taken along the search direction. The initial starting point is important for convergence. The next point is given by

$$X_{new} = X_{old} + S\lambda \tag{15.116}$$

where S is the search direction and λ is the step length.

The optimization can be based on the following:

- Minimum harmonic levels, that is, lowest value of THD
- Minimum filter losses
- Minimum cost.

The impedance of an ST filter as a function of Q at any frequency is given by

$$Z^2(Q) = R^2 \left(1 + Q^2 \delta^2 \left(\frac{\delta + 2}{\delta + 1} \right) \right) = R^2(1 + K^2 Q^2) \tag{15.117}$$

where

$$K = \delta \sqrt{\frac{\delta + 2}{\delta + 1}} \tag{15.118}$$

$Z^2(Q)$ is the optimization function for an ST filter and Q lies in some range between $a < Q < b$. This range can be treated as a constraint for optimizing variable Q. Then the penalty term is given by

$$r_k \sum_{j=1}^{m} G_j(g_j(Q)) = r_k[\log(Q - a) + \log(b - Q)] \tag{15.119}$$

Then ϕ, unconstrained function, is given by

$$\phi(Q, r_k) = Z^2(Q) + r_k[\log(Q - a) + \log(b - Q)] \tag{15.120}$$

Thus, ϕ is optimized using steepest gradient method, also see equation for optimum Q.

Similarly, the total loss in an ST filter is the sum of losses in the resistance, and the capacitor can be optimized.

There are many optimization techniques that can be applied similar to optimal load flow programs.

15.13.2   Interior Point Methods and Variants

In 1984, Karmarkar [22] announced the polynomially bounded algorithm, which was claimed to be 50 times faster than the simplex algorithm. Consider the linear programming problem defined as

$$\min c^{-t}\overline{x}$$

$$\text{st } \overline{A}\overline{x} = \overline{b}$$

$$\overline{x} \geq 0 \tag{15.121}$$

where $\overline{c}$ and $\overline{x}$ are n-dimensional column vectors, $\overline{b}$ is an m-dimensional vector, and $\overline{A}$ is an $m \times n$ matrix of rank m, $n \geq m$. The conventional simplex method requires 2^n iterations to find the solution. The polynomial time algorithm is defined as an algorithm that solves the LP problem in $O(n)$ steps. The problem of Eq. (15.121) is translated into

$$\min c^{-t}\bar{x}$$

$$\text{st } \bar{A}\bar{x} = 0$$

$$e^{-t}\bar{x} = 1$$

$$\bar{x} \geq 0 \qquad\qquad (15.122)$$

where $n \geq 2, \bar{e} = (1, 1, \ldots, 1)^t$, and the following holds:

- The point $x^0 = (1/n, \ 1/n, \ \ldots, \ 1/n)^t$ is feasible in Eq. (15.122)
- The objective value of Eq. (15.122) $= 0$
- Matrix $\bar{A}$ has full rank of m.

The solution is based on the projective transformations followed by optimization over an inscribed sphere, which creates a sequence of points converging in polynomial time. A projective transformation maps a polytope $P\epsilon R^n$ and a strictly interior point (IP) $a \epsilon P$ into another polytope P' and a point $a' \epsilon P'$. The ratio of the radius of the largest sphere contained in P' with the same center a' is $O(n)$. The method is commonly called an IP method due to the path it follows during solution. The numbers of iterations required are not dependent on the system size.

15.13.3   Karmarkar Interior Point Algorithm

The algorithm creates a sequence of points $x^0, x^1, \ldots, x^k$ in the following steps:

1. Let x^0 be the center of simplex.
2. Compute next point $x^{k+1} = h(x^k)$. Function $\phi = h(a)$ is defined by the following steps:
3. Let $\bar{D} = \text{diag}(a_1, a_2, \ldots, a_n)$ be the diagonal matrix.
4. Augment $\bar{A}\,\bar{D}$ with rows of 1 s:

$$\bar{B} = \begin{vmatrix} \bar{A}\bar{D} \\ \bar{e}^t \end{vmatrix} \qquad\qquad (15.123)$$

5. Compute orthogonal projection of DC into the null space of B:

$$\bar{c}_p = [1 - \bar{B}^t(\bar{B}\bar{B}^t)^{-1}\bar{B}]\bar{D}c \qquad\qquad (15.124)$$

6. The unit vector in the direction of $\bar{c}_p$ is

$$\bar{c}_u = \frac{\bar{c}_p}{|c_p|} \qquad\qquad (15.125)$$

7. Take a step of length ωr in the direction of $\bar{c}_u$:

$$Z = \bar{a} - \omega r \bar{c}_u, \text{ where } r = \frac{1}{\sqrt{n(n-1)}} \qquad (15.126)$$

8. Apply reverse protective transformation to z:

$$\bar{\phi} = \frac{\overline{D}\bar{z}}{e^t \overline{D}\bar{z}} \qquad (15.127)$$

Return ϕ to x^{k+l}.

The potential function is defined as

$$f(\bar{x}) = \sum_{i=1}^{n} \ln \left(\frac{c^t x}{x_i} \right) \qquad (15.128)$$

Check for infeasibility.

There should be certain improvement in the potential function at each step. If the improvement is not achieved, it can be concluded that the objective function must be positive. This forms a test of the feasibility.

15.13.4 Barrier Methods

A year after Karmarkar announced his method, Gill et al. [23] presented an algorithm based on projected Newton logarithmic barrier methods. Gill's [24] work showed that Karmarkar work can be viewed as a special case of a barrier function method for solving nonlinear programming problems. One drawback of the Karmarkar algorithm is that it does not generate dual solutions, which are of economic significance. Todd's [25] work, an extension of Karmarkar's algorithm, generates primal and dual solutions with objective values converging to a common optimal primal and dual value. Barrier function methods treat inequality constraints by creating a barrier function, which is a combination of original objective function and weighted sum of functions with positive singularity at the boundary. As the weight assigned to the singularities approaches zero, the minimum of barrier function approaches the minimum of original function.

The following are variants of Karmarkar's IP method:

- Projective scaling methods
- Primal and dual affine methods [26]
- Barrier methods
- Extended quadratic programming using IP.

Reference [27] describes a passive filter optimization function based on the following:

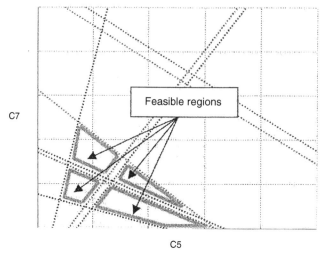

C7

C5

Figure 15.22   Optimization of filter capacitors for fifth and seventh harmonic filters with given constraints. The feasible regions bounded by constraint lines.

- Rms value of source current
- THD of source current
- THD of load voltage
- Constraints generated by harmonic resonances. The tuned frequencies where harmonic resonance is a possibility can be avoided. The constraints create an exclusion band where the capacitance of the filters cannot be placed
- Reactive power compensation
- Immunization against parameter changes.

The problem is solved using dual-logarithmic barrier interior point method. Figure 15.22 shows the feasibility regions for two filter cases: fifth and seventh harmonics [27]. The dotted lines show excluded regions because of imposed constraints. MATLAB is another powerful tool for optimization of filter parameters (also see Refs [28–33]).

15.14   GENETIC ALGORITHMS FOR FILTER DESIGNS

The framework of GA was proposed by Dr. J. Holland in 1975 [34]. It is based on the principle of natural selection and population genetics. It has been applied to applications in many different fields such as communications, computer networks, reliability, pattern recognition, neural networks, and so on.

At first, the representations of possible solutions that are called *chromosomes* or individuals must be developed. A chromosome consists of number of binary genes, which can be represented by binary codes. Different combinations of genes form different chromosomes. *Each chromosome is a possible solution of the problem.*

The set of chromosomes is called the *population of the generation*. Chromosomes in a generation are forced to evolve into better ones in the next generation by three basic GA operators:

- Reproduction
- Crossover
- Mutation
- And the problem-specified fitness function also called the objective function.

In reproduction, a number of selected exact copies of chromosomes in the current population become the offsprings. In crossover, randomly selected crossovers of two individual chromosomes are swapped to produce the offspring. In mutation, randomly selected genes in chromosomes are altered by a probability equal to the mutation rate. For binary coding, it means that digit 1 becomes 0 and vice versa. Figure 15.23 shows (a) reproduction, (b) crossover, and (c) mutation.

(a)

Current generation		Next generation
01011101	→	01011101
11000110	→	11000110

(b)

Current generation		Next generation
01011101	→	01000101
11000110	→	01011101

(c)

Current generation		Next generation
01011101	→	01001101
11000110	→	11000111

Figure 15.23   Basic operations of GAs: (a) reproduction, (b) crossover, and (c) mutation. Source: Ref. [35].

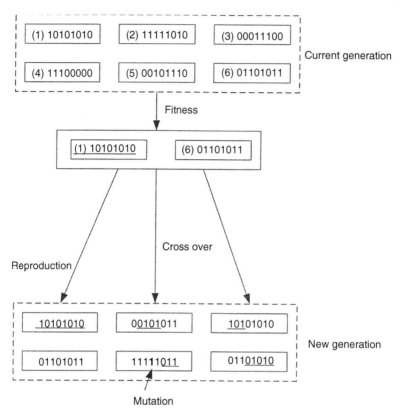

Figure 15.24   Standard procedure for GA operations from one generation to another. Source: Ref. [35].

The fitness function (objective function) plays the role of environment to distinguish between good and bad chromosomes. The conceptual procedure from one generation to another is shown in Fig. 15.24. Each chromosome is evaluated by objective function and some good chromosomes are selected [35,36].

GAs do not require the use of derivatives and offer parallel searching in a solution space rather than point-by-point search. Hence, GA can provide an optimal solution for a complex problem quickly.

The GA can be applied to different passive LC filter topologies.

15.14.1   Particle Swarm Optimization (PSO)

An approach based on particle swarm optimization (PSO) can be used, and the authors in Ref. [37] state that it takes lesser time to converge as compared to GA. It is a population-based stochastic search algorithm. It mimics the natural process of group communication of swarm of animals, for example, insects or birds. If one member finds a desirable path, the rest of the swarm follows it. In PSO, the behavior is imitated by particles with certain positions and velocities in a search space, wherein the population is called a swarm, and each member of the swarm is called a particle. Each particle flies through the search space and remembers the best position it has seen. Members of swarm communicate these positions to each other and adjust their positions and velocities based on these best positions. The velocity adjustment is based on historical behavior of the particles as well as their companions. In this way, the particles fly to the optimum position.

The design objective of filters is formed as a weighted optimization problem. PF, THD of main AC supply current, the reduction in THD are modeled for maximization of objective by considering a variable:

$$\text{Opt1} = 1 - \text{THD} \tag{15.129}$$

Thus, with weights w_1 and w_2, the main objective function is

$$\text{Obj1} = w_1 \text{PF} + w_2 \text{Opt1} \tag{15.130}$$

The composite objective function can be written as

$$\text{obj} = Max \left[\left\{ \sum \text{obj} I_k \right\} / M \right] \quad k = 2, 3, 4, \dots, M \tag{15.131}$$

where M is the number of loading positions $= 10$, and $k = 2$ represents 20% load. The constraints can be written as

$$\text{THD}_i \le \text{THD}_{i,\text{permissible}}$$

$$Q_{\min} \le Q \le Q_{\max} \tag{15.132}$$

Figure 15.25 shows a flowchart of calculations. Reference [38] describes the results of passive filter design for a 12-pulse converter-fed LCI synchronous motor drive of 60 kVA, four-pole, 400 V, 50 Hz. For example, for constant torque variable speed loads, the second-order high-pass filter parameters are $C = 1000 \ \mu\text{F}$, $L = 0.06$ mH, and $R = 0.245$ Ohms.

15.15   HVDC–DC FILTERS

DC harmonic filters are shunt filters connected between DC pole bus and the ground bus. They are tuned to DC harmonic frequencies so that these harmonics are diverted to earth and do not pass onto the DC lines. The harmonics on the DC side were discussed in Chapter 4 and the necessity of providing filters on the DC side for HVDC transmission was discussed in Chapter 13. The DC smoothing reactor is a must. The DC filters do not contribute to the shunt compensation as none is needed.

From the DC filtration point of view, Fig. 15.26 shows a schematic representation. The rectifier- and inverter-generated voltage harmonics generate the associated harmonic currents. The HVDC line acts as distributed series and shunt harmonic impedances, and the equivalent impedance at each harmonic frequency is different.

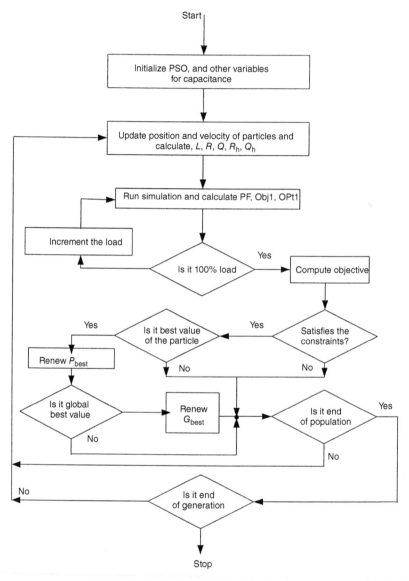

Figure 15.25   Flowchart for optimization, PSO. Source: Ref. [37].

The harmonics generated by the rectifier and inverter travel over the DC line and get superimposed. The main problem caused by DC harmonics is telephone interference due to electromagnetic induction at harmonic frequencies.

As stated in Chapter 4, the harmonic voltages are a function of pulse numbers, overlap angle, delay angle and extinction angles, type of operation – monopolar or bipolar, impedance of DC circuit to the particular harmonic, DC smoothing reactors, and damping.

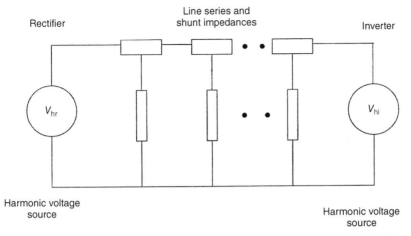

Figure 15.26   Harmonic interaction in HVDC, equivalent circuit of transmission line, and rectifier and inverter.

The design procedure is similar to AC filters. The reactive power of the filter is not significant. The ratings are determined by maximum DC voltage required capacitance and thermal rating due to harmonic.

Figure 15.27 shows a DC filter configuration with typical values for one pole. The smoothing reactors of high inductance reduce DC harmonic currents substantially.

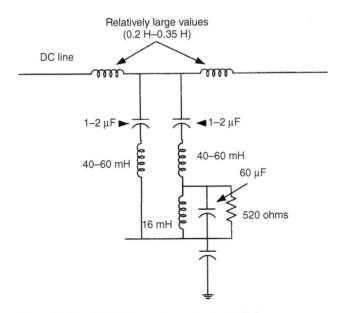

Figure 15.27   A DC filter configuration for HVDC.

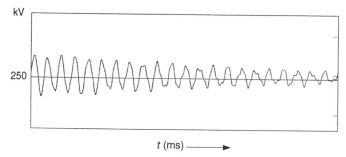

Figure 15.28   Reduction in ripple content and frequency with DC filter and EMTP simulation, see text.

Example 15.2:   The DC-side transients are simulated in Example 13. The effect of providing a DC filter of configuration in Fig. 15.27 is illustrated with an EMTP simulation. The results of this simulation are shown in Fig. 15.28. The voltage ripple and its frequency are much reduced and compare these with the simulations shown in Fig. 13.22(a) and (b).

15.16   LIMITATIONS OF PASSIVE FILTERS

Passive filters have been widely applied to limit harmonic propagation, improve power quality, reduce harmonic distortion, and provide reactive power compensation simultaneously. These can be designed for large current applications and high voltages. Many such filters are in operation for HVDC links, ASDs, and industrial and commercial power systems. Passive filters are still the only choice when high voltages and currents are involved.

Some of the limitations of the passive filters [39,40] are apparent and can be summarized as follows:

- Passive filters are not adaptable to the changing system conditions and once installed are rigidly in place. Neither the tuned frequency nor the size of the filter can be changed so easily. The passive elements in the filters are close tolerance components.

- A change in the system or operating condition can result in detuning and increased distortion. This can go undetected, unless there is online monitoring equipment in place.

- The design is largely affected by the system impedance. To be effective, the filter impedance must be less than the system impedance, and the design can become a problem for stiff systems (see Section 15.2). In such cases, a very large filter will be required. This may give rise to overcompensation of reactive power, and overvoltages on switching and undervoltages when out of service.

- Often, passive filters will require a number of parallel shunt branches. Outage of a parallel unit totally alters the resonant frequencies and harmonic current flows. This may increase distortion levels beyond permissible limits.

- Power losses in the resistance elements of passive filters can be very substantial for large filters.

- The parallel resonance between filter and the system (for single- or double-tuned filters) may cause amplification of currents of a characteristic or noncharacteristic harmonic. A designer has a limited choice in selecting the tuned frequency to avoid all possible resonances with the background harmonics. System changes will alter this frequency to some extent; however, carefully the initial design might have been selected.

- Damped filters do not give rise to a system parallel resonant frequency; however, these are not so effective as a group of ST filters. The impedance of a high-pass filter at its notch frequency is higher than the corresponding ST filter. The size of the filter becomes large to handle the fundamental and the harmonic frequencies.

- The aging, deterioration, and temperature effects detune the filter in a random manner (though the effect of maximum variations can be considered in the design stage).

- The passive filters may prove ineffective for cycloconverters, see Chapters 4 and 6.

- Definite-purpose breakers are required. To control switching surges, special synchronous closing devices or resistor closing is required.

- The grounded neutrals of wye-connected banks provide a low-impedance path for the third harmonics. Third-harmonic amplification can occur in some cases.

- Special protective and monitoring devices (not discussed) are required.

15.17   FLOWCHART FOR DESIGN OF FILTERS

Figure 15.29 shows a flowchart for the design of passive filters. For design of the minimum filter, it is assumed that no reactive power compensation is required in the system and bus voltages can be maintained close to their rated voltages by other means such as ULTC (under load tap changing or voltage regulators. Chapter 16 provides some practical designs of filters.

15.18   FILTER COMPONENTS

15.18.1   Filter Reactors

The filter reactors can be the following:

- Dry-type air core, generally used in medium-voltage and HV applications
- Dry-type iron core reactors, generally used in low- and medium-voltage applications, say up to voltage levels of 30 kV
- Fluid-filled iron-core reactors used for medium-voltage applications.

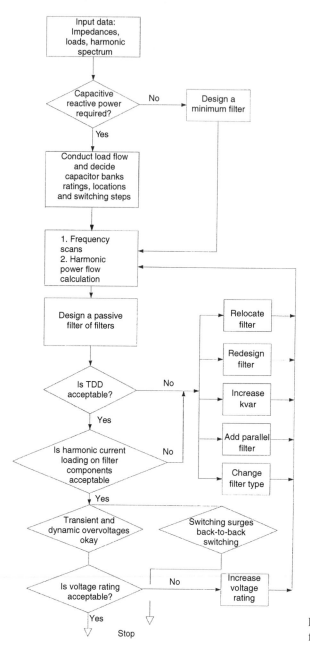

Figure 15.29 Flowchart for design of passive filters.

No standard exists on filter reactors, though IEEE Standard C57.16 [41] is often referenced.

For voltages above 30 kV, air-core reactors are most popular. Air-core reactors can be designed for higher voltages and better tolerances as compared to iron-core reactors. For medium-voltage applications, single-phase iron-core reactors can be applied, and these have smaller dimensions and require much smaller magnetic clearances as compared to air-core reactors. Three-phase iron-core reactors are not generally used as it is difficult to adjust the reactance of one phase without impacting the reactance of the other phases. The change in magnetic material properties can give rise to wider fluctuations in the reactance value in iron-core reactors, though these are designed with lower flux densities with a series air gap and are smaller in dimensions. Reactors for filter applications are subjected to high harmonic frequencies. A harmonic current flow spectrum, based on the worst-case operation, is usually required by a manufacturer for an appropriate design. The rated steady-state voltage is calculated as arithmetic sum of fundamental and harmonic voltages similar to the capacitor:

$$V_r = \sum_{h=1}^{h=\infty} I_h X_{R(h)} \tag{15.133}$$

where $X_{R(h)}$ is the reactance at the given harmonic order.

The earlier construction of air-insulated reactors, consisting of large conductors restrained in polyester or poured-in concrete, has given way to small parallel conductors, epoxy insulated, and encapsulated. Figure 15.30 shows a construction that is practically obsolete now and is replaced with construction shown in Fig. 15.31 [42]. Referring to this figure, the parallel conductors share the current path, and it is possible to use fiberglass epoxy composite to encapsulate the windings. The windings are, thus, completely supported along their lengths and can better withstand compressive and tensile short-circuit forces. The filament fiberglass utilized in conjunction with epoxy resin to encapsulate has better tensile strength than steel. The reactors must withstand system-through fault symmetrical amperes and time duration of 3 seconds is generally specified. These should also withstand the mechanical stresses brought about by asymmetrical short-circuit currents. Dynamic stresses due to switching and transformer inrush currents have to be considered.

At medium-voltage levels, the harmonic filter is generally connected in ungrounded configuration and when the reactor is located on the source side of the capacitor, the available fault current is limited; while an iron-core reactor in the same situation may saturate and will not decrease the short-circuit current. For HV grounded banks, the filter reactor may be located on the neutral side. This may allow BIL of the reactor to be less than that of the system. The air-core coils of even large reactors can be vertically stacked, resulting in space saving. The middle phase is reverse wound to balance the magnetic fluxes.

The magnetic clearance required for air-core reactors is generally large, at least equal to half the diameter of the coil on the sides of the reactor and one-fourth the diameter of the coil for any magnetic material in the foundations, provided no close loops of the magnetic materials exit. Foundation pedestals for the reactors can be designed with fiberglass reinforcements. Sometimes, where the required clearances cannot be maintained, ½ in. aluminum plates can be provided at the base and sides.

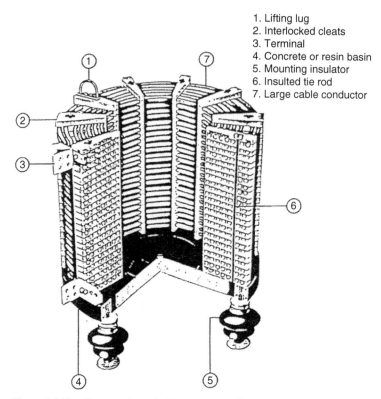

1. Lifting lug
2. Interlocked cleats
3. Terminal
4. Concrete or resin basin
5. Mounting insulator
6. Insulted tie rod
7. Large cable conductor

Figure 15.30   Construction of a filter reactor coil, with large conductors restrained in blocks, no more in use.

The reactors can be provided in metal enclosure for indoor and outdoor applications. Fiberglass enclosures are commercially available for highly corrosive atmospheres and ambient conditions. In metal enclosures, to prevent circulating currents, insulating strips and bolts are used to break the continuous magnetic path.

15.18.2   Filter Resistance Assemblies

At present, there are no specific standards for harmonic resistor assemblies, and IEEE Standard 32 [43] may be generally applied. According to IEEE 32, the time rating and permissible temperature rises on grounding resistors are the following:

Ten seconds or 1 minute short-time rating = 760 °C

Ten minutes or extended time ratings = 610 °C

Continuous rating = 385 °C.

For filter resistor assemblies, short-time ratings are not applicable, and even the permissible temperature rise can be reduced for conservatism. The tolerances on resistors with rise of temperature are another consideration. Table 15.2 shows this

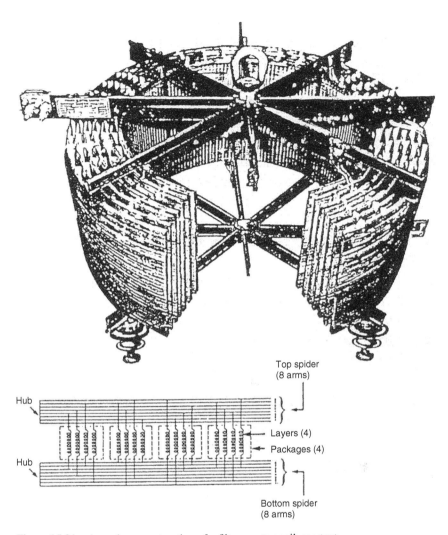

Figure 15.31   A modern construction of a filter reactor coil, see text.

TABLE 15.2   Variation in Resistance with Temperature

	AISI 304 Nickel Chrome	Aluminum Chrome Steel
Temperature coefficient	0.001 ohms/°C	0.00012 ohms/°C
Ohms at ambient	8 ohms	8 ohms
After 10 seconds	14.08 ohms	8.7 ohms
Change	43.2%	8.1%

data for two different materials: Filter resistors should be sized for continuous duty, considering the worst operating condition that gives the maximum loading.

To avoid large change in the grounding current, a temperature coefficient of no more than 0.0002 ohms/°C should be specified.

Table 15.3 shows the required major specifications for the reactors and resistors. The specifications of the capacitors are in Chapter 11.

TABLE 15.3 Specifications of Harmonic Filter Reactors and Resistors

Parameter	Applicable to Harmonic Filter Reactor	Applicable to Harmonic Resistor
Maximum system line-to-line voltage, frequency	x	x
BIL (across reactor coil and to ground may be different)	x	x
Installation: indoor, outdoor	x	x
Enclosure, that is, metal enclosure, ventilated	x	x
Environmental conditions, ambient temperature, industrial pollution, creepage required, wind velocity, ice loading, altitude, and so on	x	x
Maximum short-circuit current and its duration through the reactor, asymmetrical short-circuit current for mechanical bracing, insulation supports	x	x
A chart showing all the harmonic current spectrum. The harmonic current can vary under various operations and specify the maximum currents	x	x
Tolerances on inductance at each tap	x	x
The duty cycle of the duration of currents and voltages	x	x
Reactor inductance in millihenry. Specify taps if required	x	
Transient and dynamic voltage peaks	x	x
Limitation on magnetic clearances – a structural plan with steel sections defined may be required for proper evaluation. The depth of steel reinforcement in existing concrete pedestal is required. New foundations may have to be constructed with fiberglass reinforcements		
Limitations of height and coil dimensions	x	
Resistance value, tolerances, temperature coefficient, kilowatt rating		x
Some applications may require a resistor of low series inductance		x
Constructional features, screens, enclosures, insulators, stainless steel edge – wound or cast grid, and so on		x

15.19   FAILURE OF HARMONIC FILTERS

The major causes of failure of harmonic filters in order of importance are the following:

- Improper selection of the capacitor element voltage rating. Experience shows that this is the most important parameter. Selection of rated voltage without due considerations of all the related parameters as discussed in this chapter, and Chapter 11 is one major reason for premature failures.

- Improper fuse protection of the individual capacitor elements in externally fused banks. Better coordination is obtained with zero probability failure curves of the capacitor elements.

- Improper overall protection, especially the unbalance protection and its settings.

- Improper selection of filter reactor ratings without regard to harmonic loading, though the failure rate due to this cause is minimal. Air-core reactors with construction features shown in Fig. 15.31 are a better choice, especially for high voltages.

- Improper specification and construction of filter resistor assemblies. The resistance can get open circuit or short circuit, and both these conditions can be relayed.

The filter banks are practically maintenance free and have replaced synchronous rotating machines. However, the useful life of capacitor elements is around 20–25 years. It is necessary that after this period, *all* capacitor elements in a bank are replaced. An overall check with capacitance measurements in each phase of the bank is also required from time to time.

REFERENCES

1. C. Khun, V. Tarateeraseth, W. Khan-ngern and M. Kando. "Design procedure for common mode filter for induction motor drive using Butterworth function," IEEE Power Conversion Conference, Nagoya, pp. 417–422, 2007.
2. IEEE 519. Recommended practice and requirements for harmonic control in electrical systems, 1992.
3. E.W. Kimbark. Direct Current Transmission, Chapter 8. John Wiley, New York, 1971.
4. IEEE Standard 1531. IEEE guide for application and specifications of harmonic filters, 2003.
5. IEEE Standard C37.99. Guide for protection of the capacitor banks, 2000.
6. J.D. Anisworth. Filters, Damping Circuits and Reactive Voltamperes in HVDC Converters, in High Voltage Direct Current Converters and Systems. Macdonald, London, 1965.
7. M. M. Swamy, S. L. Rossiter, M.C. Spencer and M. Richardson. "Case studies on mitigating harmonics in ASD systems to meet IEEE-519-1992 standards," IEEE-IAS Conference Record, vol. 1, pp. 685–692, 1994.
8. H.M. Zubi, R.W. Dunn, F.V.P. Robinson and M.H. El-werfelli. "Passive filter design using genetic algorithms for adjustable speed drives," IEEE Power and Energy Society General Meeting, Minneapolis, 2010.
9. B.J. Abramovich and G.L. Brewer. "Harmonic filters for Sellindge converter station," GEC Journal of Science and Technology, vol. 48, no. 1, 1982.

10. M.A. Zamani, M. Moghaddasian, M. Joorabian, S.Gh. Seifossadat and A. Yazdani. "C-type filter design based on power-factor correction for 12-pulse HVDC converters," IEEE Industrial Electronics, IECON, 34th Annual Conference, pp. 3039–3044, 2008.

11. CIGRE Working Group 14.03. "AC harmonic filters and reactive power compensation for HVDC," CIGRE Report, 1990.

12. C.O. Gercek, M. Ermis, A. Ertas, K.N. Kose and O. Unsar. "Design implementation and operation of a new C-type 2nd order harmonic filter for electric arc and ladle furnaces," IEEE Transactions of Industry Applications, vol. 47, no. 4, pp. 1545–1557, 2011.

13. S.H.E.A. Aleem, A.F. Zobaa and M.M.A. Aziz. "Optimal C-type filter based on minimization of voltage harmonic distortion for non-linear loads," IEEE Transactions of Industrial Electronics, vol. 59, no. 1, pp. 281–289, 2012.

14. S. Goldman. Laplace Transform Theory and Electrical Transients, Dover Publications, New York, 1966.

15. Y.-S. Cho, B.-Y. Kim and H. Cha. "Transfer function approach to a passive filter design for industrial process application," Proceedings of the 2010 IEEE International Conference on Mechatronics and Automation, Xian, China, pp. 963–968, 2010.

16. J.K. Phipps. "Transfer function approach to harmonic filter design," IEEE Industry Applications Magazine, pp. 175–186, vol. 3, 1973.

17. P.E. Gill, W. Murray and M.H. Wright. Practical Optimization. Academic Press, New York, 1984.

18. D.G. Lulenberger. Linear and Non-Linear Programming, Addison Wesley, Reading, MA, 1984.

19. G.B. Dantzig. Linear Programming and Extensions, Princeton University Press, Princeton, NJ, 1963.

20. R. Fletcher and M.J.D. Powell. "A rapidly convergent descent method for minimization," Computer Journal, vol. 5, no. 2, pp. 163–168, 1962.

21. R. Fletcher and C.M. Reeves. "Function minimization by conjugate gradients," Computer Journal, vol. 7, no. 2, pp. 149–153, 1964.

22. N. Karmarkar. "A new polynomial time algorithm for linear programming," Combinatorica, vol. 4, no. 4, pp. 373–395, 1984.

23. P.E. Gill, W. Murray, M.A. Saunders, J.A. Tomlin and M.H. Wright. "On projected Newton barrier methods for linear programming and an equivalence to Karmarkar's projective method," Math Program, vol. 36, pp. 183–209, 1986.

24. B. Stott and L.J. Marinho. "Linear programming for power system security applications," IEEE Transactions of Power and Systems, vol. PAS 98, pp. 837–848, 1979.

25. M.J. Todd and B.P. Burrell. "An extension of Karmarkar's algorithm for linear programming using dual variables," Algorithmica, vol. 1, pp. 409–424, 1986.

26. R.E. Marsten. "Implementation of dual affine interior point algorithm for linear programming," ORSA Computing, vol. 1, pp. 287–297, 1989.

27. J.C. Churio-Barboza, J.M. Maza-Ortega and M. Burgos-Payan. "Optimal design of passive filters for time-varying non-linear loads," Proceedings of International Conference on Power Engineering, Energy and Electrical Drives, Malaga, Spain, 2011.

28. R.D. Koller and B. Wilamowski. "LADDER-A microprocessor tool for passive filter design and simulation," IEEE Transactions on Education, vol. 29, no. 4, pp. 478–487, 1996.

29. J.C. Churio-Barboza, J.M. Maza-Ortega and M. Burgos-Payan. "Optimal design of passive tuned filters for time varying non-linear loads," Proceedings of the 2011 International Conference on Power Engineering, Energy and Electrical Drives, Torremolinos, Spain, 2011.

30. D. Wu, Y. Chen, S. Hong, X. Zhao, J. Luo and Z. Gu. "Mathematical model analysis and LCL filter design of VSC," IEEE 7th International Power Electronics and Motor Control Conference, Harbin, China, 2012, pp. 2799–2804.

31. C. Fu and H. Wang. "An efficient optimization of passive filter design," IEEE International Conference on Industrial Informatics, Daejeon, Korea, pp. 631–634, 2008.

32. M. Azri and N.A. Rahim. "Design analysis of low-pass passive filter in single-phase grid connected transformer less inverter," IEEE Conference on Clean Energy and Technology (CET), Kula Lumpur, pp. 349–353, 2011.

33. J.M. Maza-Ortega, J.C. Churio-Barboza, J.M. Burgos-Payan. "A software-based tool for optimal design of passive tuned filters," IEEE International Symposium on Industrial Electronics (ISIE), Bari, pp. 3273–3279, 2010.

34. D.E. Goldberg. Genetic Algorithms in Search, Optimization and Machine Learning, Addison Wesley, Reading MA, 1989.

35. Y.-M. Chen. "Passive filter design using genetic algorithms," IEEE Transactions on Industrial Electronics, vol. 50, no. 1, pp. 202–207, 2003.

36. J.F. Frenzel, "Genetic algorithms," IEEE Potentials, vol. 12, pp. 21–24, 1998.

37. J. Kennedy and R. Eberhart. "Particle swarm optimization," International Conference on Neural Networks, vol. 4, pp. 1942–1948, 1995.

38. S. Singh and B. Singh. "Passive filter design for a 12-pulse converter fed LCI-synchronous motor drive," Joint Conference on Power Electronics, Drives and Energy Systems and 2010 Power India, New Delhi, India, pp. 1–8, 2010.

39. J.C. Das. "Analysis and control of harmonic currents using passive filters," TAPPI Proceedings, Atlanta Conference, pp. 1075–1089, 1999.

40. J.C. Das. "Passive filters-potentialities and limitations," IEEE Transactions on Industry Applications, vol. 40, no. 1, 232–241, 2004.

41. IEEE Standard C57.16. IEEE standard for requirements, terminology and test code for dry type air-core series connected reactors, 2011.

42. J.C. Das, W.F. Robertson and J. Twiss. "Duplex reactor for large cogeneration distribution system-an old concept reinvestigated," TAPPI Engineering Conference, Nashville, TN, pp. 637–648, 1991.

43. IEEE Standard 32. IEEE standard requirements, terminology and test procedures for neutral grounding devices, 1972.

CHAPTER 16

PRACTICAL PASSIVE FILTER DESIGNS

In this chapter, some case studies of practical harmonic analysis and passive filter designs are discussed.

16.1 STUDY 1: SMALL DISTRIBUTION SYSTEM WITH MAJOR SIX-PULSE LOADS

Figure 16.1 shows a simple system, where the six-pulse drive system loads are 77% of the total load. When the nonlinear load is more than 30% of the total load demand, a careful analysis is required for the control of total demand distortion (TDD). The 4.16-kV bus 2 is the point of common coupling (PCC).

Step 1: *Estimate harmonic current injection*

Estimation of correct harmonic emission from the nonlinear loads is of importance, the first step in the analyses. Reiterating what has been said in previous chapters:

o It varies with the operation and the drive system load, and the short-circuit level at the utility source having a profound effect. Harmonic analysis can be carried out with the maximum and minimum levels of the short-circuit currents.

o In this example, as there is only one harmonic source, the angle of harmonics need not be considered. Where more than one source is present, the angle of each harmonic should be modeled.

o A worst-case scenario is chosen for the analysis. For the example, a gating angle $\alpha = 15°$ is considered. Then using Eq. (4.40), repeated below, the overlap angle is $12.25°$:

$$\mu = \cos^{-1}[\cos\alpha - (X_s + X_t)I_d] - \alpha \qquad (16.1)$$

where X_s and X_t are the system and transformer reactances in per unit on converter base and I_d is per unit current on a converter base.

Power System Harmonics and Passive Filter Designs, First Edition. J.C. Das.
© 2015 The Institute of Electrical and Electronics Engineers, Inc. Published 2015 by John Wiley & Sons, Inc.

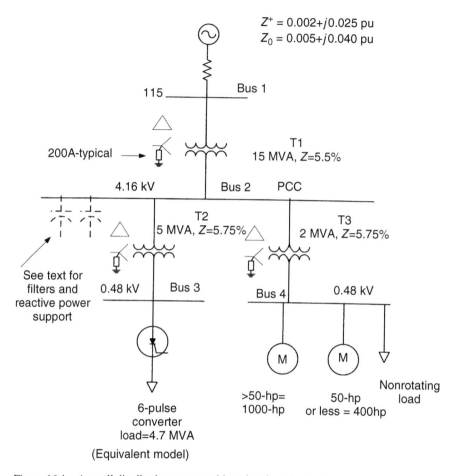

Figure 16.1   A small distribution system with major six-pulse loads.

TABLE 16.1   Estimation of Harmonic Emission

h	5	7	11	13	17	19	23	25	29	31
%	18	13	6.5	4.8	2.8	1.5	0.5	0.4	0.3	0.2

On a simplified basis, assuming trapezoidal waveform, large delay angle and small overlap angle; the harmonic spectrum can be calculated from Eqs. (4.37) to (4.39). The resulting harmonic current spectrum in terms of fundamental is shown in Table 16.1, which terminates at 31st harmonic. Harmonic analysis can be carried out to 49th harmonic or more, especially for higher pulse numbers. A small percentage of noncharacteristic harmonics, that is, second, third, fourth, and so on, are also generated, which have been ignored in Table 16.1 for this study.

Step 2: *Conduct load flow and establish need for reactive power compensation*

For conducting the load flow study, an estimate of the power factor of the nonlinear and linear loads is required. Computer-based fundamental frequency load flow calculations show that operating power factor is 0.82. It is required to control the power factor at PCC to >0.9 and also the TDD at PCC within IEEE 519 limits.

The load flow shows a demand of 5.279 MW and 3.676 Mvar from the 115-kV source, including system losses. A 1200 kvar capacitor bank at 4.16-kV bus will reduce the reactive power input from the 115-kV source to 2.44 Mvar and give an overall power factor of 0.91, approximately.

Step 3: *Ascertain short-circuit level and load demand at PCC*

To calculate the permissible TDD, short-circuit level at PCC and the load demand over a period of 15 min or 30 min is required. The short-circuit level at 4.16-kV bus, PCC, is 36.1 kA, the load demand = 800 A, the ratio $I_s/I_r = 45$, and the permissible IEEE TDD limits are as shown in Table 16.2, first row.

All the results of calculations in this study case are compiled in Table 16.2, which are referred at each step of the calculation. The first column shows the identification of the case number followed by the parameters of study and then the results of the study.

Step 4: *Conduct harmonic analysis study*

The modeling of components as discussed in Chapter 13 is applicable. There are no nearby harmonic loads, all studies are conducted with a single value of utility short-circuit level = 3987 MVA (20 kA, $X/R = 10$).

4a Harmonic analysis without capacitor bank

The harmonic current distortion without capacitors is shown in row 2 of Table 16.2. The harmonic distortion limits at 5th, 7th, 11th, and 13th harmonics exceed the permissible limits, and also the overall TDD is 19.57% versus the maximum permissible value of 8%. This result could be expected because of higher percentage of nonlinear load. Figure 16.2 depicts the harmonic spectrum at PCC and Fig. 16.3 its waveform.

4b Harmonic analysis with capacitor bank

In step 2, 1200 kvar capacitors are required for reactive power compensation at bus 2 in Fig. 16.1. A capacitor bank is formed from the individual capacitor cans of certain standard sizes in series and parallel combinations (Chapter 11). A higher than the rated voltage is selected on the capacitor units (Chapter 11).

Consider that cans of 200 kvar, rated voltage 2.77 kV, are selected in ungrounded wye configuration, one series group per phase. Then three cans per phase give a kvar of 450.42 per phase, that is, three-phase kvar at the operating voltage of 4.16 kV is

$$\text{kvar}_{4.16 \text{ kV}} = 3 \times 600 \left(\frac{2.4}{2.77} \right)^2 = 1351.3 \text{ kvar}$$

The capacitive reactance is 12.788 ohms per phase and the capacitance = 2.074E-4 μF. The results of harmonic analysis are shown in row 3 of Table 16.2.

TABLE 16.2 Calculation Steps for Design for Harmonic Filters, Study 16.1

Case #		Harmonic	5	7	11	13	17	19	23	25	29	31	TDD
1	IEEE TDD Limits		7.0	7.0	3.5	3.5	2.5	2.5	1.0	1.0	1.0	1.0	8
	No capacitor	I_h	118	85.6	42.7	31.5	18.4	9.8	3.3	2.6	2.0	1.3	–
		HD	14.75	10.70	5.34	3.94	2.30	1.23	0.41	0.33	0.25	0.16	19.57
2	600 kvar/p	I_h	137	117	121	402	31.9	10.1	1.7	1.1	0.6	0.3	
		HD	17.13	14.60	15.10	50.50	3.99	1.26	0.21	0.14	0.08	0.04	57.23
3	500 kvar/p	I_h	133	110	94.9	136	58.6	15.3	2.33	1.41	0.7	0.4	
		HD	16.63	13.75	11.86	17.00	7.32	1.91	0.29	0.18	0.09	0.05	30.86
4	5th ST filter, 600 kvar/phase n =4.6	I_h	58	70.2	37.2	27.7	16.2	8.7	2.9	2.3	1.8	1.2	
		HD	7.25	8.77	4.65	3.46	2.03	1.09	0.36	0.29	0.23	0.15	12.99
5	5 & 7 ST filters, 300/300 Kvar/p, n=4.6,6.7	I_h	86.6	33.3	33.7	25.5	15.2	8.1	2.7	2.2	1.7	1.1	
		HD	10.85	4.16	4.21	3.18	1.90	1.01	0.34	0.28	0.21	0.14	12.93
6	5 & 7 ST filters, 300/300 Kvar/p, n=4.85,6.7	I_h	60.4	33.1	33.5	25.4	15.1	8.1	2.7	2.2	1.7	1.1	
		HD	7.56	4.14	4.19	3.18	1.89	1.01	0.34	0.28	0.21	0.14	10.25
7	5 & 7 ST filters, 400/300 Kvar/p, n=4.85.6.7	I_h	50.9	32.6	32.8	24.9	14.8	8.0	2.7	2.1	1.6	1.1	
		HD	6.36	4.08	4.10	3.11	1.86	1.00	0.34	0.28	0.21	0.14	9.39

8	5,7&11ST filters, 400/300/300 kvar/p, n = 4.85,6.7,10.6	I_h	52.9	35.8	7.5	14.4	10.5	5.8	2.0	1.6	1.3	0.8	
		HD	6.61	4.48	0.94	1.80	1.31	0.73	0.25	0.20	0.16	0.10	8.38
9	5,7&11ST filters, 500/400/300 kvar/p, n = 4.85,6.7,10.6	I_h	46.5	29.3	7.4	13.9	10.1	5.6	1.9	1.6	1.3	0.8	
		HD	5.81	3.66	0.93	1.74	1.26	0.73	0.24	0.20	0.16	0.10	7.3
10	As in row 9, but with tolerances	I_h	77.00	42.2	11.7	14.8	10.4	5.71	2.0	1.6	1.3	0.8	
		HD	9.63	5.28	1.46	1.85	1.30	0.71	0.25	0.20	0.16	0.10	11.33
11	5,7,11 ST filters, 900/600/300 Kvar/p, n = 4.85,6.7,10.6	I_h	30.6	21.2	7.22	12.8	9.2	5.1	1.8	1.4	1.1	0.73	
		HD	3.83	0.65	0.90	1.60	1.15	0.64	0.23	0.18	0.14	0.09	5.18
12	As in row 11, but with tolerances	I_h	51.1	28	10.5	13.5	9.4	5.2	1.8	1.4	1.1	0.73	
		HD	6.39	3.50	1.31	1.69	1.18	0.65	0.23	0.18	0.14	0.09	7.71
13	As row 11 but 5th bank out of service	I_h	196	23.3	7.5	14.0	10.3	5.7	2.0	1.6	1.2	0.82	
		HD	24.50	2.91	0.94	1.75	1.29	0.71	0.25	0.20	0.15	0.10	24.80
14	As row 11 but 7th bank out of service	I_h	28.4	75.5	7.73	14.8	10.7	5.9	2.1	1.7	1.3	0.85	
		HD	3.55	9.44	0.97	1.85	1.34	0.74	0.26	0.21	0.16	0.11	10.42
15	As row 11 but 11th bank out of service	I_h	30	20	26.2	20.3	12.2	6.6	2.2	1.8	1.4	0.9	
		HD	3.75	2.50	3.28	2.54	1.53	0.83	0.28	0.23	0.18	0.11	6.38

Harmonics shown in actual amperes. HD: Distortion at individual harmonic.

749

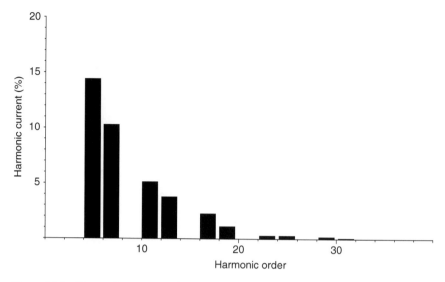

Figure 16.2 Harmonic spectrum of current at PCC without capacitors.

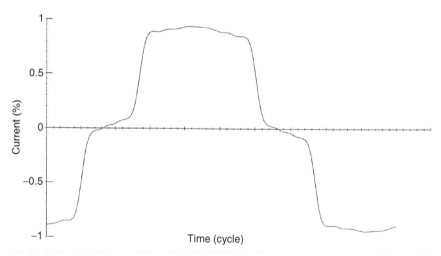

Figure 16.3 Current waveform at PCC without capacitors.

The calculated parallel resonance frequency is 822–825 Hz (a step of 3 Hz was used in the calculation), maximum impedance angle 89.64°, and minimum impedance angle −85.34°. Thus, the amplification of 13th harmonic current is apparent in Table 16.2. The distortion at this harmonic is 50.5% and the overall TDD = 57.25%. Figure 16.4 shows the spectrum at PCC and Fig. 16.5 shows its waveform, which is highly distorted. Thus, adding the capacitors has increased the harmonic distortion, and a resonant condition exists around 13th harmonic.

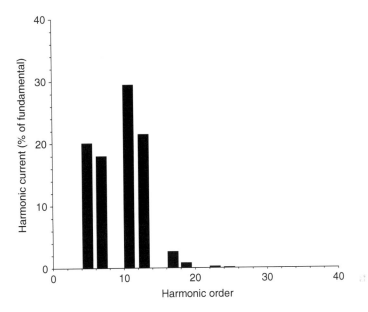

Figure 16.4   Harmonic current spectrum at PCC with capacitors.

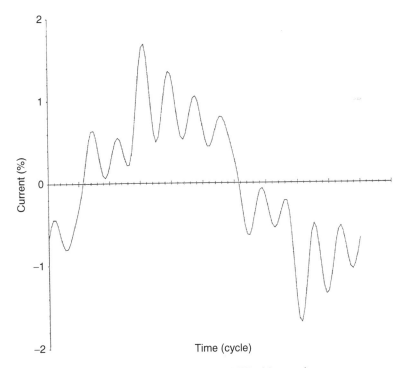

Figure 16.5   Harmonic current waveform at PCC with capacitors.

4c Harmonic analysis with capacitors sized to eliminate resonance

Sometimes, an attempt is made to size and locate the capacitors so as to eliminate resonance. This was discussed in Chapter 11 with the limitations of this approach.

If the size of the capacitor bank is reduced, the resonant frequency will shift upward. A capacitor bank of 1192 kvar at 4.16 kV (500 kvar per phase formed out of 2.77 kV individual capacitor cans) will shift the resonance to around 900 Hz. As the load does not generate this frequency, resonance can be escaped. The harmonic analysis in row 3 of Table 16.2 confirms this. Though the distortion at 13th harmonic is reduced, yet the distortion at the number of lower harmonics exceeds the permissible limits and the overall TDD is at unacceptable level of 30.86%. Thus, sizing or relocating the capacitors in a distribution system will impact the initial results, but the problem lies in much wider swings with the system changes.

Step 5: *Design a harmonic filter*

5a Form an ST fifth harmonic filter

Form a fifth harmonic filter by tuning to 4.7th harmonic. A tuning frequency below 3–10% of the harmonic to be filtered is selected to consider detuning effects as explained in Chapter 14. Then, a series reactor of $L = 1.53$ mH is required. Arbitrarily choose $Q = 40$ at fundamental frequency. The results of the calculation are shown in row 4 of Table 16.2. The 7.25% distortion at fifth harmonic almost meets the requirements of 7%, but the distortion at 7th and 11th harmonics and the overall TDD exceed IEEE limits. The parallel resonant frequency is between 266 and 268 Hz, and the series resonant frequency is between 282 and 284 Hz. *The resonance is not eliminated*, it shifts to a frequency below the tuned frequency as discussed in Chapter 14.

5b Add seventh harmonic ST filter

The splitting of 1350-kvar capacitor bank is tried to form two equal parallel ST filters, one tuned to $n = 4.7$ as before and the other tuned to $n = 6.7$. Table 16.2, row 5, shows that seventh harmonic distortion is reduced, while the fifth harmonic current flow increases, giving rise to increased distortion. This can be expected, as the size of the fifth filter has been reduced. TDD exceeds the permissible limits.

5c Effect of tuning frequency

A sharper tuning closer to fifth harmonic is tried and the filter is reformed for the same capacitor size, and $n = 4.85$. The results are shown in row 6 of Table 16.2. The fifth harmonic distortion is considerably reduced for the same size of filter bank.

5d Increase fifth filter size

As the fifth harmonic is still high, the fifth ST filter size is increased, formed with 400 kvar of 2.77-kV capacitors per phase. This brings the fifth harmonic distortion within permissible limits, but the 11th and total distortion is still high, Table 16.2, row 7.

5e Add 11th harmonic filter

An 11th ST filter formed with 300-kvar capacitors/phase, $n = 10.6$, reduces the 11th harmonic distortion. Table 16.2, row 8, shows that harmonic distortion at all harmonics is within acceptable limits, but the overall TDD is 8.38, that is, slightly higher than the permissible value of 8%.

5f *Increase size of fifth and seventh ST filters*

The fifth and seventh harmonic filters are reformed with 500-kvar and 400-kvar capacitors per phase. The results in Table 16.2, row 9, shows that the permissible distortion limits are met throughout the harmonic spectrum, and the total THD at PCC is reduced by 7.3%.

Thus, a total of 1200 kvar of capacitors per phase at 2.77 kV are required. This means that installed rating at the operating voltage is 2862 kvar, while only 1200 kvar was required for the reactive power compensation for power factor improvement. Yet, the design of the filter is not final.

Step 6: *Consider detuning effects*

The reactors and capacitors for filters are specified to be of close tolerance components to limit frequency drift; consider following tolerances as discussed in Chapter 14:

o Capacitors: +5% and no negative tolerance

o Reactors: ±2%

Also, the capacitor bank in each phase is so formed that the difference between the capacitance of phases is limited. The same tolerance as on individual capacitor cans can also be applied to the overall assembly, the total capacitance in each phase having the same tolerance with respect to other two phases as individual capacitor units. Unbalances of the capacitance between each phase give rise to unbalance currents.

Consider that capacitance of each bank increases by 5% and that of each filter reactor also increases by 2%. This is quite a conservative assumption for checking the detuning effects on the 5th, 7th, and 11th filter banks.

The results of the calculation are shown in row 10 of Table 16.2. The harmonic distortion at fifth harmonic and the overall TDD of 11.3% exceed the permissible limits. An obvious solution is to *lower* the design values of distortion, so that the increase in distortion due to detuning is still within IEEE limits. *This means that the size of the filters should be further increased.*

An iteration with various sizes shows that the 5th, 7th, and 11th harmonic filters should be formed with 900-, 600-, and 300-kvar capacitors per phase (rated voltage 2.77 kV). The results of the calculations with no tolerance and with positive tolerance of 5% on the capacitors and 2% on the reactors are shown in rows 11 and 12 of Table 16.2.

Therefore, a much larger size is required following the earlier steps. Total three-phase kvar of the filter capacitors at system voltage of 2.4 kV is 4054 kvar, while only 1200 kvar is required for reactive power compensation. A filter intended to meet only the requirements of harmonic distortion control is called a *minimum filter*, Chapter 14. In a way, this term can be misleading; the example illustrates that the requirements of controlling

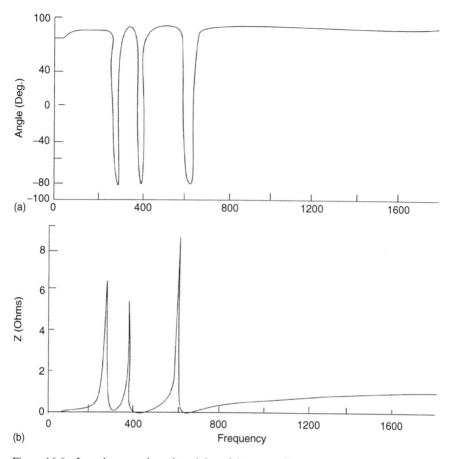

Figure 16.6   Impedance angle and modulus of three-step filter designs.

the harmonic distortion require a filter of much larger size than the reactive power compensation for the required power factor improvement. The converse can also be true.

The installation of the finally designed filter will make the overall power factor at 115-kV side approximately equal to unity.

Figure 16.6(a) and (b) shows the impedance angle and modulus of the final three-step filter.

Step 7:   *Consider outage of one of the parallel banks*

Well-designed and protected capacitor filters are practically trouble free and require little maintenance. Yet an outage cannot be ruled out. For a continuous process plant, shutting down the process because of failure of a filter may not be warranted and the effect on harmonic distortion, when one of the parallel filter banks is out of service must be considered. IEEE Standard 519 allows 50% increase in the distortion limits on a short-time basis. This increased TDD limits during the repair time of the combination filter are considered in the following calculations, Table 16.2, rows 13–15.

The calculations show that

1. On outage of 5th filter bank, the 5th harmonic distortion is increased by 24.5%, and the overall TDD is 24.8%. Also the distortion at 5th harmonic exceeds the IEEE limits.

2. On outage of 7th filter bank, the TDD on all harmonics and the over-all TDD of 10.42% are acceptable, provided the normal situation is restored quickly.

3. On outage of 11th filter bank, the overall TDD is 6.38 and acceptable.

Thus, it is only the outage of 5th ST filter, which is of concern. An online standby 5th filter can be installed and switched automatically. This could be a more economical solution, rather than shutting the process facility.

Step 8: *Consider the shifted resonant frequencies*
The shifted frequencies are as follows:

1. 5th ST filter: 260–262 Hz

2. 7th ST filter: 368–370 Hz

3. 11th ST filter: 584–586 Hz

If the shifted frequencies coincide with one of the characteristics, non-characteristics, or triplen harmonics present in the system, current magnifications at these frequencies can occur, see Chapter 15 that the shifted resonant frequencies should be at least 30 cycles apart from the adjacent odd and even harmonics. An examination of the above shifted frequencies shows that this criterion is not met in every case.

Consider overvoltage capability
Figures 11.3–11.5 show the short time and transient overvoltage and overcurrent capabilities. To apply these curves, a system study of transients is required and that can become fairly involved.
Consider that 100 switching operations are required in 1 year. Then from Fig. 11.4, $k = 2.6$ and from Eq. (11.22), for transient events such as capacitor bank switching, circuit breaker restrikes:

$$V_{tr} = \sqrt{2} \times 2.6 \times 2.77 = 10.18 \ \text{kV peak}$$

Thus, by calculation, it should be proved that this voltage or the number of switching operations is not exceeded. Now consider that the number of switching operations is reduced to 10 only, then from the same figure $k = 3.2$ and this raises the 10.18 kV to 12.53 kV.

When considering detuning, it is necessary to check the efficacy of the final filter design for various switching conditions of the plant. These can also cause detuning. The filter design is sensitive to the utility's source impedance. In the system configuration shown, some motor loads may be out of service, which will increase TDD at PCC.

Step 9: *Consider alternate filter designs*
It is possible to design a single fifth ST filter to meet the IEEE distortion limits; however, this filter will be of abnormally larger size. Similarly, the

high-pass filters with different values of Q can be tried; however, these will of still larger size.

Step 10: *Consider harmonic loading of power capacitors*

The harmonic loading on the power capacitors should be calculated and verified as discussed in Chapter 11.

Most computer programs will calculate the loadings on the harmonic filters and capacitors and flag overload conditions. These can be calculated by longhand calculations also. None of the above limits are exceeded in the final design of this study. (A reader could perform hand calculations and verify this statement.)

16.2 STUDY 2: FILTERS FOR ARC FURNANCE LOADS

Figure 16.7 shows an arc furnace installation. The total operating load is 150 MVA. 34.5-kV bus is considered a PCC. The furnace loads are connected through step-down transformers of 34.5–7.2 kV as shown in this figure. A reactive power compensation of 98 Mvar is required, which is provided by three ST filters formed with 49, 24.5, and 24.5 Mvar capacitors, respectively, for second, fourth, and fifth harmonics. These are connected at the main 34.5-kV bus. This reactive power compensation is calculated at full load. As stated in Chapter 6, arc furnace loads give rise to flicker due to variations

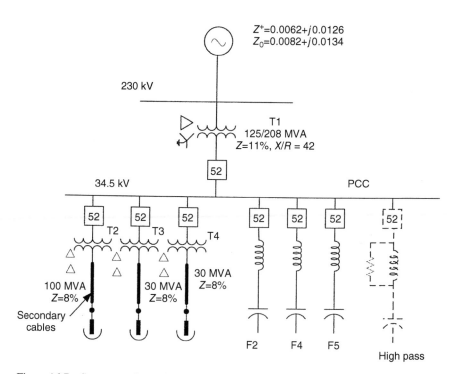

Figure 16.7   System configuration with arc furnace loads.

in reactive power demand and erratic load patterns, and normally a TCR, SVC, or similar fast response device is required.

(a) *Model with current injection*

Table 4.6 shows the harmonic current spectrum during melting and refining from IEEE 519. The melting period gives higher magnitude of harmonics, which are considered in this simulation.

The third harmonic is generated; however, no filter is provided for it because the delta windings of the transformers will filter out the third harmonics and no third harmonics appear in the lines. The three ST filters are formed as follows:

o Second harmonic ST filter: *double-wye*, ungrounded, three-series groups, each group containing eight units of 400-kvar capacitors of rated voltage 7.2 kV. This gives a three-phase Mvar of 49; see Fig. 11.14(d) for a double-wye ungrounded bank configuration.

o Fourth and fifth ST filters: single-wye ungrounded, three-series groups, each group containing eight units of 400-kvar capacitors of rated voltage 7.2 kV. This gives a three-phase Mvar of 24.5.

The tuning frequencies for second, fourth, and fifth ST filters are 1.95, 3.95, and 4.95 times the fundamental frequency, respectively. This gives series reactors of 1.694E-2 mH, 8.259E-3 mH, and 5.260E-3 mH for the second, fourth, and fifth filters, respectively. Q factors are second harmonic filter = 60 and fourth and fifth harmonic filters = 40.

The system is impacted with a harmonic spectrum during the melting cycle of the furnaces. We observe that each ST filter operates effectively providing a low impedance path for the harmonic it is intended to shunt away.

The load demand is 319.7 A at 230 kV, the short-circuit current at PCC = 20 kA, ratio I_s/I_r = 62.5. The permissible total TDD = 3.5% at 230-kV bus. And at 34.5 kV, it is 12%. Also the limits of each of the harmonics as per IEEE 519 should be met. The results of harmonic flow calculations and TDD at the PCC are shown in Table 16.3, which shows that when 34.5-kV bus is declared as PCC, all the harmonic emission limits are met, *but not so if we declare 230-kV bus as the PCC*. The permissible second harmonic limit at 230 kV is 0.75% while the actual calculated is 2.20%. *To control even harmonics in arc furnace installations as per permissive limits requires much larger second harmonic filter.* The second harmonic filter size will require to be approximately doubled to reduce it to 0.75%. This will be indeed a very large filter. The voltage distortion at 230 kV is 0.16%, and at 34.5-kV bus it is 1.27%. IEEE limits are shown.*

The impedance modulus and impedance angle are shown in Figs 16.8 and 16.9. The impedance modulus and the shifted resonance frequencies are shown in Table 16.4.

The voltage waveforms at 34.5-kV bus and 230-kV bus are shown in Fig. 16.10, current waveforms through filters in Fig. 16.11, and current

---

*Only in Table 16.3, as these are not of consideration for other cases of this study.

TABLE 16.3 Calculations of Harmonic Emissions, Arc Furnace, Three ST Filters (49, 24.5, 24.5 Mvar), IEEE 519 Current Injection Model

Harmonic	Current through Filters			TDD Calculations			Harmonic Voltage Distortion			
	Second ST filter	Fourth ST filter	Fifth ST filter	TDD At 230 kV or 34.5 kV bus	TDD IEEE limits 34.5 kV	TDD IEEE limits 230 kV	34.5-kV PCC		230-kV bus	
							IEEE limits	34.5 kV	IEEE limits	230 kV
Fundamental A or V	1117.5A	439.9A	429.3A	310.76A, 230 kV 2071.7A, 34.5 kV	2071.7	310.76		34.5 kV		230 kV
$h = 2$	15.39	1.38	1.25	**2.20**	2.5	0.75	3.0	0.54	1.0	0.07
$h = 4$	0.09	15.11	1.07	0.20	2.5	0.75	3.0	0.10	1.0	0.01
$h = 5$	0.06	0.75	24.32	0.15	10	3	3.0	0.09	1.0	0.01
$h = 7$	0.51	3.58	7.86	1.34	10	3	3.0	1.14	1.0	0.14
Total TDD, THDV			7.86	2.54	12	3.75	5.0	1.27	1.5	0.16

All harmonic currents and voltages are in percentage of the fundamental currents and voltages.

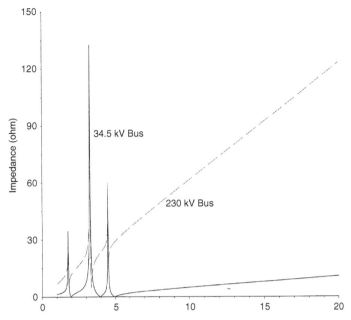

Figure 16.8   Impedance modulus with second, fourth, and fifth harmonic filters and typical current injection model.

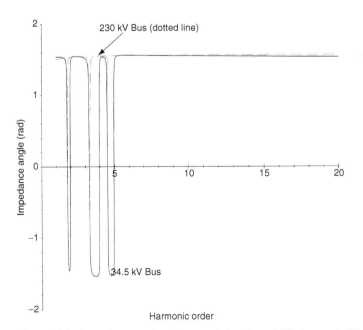

Figure 16.9   Impedance angle with second, fourth, and fifth harmonic filters and typical current injection model.

TABLE 16.4 Impedance Modulus and Resonant Frequency Calculations of Harmonic Emissions, Arc Furnace, Three ST Filters (49, 24.5, 24.5 Mvar)

Bus Identification	Impedance Modulus (Ω)	Resonant Harmonic	Resonant Frequency (Hz)
34.5-kV bus	34.43	1.767	106
	132.59	3.267	196
	56.75	4.5	270
230-kV bus	21.36	1.73	104
	91.39	3.27	196
	60.09	4.50	270

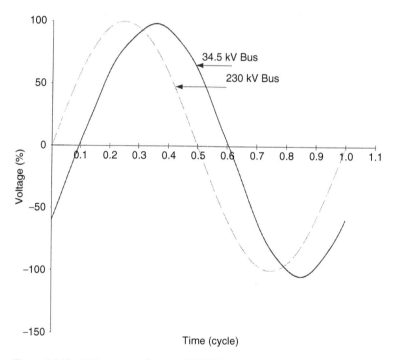

Figure 16.10 Voltage waveforms at 34.5-kV and 230-kV (PCC) buses.

waveform at PCC in Fig. 16.12, which looks almost sinusoidal. None of the components in any filters are overloaded.

(b) *Model with typical voltage harmonics*

Generally, it is not the current injection model but harmonic voltage injection model that is used for the arc furnace harmonic studies. The voltage harmonics are shown in Table 16.5 in normal and worst-case scenarios. The voltage harmonics are applied through the furnace transformer secondary leads, which should be properly modeled.

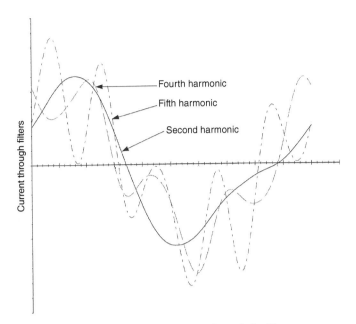

Figure 16.11   Waveforms of the current through the filters.

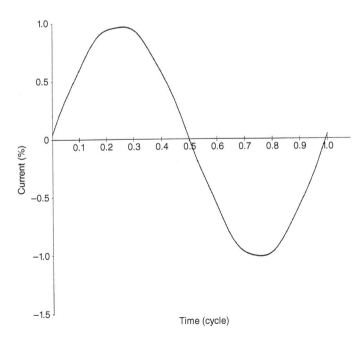

Figure 16.12   Waveform of the current at PCC.

TABLE 16.5   Voltage Harmonics for Harmonic Analysis of Arc Furnace

Harmonic Order	Maximum Voltage Distortion	Typical Voltage Distortion (Percentage of Fundamental Voltage)
2	17	5.0
3	29	20
4	7.5	3.0
5	10	10
6	3.5	1.5
7	8	6
8	2.5	1
9	5	3

Table 16.6 shows the tabulations. It is noted that the pattern of harmonic loading in filters changes and also current and voltage distortion levels change. There is not much difference with respect to the current model. Figure 16.13 shows the frequency scan plot.

(c) *Model with maximum voltage harmonics*

If the worst-case harmonics shown in Table 16.5 are modeled, the resulting harmonic voltage waveform at 34.5-kV bus is shown in Fig. 16.14. With this model, the second harmonic ST filter is not even adequate to control the second harmonic to acceptable level of 2.5% for 34.5-kV bus as the PCC. The distortion at second harmonic becomes 3.5%, though all the other harmonic indices meet IEEE requirements at 34.5-kV bus. It is necessary to raise the size of this filter from 49 Mvar to 72 Mvar to control the second harmonic. It will be a very large filter, double-wye ungrounded connection, three series groups of 7.2 kV capacitors, each rated for 400 kvar and 12 units in parallel per series group. This will give 73.5 Mvar at 34.5 kV.

The results of a study with this configuration are shown in Table 16.7. Even with increased filter size, the TDD and voltage distortion are slightly more than that in case 2.

(d) *Study with second harmonic filter as an ST filter and fourth and fifth harmonic filters as second-order high-pass damped filters.*

Damped filters for arc furnace installations are shown in Fig. 5.10. As we discussed in Chapter 5, arc furnace loads give rise to interharmonics, and with ST filters, there is a possibility of their magnification (though the interharmonics are not modeled in this study case). Damped filters do not give rise to shifted resonant frequencies. In this case study, the second harmonic filter is still retained as an ST filter, while fourth and fifth harmonic filters are turned into second-order high-pass filter. In this case, also the worst-case voltage harmonic model is considered.

The results of this case are shown in Table 16.8. If 230-kV bus is considered as a PCC, not only the second harmonic but also the overall TDD of 4.73 exceeds IEEE requirements. *This amply shows the vast differences in decaling*

TABLE 16.6 Calculations of Harmonic Emissions, Arc Furnace, Three ST Filters (49, 24.5, 24.5 Mvar), Typical Voltage Injection Model

Harmonic	Current through Filters			TDD Calculations			Harmonic Voltage Distortion	
	Second ST filter	Fourth ST filter	Fifth ST filter	TDD at 230-kV or 34.5-kV bus	TDD IEEE limits 34.5 kV	TDD IEEE limits 230 kV	34.5-kV PCC	230-kV bus
Fundamental A or V	1117.5A	439.9A	429.3A	310.76A, 230 kV 2071.7A, 34.5 kV	2071.7	310.76	34.5 kV	230 kV
$h = 2$	7.99	0.71	0.65	1.14	2.5	0.75	0.29	0.04
$h = 4$	0.08	13.69	0.97	0.18	2.5	0.75	0.09	0.01
$h = 5$	0.11	1.32	43.31	0.27	10	3	0.17	0.02
$h = 7$	0.61	4.28	9.41	1.61	10	3	1.40	0.17
$h = 8$	0.10	0.67	1.31	0.28	2.5	0.75	0.28	0.03
Total TDD, THDV				2.02	12	3.75	1.46	0.18

All harmonic currents and voltages are in percentage of the fundamental currents and voltages.

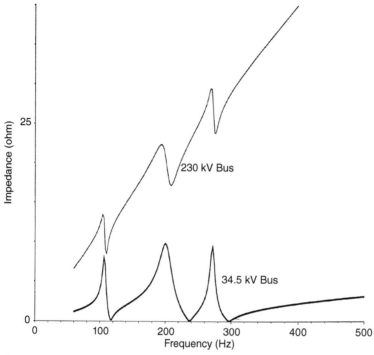

Figure 16.13 Impedance modulus with second, fourth, and fifth harmonic filters and typical voltage harmonics model.

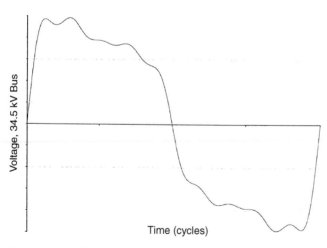

Figure 16.14 Voltage waveform, worst-case voltage harmonics, arc furnace models.

TABLE 16.7 Calculations of Harmonic Emissions, Arc Furnace, Three ST Filters (73.5, 24.5, 24.5 Mvar), Worst-Case Voltage Harmonic Model

Harmonic	Current through Filters			TDD Calculations			Harmonic Voltage Distortion	
	Second ST filter	Fourth ST filter	Fifth ST filter	TDD at 230-kV or 34.5-kV bus	TDD IEEE limits 34.5 kV	TDD IEEE limits 230 kV	34.5-kV PCC 34.5 kV	230-kV bus 230 kV
Fundamental A or V	1714	462.6	451.3	341.9A, 230 kV 2279.6A, 34.5 kV	2279.6	341.9		
$h = 2$	18.47	1.58	1.44	**2.37**	2.5	0.75	0.565	0.08
$h = 4$	0.19	31.41	2.31	0.40	2.5	0.75	0.22	0.03
$h = 5$	0.10	1.21	39.93	0.23	10	3	0.16	0.02
$h = 7$	0.73	5.17	11.38	1.82	10	3	1.75	0.22
$h = 8$	0.23	1.51	2.98	0.59	2.5	0.75	0.65	0.08
Total TDD, THDV				3.08	12	3.75	1.92	0.25

All harmonic currents and voltages are in percentage of the fundamental currents and voltages.

TABLE 16.8 Calculations of Harmonic Emissions, Arc Furnace, Second Harmonic Filter of 73.5-Mvar ST Filter, 24.5 Mvar Fourth and Fifth Harmonic Filters, Second-Order Damped Filters, Worst-Case Voltage Harmonic Model

Harmonic	Current through Filters			TDD Calculations			Harmonic Voltage Distortion	
	Second ST filter	Fourth ST filter	Fifth ST filter	TDD at 230-kV or 34.5-kV bus	TDD IEEE limits 34.5 kV	TDD IEEE limits 230 kV	34.5-kV PCC 34.5 kV	230-kV bus 230 kV
Fundamental A or V	1714	462.6	451.3	341.9A, 230 kV 2279.6A, 34.5 kV	2279.6	341.9		
$h = 2$	18.47	1.57	1.43	**2.37**	2.5	0.75	0.65	0.08
$h = 4$	1.11	22.14	9.46	2.26	2.5	0.75	1.24	0.15
$h = 5$	1.23	13.67	20.24	2.79	10	3	1.91	0.24
$h = 7$	0.76	5.63	11.82	1.89	10	3	1.82	0.23
$h = 8$	0.23	1.57	3.31	0.57	2.5	0.75	0.63	0.08
Total TDD, THDV				**4.73**	12	3.5	2.95	0.38

All harmonic currents and voltages are in percentage of the fundamental currents and voltages.

TABLE 16.9 Calculations of Harmonic Emissions, Arc Furnace, Second Harmonic filter of 73.5-Mvar Type-C Filter, 24.5-Mvar Fourth and Fifth Harmonic Filters, Second-Order Damped Filters, Worst-Case Voltage Harmonic Model

Harmonic	Current through Filters			TDD Calculations			Harmonic Voltage Distortion	
	Second ST filter	Fourth ST filter	Fifth ST filter	TDD at 230-kV or 34.5-kV bus	TDD IEEE limits 34.5 kV	TDD IEEE limits 230 kV	34.5-kV PCC	230-kV bus
Fundamental A or V	1714	462.6	451.3	388.9A, 230 kV 2593A, 34.5 kV	2593	388.9	34.5 kV	230 kV
$h = 2$	25.62	2.10	1.91	**2.72**	2.5	0.75	0.85	0.11
$h = 4$	1.40	21.10	9.01	1.86	2.5	0.75	1.16	0.14
$h = 5$	1.86	13.20	19.57	2.32	10	3	1.81	0.23
$h = 7$	1.55	5.47	11.48	1.58	10	3	1.73	0.21
$h = 8$	0.52	1.52	3.19	0.48	2.5	0.75	0.59	0.07
Total TDD, THDV				**4.35**	12	3.5	2.91	0.37

All harmonic currents and voltages are in percentage of the fundamental currents and voltages.

very high bus voltages above 161 kV as the PCC versus the PCC being at the secondary side of the utility interconnecting transformer. The TDD and voltage distortion are higher than case (c) because damped filters are not so effective as ST filters. If we increase the size of the damped filters, the distortions can be reduced.

(e) *As case (c) except that the second harmonic filter is turned into a type C filter.*

Chapter 15 shows the advantages of a C type filter for arc furnace installations for controlling the transients. The parameters of C type filter are ascertained using equations in Chapter 15.

- $C1 = 161\ \mu F$
- $C = 485\ \mu F$
- $L = 14.62\ mH$
- $R = 18\ ohms$

The study results of this case are shown in Table 16.9. This shows almost identical results to case (c).

We may conclude that all the designs meet the requirement of IEEE harmonic limits when 34.5-kV bus is declared as the PCC, but not so when 230-kV bus is declared as PCC. Limiting the second harmonic distortion to 0.75% at 230-kV bus will require almost 150 Mvar filter at 34.5 kV, which will bring down the overall TDD also to less than 3.5% in all the cases. However, this is a large filter and will give rise to overvoltages, and two double-wye connected filters in parallel will need to be provided. The overcompensation of reactive power will not be acceptable and a STATCOM, SVC, or TCR should be provided (see Fig. 5.11).

(f) *Switching transient considerations*

Further considerations for design of filters for arc furnace are the inrush current of transformers, which is rich in harmonics. A furnace transformer may be switched often with the furnace loads. This requires that

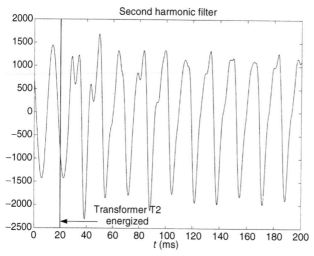

Figure 16.15   Inrush current 49-Mvar second harmonic filter on switching of 100-MVA transformer at 20 ms.

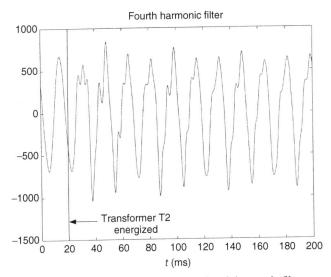

Figure 16.16   Inrush current 24.5-Mvar fourth harmonic filter on switching of 100-MVA transformer at 20 ms.

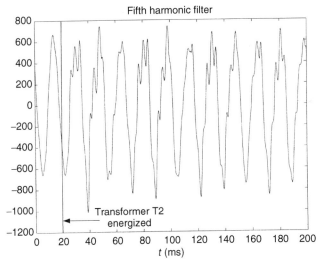

Figure 16.17   Inrush current 24.5-Mvar fifth harmonic filter on switching of 100-MVA transformer at 20 ms.

- The transient loading of the filter capacitors should be calculated.
- The capacitors should not get overloaded beyond their transient limits.
- The dynamic stresses of frequent switching should be considered. A furnace transformer may be switched 100–150 times per day.
- Other transients as discussed in Chapter 5 should also be considered.
- The TDD at the PCC should remain within acceptable limits.

    Figure 16.7 shows that each filter bank is individually controlled through a dedicated definite-purpose breaker to reduce the transient inrush currents.

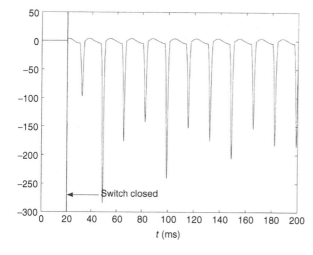

Figure 16.18 Inrush current of 100-MVA transformer switch closed at 20 ms.

The filters are switched in a sequence from lower harmonic to the higher harmonic.

Consider that 100 MVA transformer is switched, while 49, 24.5, 24.5 Mvar for second, fourth, and fifth harmonic filters are in service. Simulations of filter transient inrush currents using EMTP/ATP transient inrush currents are depicted as follows:

Figures 16.15–16.17 show that the inrush currents impact on the filters, and the 100 MVA transformer is switched at 20 ms. The transient lasts for a short duration of couple of cycles.

From Fig. 11.4, the peak transient voltage withstand for 10,000 operations a year is $2\sqrt{2}$ times the rated voltage. From Fig 11.5, the frequency of transient overcurrents is not specific though 12 times the peak currents can be withstood. Figure 16.15 for transient currents in second harmonic filter illustrates that the current transients last for a short duration and the peak is approximately 1.7 times the normal current flow.

Figure 16.18 is a simulation of 100 MVA transformer inrush current.

16.3 STUDY 3: FILTERS FOR TWO 8000-HP ID FAN DRIVES

A basic system configuration is shown in Fig. 16.19. Note that all the details of the distribution from the 13.8-kV buses are not shown. In any harmonic study, it is imperative that all the distribution system is modeled accurately and lumping of loads is done carefully (Chapter 14).

It is assumed that there are no nearby harmonic producing loads connected to the utility systems, and the utilities can be represented by a single Thévenin impedance representative of the short-circuit currents.

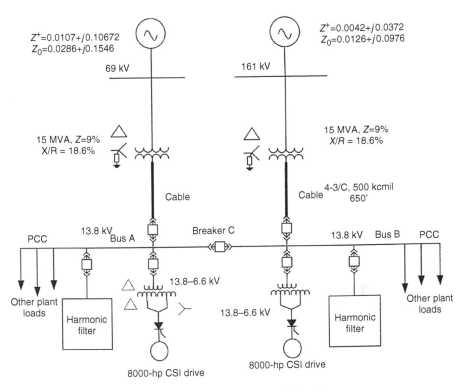

Figure 16.19 A system configuration with two 8000-hp ID fan drives.

The ASDs for the two ID fan motors are the major nonlinear loads. The complete distribution modeled for the study, medium-voltage and low-voltage loads, motor control centers, rotating and static loads modeled for the study (approximately 60 buses) are not shown. The system can be operated with either of the two utility sources, 69 kV or 161 kV. Usually, the bus section breaker C is kept open and 13.8-kV bus A loads are served from 69 kV utility source, while 13.8-kV bus B loads are served from 161 kV utility source. In case of outage of a utility source, the bus section breaker C is closed and the entire loads are served from either 69 kV or 161 kV. The 13.8-kV buses are the PCC.

Some problems with the operation of low-voltage systems and failure of components were noted, also some electronic circuit boards in the drive systems failed, which led to the investigations of the harmonic problems.

The 8000-hp drive systems are 18-pulse systems with front-end SCRs, served through three-winding transformers. These can be modeled as equivalent two winding transformers (Chapter 12). The capacitance of cables should be carefully modeled, as these impact the resonant frequencies.

(a) *Measurements*

As the system was in operation, the measurements of harmonics were conducted under various scenario of operation, and all results are not shown.

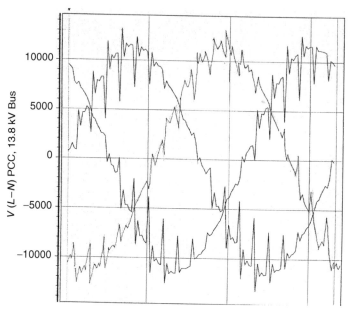

Figure 16.20   Voltage distortion at 13.8-kV bus (PCC), with an ID fan operating at 2.48 MW, $THD_V = 25.5\%$.

The measurement plots of the voltage and current distortion are shown as follows:

Figure 16.20 shows the voltage waveform distortion when the ID fans are operating at 2.48 MW (partial load), the power factor dips to 0.57. A voltage distortion of 25.5% is measured at 13.8 kV PCC.

Figure 16.21 shows the current waveform distortion, rms input current = 165 A. The current distortion is around 10%.

Figure 16.22 shows the voltage waveform distortion at a 480-V bus in the distribution. This is also approximately 26%.

The voltage distortion at 13.8-kV buses passes on to all the buses downstream in the distribution system. Note the high short-circuit impedance of the utility sources shown in Fig.16.19.

The IEEE 519 is not specific about the voltage distortion to be observed by a consumer. *Note that the source impedance of the utility systems is high representing a weak tie connection to the distribution. It is the harmonic current flow through this high impedance that gives rise to high voltage distortion.* In such situations, a consumer must not only consider the TDD but simultaneously the voltage distortion also, which can have much detrimental impact on the system, including harmonic loading and failure of system components, as in this study case.

(b) *Modeling of Harmonics*

A number of measurements at various operating points were taken, when the ID fans are operating at varying loads. The harmonic emissions are conservatively estimated from the measurement results. The harmonics modeled are shown in

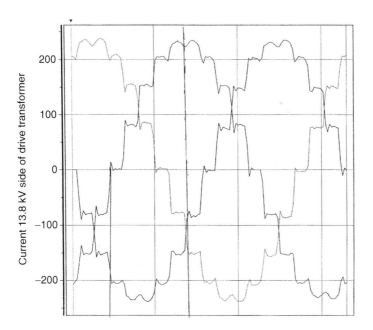

Figure 16.21   The current distortion at 13.8-kV bus (PCC), TDD = 10%.

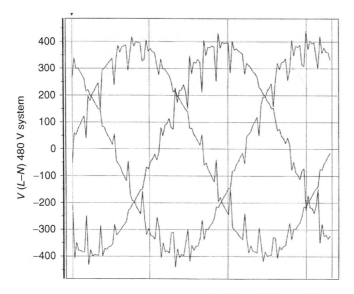

Figure 16.22   The voltage distortion at 480-V bus, THD$_V$ = 26%.

Table 16.10. Noncharacteristics and triplen harmonics are modeled based on measurements and worst operating loads. *Note that harmonics up to 73rd are modeled.* This is an 18-pulse system, therefore, the characteristic harmonics are 17, 19; 35, 37; 53, 55; 71, 73; and so on. Table 16.10 shows considerably higher magnitude of these harmonics in relation to other harmonics

TABLE 16.10   Harmonic Emission from ID Fans

Harmonic Order	Harmonic Percentage of Fundamental	Harmonic Order	Harmonic Percentage of Fundamental	Harmonic Order	Harmonic Percentage of Fundamental
2	0.07	14	0.08	49	0.2
3	0.98	15	0.86	53	3.0
4	0.06	17	8.0	55	4.0
5	0.84	19	6.0	59	0.10
6	0.12	23	0.26	61	0.10
7	0.41	25	0.28	65	0.10
8	0.06	29	0.20	67	0.10
9	0.20	31	0.20	71	2.0
10	0.08	35	4.0	73	2.0
11	0.26	37	3.0		
12	0.08	41	0.20		
13	0.26	43	0.20		

18-Pulse characteristics harmonics shown in heavy-lined blocks.

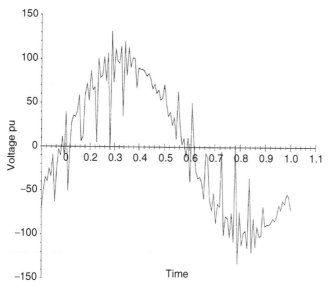

Figure 16.23   Modeled harmonic voltage distortion at 13.8-kV bus = 42.41%, with ID fans operating at 6 MW.

(c) *Study without Harmonic Filters*

With the harmonic injection as in Table 16.10 is applied, the voltage distortion at PCC, that is, at 13.8-kV buses rises to 42.41%. The harmonic voltage waveform and its spectrum are plotted in Figs 16.23 and 16.24, respectively.

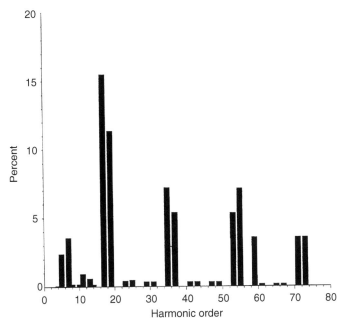

Figure 16.24   Harmonic distortion spectrum (voltage) at 13.8-kV buses, with bus section closed.

These show high levels of distortions and require careful analyses for the filter designs.

(d) *Selection and Design of Filters*

Looking at the harmonic spectrum, it is obvious that ST filters cannot be selected. These will give rise to resonant frequencies below their notch frequencies and it is practically impossible to place these so that these do not coincide wit1h the load-generated harmonics under various switching conditions. Thus, second-order high-pass damped filters are obvious choice. As discussed in Chapter 15, these do not give rise to shifted resonant frequencies but require much larger capacitors. Another advantage that is of much importance in this study case is that these are more tolerant to variations in the tolerances of the filter components.

The iterative designs of the filters are not discussed nor are their equations for calculations are repeated here (see Chapter 15). The design selected is two identical filters *one on each side of the bus section*, second-order high-pass filters with the following component ratings:

o 10 units per phase, 400 kvar, rated voltage 9960V, two bushing, 95 kV BIL, low-loss type. connected in ungrounded wye configuration. Each capacitor can is individually fused with 65A current limiting fuse. The three-phase kvar at operating voltage of 13.8 kV = 7679. The 400 kvar capacitor units are selected with no negative tolerance and plus tolerance limited to 5%. Note that a rated voltage of 9960 V is selected, while the system line-to-neutral voltage is 7967 V (Chapter 15).

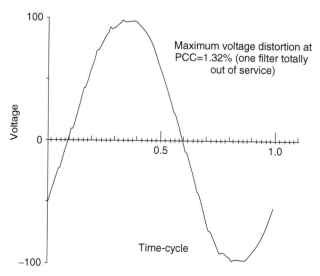

Figure 16.25   Voltage waveform at 13.8-kV PCC, the voltage distortion reduced from 42.4% to 1.32%.

- 13.8 kV, air-cooled, three-phase, vertically stacked coils, center phase reverse wound, 0.80 ohms (2.122 mH), 350 A rms continuous, harmonic spectrum through the reactor is specified based on study results, tolerance ±2.5%.

- Three-phase 0.25 ohms, 350 A rms continuously rated, tolerance ±10%, temperature rise limits as specified in Section 15.

- Ungrounded wye connection.

(e) *Effectiveness of Designed Filters*

The effectiveness of the filters is illustrated in Fig. 16.25, which shows that the voltage distortion of 42.41% in Fig. 16.23 is reduced by 1.3%.

The harmonic spectrum of the currents through filters is shown in Table 16.11 as the percentage of the fundamental frequency current. The spectrum of the harmonic current through the filter is illustrated in Fig. 16.26 and its waveform is plotted in Fig. 16.27. It is seen that the designed second-order high-pass filter shunts out a number of harmonics from the low order to high order, which is not possible with an ST filter.

(f) *Study of the Various Switching Conditions*

It is necessary to study a number of switching conditions, which are shown in Table 16.12.

- Cases 1–3 explore varying ID fan operations, with 13.8-kV bus section breaker C closed and system connected to 161 kV.

- Case 4 is a repeat of case 3, but system connected to 69 kV.

TABLE 16.11   Harmonic Current Spectrum through the Filter (Percentage of Fundamental)

Harmonic Order	Harmonic Current	Harmonic Order	Harmonic Current	Harmonic Order	Harmonic Current
2	3.17692E-02	19	22.7226	49	0.713385
4	0.173867	23	0.84045	53	10.6922
5	4.78747	25	0.965179	55	14.2526
7	7.12484	29	0.726265	59	7.12395
8	0.409609	31	0.723393	61	0.356168
10	0.403888	35	14.386	65	0.356155
11	1.86489	37	10.7676	67	0.356166
13	1.18739	41	0.715698	71	7.12433
14	0.329781	43	0.714922	73	7.12506
17	31.0012	47	0.713789		

- Cases 5–7 are with bus section breaker open and both the utility sources serving the loads.
- Case 8 investigates the detuning due to tolerance on filter reactors and capacitors.
- Case 9 investigates detuning when one capacitor can in the filter bank goes out of service.
- Case 10 investigates complete outage of a filter bank.

(g) *Study results*

The results of these calculations are shown in Table 16.13. In this table, cases A and B are operations without any filters, which are described as follows:

- Case A: ID fans operate at 2.48 MW, PF = 0.57, 161 kV source, 13.8-kV bus section breaker C closed.
- Case B: ID fans operate at 6 MW, PF = 0.84, 161 kV source, 13.8 kV bus section breaker C closed.

According to IEEE 519, the maximum permissible TDD for the $I_s/I_r < 20$ is 5%. Furthermore, not only the TDD but also the maximum distortion at each of the harmonics must be limited. Table 16.14 shows that this TDD limit is not exceeded in any of the study cases. The maximum TDD of 1.41% occurs for case 5. A tabulation of IEEE limits versus the calculated harmonic distortion at each harmonic to 73rd is shown in Table 16.5. Note that noncharacteristic harmonics are reduced by 25% and the characteristic harmonics are increased by $\sqrt{p/6} = \sqrt{3}$ for the 18-pulse system.

The calculations show that IEEE 519 harmonic limits for this worst-case scenario for fifth harmonic are slightly above IEEE limits. However, this is acceptable as this is not the normal operating mode of the system. If this is

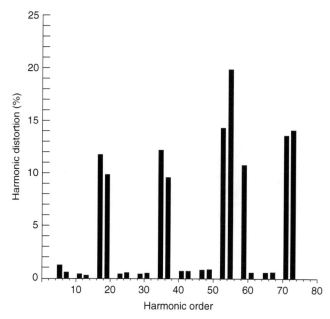

Figure 16.26   Spectrum of harmonic current loading of the filters in percentage of fundamental current.

TABLE 16.12   Switching Conditions for Analyses

Study Case	Description	Bus Section Breaker C	Source
1	ID fans operating at 1.7 MW, 0.4 PF	Closed	161 kV
2	ID fans operating at 2.48 MW, 0.57 PF	Closed	161 kV
3	ID fans operating at 6 MW, 0.84 PF	Closed	161 kV
4	ID fans operating at 6 MW, 0.84 PF	Closed	69 kV
5	ID fans operating at 1.7 MW, 0.4 PF	Open	161 kV and 69 kV
6	ID fans operating at 2.48 MW, 0.57 PF	Open	161 kV and 69 kV
7	ID fans operating at 6 MW, 0.84 PF	Open	161 kV and 69 kV
8	As case 7, filters of bus A with +10% tolerance on capacitors and +2.5% tolerance on reactor	Open	161 kV and 69 kV
9	As in case 7, one capacitor can of 300 kvar out of service on bus A	Open	161 kV and 69 kV
10	As in case 3, filter at bus 1 totally out of service	Closed	161 kV

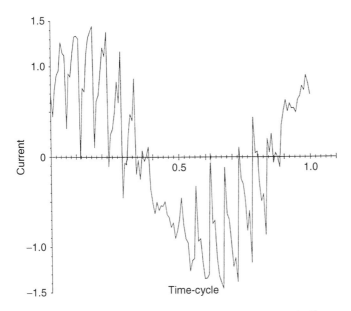

Figure 16.27   Waveform of the current through the harmonic filter.

TABLE 16.13   Summary Results of the Harmonic Analysis Study, Various Operating Conditions

Filters	Case #	THD$_V$ 13.8-kVBus A	THD$_V$ 13.8-kVBus B	TDD Bus A	TDD Bus B	THD$_V$ 480-V System
No	A	25.42	25.42	10.06	10.06	25.42
No	B	42.41	42.41	10.95	10.95	42.41
Yes	1	0.59	0.59	0.6	0.6	0.51
Yes	2	0.56	0.56	0.47	0.47	0.40
Yes	3	0.68	0.68	0.46	0.46	0.56
Yes	4	0.68	0.68	0.41	0.41	0.55
Yes	5	0.59	0.59	1.41	1.40	0.49
Yes	6	0.56	0.56	0.97	1.18	0.47
Yes	7	0.61	0.61	0.87	1.05	0.57
Yes	8	0.67	0.69	0.76	1.05	0.55
Yes	9	0.72	0.69	1.01	1.05	0.59
Yes	10	1.32	1.32	0.92	0.92	1.09

of concern, the filter size has to be slightly increased. Table 16.14 shows that triplen harmonics are zero; this is because of the delta-connected windings, which act as sink to the zero sequence third harmonics.

(h) *Practical Installations*
A practical filter installation is shown in Fig. 16.28. This illustrates that

TABLE 16.14   IEEE Permissible Distortion Limits versus Calculated Distortion at PCC for Case 5, Bus B, Table 16.13

h	IEEE Limits	Design Value	h	IEEE Limits	Design Value
2	1	0.07	31	0.15	0.01
3	1	0	**35**	**0.52**	**0.13**
4	1	0.12	**37**	**0.52**	**0.09**
5	1	1.06	41	0.15	0.01
6	1	0	43	0.15	0
7	1	0.18	45	0.15	0
8	1	0.02	47	0.15	0
9	1	0	49	0.15	0
10	1	0.02	**53**	**0.52**	**0.05**
11	0.5	0.05	**55**	**0.52**	**0.06**
12	0.5	0	59	0.15	0
13	0.5	0.04	61	0.15	0
14	0.5	0.01	65	0.15	0
15	0.5	0	67	0.15	0
17	**2.598**	**0.75**	**71**	**0.52**	**0.02**
19	**2.598**	**0.48**	**73**	**0.52**	**0.02**
23	0.15	0.02			
25	0.15	0.02			
29	0.15	0.01			

○ Resistance switching may be used to control the switching transients as an option. This requires two circuit breakers (Chapter 11). However, the results of switching studies showed that this is not necessary for the configuration under study.

○ MMPR (Microprocessor-based multifunction relays) provide bank protection.

○ The capacitor bank, consisting of individual-fused capacitor units in parallel and other devices shown in Fig. 16.28, is located in a metal enclosure suitable for indoor or outdoor installations. Alternatively, all devices can be outdoor rack mounted.

○ A three-phase disconnect and four-pole grounding switch is provided, inter-locked with each other. Disconnect switch is for local isolation and cannot be operated unless the upstream breakers are first opened and locked. The grounding switch can only be operated if the breakers and the incoming dis-connect switch are opened and locked. It is required for safety and connects the ungrounded capacitors to ground when the equipment is under mainte-nance, though the capacitors are provided with internal discharge resistances.

○ The disconnect switch does not have a load current or short-circuit withstand capability, it must withstand the momentary duty (fist cycle) short-circuit currents of the system for 10 cycles.

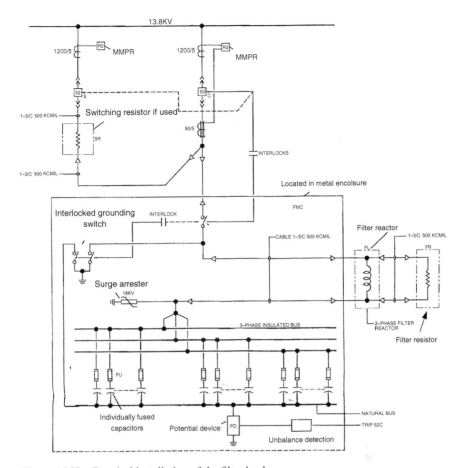

Figure 16.28   Practical installation of the filter bank.

o Surge arresters of appropriate rated voltage and maximum continuous operating voltage (MCOV) are provided at the junction of capacitors and filter reactor as shown. Surge protection is not discussed, see Chapter 11.

o A voltage is developed across the potential device (PD), provided with surge protection, if a fuse operates in capacitor can. The unbalance detection device will detect a single fuse failure and provide an alarm, and trip in case more than one fuse operate. The device is programmable. The unbalance protection is discussed in Chapter 11.

o The isolated neutral bus of the capacitor bank carries the discharge current in case of failure and is generally sized equal to the phase buses.

o The cable connections to the filter reactor and resistor are shown in this figure, and all cable sizes are conservatively selected to carry the required rms current plus a factor of 1.35.

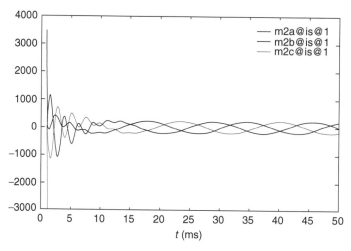

Figure 16.29   EMTP simulation of switching inrush current.

- o The filter reactors are air core, dry type, vertically stacked, and of the required Q factor and tolerances. Iron core reactors, though smaller in dimensions, are a possibility for voltages up to 30 kV (Chapter 15).

(i) *Study of switching transients*

An EMTP simulation of the switching transient on energization of a filter bank without *resistance switching* is shown in Fig. 16.29. This shows that the transients do not last for three-fourths of a cycle. The resistor in the filter somewhat helps to dampen the transient, though these can be further reduced by resistance switching. Thus, no control of the switching transients was implemented.

16.4   STUDY 4: DOUBLE-TUNED FILTER ON A THREE-WINDING TRANSFORMER

Figure 16.30 depicts a 50-MVA three-winding transformer, serving buses 3 and 4 at 4.16 kV and 13.8 kV, respectively. The 4.16-kV bus carries a 5-MVA 12-pulse ASD load and a 2 MVA of static load, while 13.8-kV bus carries a 6-pulse ASD load of 20 MVA and static load of 10 MVA. The harmonic spectra for these drive systems have been shown in previous chapters. To obtain close to rated operating voltages on load, the three-winding transformer primary (138 kV) taps are set to provide 2.5% voltage boost at 4.16 kV and 13.8 kV, and a 15-Mvar capacitive power compensation is required at 13.8-kV bus.

The harmonic analysis with the 15-Mvar capacitor bank shows that the distortion levels increase above the ambient level when no reactive power compensation is provided. This suggests that

- • A double-tuned filter 5/7 harmonics can be provided at 13.8-kV bus, sized 9 Mvar for the fifth harmonic and 6 Mvar for the seventh harmonic as these are the dominant harmonics of 6-pulse system.

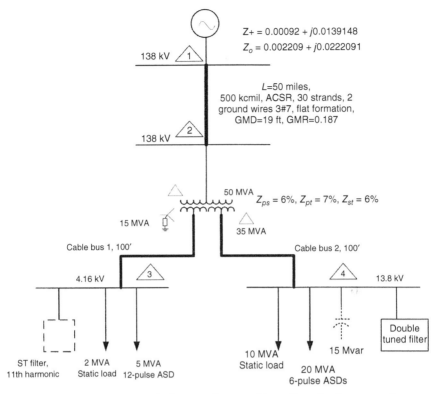

Figure 16.30   50-MVA three-winding transformer, with nonlinear loads on secondary windings of 4.16 and 13.8 kV.

- Simultaneously, an 11th harmonic single-tuned filter can be provided at 4.16-kV bus, which will control the 11th and 13th dominant harmonics of 12-pulse load on this bus.

With this objective, the results of harmonic flow are shown in Figs 16.31 and 16.32, which illustrate the current spectra and waveforms of the currents in cable buses 1 and 2. The results are also tabulated in Table 16.15. While 13.8-kV bus double-tuned filter behavior can be expected and the harmonics on this 13.8-kV bus are effectively reduced, the harmonic amplification occurs at 4.16-kV bus. Note the distortion of 11.29% on 11th harmonic, though this bus is having a 6-Mvar 11th harmonic filter.

Next, the 11th harmonic filter at 2.4-kV bus is removed.

The result of this calculation is shown in Figs 16.33 and 16.34, which depict that the current distortion through cable bus 1 is much reduced, though the distortion through cable bus 2 is slightly increased. These results are also plotted in Table 16.15.

It is an unusual case of coupling of two buses, one at 13.8 kV and the other at 4.16 kV, through the three-winding transformer and the harmonic distortion *brought down by not adding but removing an ST filter. Yet, the distortion of the current through 138-kV line increases slightly.*

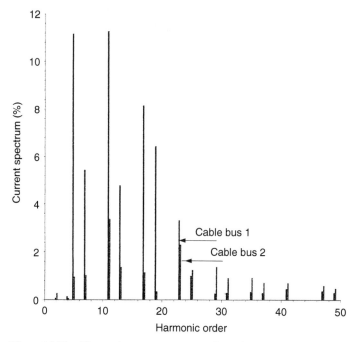

Figure 16.31   Harmonic current spectrum through cable buses 1 and 2 in Fig. 16.30, with 11th harmonic filter on 4.16-kV bus.

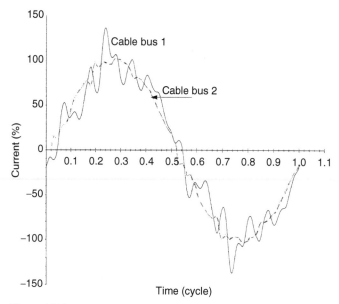

Figure 16.32   Waveform of the current through cable buses 1 and 2 in Fig. 16.30, with 11th harmonic filter at 4.16-kV bus.

TABLE 16.15   Calculations of Harmonic Distortions, Section 16.4

ID	With 11th Harmonic Filter			Without 11th Harmonic Filter		
	2.4-kV cable bus 1	13.8-kV cable bus 2	138-kV line L1	2.4-kV cable bus 1	13.8-kV cable bus 2	138-kV line L1
If = 100%	922.68	1056.29	136.05	961.97	1039.12	130.97
2	0.08	0.28	0.23	0.04	0.28	0.22
4	0.14	0.07	0.09	0.03	0.13	0.11
5	**11.18**	**0.96**	**3.27**	1.73	**2.59**	**2.76**
7	5.45	1.04	0.43	0.79	0.02	0.23
11	**11.29**	**3.38**	**0.89**	3.45	**0.48**	2.85
13	4.79	1.37	0.72	2.05	0.31	1.93
17	**8.17**	**1.15**	**1.83**	2.42	2.17	3.76
19	6.45	0.36	1.89	1.36	2.19	2.33
23	3.32	2.31	1.76	0.66	2.28	1.40
25	1.0	1.25	0.68	0.57	1.25	0.65
29	0.27	1.37	0.44	0.08	1.37	0.46
31	0.30	0.92	0.25	0.10	0.91	0.27
35	0.34	0.92	0.21	0.26	0.90	0.25
37	0.29	0.72	0.16	0.20	0.69	0.18
41	0.47	0.71	0.14	0.24	0.69	0.16
43	0.01	0	0	0.02	0	0
47	0.37	0.60	0.15	0.25	0.57	0.18
49	0.31	0.48	0.15	0.17	0.46	0.17
THD %	20.65	5.38	4.81	5.33	5.33	6.41

These results could not be intuitively guessed. An investigation is carried out by $Z - \omega$ plots of 138-kV bus 2, 4.16-kV bus 3, and 13.8-kV bus 4, illustrated in Fig. 16.35(a)–(c). An interaction is noted between the filters, which changes the frequency plots of these buses with the 11th harmonic filter at 2.4-kV bus 3 in service and out of service. This can be attributed in part to the low impedance between buses 3 and 4 through the three-winding transformer coupling.

Damped filters could be used that will minimize this interaction.

16.5   STUDY 5: PV SOLAR GENERATION PLANT

We have not discussed solar generation. Solar energy magnitude is reflected in a radiation level of some 400 BTU/ft². The solar energy can be used for water heating, industrial process heating, homes, fans pumps, traffic lights, and hybrid PV powered homes. The two technologies for converting solar energy into electricity are as follows:

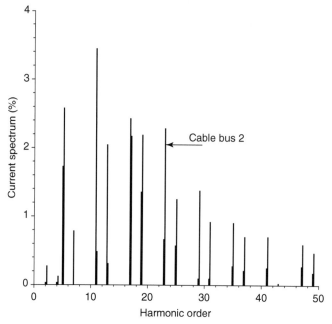

Figure 16.33　Harmonic current spectrum through cable buses 1 and 2 in Fig. 16.30, *without* 11th harmonic filter on 4.16-kV bus.

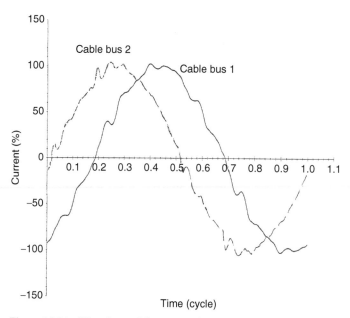

Figure 16.34　Waveform of the current through cable buses 1 and 2 in Fig. 16.30, *without* 11th harmonic filter on 4.16-kV bus.

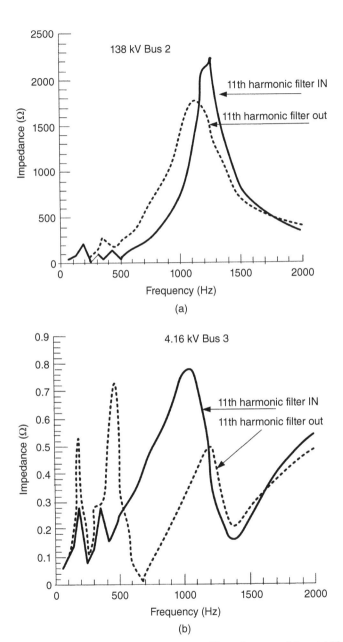

Figure 16.35   (a)–(c) Frequency scan of impedance modulus at 138, 4.16, and 13.8 buses, with and without 11th harmonic filter, Fig. 16.30.

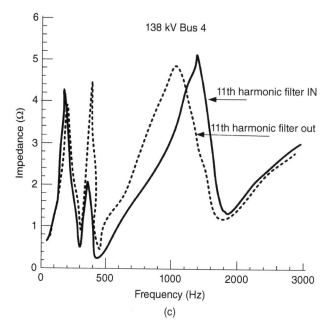

Figure 16.35 (*Continued*)

- Solar thermomechanical systems, in which the solar radiation is used to heat a working fluid that runs the turbines. The solar energy is collected through mirrors which track the sun and focus the energy to heat a working fluid.
- Solar photovoltaic cells, which directly convert solar radiation to electric current.

PV cells were first used to power the satellites. PV is a device that converts sunlight directly into electricity. The basic structure of a PV cell is shown in Fig 16.36. The incident photons cause generation of electron–hole pairs in both p and n-layers. This photon-generated minority carries freely cross the junction, which increases minority flow many times. The major component is the light-generated current when load is connected to the cell. There is also thermally generated reverse current, also called dark current, as it flows even in the absence of light. The light current flows in the opposite direction to the forward diode current of the junction.

A PV cell is classified as follows:

- In terms of material, such as noncrystalline silicon, polycrystalline silicon, amorphous silicon, gallium arsenide, cadmium telluride, and cadmium sulfide.
- In terms of technology, such as single crystal bonds and thin film.

The conversion efficiency varies. Under normal temperature of 25°C and illumination level of $100mW/cm^2$, amorphous silicon has an efficiency of 5–6% while gallium arsenide has an efficiency of 20–25%.

The capacity to produce solar cells in the United States is increased by 20%, with the official inauguration of new thin film facility, United Solar Systems, which

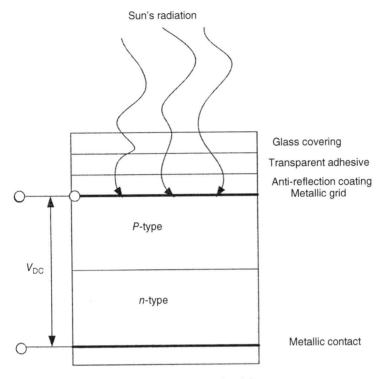

Figure 16.36   PV solar cell, basic constructional features.

can produce 30 MW of solar cells per year. The PV panels are installed at the top of CN tower in Toronto, the tallest such tower in the world. Most of the solar capacity of 6.5 MW in Canada alone is for off-grid applications for light houses, homes, navigations buoys, remote telecommunication towers, and so on. There are an estimated 12,000 residential solar water systems, seasonal pool heaters, and hot water units in Canada alone.

Till the mid-1990s, the applications were standalone systems located where connections to utility grids were impractical. Utility interactive PV cells are classified by IEEE Standard 929. Generally, utility interactive systems do not incorporate any form of energy storage and supply power to the grid when operating. As the output is DC, it is necessary to convert it to AC before interconnection with utility. An inverter is used, called here power conditioning unit (PCU). Automatic disconnects are provided in the event of loss of utility voltage. PCU is synchronized with the utility voltage. Figure 16.37 shows solar resources in the United States.

16.5.1   Solar Plant Considered for Harmonic Analysis

The harmonic emission from large solar plant is modeled in this study. There has been an attempt to obtain larger currents and increase voltage from interconnections of

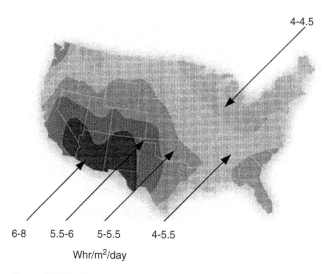

Figure 16.37   Solar energy resources in the United States.

solar cells and innovations in the design of inverters. Figure 16.38 shows a 2.5-MVA, 13.8 kV–0.8 kV, three-phase transformer connected to two solar inverters, which in turn connect to a number of solar cells in series and parallel configurations for the required output of 900 A maximum current and three-phase inverter output at 800 V.

This figure shows that 10 such 13.8–0.8 kV transformers, primary windings in delta connection, and the secondary windings in wye connection, solidly grounded, form "one chain" served from a 1200-A, 13.8-kV circuit breaker. Thus, the maximum possible output from one chain is approximately 75 MW. Three such chains are served from 3000-A, 13.8-kV bus 1 (Fig. 16.39). The operating power factor varies between 0.95 lagging to 0.95 leading. When operating at lagging power factor, a 6 Mvar of compensation is required through voltage-controlled auto-switched capacitor bank.

Figure 16.39 illustrates that three such 13.8-kV buses, bus 1, bus 2, and bus 3, are each connected to 55/92 MVA, 13.8–230 kV step-up transformer. Thus, the total installed capacity of 9 chains, each chain consisting of ten 2.5-MVA transformers (a total of 90 transformers), is approximately 225 MW of power generation at peak hours.

The objective of the study is to ascertain

- If there are harmonic distortion problems due to 180 solar inverters, with two inverters on the secondary of each transformer?

- When 6-Mvar capacitor banks are switched at each bus, will any harmonic resonance problem occur?

- Whether the switching of capacitors will create any transient overvoltages, which could be damaging to the inverters?

We will not discuss the aspect of switching transients. The references quoted in Chapter 11 may be seen.

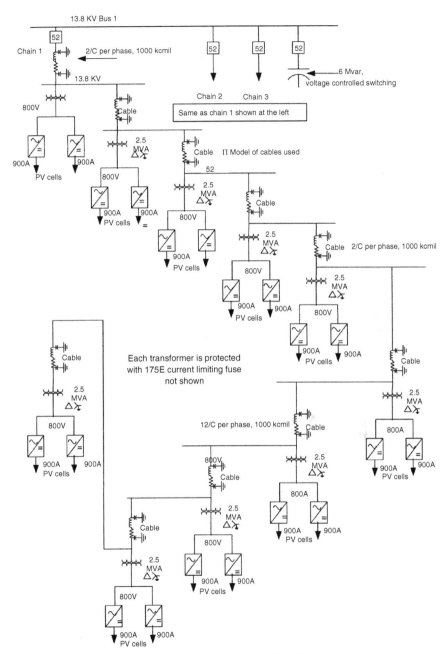

Figure 16.38   A system configuration for solar-generating plant, three chains, each chain consisting of ten 2.5-MVA step-up transformers connected to a 13.8-kV bus.

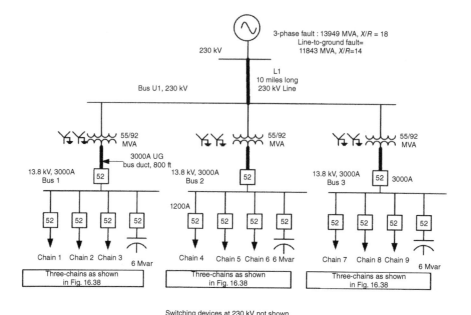

Figure 16.39   Overall configuration of a 225-MW solar generating plant.

The results of the study are shown in the following figures and tables with explanations:

Table 16.16 shows the current harmonic emission at the inverters connected to the secondary side of the 13.8–0.8 kV, delta–wye grounded 2.5-MVA transformers in terms of percentage of the fundamental current. The odd and even harmonics are separately listed. These are low, the inverters have line filters. There is some variation in the harmonic emission between phases, and Table 16.16 shows the maximum values. The harmonic emission will vary from manufacturer to manufacturer, and ascertaining a correct spectrum is important.

Figure 16.40 depicts the impedance modulus and phase angle; frequency scan on a 13.8-kV bus. The results are the same on any 13.8-kV bus 1, 2, or 3 due to symmetry. As expected when the capacitors (total of 18 Mvar) are in service, the resonant frequency is lowered. Note the resonant condition when the capacitors are not in service, which occurs above 40th harmonic. As harmonics above 40th are not modeled, this has no impact.

Figure 16.41 shows the harmonic spectrum of voltages at 13.8-kV buses, with capacitors out of service and with capacitors in service. It is seen that when the capacitors are in service, the higher order harmonics are much reduced, and there is reduction of three to four times in high-order harmonics. For example, the 23rd harmonic is reduced by a factor of 2.3 when capacitors are in service, and 29th harmonic is reduced by a factor of 3.5. The total voltage harmonic distortion at 13.8-kV buses *is 0.14% when capacitors are out of service and reduces by 0.08% with capacitors in*

TABLE 16.16   Harmonic Emission, Solar Power Plant

Harmonic Order Odd Harmonics	Percentage of Fundamental	Harmonic Order Even Harmonics	Percentage of Fundamental
3	0.14	2	0.24
5	0.18	4	0.30
7	0.09	6	0.05
9	0.06	8	0.11
11	0.09	10	0.13
13	0.04	12	0.05
15	0.06	14	0.12
17	0.12	16	0.15
19	0.30	18	0.06
21	0.12	20	0.16
23	0.25	22	0.20
25	0.10	24	0.12
27	0.07	26	0.12
29	0.11	28	0.13
31	0.24	30	0.11
33	0.11	32	0.19
35	0.11	34	0.22
37	0.12	36	0.09
39	0.07	38	0.14
		40	0.14

service. Usually, the situation is otherwise, the presence of capacitors giving higher distortion.

Figure 16.42 shows similar harmonic spectrum of voltages for 230-kV bus. It has the same pattern as the 13.8-kV buses in Fig. 16.41. The total current harmonic distortion at 230-kV bus is 0.13% when capacitors are out of service and reduces by 0.08% when the capacitors are in service.

Table 16.17 plots the amplitude of individual harmonics. The triplen harmonics are eliminated due to delta connection of the transformer windings.

The current harmonic spectrum through 230 line L1, with capacitors out of service and with capacitors in service, is tabulated in Table 16.18. It follows the pattern of voltage spectra.

No study should be considered to be applicable to all situations. The solar plants have used a variety of voltage control devices, such as STATCOM, shunt reactors, and filters. The harmonic emissions from the PV cell inverters can vary and the reactive power compensation can vary, which can give entirely different results.

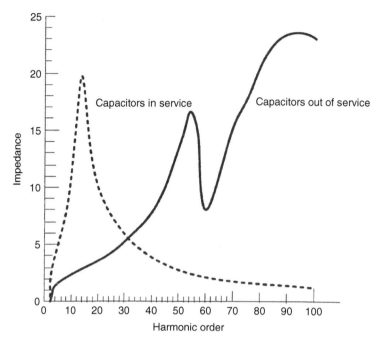

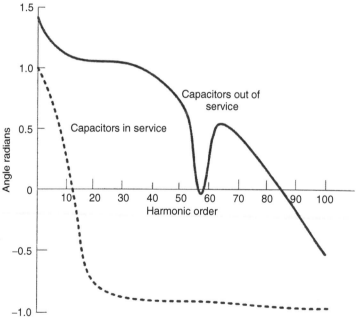

Figure 16.40   Frequency scan, impedance angle and modulus at a 13.8-kV bus.

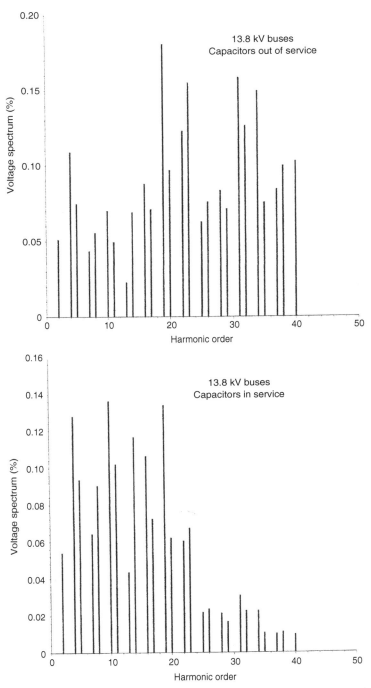

Figure 16.41   Spectrum of harmonic voltages at a 13.8-kV bus, with and without capacitors (shown in Figure 16.39).

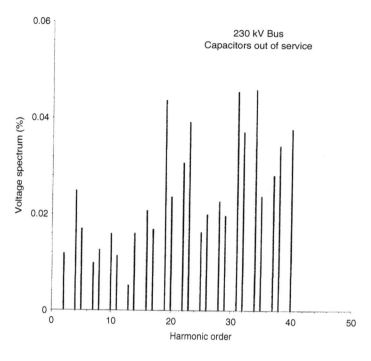

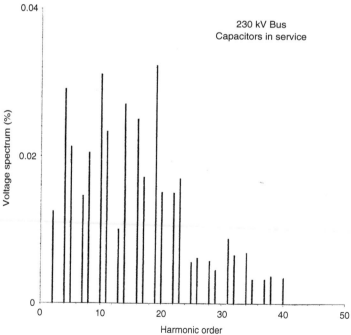

Figure 16.42 Spectrum of harmonic voltages at a 230-kV bus, with and without capacitors (shown in Figure 16.39).

TABLE 16.17 Harmonic Voltage Distortion with and without Capacitor Banks

Harmonic Order Odd Harmonics	All 13.8-kV Buses Capacitors Out of Service	230-kV Bus Capacitor Out of Service	All 13.8-kV Buses Capacitors in Service	230-kV Bus Capacitors in Service	Harmonic Order Even Harmonics	All 13.8-kV Buses Capacitors Out of Service	230-kV Bus Capacitor Out of Service	All 13.8-kV Buses Capacitors in Service	230-kV Bus Capacitors in Service
3	0	0	0	0	2	0.05	0.01	0.05	0.01
5	0.07	0.02	0.09	0.02	4	0.11	0.02	0.13	0.03
7	0.04	0.01	0.06	0.01	6	0	0	0	0
9	0	0	0	0.	8	0.06	0.01	0.09	0.02
11	0.05	0.01	0.10	0.02	10	0.07	0.02	0.14	0.03
13	0.02	0.01	0.04	0.01	12	0	0	0	0
15	0	0	0	0	14	0.07	0.02	0.12	0.03
17	0.07	0.02	0.07	0.02	16	0.09	0.02	0.11	0.03
19	0.18	0.04	0.13	0.03	18	0	0	0	0
21	0	0	0	0	20	0.10	0.02	0.06	0.02
23	0.16	0.04	0.07	0.02	22	0.12	0.03	0.06	0.02
25	0.06	0.02	0.02	0.01	24	0	0	0	0
27	0	0	0	0	26	0.08	0.02	0.02	0.01
29	0.07	0.02	0.02	0	28	0.08	0.02	0.02	0.01
31	0.16	0.05	0.03	0.01	30	0	0	0	0
33	0	0	0	0	32	0.13	0.04	0.02	0.01
35	0.08	0.02	0.01	0	34	0.15	0.05	0.02	0.01
37	0.08	0.03	0.01	0	36	0	0	0	0
39	0	0	0	0	38	0.10	0.03	0.01	0
					40	0.10	0.04	0.01	0

THDv with capacitors in service

230-kV bus: 0.08%

13.8-kV buses: 0.08%

THDv with capacitors out of service

230-kV bus: 0.13%

13.8-kV buses: 0.14%

TABLE 16.18 Harmonic Current Distortion 230-kV Line, with and without Capacitor Banks

Harmonic Order Odd Harmonics	Harmonics Injected On Sec of 2500 kVA Transformer	Harmonics In 230-kV Line, Capacitors Out of Service	Harmonics In 230-kV Line, Capacitors in Service	Harmonic Order Even harmonics	Harmonics Injected On Sec of 2500 kVA Transformer	Harmonics In 230-kV Line, Capacitors Out of Service	Harmonics In 230-kV Line, Capacitors in Service
3	0.14	0	0	2	0.24	0.20	0.22
5	0.18	0.12	0.15	4	0.30	0.22	0.26
7	0.09	0.05	0.08	6	0.05	0	0
9	0.06	0	0	8	0.11	0.06	0.09
11	0.09	0.04	0.08	10	0.13	0.06	0.11
13	0.04	0.01	0.03	12	0.05	0	0
15	0.06	0	0	14	0.12	0.04	0.07
17	0.12	0.04	0.04	16	0.15	0.05	0.06
19	0.30	0.09	0.06	18	0.06	0	
21	0.12	0	0	20	0.16	0.04	0.03
23	0.25	0.06	0.03	22	0.20	0.05	0.03
25	0.10	0.02	0.01	24	0.12	0	
27	0.07	0	0	26	0.12	0.03	0.01
29	0.11	0.03	0.01	28	0.13	0.03	0.01
31	0.24	0.06	0.01	30	0.11	0	0
33	0.11	0	0	32	0.19	0.05	0.01
35	0.11	0.03	0	34	0.22	0.05	0.01
37	0.12	0.03	0	36	0.09	0	0
39	0.07	0	0	38	0.14	0.04	0
				40	0.14	0.04	0

THDI = 0.13% capacitors out of service, THDI = 0.08 % capacitors in service

16.6   STUDY 6: IMPACT OF HARMONICS AT A DISTANCE

Figure 16.43 shows generation/transmission/subtransmission/utility service to an industrial consumer and small commercial load.

There are a total of four utility generators, G1, G2, G3, and G4 connected to 400 kV; two utility generators G5 and G6 connected to 230 kV, and one industrial generator of 30 MVA connected to 13.8-kV bus B-20, industrial plant load. All utility generators are high resistance grounded through a distribution transformer, and the industrial generator of 30 MVA is low resistance grounded.

A total of 15 transmission lines of which 5 at 400 kV, 9 at 230 kV, and 1 at 115 kV, and 1 distribution line at 12.47 kV are modeled; the length of the lines is shown in Fig. 16.43.

A total of 17 transformers, ratings, and voltage transformation ratios as shown are modeled. All generator step-up transformers have primary windings in

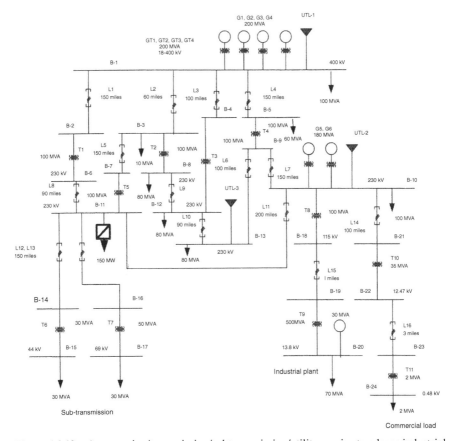

Figure 16.43   A generation/transmission/subtransmission/utility service to a large industrial and commercial consumer to study distance effect of harmonics.

delta-connected and high-voltage wye-connected winding solidly grounded. All other transformers have delta primary and solidly grounded secondary windings, except 50-MVA transformer serving industrial loads, which has low-resistance grounded wye windings.

The total load served from the system is 731.5 MW and 360 Mvar. Some loads are composite loads, that is, static and AC motor loads, not indicated in Fig. 16.43. The total generation is 1020 MW and 73 Mvar. Approximately, 16 MW of active power losses occur in the system.

There is only one 150 MW converter load at bus B-11 in the entire system. There are no other shunt capacitors or reactive power compensation devices.

No reactive power compensation is required when the system is loaded as shown. Approximately, 341 Mvar is supplied into the utility systems apart from meeting the load-reactive power demand and the system losses. This is so, because the shunt capacitance of many long transmission lines generates leading Mvar. Note that when the system is unloaded, *lagging* reactive power compensation in the controlled and noncontrolled form of shunt reactors, SVCs, or other devices will be required.

With appropriate off-load tap settings on the transformers, the load voltages can be maintained on all buses within acceptable limits, without any other voltage regulation. Practically to account for voltage variations under various loading conditions, controllable and fixed reactive power compensation will be required.

Three utility sources, one at 400 kV and the other two at 230 kV, U-1, U-2 and U-3, are represented by their three phase and single line to ground fault impedances at fundamental frequency, with no external harmonics.

The objective of this study is to demonstrate that only one 150 MW harmonic producing load results in harmonics in all sections of the power system including the small 2-MVA commercial load.

The study results are shown as follows.

- Figures 16.44 and 16.45 show the harmonic current flow into the three utility sources, spectrum, and current waveform, respectively. Figure 16.46 is a frequency scan of the buses having utility interconnection. The harmonic currents injected into the utility systems are also tabulated in Table 16.19. This means that harmonic amplification is occurring in the system, *though there are no shunt capacitors.*

- Figures 16.47 and 16.48 show the harmonic current spectra and waveforms at the far ends of the system, in transformers T6 (subtransmission system), in transformer T9 (industrial distribution system), and in 2-MVA transformer T11 (commercial distribution). It is seen that these systems are impacted with considerable harmonics (Table 16.20).

- Amplification and attenuation of harmonic voltages occur through the transmission lines. Figure 16.49 shows the harmonic voltage profiles at bus B-11, at B-6, and at B-14. Note the variations of harmonic voltages. A similar plot for bus B-1, bus B-2, and bus B-5 is shown in Fig. 16.50. The harmonic voltages are function of harmonic impedances at the particular harmonic current flows, which show resonances.

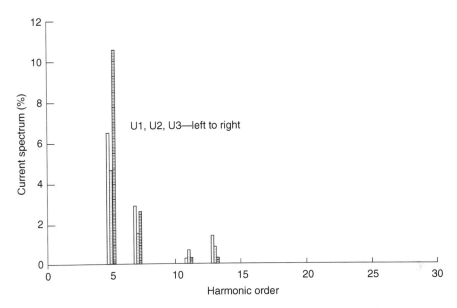

Figure 16.44   Spectrum of harmonic current flows in three utility sources, Fig. 16.43.

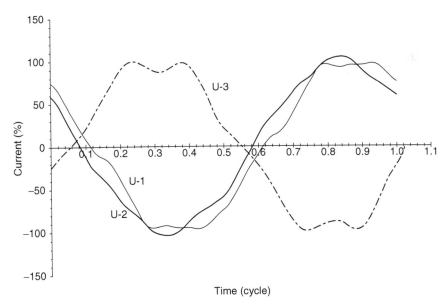

Figure 16.45   Waveform of the harmonic current flows in three utility sources, Fig. 16.43.

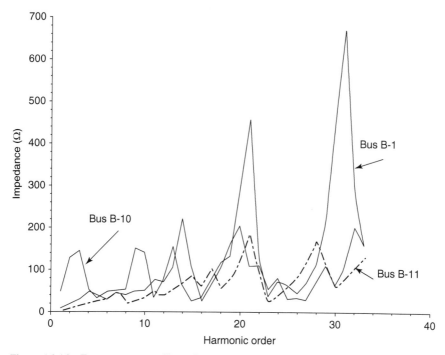

Figure 16.46   Frequency scan of impedance modulus at a number of buses connected to utility sources in Fig. 16.43, showing number of resonant frequencies.

TABLE 16.19   Harmonic Current Flow into the Utility Sources

Harmonic Order	U1, 400 kV	U2, 230 kV	U3, 230 kV
Fundamental current	587.82	372.46	295.01
5	6.5	4.68	10.71
7	2.94	1.59	2.60
11	0.15	0.66	0.09
13	1.4	0.76	0.10
17	0.16	0.16	0.20
19	0.04	0.05	0.23
23	0.02	0.01	0.03
25	0.01	0.01	0.01

A single harmonic source has far-reaching impacts spread over miles apart. This is because of resonances and partial resonances that can occur in the transmission systems due to longline models. See the capacitance charging current estimates of high-voltage transmission lines provided in Chapter 13.

The study shows that mitigation of harmonics at bus B-11 is required by provision of appropriate filters. Mitigating the lower order harmonics of fifth and seventh will curtail their amplitudes at a distance, but not entirely eliminate it. Partial resonances will still occur.

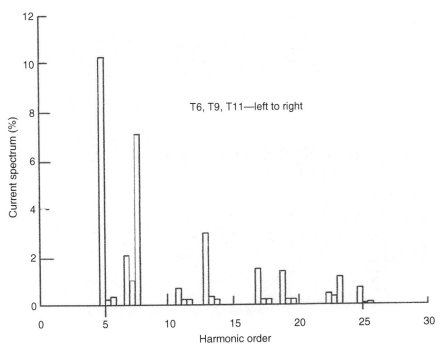

Figure 16.47   Spectra of harmonic current flows in transformers at remote ends of the system configuration shown in Fig. 16.43.

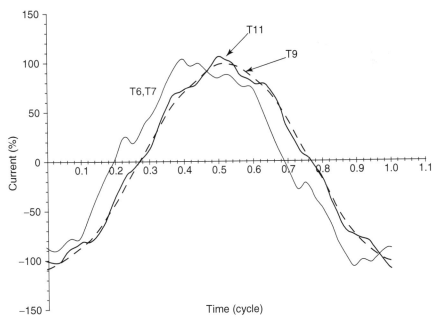

Figure 16.48   Waveforms of the harmonic current flows in transformers at remote ends of the system configuration shown in Fig. 16.43.

TABLE 16.20   Harmonic Current Flow, Transformers T6, T9, and T11

Harmonic	T6, 50 MVA Subtransmission	T9, 50 MVA Industrial Load	T11, 2 MVA Commercial Load
Fundamental	75.33A = 100%	179.01A = 100%	96.92A = 100%
5	10.08	0.11	0.15
7	2.06	1.05	7.06
11	0.66	0.07	0.10
13	2.91	0.25	0.27
17	1.36	0.16	0.16
19	1.36	0.15	0.19
23	0.37	0.07	1.05
25	0.71	0.02	0.04

Harmonic currents shown in percentage of fundamental currents.

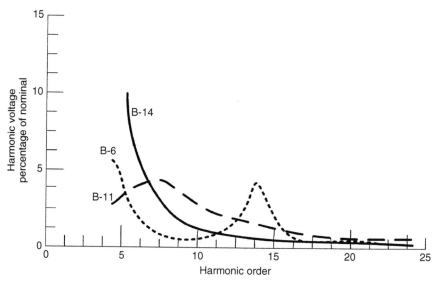

Figure 16.49   Harmonic voltages versus frequency, buses B-6, B-11, and B-14, showing amplifications and attenuations, configuration shown in Fig. 16.43.

16.7   STUDY 7: WIND GENERATION FARM

In Section 4.15.4, harmonic emission from wind power generation is explained. More than 94 GW of wind power generation has been added worldwide by end of 2007, out of which over 12 GW is in the United States and 22 GW in Germany alone. Looking at energy penetration levels (ratio of wind power delivered by total energy delivered), Denmark leads, reaching a level of 20% or more, followed by Germany. In some hours of the year, the wind energy penetration exceeds 100%, with excess sold to Germany

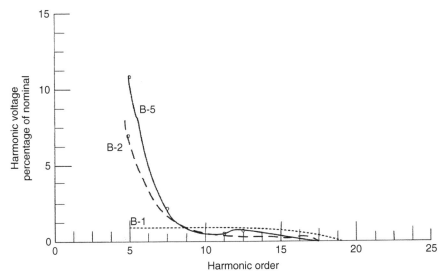

Figure 16.50   Harmonic voltages versus frequency, buses B-1, B-2, and B-5, showing amplifications and attenuations, configuration shown in Fig. 16.43.

and NordPool. Nineteen offshore projects operate in Europe producing 900 MW. The US offshore wind energy resources are abundant.

In the United States, wind power generation accounts for approximately 0.6% of the total, and Renewable Energy Laboratory (DOE/NREL) took an investigation on how 20% of energy from wind will look like in 2030. Some considerations are as follows:

- Power flow control
- Reactive power compensation
- Congestion management
- Long-term and short-term voltage stability
- Transient stability and low-voltage ride through capability
- Green house gas reduction
- Harmonics and resonances

With respect to single-unit wind turbines in the United States, there has been a progressive increase in the ratings from 100 kW in the 1980s to 5 MW planned for offshore units. Worldwide, an 8-MW geared single unit with 80 m long blades has been commissioned in January 2014.

An induction motor will act as an induction generator when driven above its synchronous speed. At $s = 0$, the induction motor torque is zero, and if it is driven above its synchronous speed, the slip becomes negative and generator operation results. The torque–speed characteristics of the machine above synchronism are similar to that for running as an induction motor (Fig. 16.51). Also, the equivalent circuit can be drawn akin to an induction motor (Fig. 16.52).

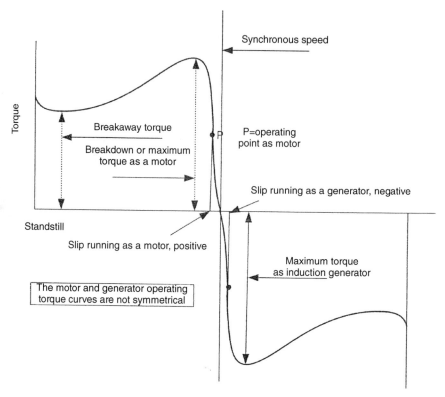

Figure 16.51   Torque–speed characteristics of an induction motor and induction generator.

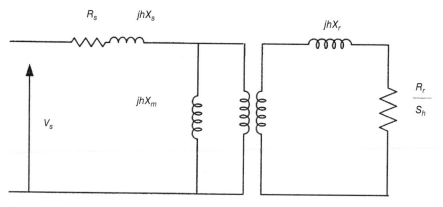

Figure 16.52   Equivalent circuit model of an induction generator.

The maximum torque can be written as

$$T_m = \frac{v_s^2}{2\left[\sqrt{\left(R_s^2 + (X_s + X_r)\right)^2} \pm R_s\right]} \tag{16.2}$$

The negative R_s represents the maximum torque required to drive the machine as generator. The maximum torque is independent of the rotor resistance. For supersynchronous speed operation (generator operation), the maximum torque is independent of R_r (rotor resistance), same as for motor operation, but increases with reduction of both stator and rotor reactances.

Induction generators do not need synchronizing and can run in parallel without hunting and at any frequency, the speed variations of the prime mover are relatively unimportant. Thus, these machines are applied for wind power generation.

An induction generator must draw its excitation from the supply system, which is mostly reactive power requirement. An induction generator can be self-excited through a capacitor bank, without external DC source, but the frequency and generated voltage will be affected by speed, load, and capacitor rating. For an inductive load, the magnetic energy circulation must be dealt with by the capacitor bank as induction generator cannot do so. On a sudden short circuit, the excitation fails, and with it the generator output; so in a way the generator is self-protecting.

As the rotor speed rises above synchronous speed, the rotor EMF becomes in phase opposition to its subsynchronous position, because the rotor conductors are moving faster than the stator rotating field. This reverses the rotor current also and the stator component reverses. The rotor current locus is a complete circle. The stator current is clearly a leading current of definite phase angle. *The output cannot be made to supply a lagging load.* The presence of capacitors can lead to harmonic resonance and self-excitation. Induction generators produce harmonic and synchronous pulsating torques, akin to induction motors (Chapter 3).

Strict regulations apply to the connections of wind power plants to the utility grids, with respect to voltage control, fault clearance times, and the time duration of voltage dips: according to recommendations of WECC (Western Electricity Coordinating Council) Wind Generation Task Force (WGTF); not completely discussed here. Thus, a prior reactive power compensation study is undertaken. As we have previously seen, the reactive power compensation, power factor, and voltage profiles are interrelated, the system impedance playing an important role. With reference to Fig. 16.53, equivalent circuit, the reactive power required by an induction generator, can be written as

$$Q = \frac{-b}{2a}V_1^2 + \frac{\sqrt{(b^2 - 4ac)V_1^4 + 4aPV_1^2}}{2a} \tag{16.3}$$

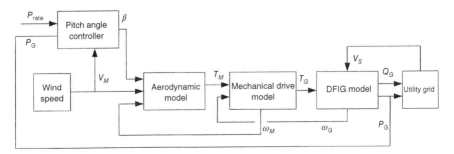

Figure 16.53 Components of main control circuit of wind power generation.

where

$$X_{ss} = X_s + X_m, \quad a = \frac{R_r X_{ss}^2}{X_m \sin^2\varphi}, \quad b = \frac{2R_r X_{ss}}{X_m^2} + \frac{1-s}{\tan\varphi}, \quad c = \frac{R_r}{X_m^2} \quad (16.4)$$

P is the active power and φ is the power factor angle.

The wind turbine dynamic models consist of pitch angle control, active and reactive power control, the drive train model, and the generator model. These models are considered the proprietary of the turbine manufacturer and can be obtained only under confidentiality agreements. Some efforts have been directed by WECC toward generic models, which are now available in some commercial software packages.

The overall control system diagram of a wind generation using DFIG is shown in Fig. 16.53. It has four major control components:

- Pitch angle control model
- Vector decoupling control system of DFIG
- Grid VSC control model
- Rotor VSC control system

Wind turbines are not able to maintain the voltage level and the required power factor. According to one regulation, it should be capable of supplying rated MW at any point between 0.95 power factor lagging to leading at the PCC. The reactive power limits defined at rated MW at lagging power factor will apply to all active power output levels above 20% of the rated MW output. Also the reactive power limits defined at rated MW at leading power factor will apply to all active power output levels above 50% of the rated MW output. See Fig. 16.54 for further details; this

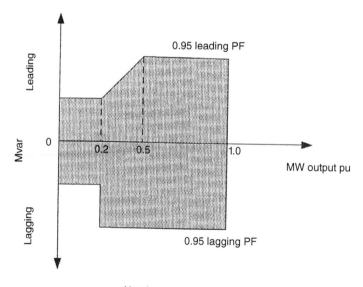

Not drawn to scale

Figure 16.54 Operation requirements of a wind generating farm for connections at the utility source, output verses power factor.

figure is for interconnection at the grid, PCC, *and not for individual operating units.* Thus, the reactive power compensation, fault levels and short-circuit analysis, the variations in the active power due to wind speed in the particular area over the course of a day, month-to-month, peak, and lowest raw electricity that will be generated are the first set of studies performed in the planning stage. These allow fundamental equipment ratings to be selected and protection system designs. Considerations such as cables versus overhead lines for the connection of collector buses to grid step-up transformer also arise. Further need for dynamic studies, voltage profiles, fault clearance times, studies documenting the grid connection requirements also arise. In fact, wind power generation and utility interconnections required extensive studies apart from harmonic considerations.

Due to the stochastic nature of wind turbine harmonics, probability concepts have been applied. Autoregressive moving average (ARMA) model is the statistical analysis of time series and provides parsimonious description of a stationary stochastic process in terms of two polynomials, one for autoregression and the other for moving average.

The mathematical model of a variable-speed constant-frequency generator under a *dq* synchronous rotating coordinate system is given by

$$
\begin{vmatrix} v_{sd} \\ v_{sq} \\ v_{rd} \\ v_{rq} \end{vmatrix} = \begin{vmatrix} pL_s + R_s & -\omega_1 L_s & pL_m & -\omega_1 L_m \\ \omega_1 L_s & pL_s + R_s & \omega_1 L_m & pL_m \\ pL_m & -(\omega_1 - \omega_r) L_m & pL_r + R_r & -(\omega_1 - \omega_r)L_r \\ (\omega_1 - \omega_r)L_m & pL_m & (\omega_1 - \omega_r)L_r & pL_r - R_r \end{vmatrix} \begin{vmatrix} i_{sd} \\ i_{sq} \\ i_{rd} \\ i_{rq} \end{vmatrix}
$$
(16.5)

The subscripts s and r denote stator and rotor, and p is the differential operator.

Generator flux is given by

$$
\begin{vmatrix} \varphi_{sd} \\ \varphi_{sq} \\ \varphi_{rd} \\ \varphi_{rq} \end{vmatrix} = \begin{vmatrix} L_s & 0 & L_m & 0 \\ 0 & L_s & 0 & L_m \\ L_m & 0 & L_r & 0 \\ 0 & L_m & 0 & L_r \end{vmatrix} \begin{vmatrix} i_{sd} \\ i_{sq} \\ i_{rd} \\ i_{rq} \end{vmatrix}
$$
(16.6)

If stator flux linkage is in the same direction as *d*-axis of rotating coordinate system, then $\varphi_{sd} = 0$. Thus,

$$
\phi_{sd} = \phi_s
$$

$$
\phi_{sq} = 0
$$
(16.7)

If stator coil resistance is ignored,

$$
v_{sd} = 0
$$

$$
v_{sq} = |v_s|
$$
(16.8)

The generator-side active and reactive powers are

$$
P_s = v_{sd} i_{sd} + v_{sq} i_{sq} = v_{sq} i_{sq}
$$

$$
Q_s = v_{sq} i_{sd}
$$
(16.9)

The P_s and Q_s can be decoupled using above equations.

The crowbar protection is specific to DFIG. The rotor-side converter must be protected in case of nearby faults. When the currents exceed a certain limit, the rotor-side converter is bypassed to avoid any damage.

16.7.1 Model for Harmonic Studies

For harmonic power flow analysis, the harmonic currents are modeled in parallel with the asynchronous machine model. A resistive load in parallel with the generator and current source is placed to model turbine auxiliary loads.

Considerable lengths of cables are involved and modeling of cable capacitance is important. Consider

- Capacitance of the turbine capacitor bank, if provided
- Capacitance of the collector cables
- Capacitance of the substation capacitors

Some simplifying assumptions can be made, that is, triplen harmonics will be trapped by transformer windings, even harmonics are eliminated due to waveform symmetry, the harmonic currents produced by all turbines can be assumed to have same phase angles.

For harmonic analysis, the harmonic spectra are best obtained from a manufacturer for the specific installation. The output filters (see Fig. 4.35) impact the harmonic emission passed on to the AC lines.

Figure 16.55 shows a 100-MW wind farm generation for harmonic analysis (1000-MW wind generation in a single location is being planned). The length and sizes of cables are indicated and also the source impedance and transformer impedances.

Table 16.21 shows the planning level of harmonics recommended by Energy Network Association, United Kingdom. The harmonic current spectra are usually supplied by the manufacturer, Table 16.22 shows harmonic emissions from a WTG. Third harmonics need not be modeled.

Also some harmonic pollution in the grid connection itself is modeled:

- 5th harmonic voltage $= 3\%$
- 7th harmonic voltage $= 2.0\%$
- 11th harmonic voltage $= 1.5\%$
- 13th harmonic voltage $= 10.\%$
- 17th harmonic voltage $= 0.5\%$

This gives an ambient harmonic distortion at the utility source $= 4.06\%$ and makes the design of filters difficult.

The capacitive reactive power demand is a total of 30 Mvar, shown in the form of shunt capacitors at main substation bus. There are no additional capacitors at the collector bus.

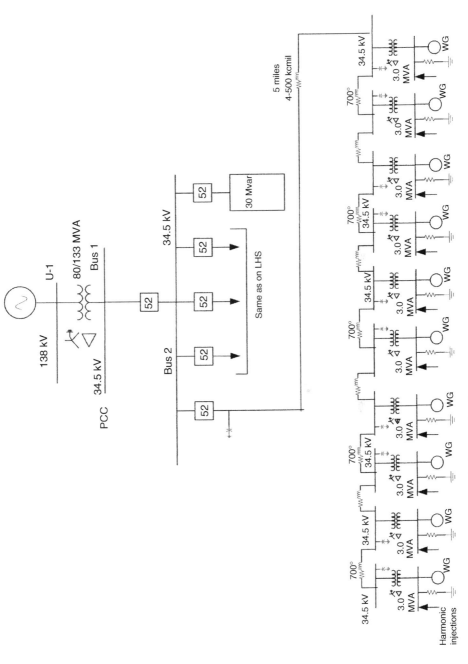

Figure 16.55 Configuration of a 100-MW wind generation farm.

811

TABLE 16.21   Planning Levels for Harmonic Voltages in Systems > 20 kV and < 145 kV

Odd Harmonics		Triplen Harmonics		Even Harmonics	
Order	Voltage (%)	Order	Voltage (%)	Order	Voltage (%)
5	2.0	3	2.0	2	1.0
7	2.0	9	1.0	4	0.8
11	1.5	15	0.3	6	0.5
13	1.5	21	0.2	8	0.4
17	1.0	> 21	0.2	10	0.4
19	1.0			12	0.2
23	0.7			> 12	0.2
25	0.7				
> 25	$0.2 + 0.5(25/h)$				

TABLE 16.22   Harmonic Emission from a typical DFIG

Harmonic Order	Harmonic Current Percentage of Fundamental	Harmonic Order	Harmonic Current Percentage of Fundamental
2	1.0	17	0.76
3	0.51	19	0.42
4	0.43	22	0.33
5	1.32	23	0.41
6	0.42	25	0.24
7	1.11	26	0.2
8	0.42	28	0.15
10	0.61	29	0.27
11	1.52	31	0.24
13	1.91	35	0.35
14	0.50	37	0.26
16	0.37		

To avoid resonance problems, ST, DT, LP, and type C filters have been used in the wind farms. Also STATCOM and SVCs are employed – these give less of a harmonic pollution and less of resonance problems.

The study is conducted in the following steps:

- Establish ambient harmonics without 30-Mvar capacitor bank.
- Ascertain the harmonic resonance when 30-MVA banks are switched in service.
- Provide filters to bring harmonic distortion levels to acceptable limits.
- Is 30-Mvar capacitor bank/filters a proper choice for this application?

The study results with ambient harmonics and with 30 Mvar capacitors are shown in following figures and tables:

Figure 16.56(a) and (b) is frequency scans impedance angle and modulus without 30-Mvar capacitor bank.

Figure 16.56(c) and (d) is frequency scans impedance angle and modulus with 30-Mvar capacitor bank.

Figure 16.57(a) and (b) is voltage harmonic spectrum and its waveform at PCC without 30-Mvar capacitor bank.

Figure 16.57(c) and (d) is voltage harmonic spectrum and its waveform at PCC with 30-Mvar capacitor bank.

Figure 16.58(a) and (b) is current distortion at PCC and its waveform at PCC without 30-Mvar capacitor bank.

Figure 16.58(c) and (d) is current distortion at PCC and its waveform at PCC with 30-Mvar capacitor bank.

Also the results of current and voltage distortions are plotted in Tables 16.23 and 16.24. This shows that with 30-Mvar capacitor bank in service resonance occurs at fifth harmonic, it is 73.70% of nominal voltage and 43.82% of fundamental current. The distorted waveforms in Figs 16.57(d) and 16.58(d) illustrate this situation.

From this study, it seems that a single fifth harmonic filter will bring down the fifth harmonic distortion. If the 30-Mvar capacitor bank is turned into fifth harmonic ST filter, the results of current distortion at PCC are shown in Fig. 16.59(a). Considerable harmonic amplification occurs at second and fourth harmonics. Change in tuning frequency controls these harmonics only slightly.

The various strategies of sizing the filters and their types are tried. Two ST filters with a damped filter give the best results. However, the size of the three filters amounts to 60 Mvar. The design of filters is also impacted by harmonic voltages modeled in the 138-kV source. The current distortion at PCC with these filters in place is illustrated in Fig. 16.59(b).

Apart from the control of harmonics, we cannot lose sight of the wind farm operation. Referring to Fig. 16.54, even with 30-Mvar filters, some inductive reactive power compensation is required at low loads.

Study with STATCOM   This shows that selection of a STATCOM as a variable reactive power compensation device will be an appropriate choice. It can be controlled with respect to desired power factor and power output.

Table 6.1 shows that STATCOM does give rise to some harmonics. These harmonics should be modeled for accuracy.

The results of the study with STATCOM supplying 30 Mvar of capacitive reactive power into the system are shown in Fig. 16.60. Figure 16.60(a) and (b) shows voltage spectrum and its waveform, and Fig. 16.60(c) and (d) illustrates the current spectrum and its waveform at PCC. It is observed that the harmonics are reduced to almost ambient level. The summary results are also shown in Table 16.25. It is interesting to note that

- Eighth harmonic is amplified, showing resonance at this frequency. This exceeds IEEE limits. Note that for generating plants, the IEEE limits correspond to $I_s/I_r < 20$, *irrespective of actual* I_s/I_r.

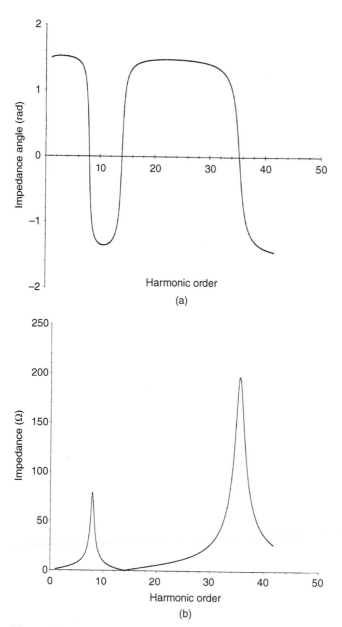

Figure 16.56   (a) and (b) Frequency scan phase angle and impedance modulus, ambient harmonics; (c) and (d) with 30-Mvar capacitor bank connected in service at 34.5 kV, Fig. 16.56.

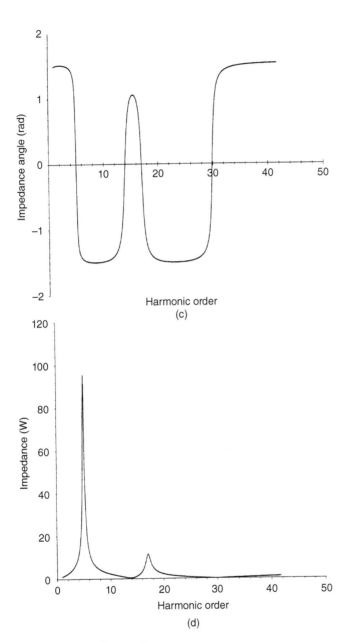

Figure 16.56   (*Continued*)

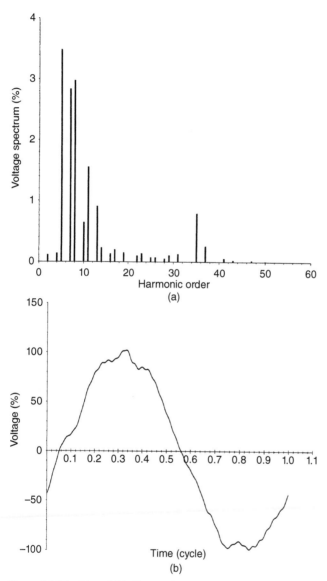

Figure 16.57   (a) and (b). Harmonic voltage spectrum and voltage waveform at PCC, ambient harmonics; (c) and (d) with 30-Mvar capacitor bank connected in service at 34.5 kV, Fig. 16.55.

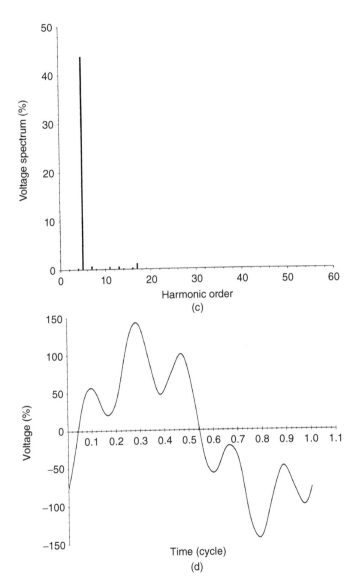

Figure 16.57   (*Continued*)

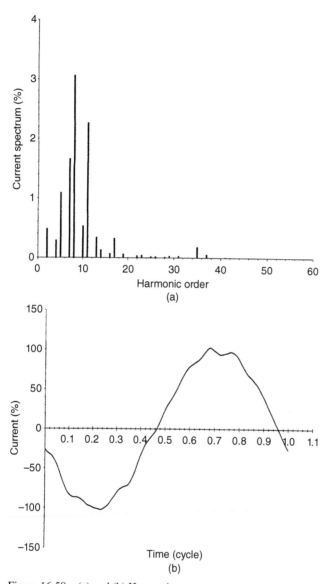

Figure 16.58 (a) and (b) Harmonic current spectrum and current waveform at PCC, ambient harmonics; (c) and (d) with 30-Mvar capacitor bank connected in service at 34.5 kV, Fig. 16.55.

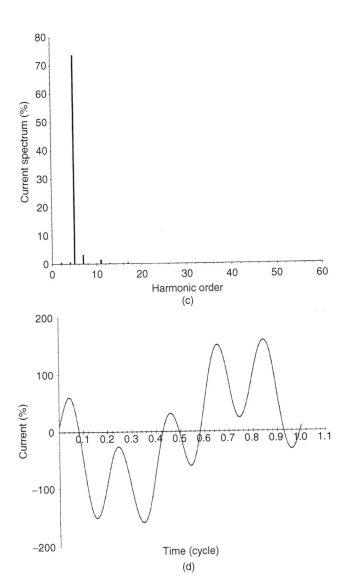

Figure 16.58   (*Continued*)

TABLE 16.23 Harmonic Current Distortion at PCC, 34.5 kV

Harmonic Order	30-Mvar Capacitor Out of Service	30-Mvar Capacitor in Service	IEEE Limits
Fundamental	1572.3A (100%)	1533.8A (100%)	
2	0.49	0.58	1.0
4	0.29	0.66	1.0
5	1.09	**73.70**	**4.0**
7	1.65	3.27	4.0
8	**3.06**	0.19	**1.0**
10	0.53	0.15	1.0
11	**2.26**	1.53	**2.0**
13	0.34	0.30	2.0
14	0.13	0.11	0.5
16	0.07	0.15	0.5
17	0.32	0.45	1.5
19	0.07	0.07	1.5
22	0.04	0.01	0.375
23	0.05	0.01	0.6
25	0.03	0	0.6
26	0.02	0	0.15
28	0.02	0	0.15
29	0.03	0	0.6
31	0.03	0	0.6
35	0.17	0	0.3
37	0.05	0	0.3
41	0.01	0	0.3
43	0	0	0.3
47	0	0	0.3
50	0	0	0.075

Total permissible TDD = 5%.

Without 30-Mvar capacitor = 4.39%. The bold numbers show that harmonics exceed permissible values.

With 30-Mvar capacitor = 73.80%.

- The importance of modeling even harmonics is illustrated. Resonance at even-order harmonics is very possible, and such cases are on record.
- The total current distortion at PCC is 3.86% versus permissible level of 5%.
- The ambient harmonic voltage distortion modeled at 138-kV utility source is 4.06%. If the voltage distortion at PCC is to be limited to 5% it is a constraint, and the harmonics at PCC should be further reduced.

The harmonic distortions at PCC are further reduced by reducing the capacitive reactive power output to 20 Mvar and providing two ST filters of 5 Mvar each.

TABLE 16.24   Harmonic Voltage Distortion, PCC, 34.5 kV

Harmonic Order	30-Mvar Capacitor Out of Service	30-Mvar Capacitor in Service
2	0.12	0.14
4	0.14	0.31
5	3.49	**43.82**
7	2.84	0.72
8	2.98	0.18
10	0.65	0.18
11	1.56	0.55
13	0.91	0.58
14	0.23	0.19
16	0.21	0.30
17	0.35	1.20
19	0.11	0.16
22	0.14	0.04
23	0.08	0.04
25	0.08	0.01
26	0.06	0.01
28	0.11	0
29	0.13	0.01
31	0.79	0.01
35	0.26	0
37	0.06	0
41	0.03	0
43	0.02	0
47	0.01	0
50	0.01	0
THD_V	5.90	44.25

Various tuning frequencies and filter types are tried. The tuning frequencies selected are 147 Hz and 602 Hz, respectively, which may seem to be rather an odd choice, but give the least harmonic distortions. Due to harmonics of 5th, 7th, 11th, 13th, and 17th at the source, the filters tuned to these frequencies will draw harmonic currents from the source. Figure 16.61(a) and (b) shows the voltage spectrum and its waveform, while Fig. 16.61(c) and (d) illustrates the current spectrum and its waveform at PCC.

Before finalization, the study should also consider lower load operations, outage of a filter or filter component as illustrated in some other studies in this chapter (Table 16.26).

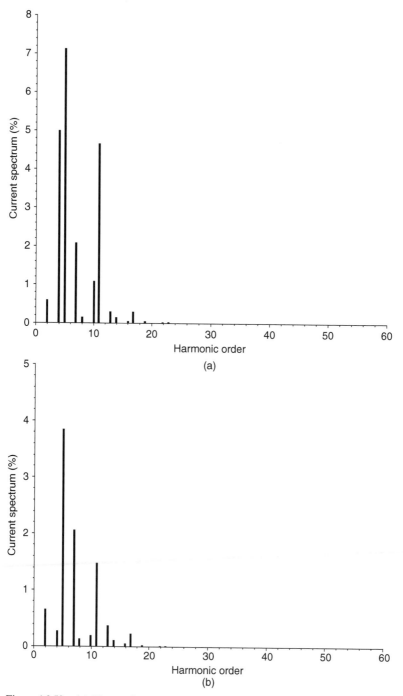

Figure 16.59   (a), Harmonic current spectrum at PCC with one 30Mvar ST fifth harmonic filter, (b) with multiple filters of 60 Mvar.

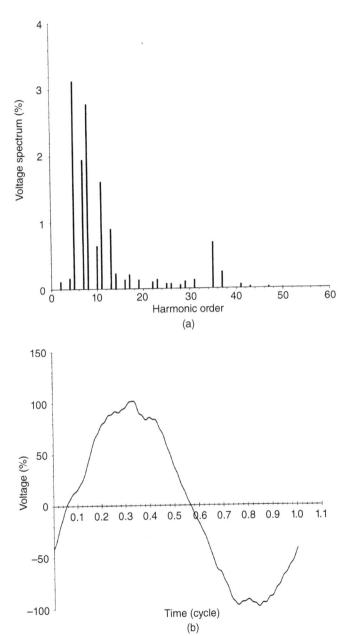

Figure 16.60   (a)–(d) Harmonic voltage spectrum, voltage waveform, current spectrum, and current waveform at PCC with STATCOM, output 30 Mvar (capacitive).

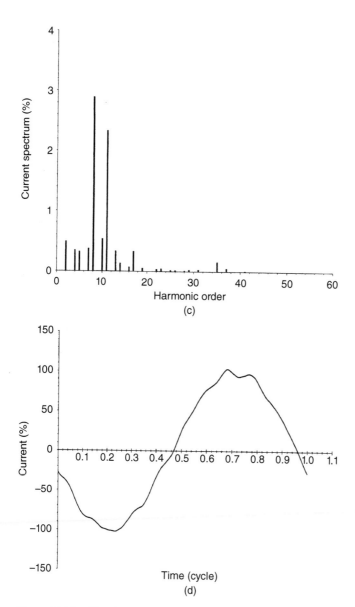

Figure 16.60   (*Continued*).

TABLE 16.25   Harmonic Current Distortion at PCC with STATCOM, 34.5 kV

Harmonic Order Fundamental	With STATCOM 1572.3A (100%)	IEEE Limits	Voltage Distortion
2	0.49	1.0	0.12
4	0.35	1.0	0.17
5	0.32	4.0	3.13
7	0.38	4.0	1.95
8	**2.88**	1.0	2.78
10	0.53	1.0	0.64
11	**2.33**	2.0	1.61
13	0.34	2.0	0.90
14	0.13	0.5	0.23
16	0.07	0.5	0.14
17	0.32	1.5	0.21
19	0.07	1.5	0.13
22	0.04	0.375	0.11
23	0.05	0.6	0.14
25	0.03	0.6	0.08
26	0.02	0.15	0.08
28	0.02	0.15	0.06
29	0.03	0.6	0.11
31	0.03	0.6	0.13
35	0.17	0.3	0.69
37	0.05	0.3	0.25
41	0.01	0.3	0.06
43	0	0.3	0.03
47	0	0.3	0.02
50	0	0.075	0.01

Total permissible $THD_I = 5\%$.

Calculated TDD $THD_I = 3.86$. The values shown in bold exceed IEEE limits.

$THD_V = 5.18$.

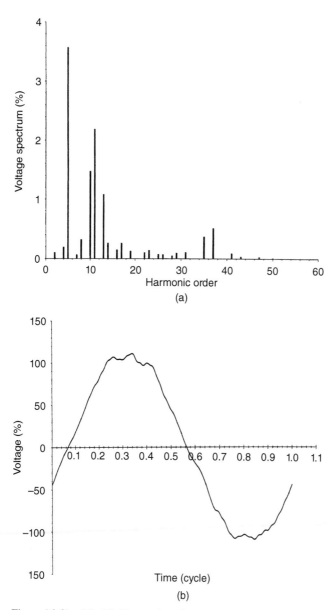

Figure 16.61   (a)–(d). Harmonic voltage spectrum, voltage waveform, current spectrum, and current waveform at PCC with STATCOM, output 20 Mvar (capacitive), and two ST filters of 5 Mvar each (see text).

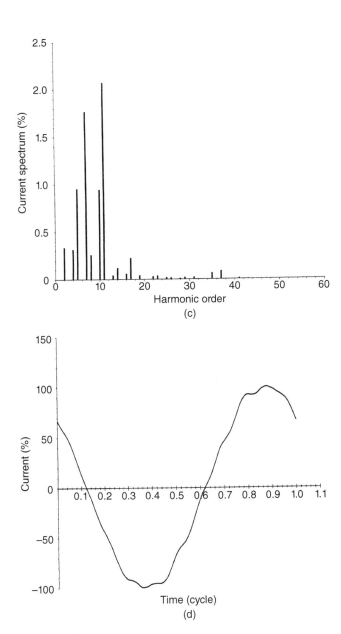

Figure 16.61   (*Continued*)

TABLE 16.26   Harmonic Current Distortion at PCC with STATCOM and Filters, 34.5 kV

Harmonic Order Fundamental	With STATCOM and Filters 1572.3A (100%)	IEEE Limits	Voltage Distortion
2	0.34	1.0	0.10
4	0.31	1.0	0.20
5	0.96	4.0	3.57
7	1.77	4.0	0.07
8	0.26	1.0	0.32
10	0.95	1.0	1.48
11	1.98	2.0	2.19
13	0.04	2.0	1.08
14	0.12	0.5	0.26
16	0.06	0.5	0.15
17	0.23	1.5	0.27
19	0.04	1.5	0.13
22	0.03	0.375	0.11
23	0.04	0.6	0.14
25	0.02	0.6	0.08
26	0.02	0.15	0.07
28	0.01	0.15	0.05
29	0.02	0.6	0.10
31	0.02	0.6	0.11
35	0.07	0.3	0.37
37	0.09	0.3	0.51
41	0.01	0.3	0.09
43	0.01	0.3	0.04
47	0	0.3	0.03
50	0	0.075	0.01

Total permissible THD_I = 5%.

Calculated THD_I = 3.09.

THD_V = 4.69.

INDEX

Power System Harmonics and Passive Filter Designs, First Edition. J.C. Das.
© 2015 The Institute of Electrical and Electronics Engineers, Inc. Published 2015 by John Wiley & Sons, Inc.

IEEE Press Series
on Power Engineering

Series Editor: M. E. El-Hawary, Dalhousie University, Halifax, Nova Scotia, Canada

The mission of IEEE Press Series on Power Engineering is to publish leading-edge books that cover the broad spectrum of current and forward-looking technologies in this fast-moving area. The series attracts highly acclaimed authors from industry/academia to provide accessible coverage of current and emerging topics in power engineering and allied fields. Our target audience includes the power engineering professional who is interested in enhancing their knowledge and perspective in their areas of interest.

1. *Principles of Electric Machines with Power Electronic Applications, Second Edition*
M. E. El-Hawary

2. *Pulse Width Modulation for Power Converters: Principles and Practice*
D. Grahame Holmes and Thomas Lipo

3. *Analysis of Electric Machinery and Drive Systems, Second Edition*
Paul C. Krause, Oleg Wasynczuk, and Scott D. Sudhoff

4. *Risk Assessment for Power Systems: Models, Methods, and Applications*
Wenyuan Li

5. *Optimization Principles: Practical Applications to the Operations of Markets of the Electric Power Industry*
Narayan S. Rau

6. *Electric Economics: Regulation and Deregulation*
Geoffrey Rothwell and Tomas Gomez

7. *Electric Power Systems: Analysis and Control*
Fabio Saccomanno

8. *Electrical Insulation for Rotating Machines: Design, Evaluation, Aging, Testing, and Repair, Second Edition*
Greg Stone, Edward A. Boulter, Ian Culbert, and Hussein Dhirani

9. *Signal Processing of Power Quality Disturbances*
Math H. J. Bollen and Irene Y. H. Gu

10. *Instantaneous Power Theory and Applications to Power Conditioning*
Hirofumi Akagi, Edson H. Watanabe, and Mauricio Aredes

11. *Maintaining Mission Critical Systems in a 24/7 Environment*
Peter M. Curtis